Human Genes and Diseases

Horizons in Biochemistry and Biophysics Series

Editorial Board

Horizons in Biochemistry
and Biophysics

Volume 8

Human
Genes and Diseases

Volume Editor
F. Blasi

Istituto Internazionale
di Genetica e Biofisica
Naples, Italy

Series Editors
E. Quagliariello, *Editor-in-Chief*
and
F. Palmieri, *Managing Editor*
Department of Biochemistry, University of Bari

A Wiley–Interscience Publication

JOHN WILEY & SONS
Chichester · New York · Brisbane · Toronto · Singapore

Library of Congress Cataloging in Publication Data:
Main entry under title:

Human genes and diseases.
   (Horizons in biochemistry and biophysics; v. 8)
   'A Wiley–Interscience publication.'
   Includes index,
   1. Medical genetics.   2. Molecular genetics.
I. Blasi, F.   II. Title.   III. Series. [DNLM:
1. Genetics, Medical.   2. Hereditary Diseases.
W1 HO596T v. 8/QZ 50 H0164]
QH345.H66   vol. 8   574.19 s   [616'.042]   85-32313
[RB155]

ISBN 0 471 91167 4

British Library Cataloguing in Publication Data:
Human genes and diseases.—(Horizons in biochemistry
   and biophysics; v. 8)
   1. Medical genetics
   I. Blasi, F.   II. Series
   616'.042   RB155

ISBN 0 471 91167 4

Printed in Great Britain

Contents

List of Contributors

Preface

1   **Towards a complete linkage map of the X-chromosome**   1
K. E. DAVIES AND R. WILLIAMSON

2   **From Hemophilia B to hemophilia A *via* the fragile X locus: genes and recombination in the distal region of the human X-chromosome' long arm**   51
I. OBERLÉ AND MANDEL

3   **Mapping of rare X-linked genes through DNA polymorphism and identification of crossover points**   91
G. ROMEO

4   **The role of HPRT genes in human disease**   123
D. J. JOLLY

5   **The utilization of the human phosphoglycerate kinase gene in the investigation of X-chromosome inactivation**   169
M. A. GOLDMAN, S. M. GARTLER, E. A. KEITGES AND D. E. RILEY

6   **Metallothionein gene regulation in Menke's Disease**   207
A. LEONE

7   **Beta-globin gene disorders in Italy and in the Mediterranean area**   257
S. OTTOLENGHI AND C. CARESTIA

8   **The molecular genetics of hyperlipidemia**   299
C. C. SHOULDERS AND F. E. BARALLE

9   **Human collagens: biochemical, molecular and genetic features in normal and diseased states**   341
F. RAMIREZ, F. O. SANGIORGI AND P. TSIPOURAS

10   **Human plasminogen activators. Genes and proteins structure**   375
F. BLASI, A. RICCIO AND G. SEBASTIO

11   ***In vitro* transformation of epithelial cells by acute retroviruses**   415
G. VECCHIO, P. P. DiFIORE, A. FUSCO, G. COLETTA, B. E. WEISSMAN AND S. A. AARONSON

12   **Monoclonal antibodies to the insulin receptor as probes of insulin receptor structure and function**   471
I. D. GOLDFINE AND R. A. ROTH

13 Cellular oncogenes and the pathogenesis of human cancer   503
    R. DALLA FAVERA AND E. CESARMAN

14 Molecular genetics of human B cell neoplasia   545
    C. M. CROCE

15 The erb-B related growth factors receptors   571
    T. YAMAMOTO, K. SEMBA AND K. TOYOSHIMA

16 Aldolase gene and protein families: structure, expression and
    molecular pathophysiology   611
    F. SALVATORE, P. IZZO AND G. PAOLELLA

Index   667

List of Contributors

ARONSON, S. A.
Laboratory of Cellular and Molecular Biology, National Cancer Institute, National Institutes of Health, Bethesda, Maryland 20205, U.S.A.

BARALLE, F. E.
Sir William Dunn School of Pathology, University of Oxford, South Parks Road, Oxford OX12 3RE, United Kingdom

BLASI, F.
International Institute of Genetics and Biophysics, Consiglio Nazionale delle Ricerche, via Marconi, 12, 80123 Naples, Italy

CARESTIA, C.
International Institute of Genetics and Biophysics, Consiglio Nazionale delle Ricerche, via Marconi, 12, 80123 Naples, Italy

CESARMAN, E.
Department of Pathology and Kaplan Cancer Center, New York University, 550 First Avenue, 10016 New York, U.S.A.

COLLETTA, G.
Dipartimento di Biologia e Patologia Cellulare e Molecolare, II Faculty of Medicine, University of Naples, via S. Pansini 5, 80131 Naples, Italy; Centro di Endocrinologia e Oncologia Sperimentale del CNR, Naples, Italy

CROCE, C. M.
The Wistar Institute, 36th at Spruce Street, Philadelphia, Pennsylvania 19104, U.S.A.

DALLA FAVERA, R.
Department of Pathology and Kaplan Cancer Center New York University, 550 First Avenue, 10016 New York, New York, U.S.A.

DAVIES, K. E.
Nuffield Department of Clinical Medicine, John Radcliffe Hospital, Headington, Oxford, OX3 9DU, United Kingdom

DI FIORE, P.
Laboratory of Cellular and Molecular Biology National Cancer Institute, National Institutes of Health, Bethesda, Maryland 20205, U.S.A.

FUSCO, A.   *Dipartimento di Biologia e Patologia Cellulare e Molecolare, II Faculty of Medicine, University of Naples, via S. Pansini 5, 80131 Naples, Italy; Centro di Endocrinologia e Oncologia Sperimentale del CNR, Naples, Italy*

GARTLER, S. M.   *Departments of Genetics and Medicine SK50, and the Center for Inherited Diseases, University of Washington, Seattle, Washington 98195, U.S.A.*

GOLDFINE, I. D.   *Cell Biology Laboratory, Mount Zion Hospital and Medical Center, P. O. Box 7921, San Francisco, California 94120, U.S.A.*

GOLDMAN, M. A.   *Department of Genetics and Medicine SK50, and the Center for Inherited Diseases, University of Washington, Seattle, Washington 98195, U.S.A.*

JOLLY, D. J.   *Department of Pediatrics, M-009H, University of California, San Diego, La Jolla, California 92093, U.S.A.*

KEITGES, E. A.   *Departments of Genetics and Medicine SK50, and the Center for Inherited Diseases, University of Washington, Seattle, Washington 98195, U.S.A.*

IZZO, P.   *Istituto di Scienze Biochimiche, University of Naples, 2nd Medical School, via S. Pansini 5, 80131 Naples, Italy*

LEONE, A.   *Laboratory of Biochemistry, National Cancer Institute, National Institutes of Health, Bethesda, Maryland 20205, U.S.A.*

MANDEL, J-L.   *Inserm U-184, Faculté de Medicine, 11 rue Humann, Strasbourg-Cedex 67085, France*

OBERLÉ, I.   *Inserm U-184, Faculté de Médicine, 11 rue Humann, Strasbourg-Cedex 67085, France*

OTTOLENGHI, S.   *Cattedra Di Patologia Generale, Facoltà di Farmacia, University of Milan, and Centro per lo studio della Patologia Cellulare del CNR, Milan, Italy*

PAOLELLA, G.   *Istituto di Scienze Biochimiche, University of Naples 2nd Medical School, via S. Pansini 5, 80131 Naples, Italy*

RAMIREZ, F.
Department of Obstetrics and Gynecology, University of Medicine and Dentistry of New Jersey, Rutgers Medical School, Piscataway, New Jersey 08854, U.S.A.

RICCIO, A.
International Institute of Genetics and Biophysics, Consiglio Nazionale delle Ricerche, via Marconi 12, 80123 Naples, Italy

RILEY, D.E.
Departments of Genetics and Medicine SK50, and the Center for Inherited Diseases, University of Washington, Seattle, Washington 98195, U.S.A.

ROMEO, G.
Laboratory of Genetics, Institute of Neurology, University of Bologna, via Foscolo 7, 40123 Bologna, Italy

ROTH, R. A.
Department of Pharmacology, Stanford University School of Medicine, Stanford, California 94305, U.S.A.

SALVATORE, F.
Istituto di Scienze Biochimiche, University of Naples 2nd Medical School, via S. Pansini 5, 80131 Naples, Italy

SANGIORGI, F. O.
Department of Obstetrics and Gynecology, University of Medicine and Dentistry of New Jersey, Rutgers Medical School, Piscataway, New Jersey 08854, U.S.A.

SEBASTIO, G.
International Institute of Genetics and Biophysics, Consiglio Nazionale delle Ricerche, via Marconi 12, 80123 Naples, Italy

SEMBA, K.
Institute of Medical Sciences, The University of Tokyo, 4-6-1 Shirokanedai Minato-Ku, Tokyo 108, Japan

SHOULDERS, C. C.
Sir William Dunn School of Pathology, University of Oxford, South Parks Road, Oxford OX1 3RE, United Kingdom

TOYOSHIMA, K.
Institute of Medical Sciences, The University of Tokyo, 4-6-1 Shirokanedai Minato-Ku, Tokyo 108, Japan

TSIPOURAS, P.   *Department of Pediatrics, University of Medicine and Dentistry of New Jersey, Rutgers Medical School, Piscataway, New Jersey 08854, U.S.A.*

VECCHIO, G.   *Dipartimento di Biologia e Patologia Cellulare e Molecolare, II Faculty of Medicine, University of Naples, via S. Pansini 5, Naples 80131, Italy; Centro di Endocrinologia e Oncologia Sperimentale del CNR, via S. Pansini 5, Naples, Italy*

YAMAMOTO, T.   *Institute of Medical Sciences, The University of Tokyo, 4-6-1 Shirokanedai Minato-Ku, Tokyo 108, Japan*

WEISSMAN, B. E.   *Laboratory of Cellular and Molecular Biology, National Cancer Institute, National Institutes of Health, Bethesda Maryland 20205, U.S.A.*

WILLIAMSON, R,   *Department of Biochemistry, St. Mary's Hospital Medical School, Paddington, London W2 1PG, United Kingdom*

Preface

Molecular genetics has reached a degree of refinement such that its contributions have now become an absolute requirement in fields like biochemistry and pathology. This is even more true for human molecular genetics where information gained is of value in basic genetics, biochemistry and pathology. This volume of Horizons in Biochemistry and Biophysics is therefore devoted to the molecular genetic approach to human diseases. The volume is dedicated not just to the experts in the field; medicine and biology students, physicians and biochemists, should take advantage from it. Each chapter contains a comprehensive introduction that facilitates the subsequent reading of its content.

The content of the book is essentially divided into two parts; the first dealing with X-chromosome linked genes, the second with autosomal genes. The latter moves on from an investigation of the more classical autosomal genes to a discussion of other possibly pleiotropic genes, whose function has been connected with several types of diseases, most notably oncogenes.

The first part of the book, dedicated to the X chromosome, starts out with two chapters on modern methods for mapping the human X chromosome, in particular with the use of the restriction fragments length polymorphism (RFLP) method, which appears to be an extremely promising technique and should allow in the near future the complete mapping of the human genome as well as a genetic approach to multi-genic diseases. Other chapters of this section deal with diagnosis, molecular genetics and future therapeutic approaches to classic X-linked diseases. Finally, a chapter is dedicated to the use of a cloned X-linked gene as a tool with which to study the phenomenon of X-chromosome inactivation..

Among the extremely large collection of autosomal diseases, we have chosen one, thalassemia, on which a great deal of information has been accumulated over the years, and others that have received relatively less attention at the genetic level, i.e. collagen diseases and hyper-lipidemias. The latter represent a group of diseases the detailed knowledge of which will lead to the understanding at the molecular level of the genetic basis of predisposition to diseases like arteriosclerosis and its effects.

The final group of chapters deals with genes that have been implicated with the genesis of cancer, or with the establishment of the malignant phenotype. These include the genes for plasminogen activators and some oncogenes. Plasminogen activators have received public attention recently mostly for their possible use as thrombolytic agents in the treatment of thrombosis and heart infarction. However, they appear to have an important role in the pathogenesis of other diseases, like pemphigus, revealing another important feature of tumor cells, i.e. the ability to dissociate from the main tumor, reach the circulation and re-implant at a distant location (metastasis). After the general chapter on transformation of epithelial cells by retroviruses, the remainder of this section is devoted to specific neoplasis (like B-cell neoplasms), oncogenes and their products with special reference to the epidermal growth factor and the insulin receptor.

F. Blasi
F. Palmieri
E. Quagliariello

Towards a Complete Linkage Map
Of The Human X Chromosome

Kay E. Davies[1] and Robert Williamson[2]

[1] Nuffield Department of Clinical Medicine, John
Radcliffe Hospital, Headington, Oxford, OX3 9DU

[2] Department of Biochemistry, St Mary's Hospital
Medical School, Paddington, London, W2 1PG

Summary

In the seven years since the first human gene was
cloned, several hundred coding sequences and many more
random single copy DNA sequences have been isolated.
Many of these show restriction fragment length
polymorphisms (RFLPs) and can be used as genetic
markers in inheritance studies. RFLPs enable the
construction of complete linkage maps of individual
human chromosomes which can then be used as a mapping
resource for other genes and disease loci. The
isolation of chromosome specific sequences has been
greatly facilitated by the purification of human
chromosomes by flow cytometry. Sub-localisation of
the polymorphic DNA probes along the chromosome can be
achieved using in situ hybridisation or rodent/human
hybrid cell lines. There are now more than one
hundred DNA probes assigned to the human X chromosome
and a preliminary genetic map sugggests that the
chromosome is at least 200cM long. Some of these DNA
sequences have been shown to be linked to disease loci

such as Duchenne and Becker muscular dystrophy, X-
linked mental retardation and retinitis pigmentosa.

Introduction

Traditionally, genetics is the study of the
inheritance of phenotypes. The mode of inheritance -
dominant, recessive or sex-linked - can be determined
by following the segregation of phenotypes in
families. Any phenotype determined exclusively by a
gene or genes on the X chromosome will exhibit sex-
linked inheritance, and in the case of alleles which
cause pathology, this appears as a sex-linked disease,
such as haemophilia or Duchenne muscular dystrophy.
Therefore, the X chromosome is unique in that genes on
it can be defined by the mode of inheritance alone;
for autosomes, other techniques, such as gene
expression in hybrid cells, must be used for
chromosome assignment. Because of this, many
phenotypes, both pathological and normal, and several
coding genes, were assigned to the X chromosome before
direct gene analysis became possible (see McKusick and
Ruddle, 1977).

Gene cloning is revolutionising human genetics,
as it allows direct gene analysis in families and
correlation of genotype and phenotype unequivocally.
Gene libraries, whether total, chromosome-specific, or
of expressed genes in a tissue, are unique to the
person whose genome is under study. Such gene
libraries are not only a resource for isolating
specific gene sequences which code for proteins, or
which are located at a particular site on a
chromosome: they also provide the raw material to
allow the study of variation, both for each individual
and also in a family. Any DNA sequence variant can be
studied for its expression in an individual, and can

be followed through a family either as a functional or
a linkage marker.

Conventional linkage markers

Before the advent of DNA recombinant technology,
two linkage groups on the human X chromosome had been
described. These were the Xg cluster on the short arm
and the colour blindness cluster on the long arm.

The precise localisation of the Xg blood group
locus was unknown for some time after its initial
discovery in 1962 (for review see Race and Sanger,
1975). The locus was eventually found to be linked
to X-linked ichthyosis (steroid sulphatase (STS)
deficiency) which was assigned to Xp22.3-Xpter by
deletion mapping (Adam et al, 1969; Tiepolo et al,
1980; Mohandas et al, 1980). Within the Xg cluster, Xg
appears to be distal to STS by deletion mapping
(Tiepolo et al, 1980). Other loci in this cluster
include the Xk blood group (XK locus), chronic
granulomatous disease (CGD), ocular albinism (OA) and
X-linked retinoschisis (RS) (see Race and Sanger, 1975
and Wolf et al, 1980).

The colour blindness cluster which includes both
protan and deutan loci, maps near the end of the long
arm of the human X chromosome. The linkage group
includes glycerol 6-phosphate dehydrogenase (G6PD),
colour blindness (CBD, CBP), haemophilia A and B (HEMA
and HEMB), adrenoleukodystrophy (ALD) and serum group
Xm (XM) (see McKusick and Ruddle, 1977 and Siniscalco,
1979). Somatic cell hybridisation studies localise the
G6PD locus to the terminal Giemsa-negative band of the
long arm, Xq28 (Pai et al, 1980). Studies of
radiation-induced gene segregation showed that the
hypoxanthine guanine phosphoribosyltransferase (HPRT)
locus is physically rather close to the G6PD locus

(Goss and Harris, 1977). However, close genetic
linkage was not observed in family studies (Francke et
al, 1974). This was the first indication that the
frequency of recombination increases towards the end
of the long arm.

It is perhaps worth noting at this point an
interesting observation made by McKusick (1980). The
human X chromosome contains two loci for the clotting
factors (HEMA and HEMB), three loci for colour
blindness (deutan, protan partial colour blindness),
three loci for muscular dystrophy (Duchenne and Becker
- although these may be allelic - and Emery types),
probably two loci for ocular albinism, four loci for
end-organ unresponsiveness, and at least four loci for
severe combined immunodeficiency (for refs see
McKusick, 1983). These may be consequences of ancient
duplications, frozen in evolution by Lyonisation. The
use of DNA recombinant technology to map the human X
chromosome and identify more X-linked functional genes
should shed light on the molecular basis of these
observations.

Restriction fragment length polymorphisms

DNA sequence variants arise as the result of
single base changes, insertions or deletions and many
of them result in the removal or creation of a
restriction enzyme site. They are therefore known as
restriction fragment length polymorphisms (RFLPs).
The first demonstration of a random RFLP, and its
usefulness in genetic studies, was by Kan and Dozy
(1978), who showed that an RFLP for the enzyme HpaI is
in disequilibrium with the gene for sickle cell beta-
globin in West Africans. Immediately afterwards,
Solomon and Bodmer (1979) remarked on the significance
of this in establishing a new class of random

polymorphic markers, relatively few of which would be required to provide a linkage map of the human genome. This concept, later extended by Botstein et al (1980), has turned out to be accurate.

The applicability of RFLPs in genetic mapping depends on how common they are in the human genome. Jeffreys (1979) studied the occurrence of variants in, and around, the human beta-globin gene region, and concluded that point mutations occurred approximately once in every hundred base pairs. Screening of individuals for polymorphisms with other DNA probes suggest that the distribution of RFLPs throughout the genome is not random. Some sequences do not detect any variants within 20kb of DNA using a panel of twenty families and twenty different restriction enzymes, whereas other loci, such as phenylalanine hydroxylase, are highly polymorphic (Woo et al, 1984). In addition, evidence is accumulating that suggests that X chromosome DNA probes are less polymorphic than other chromosome sequences (Hofker et al, 1985).

The usefulness of a polymorphism for linkage studies obviously depends on its frequency in the general population. Most of the RFLPs reported so far consist of two alleles and the frequency of informative heterozygotes is low. Highly polymorphic loci and hypervariable regions derived by insertion and deletion of DNA fragments have been reported on autosomes, but no similar sequences have been found on the human X chromosome (Wyman and White, 1980; Bell et al, 1981; Goodbourn et al, 1983; Capon et al, 1983; Jeffreys et al, 1985). Many of the single point mutations are detected by the enzymes MspI or TaqI and it has been postulated that it is due to the methylation of CpG and deamination of the C to a T

residue (Barker et al, 1984). Both of these
restriction enzymes contain CpG in their recognition
sequences.

Mapping the human genome

The number of markers which are required for a
linkage map is relatively easy to determine; it is
related to the number of cross-overs that occur during
meiosis. On average, there are two meiotic cross-
overs in males per chromosome (the number is much more
difficult to determine for females, as first meiotic
prophase occurs during embryonic development and is
difficult to study) (Hulten, 1974; Laurie et al, 1981;
Hulten et al, 1982). The total number of Morgans (one
centimorgan is the genetic distance corresponding to a
1% chance of crossover occurring per generation) for
the human genome is approximately 30; therefore, the
human X chromosome, which contains 5% of the genome
DNA, might be expected to be some 150 centimorgans in
genetic length. This is a very imperfect estimate,
because the X chromosome may differ from autosomes in
this. In addition, a higher frequency of
recombination between autosomal loci in females
relative to males is well documented (Cook, 1965;
Renwick and Schulze, 1965), and X chromosome
recombination only occurs in females.

Various calculations have been made of the number
of polymorphic DNA markers that would be required to
provide a complete human linkage map (Solomon and
Bodmer, 1979; Botstein et al, 1980; Bishop et al,
1983). Assuming a total recombination fraction of 30
Morgans, and that the human genome approximates to a
linear array of genes laid out along a single
dimension, it is obvious that 150 markers, spaced
exactly equally at 20 centimorgans distance, would

ensure that any random phenotype, or DNA sequence, would be no more than 10 cM from one marker. Of course, such a simplistic calculation is a minimum estimate, from several viewpoints. First, it applies to male meiosis, and female meiosis probably involves a higher degree of crossing over. Second, probes will be distributed randomly and not at equal spacing. Third, the human genome is divided into chromosomes and recombination is not uniform along their length. Finally, any family will only be informative for a proportion of the markers that can be assigned. Therefore, the number of random polymorphic clones that should be placed, chromosome by chromosome, in the human genome to provide a complete map is more like 1,000 than 150.

Isolation of X chromosome sequences

The first methods of obtaining X chromosome specific sequences depended on somatic cell hybrids. Schmeckpeper et al (1979) demonstrated that it is possible to isolate X-chromosome sequences by utilising nucleic acid reassociation to enrich for X-chromosomal DNA. This approach was first used to obtain Y chromosome specific DNA (Kunkel, 1976, 1977) and depends on reassociation between radiolabelled DNA from cells containing "probe" chromosome and excess of unlabelled "driver" DNA from cells devoid of probe chromosome. Olsen et al (1980), using a related approach, obtained X chromosome sequences by hybridisation of labelled human DNA with DNA from a human-mouse hybrid cell line which contains a single human X chromosome. Both of these approaches are limited to the isolation of sequences that are not homologous between mouse and man and are therefore not suitable for the construction of representative

libraries. Gusella et al (1980) have demonstrated
that somatic cell hybrids which contain the human
chromosome of interest as their only human component
can be cloned directly and the human sequences
differentiated from the mouse sequences by their
hybridisation to the human Alu repeat. This method is
very useful, particularly if a series of libraries
cloned into both cosmids and phage is required.
However, it is of limited use for library-library
comparisons such as cDNA and genomic library cross
screening (Davies et al, 1983d).

The most straightforward method of cloning a
chromosome enriched library is to purify the
chromosomes first (Young and Davies, 1983). Attempts
to fractionate metaphase chromosomes on gradients gave
disappointing results (Padgett et al, 1977; Pinaev et
al, 1979; Collard et al, 1982) because while they give
enrichment for different size classes of chromosomes,
the similar size of many human chromosomes does not
allow significant purification of individual human
chromosomes. This problem can be resolved by using
flow cytometry to sort chromosomes stained with
fluorescent dyes such as ethidium bromide or Hoescht
(Carrano et al, 1979). In the case of ethidium
bromide, the binding of the dye is directly
proportional to the DNA content and the human X
chromosome sorts between chromosomes 7 and 8. By flow
sorting chromosomes prepared from a (48,XXXX) cell
line, Davies et al (1981) constructed a library in
which 90% of the sequences were derived from the X
chromosome. More recently, sorting in two dimensions
using somatic cell hybrids and different combinations
of dyes has increased the degree of purification
attainable using this technique (Muller et al, 1983a).

Figure 1 shows the sorting profile obtained from the FACS-II cell sorter and a 48,XXXX cell line.

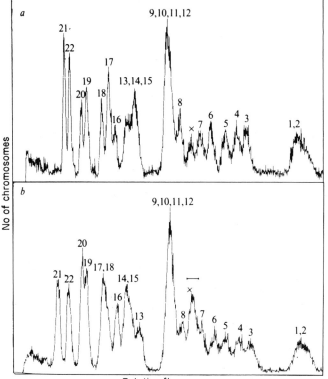

Figure 1   Flow karyotypes from: a, 46,XY diploid fibroblast cells; b, GM1416 48,XXXXX lymphoblastoid cells (reprinted by permission from Nature, 293, 374-376, copyright MacMillan Journals Ltd).

Isolation of single copy sequences for genetic mapping

The libraries constructed by the methods described above all contain inserts in phage or plasmids of 6kb or more which means that most of the clones contain repetitive sequences and cannot be used directly in genetic linkage analysis. Thus, the individual sequences need to be subcloned before use

in Southern blotting experiments. This is a rather
tedious task so the libraries are usually prescreened
with highly repetitive DNA to eliminate the most
abundant repeats and enrich for single-copy sequences
(Davies et al, 1983b). Two methods have recently been
described to reduce this problem and allow the direct
analysis of the clones without removal of repetitive
DNA. The first method relies on a prehybridisation
step which effectively competes out the repetitive DNA
and allows the analysis of the clone by Southern
blotting (Davies et al, 1983b). The second approach
relies on the hybridisation of the clones under
conditions of very high stringency so that the
repetitive DNA does not hybridise (Fisher et al,
1984). Figure 2 shows the use of the former technique
in the identification of a human X chromosome single
copy sequence. The single-copy X chromosome band can
be clearly seen in the human track after preincubation
of the probe with excess cold sheared human DNA.

Physical mapping

Once single copy sequences have been identified,
they need to be localised physically along the
chromosome length and shown to be unique in the genome
before they can be used for genetic analysis. This
can be performed by a variety of methods.

1. Chromosome sorting

DNA sequences can be assigned to a chromosome by
sorting the chromosomal DNA and by analysis using dot
hybridisation or Southern blotting (Davies et al,
1983a). The advantage of this approach is that it
assigns the sequence to a particular chromosome and
also demonstrates whether the sequence is unique by
the presence or absence of hybridisation to other

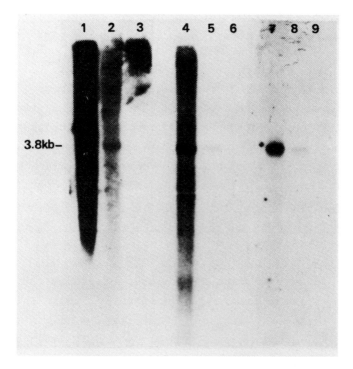

Figure 2   Characterisation of single-copy
sequences in clones containing repeats.
DNA digested with EcoR I: tracks 1, 4 and 7
human; tracks 2, 5 and 8 HORL9X DNA;
tracks 3, 6 and 9 mouse DNA.   Tracks 1-3
hybridisation with total phage clone;
tracks 4-6 hybridisation with small sub-fragment
of clone (reproduced from Davies et al (1983b)
with kind permission from Cold Spring Harbour
Laboratory).

areas of the sorting profile. In addition, the sorting
of translocations and DNA restriction analysis can be
used to localise sequences regionally (Lebo et al,
1979).

2. Dosage hybridisation

This method involves hybridising the clone of interest to a series of cell line DNAs derived from individuals deleted for part of the human X chromosome. Females with deletions in only one chromosome can be detected by the observation that they will only give half of the normal hybridisation signal if the probe lies within the deleted region. Kunkel et al (1983) localised several X chromosome probes in this manner and determined their specifity by hybridisation to cell lines with increased numbers of X chromosomes.

3. Somatic cell hybrids

Because the human X chromosome contains the gene for HPRT, its presence can be selected for in HAT medium (Littlefield et al, 1964). When a rodent-human hybrid cell line is made between a normal human cells and rodent cells which are HPRT-negative, and the fused cells are grown in HAT, the human chromosomes are gradually eliminated from the hybrid with the exception of the X chromosome which is required for cell survival. Somatic cell hybrids have been constructed which contain the whole or translocated parts of the human X chromosome. The power of a panel of such hybrids in the regional assignment of a DNA sequence is demonstrated in figure 3 where RC8 is assigned to a region of the X chromosome (Murray et al, 1982). Somatic cell hybrids containing deleted X chromosomes can also be of great value. Ingle et al (1985) have constructed a hybrid containing a human X chromosome deleted Xp11.4-Xp21.3 to identify sequences close to the Duchenne muscular dystrophy locus.

Many human DNA sequences also cross-hybridise to rodent DNA and it is not always possible to

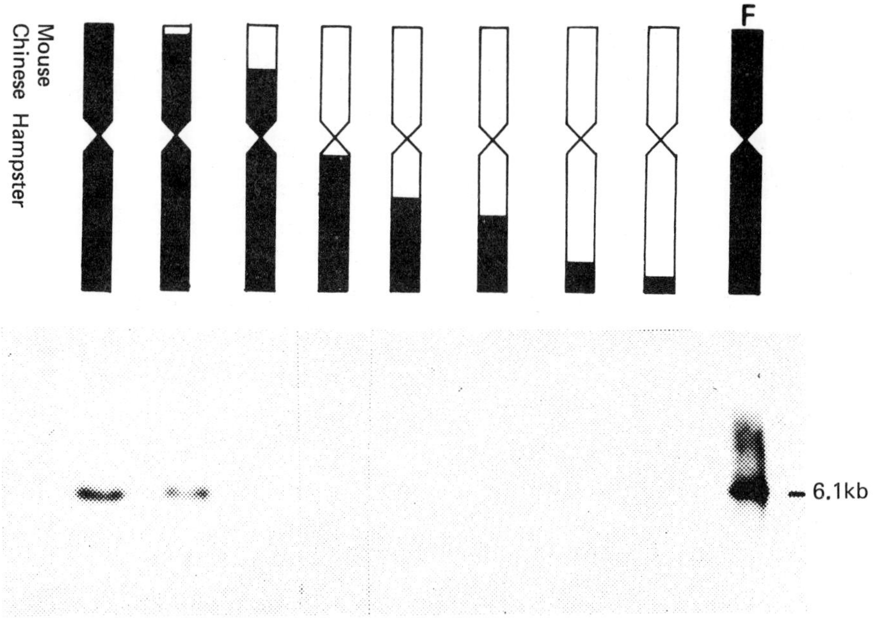

Figure 3   Localisation of an X-chromosome
probe to Xp21-Xp22.3 on a somatic cell
hybrid panel (reprinted by permission
from Nature, 300, 67-71, copyright
MacMillan Journals Ltd).

differentiate between the rodent bands and the human
bands in a Southern blot.   This is particularly
difficult with multigene families such as glycerol 3-
phosphate dehydrogenase (Benham et al, 1984).

Occasionally, when hybrid cells are grown for
many passages, the human X chromosome fragments, and
only the region near Xqter (where there is selection
pressure for the HPRT gene) is retained.   Thus,
hybrids need to be regularly karyotyped to test their
stability.

The fragmentation of the X chromosome can be
induced by irradiation of the cells with X-rays and
then exploited for mapping.   If the resultant

irradiated cell is fused with a mouse cell and HAT
selection is applied, hybrids can be constructed with
varying amounts of the terminal end of the long arm of
the human X chromosome (Goss and Harris, 1977). This
provides a panel of hybrids containing different
regions of the human X chromosome from Xqter towards
Xp.

4. In situ hybridisation

Perhaps the most useful, yet most difficult,
technique for localisation of cloned single copy
sequences is in situ hybridisation. This technique
was first used by Gall and Pardue (1969) for
repetitive sequences, and later was attempted for
single copy genes (Harper and Saunders, 1981; Malcolm
et al, 1981).

This technique has been used to assign six single
copy sequences regionally along the human X chromosome
(see figure 4).

This method is not yet very reproducible and some
probes give a much better signal than others for
reasons not yet fully understood. It is particularly
useful if the sequence of interest is thought to be
localised near a translocation breakpoint. OTC had
been localised by in situ hybridisation with a
(48,XXXX) cell line to the region Xp11.4-Xp21
(Lindgren et al, 1984). The silver grains on the slide
corresponding to hybridisation were centred in this
region but more precise localisation was not possible.
Further experiments, using a cell line from a female
Duchenne muscular dystrophy patient with a balanced
X/9 translocation (46,X,t(X;9) (p21) (p22)), localised
OTC below the Xp21 breakpoint because no grains were
observed on the derivative chromosome containing Xp21-
Xpter.

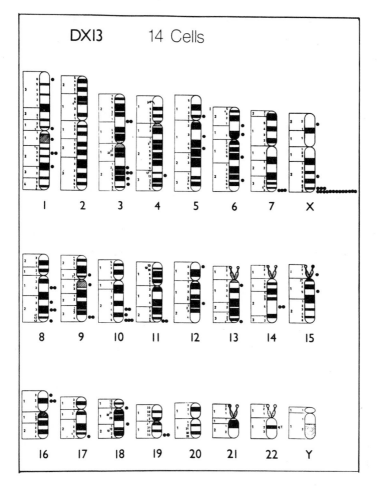

Figure 4   In situ hybridisation of X
chromosome probe DX13.   Each dot represents
the chromosomal position of a single silver
grain (reproduced from from Hartley, D.A.,
Thesis, (1983) with kind permission).

It is also possible to use non-radioactive
probes for in situ hybridisation which help the
background problem.   Biotin analogues are already
being tested for this application but only limited

success has been achieved for single copy sequences
(Leary et al, 1983; Tchen et al, 1984).

Localisation of mutant loci: X-linked muscular dystrophies

Duchenne muscular dystrophy is one of the most
common severe X-linked diseases. This progressive
dystrophy, usually fatal before the age of 25, affects
approximately one boy in 2,000 (Emery, 1980). The
biochemical defect is a complete enigma, and there
were no linkage data between DMD and any other marker
prior to 1982 (for review, see Moser, 1984b). However,
there was a clue to the location of the gene defect.
In females, one of the two X chromosomes is normally
inactive, and this inactivation occurs randomly in
normal women. When a balanced translocation occurs
between the X chromosome and an autosome, the
translocated portions of the X chromosome remain
active and the intact X becomes inactivated in nearly
all cells. Twelve females have been found who have a
DMD-like phenotype, associated with a translocation of
the X chromosome with a breakpoint at or near Xp21
(see Elejalde and Elejalde, 1983 for review).
Therefore, it was thought that this region might be
the locus of the gene which is mutated in DMD.
However, the data cannot be accepted uncritically, as
the female patients have a spectrum of symptoms
ranging from classic severe DMD through to mild
disease, more typical of Becker muscular dystrophy.
At that time, Becker muscular dystrophy was thought to
be on the long arm of the X chromosome because it
showed linkage to colour blindness.

Since DMD is not within measurable genetic
distance of any of the classical linkage groups on the
X chromosome, confirmation of the localisation of the

defect at Xp21 required the isolation of RFLPs localised close to this region. One DNA sequence isolated from the X chromosome library was localised to Xp21-Xp22.3 and revealed a TaqI polymorphism in Southern blots of normal individuals (Davies et al, 1981; Murray et al, 1982). This DNA locus, laboratory acronym RC8, was used to screen DMD carrier females for informative heterozygotes for the TaqI polymorphism. Fifteen informative kindreds were found after screening sixty known carrier females. Analysis of the segregation of RC8 and DMD in these kindreds demonstrated that RC8 is linked to DMD, at a recombination fraction of approximately 0.15 (Murray et al, 1982; Davies et al, 1983c). This was the first use of DNA linkage to prove the location of a mutation causing the disease, where the biochemical defect is unknown.

A second RFLP, detected by the probe L1.28, was also found to be linked to DMD. L1.28 is localised on the opposite side of the locus to RC8 at Xp11.4 and is genetically linked at approximately 20cM. If the genetic distance between RC8 and L1.28 is assumed to be about 30cM, then the frequency of double recombinants between them would be expected to be quite low. Therefore, used together as bridging markers for disease, they can be used for exclusion of carrier status in women at risk (Harper et al, 1983; Pembrey et al, 1983). This approach is, of course, limited by how polymorphic these markers are in the general population. For example, only 25% of mothers will be heterozygous at the RC8 locus and almost 50% at the L1.28 locus. Only 1 in 10 mothers will be informative for both. Aldridge et al (1984) estimated that between four and eight two allele loci of high frequency would be needed on each side of the DMD

locus in order to ensure double heterozygosity at the
90% level.

There are now at least 11 polymorphic loci
flanking the DMD locus available for use in carrier
status determination (Pearson and Van Ommen, 1984).
These have also been used for prenatal diagnosis where
the carrier mother was heterozygous for two markers
proximal and one marker distal to the locus (Bakker et
al, 1985.) Analysis of the segregation pattern
indicated the occurrence of a double recombinant event
between tne bridging loci. Other families have also
been studied where double recombinants occur (Davies
et al, 1985b; Bakker et al, 1985). It is not yet clear
whether this indicates a higher frequency of
recombination around the DMD region since the sample
size is small. However, the results may also be
explained if rearrangements, such as inversions, occur
in this region of the X chromosome.

As mentioned earlier, twelve female DMD patients
have now been characterised who possess balanced
X/autosome translocations with a breakpoint at Xp21.
Careful cytogenetic examination of the chromosomes
suggests that the breakpoint is not exactly the same
in all of the cases (Boyd and Buckle, 1985). This
may indicate that the locus is highly heterogeneous
or that there are position effects. Support for the
former is given by the fact that the clinically
milder Becker muscular dystrophy (BMD) is localised
within the same region of the X chromosome and may be
allelic (Kingston et al, 1983a,b). In addition, DMD
shows a very high spontaneous mutation rate (7 X 10^{-5}) (Moser, 1984a). The cloning of one of the
breakpoints in the female translocations should shed
light on this (Worton et al, 1984).

X-linked mental retardation

The most common form of X-linked mental retardation is associated with a fragile site in the band Xq27 of the human X chromosome in cells of the affected individuals (see De Arce and Kearns, 1984 for review). The fragile site is detected after culturing lymphocytes in folate-deficient media. Its actual relationship to X-linked mental retardation is unclear. Less than 1% of males with the gene express the fragile (X) and have a normal IQ (Brondum-Nielsen et al, 1981; Webb et al, 1981; Sherman et al, 1984). However, there are now many reports of males who are mentally normal and transmit the fragile (X) to their daughters (Martin and Bell, 1943; Wolff et al, 1978; Brondum-Nielsen et al, 1981; Camerino et al, 1983; Jacobs et al, 1983; Froster-Iskenius et al, 1984). Thus fragile (X) associated mental retardation does not show the normal inheritance pattern for an X-linked recessive disease. Both variable penetrance and a premutation event have been proposed to account for this unusual inheritance pattern (Pembrey et al, 1984).

Linkage studies of HPRT, which lies proximal to the fragile site at Xq26, indicate that the fragile site is not closely linked to these loci (Nussbaum et al, 1983). A random DNA probe 52A, which is distal to HPRT but proximal to the fragile site (Mattei et al, 1985) at Xq27 also shows only loose linkage with the fragile site (Davies et al, 1985b). However, Filippi et al (1983) found measurable linkage between G6PD and the fragile site. Factor IX was first reported to be very closely linked to the disease in a large French kindred (Camerino et al, 1983) but the genetic distance appears to be about 20cM in other families 'Choo et al, 1984; Davies et al, 1985b). Studies by

Purrello et al (1985) with factor VIII demonstrate
that this locus lies distal to the fragile site. These
authors postulate that the fragile site is a hot spot
for recombination such that markers bridging it
recombine freely whereas those on either side
segregate proportionally to physical distance (Purello
et al, 1985; Szabo et al, 1985). This view, however,
is not supported by the data of Davies et al (1985b)
who report a high frequency of recombination across
the fragile site and between markers lying proximal,
but close physically, to the fragile site. Studies of
the segregation of more markers bridging the Xq27
region need to be undertaken in both normal and
affected kindreds before any conclusions can be drawn.
What is abundantly clear is that there is as yet no
clinically useful closely linked probe for this
disorder (Choo et al, 1984).

Other X-linked disorders

Several other X-linked disorders have shown to be
linked to RFLPs amongst them are retinitis pigmentosa
(Bhattacharya et al, 1984), Menke's disease (MNK
(Friederich et al, 1983), retinoschisis (RS) (Wieacker
et al, 1983c) and haemophilia A (Harper et al, 1984)
(see Human Gene Mapping Conference VII, 1983; Ropers
et al, 1983).

Several gene specific probes for genetic
disorders have also been isolated, notably PGK, OTC,
HPRT, G6PD, F9 (factor IX), F8 (factor VIII), (Singh-
Sam et al, 1983; Michelson et al, 1983; Horwich et al,
1984; Brennand et al, 1982; Jolly et al, 1982; Persico
et al, 1981; Choo et al, 1982; Camerino et al, 1984;
Gitschier et al, 1984; Wood et al, 1984; Vehar et al,
1984; Toole et al, 1984). Pseudogenes have also been
identified on the X chromosome and in some instances
more than one copy is present. Examples are an actin-

like sequence and pseudo-genes for arginosuccinate synthetase and glycerol 3'-phosphate dehydrogenase (Hanauer et al, 1984; Beaudet et al, 1982; Benham et al, 1984).

Mapping with non-disjunction families

Mothers of Klinefelter's (XXY) children informative for X-linked markers can be used to map the distance and sequential order of the X-linked loci with respect to the centromere (Siniscalco et al, 1979). During the first meiosis homologous loci separate and the division is "reductional" with respect to the genes involved. However, the division will be "equational" when the sister loci separate. Since the centromere of each homolog is thought to stay undivided until the second meiosis, the centromeres will always be separated reductionally. Thus in a heterozygote for an X-linked marker the equational separation of the two alleles can be accomplished only if the original relationship between the marker and the centromere is changed by a recombinational event in the first meiotic division. Therefore the relative proportion of reductional and equational separation of X-linked loci during the first meiosis will depend upon the distance between each marker and the centromere.

Repetitive sequences on the human X chromosome

Several repetitive sequence families have been localised on the X chromosome. A 2kb tandem Bam H1 repeat consists of 5000-7000 copies and constitutes 6-10% of the X chromosome predominantly at the centromere (Yang et al, 1982; Willard and Smith, 1982). A 550bp region of this repeat DNA shares homology with a cloned alpha-dimer and also shows homology with several primate DNAs, although the X

linkage of these sequences in other primates has not
been determined (Willard and Smith, 1982; Willard et
al, 1983).

More recently, Jabs et al (1984) identified
another centromeric sequence organised as a tandem
3.0kb Bam H1 repeat. The repetitive DNA is localised
at the X chromosome centromere and at other autosomal
centromeric regions. Since some individuals possess
additional repeats of different sizes, these
polymorphic variants should be useful for linkage
studies.

X/Y homologous sequences

It has long been known that the terminal portion
of the short arm of the X chromosome undergoes
synopsis at male meiosis (Moses et al, 1975; Polani,
1981; Chandley et al, 1984). X chromosome genes
contained in this segment including Xg and STS, escape
inactivation when carried on a structurally normal X
chromosome. It has been assumed that each of these X
chromosome genes corresponds to a homologous sequence
on the Y chromosome and indeed evidence has been
presented for the existence of a regulatory "Yg" locus
and for the homology of M1C2X and M1C2Y (Goodfellow et
al, 1983a; Goodfellow et al, 1983b). However, it may
not be the coding regions but their flanking sequences
(repetitive or single-copy) that are sufficiently
homologous to enable meiotic pairing.

The cloning of single-copy DNA sequences for the
X and Y chromosomes has permitted the investigation of
homology in several regions of the X chromosome. One
sequence has been reported to be localised Xp223-Xpter
but it is not clear whether the Y site is within the
pairing region (Camerino et al, 1984). One sequence
has been localised to Xq13-Xq22 on the human X
chromosome and shows homology to a sequence on the Y

chromosome (Page et al, 1982). Ten clones from a Y
enriched cosmid library that map to both the X and Y
have been mapped to Xq (Weissenbach et al, 1984).
Homologous sequences are thus not confined to the
meiotic pairing region (Cooke et al, 1984; Rappold et
al, 1984). One repetitive sequence localised Ycen-
Yq11 from a Y enriched library hybridises to an X
chromosome sequence in Xq12-Xq21 (Muller et al, 1983).

A single obligatory crossover between X and Y
within the pairing region has been proposed in mice
since sex reversal has been shown to be due to
crossover of sequences from the Y chromosome to the X
chromosome (Singh and Jones, 1982). Some human XX
males possess extra chromosomal material but many do
not (for review see Burgoyne, 1982; Polani, 1982). The
understanding of the mechanisms underlying these
observations must await a fuller characterisation of
the X and Y chromosomes in the pairing region.

Total linkage map of the human X chromosome

A linkage map of the human X chromosome was
constructed by Drayna et al (1984) using RFLPs and
conventional markers. They estimated the total length
to be of the order of 200cM. However, as can be seen
from figure 5, genetic distances between markers are
not directly proportional to the physical distances.
Recombination increases towards the telomeres in a
very similar way to that observed for autosomes
(Hulten et al, 1982; Laurie et al, 1981). In
particular, towards the end of the long arm factor IX
(F9) and factor VIII (F8) only show very loose linkage
in spite of lying within a relatively short physical
distance at Xq26 and Xq28 respectively. Similarly, on
the short arm RC8 is not closely linked to Xg
(Sarfarazi et al, 1983) and shows only loose linkage
with STS and RS (Wieacker et al, 1983a,c).

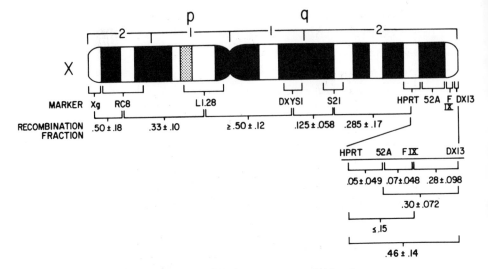

Figure 5   Genetic linkage map of the human
X chromosome using random RFLPs (reprinted
from Drayna et al (1984) with kind
permission from Proc. Natl. Acad. Sci. USA).

The direct comparison of the genetic and physical
map is presented in figure 6. The DNA sequences were
localised by in situ hybridisation (Hartley et al,
1984) and the genetic map was constructed from Drayna
et al (1984). As this map becomes more complete with
the inclusion of new markers, it should be possible to
localise any X-linked phenotype to a small region of
the X chromosome. A summary of the assigments of
cloned DNA sequences is given in figure 7.

Comparative mammalian cytogenetics has indicated
many differences in karyotype between autosomes but a
remarkable conservation of the X chromosome (Ohno,
1964). This led Ohno to suggest the hypothesis of
evolutionary conservation of X-linkage groups (Ohno,
1967). This has been tested for the enzyme loci HPRT
and GALA (alpha-galactosidase) in the mouse and human
X chromosomes where the order is found to be inverted

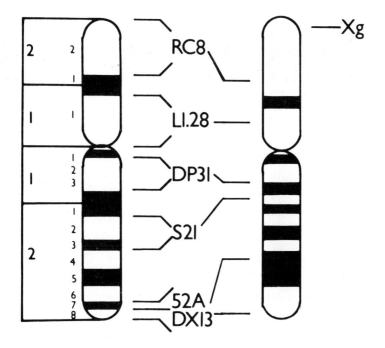

Physical   Genetic

Figure 6   Physical and genetical maps of
the human X chromosome (reproduced from
Hartley et al (1984) with kind
permission from Nucl. Acids Res).

(Francke et al, 1980).   A more detailed analysis by
Buckle et al (1983) suggests that for most genes,
while evolutionary pressure has conserved X syntenic
groups in mammals, restructuring within the X
chromosome has led to changes in their relative
positions.

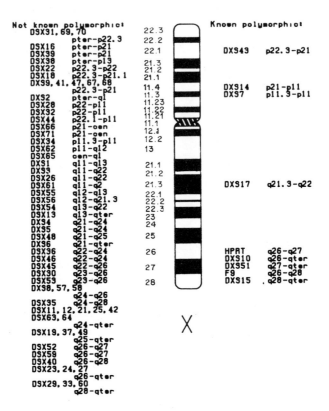

Figure 7   Mapping of cloned DNA probes on
the human X chromosome (reprinted from Human
Gene Mapping VII with kind permission from
Cytogenet. Cell Genet).

Summary
     The construction of a complete linkage map of the
human X chromosome is now within sight as more DNA
sequences are isolated and characterised from X
chromosome enriched libraries. This will provide a
valuable resource for the molecular geneticist to test
Ohno's hypothesis of evolutionary conservation of
primordial X-linkage groups and to identify the sites

of genetic recombination. Only by analysing the X chromosome at the DNA level, will the scientist be able to understand the molecular basis of X-linked disorders and pairing of the sex chromosomes at meiosis. The human X chromosome map has indeed advanced very rapidly since Sturtevant's paper first reported the linear arrangement of sex-linked factors in Drosophila (Sturtevant, 1913). The next decade should witness further significant advances in our understanding of the molecular genetics of the human X chromosome.

Acknowledgements

We are very grateful to Rachel Kitt for the patient typing of this manuscript. We thank the Medical Research Council, The Muscular Dystrophy Group of Great Britain and The Muscular Dystrophy Association of America for financial support.

References

Adam, A., Ziprkowski, L., Feinstein, A., Sanger, R.,
   Tippett, P., Gavin, J., and Race, R.R. (1969).
   'Linkage relations of X-borne icthyosis to the Xg
   blood groups and to other markers of the X in
   Israelis', J. Hum. Gen., 32, 323-332.

Aldridge, J., Kunkel, L., Bruns, G., Tantravahi, U.,
   Lalande, M., Brewster, T., Moreau, E., Wilson, M.,
   Bromley, W., Roderick, T., and Latt, S.A. (1984).
   'A strategy to reveal high-frequency RFLPs along
   the human X chromosome', Am. J. Hum. Genet., 36,
   546-565.

Bakker, E., Hofker, M.H., Goorl, N., Mandel, J.L.,
   Davies, K.E., Kunkel, L.M., Willard, H.F., Fenton,
   W.A., Sandkuyl, L., Majoor-Krakauer, D., Van
   Essen, A., Jahoda, M., Sachs, E.S., Van Ommen,
   G.J.B., and Pearson, P.L. (1985). 'Prenatal
   diagnosis and carrier detection of Duchenne
   muscular dystrophy with closely linked RFLPs',
   Lancet, i, 655-658.

Barker, D., Schafer, M., and White, R. (1984).
   'Restriction sites containing CpG show a higher
   frequency of polymorphism in human DNA', Cell, 36,
   131-138.

Beaudet, A.L., Su, T.S., O'Brien, W.E., D'Eustachio,
   P., Barker, P.E., and Ruddle, F.H. (1982).
   'Dispersion of argininosuccinate-synthetase-like
   human genes to multiple autosomes and the X
   chromosome', Cell, 30, 287-293.

Bell, G.I., Selby, M.J., and Rutter, W.J. (1982).
    'The highly polymorphic region near the human
    insulin gene is composed of simple tandemly
    repeating sequences', Nature, 295, 31-34.

Benham, F.J., Hodgkinson, S., and Davies, K.E.
    (1984). 'A glyceraldehyde-3-phosphate
    dehydrogenase pseudogene on the short arm of the
    human X chromosome defines a multigene family',
    EMBO J., 3, 2635-2640.

Bhattacharya, S.S., Wright, A.F., Clayton, J.F.,
    Price, W.H., Phillips, C.I., McKeown, C.M.E., Jay,
    M., Bird, A.C., Pearson, P.L., Southern, E.M., and
    Evans, H.J. (1984). 'Close genetic linkage between
    X-linked retinitis pigmentosa and a restriction
    fragment length polymorphism identified by
    recombinant DNA probe', Nature, 309, 253-255.

Bishop, D.T., Williamson, J.A., and Skolnick, M.H.
    (1983). 'A model for restriction fragment length
    distributions', Am. J. Hum. Genet., 35, 795-815.

Botstein, D., White, R.L., Scolnick, M.H., and Davis,
    R.W. (1980). 'Construction of a genetic linkage
    map in man using restriction fragment length
    polymorphisms', Am. J. Hum. Genet., 32, 314-331.

Boyd, Y., and Buckle, V.T. (1985). 'Cytogenetic
    heterogeneity in translocations associated with
    Duchenne muscular dystrophy', submitted for
    publication.

Brennand, J., Chinault, A.C., Konecki, D.W., Melton,
    D.W., and Caskey, C.T. (1982). 'Cloned cDNA

sequences of the hypoxanthine/guanine
phosphoribosyltransferase gene from a mouse
neuroblastoma cell line found to have amplified
genomic sequences', Proc. Natl. Acad. Sci. USA,
79, 1950-1954.

Brondum-Nielsen, K., Tommerup, N., Poulen, H., and
Mikkelsen, M. (1981). 'A pedigree showing
transmission by apparently unaffected males and
partial expression in female carriers', .
Hum. Genet., 59, 23-25.

Buckle, V.J., Edwards, J.H., Evans, E.P., Jonasson,
J., Lyon, M.F., Peters, J., Searle, A.G., and
Wedd, N.S. (1984). 'Chromosome maps of man and
mouse', Clin. Genet., 26, 1-11.

Burgoyne, P.S. (1982). 'Genetic homology and crossing
over in the X and Y chromosomes of mammals',
Hum. Genet., 61, 85-90.

Camerino, G., Grzeschik, K.H., Jaye, M., De La Salle,
H., Tolstoshev, P., and Lecocq, J.P. (1984).
'Regional localization on the human X chromosome
and polymorphism of the coagulation factor IX gene
(hemophilia B locus)', Proc. Natl. Acad. Sci. USA,
81, 498-502.

Camerino, G., Mattei, M.G., Mattei, J.F., Jaye, M.,
and Mandel, J.L. (1983). 'Close linkage of fragile
X mental retardation syndrome to hemophilia B and
transmission through a normal male', Nature, 306,
701.

Capon, D.J., Chen, E.Y., Levinson, A.D., Seeburg, P.H., and Goeddel, D.V. (1983). 'Complete nucleotide sequences of the T24 human bladder carcinoma oncogene and its normal homologue', Nature, 302, 33-37.

Carrano, A.V., Gray, J.W., Langlois, R.G., Burkhardt-Schultz, K.J., and Van Dilla, M.A. (1979). 'Measurement and purification of human chromosomes by flow cytometry and sorting', Proc. Natl. Acad. Sci. USA, 76, 1382-1384.

Chandley, A.C., Goetz, P., Hargreave, T.B., Joseph, A.M. and Speed, R.M. (1984). 'On the nature and extent of XY pairing at meiotic prophase in man', Cytogenet. Cell Genet., 38, 241-247.

Choo, K.H., George, D., Filby, G., Halliday, J.L., Leversha, M., Webb, G., and Danks, D.M. (1984). 'Linkage analysis of X-linked mental retardation with and without fragile-X using factor IX gene probe', Lancet, ii, 349.

Choo, K.H., Gould, K.G., Rees, D.J.G., and Brownlee, G.G. (1982). 'Molecular cloning of the gene for human anti-haemophilic factor IX', Nature, 299, 178-180.

Collard, J.G., Schijven, J., Tulp, A., and Meulenbrock, M. (1982). 'Localisation of genes on fractionated rat chromosomes by molecular hybridisation', Exp. Cell Res., 137, 463-469.

Cook, P.J.L. (1965). 'The Lutheran-secretor recombination fraction in man: a possible sex difference', <u>Ann. Hum. Genet.</u>, 28, 393-401.

Cooke, H.J., Brown, W.A.R., and Rappold, G.A. (1984).
'Closely related sequences on human X and Y
chromosomes outside the pairing region', <u>Nature</u>,
311, 259-261.

Davies, K.E., Briand, P., Ionasescu, V., Ionasescu,
G., Williamson, R., Brown, C., Cavard, C., and
Cathelineau, L. (1985a). 'Gene for OTC:
characterisation and linkage to Duchenne muscular
dystrophy', <u>Nucleic Acids Res.</u>, 13, 155-165.

Davies, K.E., Harper, K., Bonthron, D., Krumlauf, R.,
Polkey, A., Pembrey, M.E., and Williamson, R.
(1983a). 'Use of a chromosome 21 cloned DNA probe
for the analysis of non-disjunction in Down's
syndrome', <u>Hum. Genet.</u>, 66, 54-56.

Davies, K.E., Hartley, D.A., Murray, J.M., Harper,
P.S., Hill, M.E.E., Casey, G., Taylor, P., and
Williamson, R. (1983b). 'The characterisation of
sequences from a human X chromosome library for
the study of X-linked diseases', in <u>Banbury
Report: Recombinant DNA Applications to Human
disease</u> 14th Edn. (Eds. C.T. Caskey and R.L.
White), pp. 279-290, New York Cold Spring Harbour
Laboratory, Coldspring Harbour, New York.

Davies, K.E., Mattei, M.G., Mattei, J.F., Veenema,
H., McGlade, S., Harper, K., Tommerup, N.,
Nielsen, K.B., Mikkelsen, M., Beighton, P.,
Drayna, D., White, R., and Pembrey, M.E. (1985b).
'Linkage studies of X-linked mental retardation:

high frequency of recombination in the telomeric region of the human X chromosome', Hum. Genet, 70, 249-255.

Davies, K.E., Pearson, P.L., Harper, P.S., Murray, J.M., O'Brien, T., Sarfarazi, M., and Williamson, R. (1983c). 'Linkage analysis of the two cloned DNA sequences flanking the Duchenne muscular dystrophy locus on the short arm of the human X chromosome', Nucleic Acids Res., 11, 2303-2312.

Davies, K.E., Taylor, P., and Mueller, C.R. (1983d). 'Sex chromosome-specific DNA sequences', Differentiation, 23, 44-47.

Davies, K.E., Young, B.D., Elles, R.G., Hill, M.E., and Williamson, R. (1981). 'Cloning of a representative genomic library of the human X chromosome after sorting by flow cytometry', Nature, 293, 374-376.

De Arce, M.A., and Kearns, A. (1984). 'The fragile X syndrome: the patients and their chromosomes', J. Med. Genet., 21, 84-92.

Drayna, D., Davies, K.E., Hartley, D.A., Williamson, R., and White, R. (1984). 'Genetic mapping of the human X chromosome using restriction fragment length polymorphisms', Proc. Natl. Acad. Sci. USA, 81, 2836-2839.

Elejalde, B.R., and Elejalde, M.M. (1983). 'Phenotypic manifestations of X-autosome translocations', in Cytogenetics of the X Chromosome 2nd Edn. (Ed. A.A. Sandberg), pp. 225-

244, Alan R. Liss Inc, New York.

Emery, A.E.H. (1980). 'Duchenne muscular dystrophy: genetic aspects, carrier detection and antenatal diagnosis', Br. Med. Bull., 36, 117-122.

Filippi, G., Rinaldi, A., Archidiacono, N., Ricchi, M., Balazs, I., and Siniscalco, M. (1983). 'Linkage between G6PD and fragile-X syndrome', Am. J. Med. Genet., 15, 113-119.

Fisher, J.H., Gusella, J.F., and Scoggin, C.H. (1984). 'Molecular hybridization under conditions of high stringency permits cloned DNA segments containing reiterated DNA sequences to be assigned to specific chromosomal locations', Proc. Natl. Acad. Sci. USA, 81, 520-524.

Francke, U., Bakay, B., Connor, J.D., Coldwell, J.G., and Nyhan, W.L. (1974). 'Linkage relationships of X-linked enzymes glucose-6-phosphate dehydrogenase and hypoxanthine guanine phosphoribosyltransferase: recombination in female offspring of compound heterozygotes', Am. J. Hum. Genet., 26, 512-522.

Francke, U., and Taggart, R.T. (1980). 'Comparative gene mapping: order of loci on the X chromosome is different in mice and humans', Proc. Natl. Acad. Sci. USA, 77, 3595-3599.

Friedrich, U., Horn, N., and Stene, J. (1983). 'Close linkage of the gene for Menkes disease to the centromere region of the human X chromosome', Ann. Hum. Genet., in press.

Froster-Iskenius, U., Schulze, A., and Schwinger, E. (1984). 'Transmission of the marker X syndrome trait by unaffected males: conclusions from studies of large families', Hum. Genet., 67, 419-427.

Gall, J.G., and Pardue, M.L. (1969). 'Formation and detection of RNA-DNA hybrid molecules in cytological preparations', Proc. Natl. Acad. Sci. USA, 63, 378-383.

Gitschier, J., Wood, W.I., Goralka, T.M., Wion, K.L., Chen, E.Y., Eaton, D.H., Vehar, G.A., Capon, D.J., and Lawn, R.M. (1984). 'Characterisation of the human factor VIII gene', Nature, 312, 321-326.

Goodbourn, S.E.Y., Higgs, D.R., Clegg, J.B., and Weatherall, D.J. (1983). 'Molecular basis of length polymorphism in the human zeta-globin gene complex', Proc. Natl. Acad. Sci. USA, 80, 5022-5026.

Goodfellow, P. (1983). 'Expression of the 12E7 antigen is controlled independently by genes on the human X and Y chromosomes', Differentiation, 23, 35-39.

Goodfellow, P., Banting, G., Sheer, D., Ropers, H.H., Caine, A., Ferguson-Smith, M.A., Povey, S., and Voss, R. (1983). 'Genetic evidence that a Y-linked gene in man is homologous to a gene on the X chromosome', Nature, 302, 346-349.

Goss, S., and Harris, H. (1977). 'Gene transfer by means of cell fusion: radiation-induced gene segregation', J. Cell Sci., 25, 17-37.

Gusella, J.F., Keys, C., Varsanyi-Breiner, A., Kao,
F.T., Jones, C., Puck, T.T., and Housman, D.
(1980). 'Isolation and localization of DNA
segments from specific human chromosomes',
Proc. Natl. Acad. Sci. USA, 77, 2829-2833.

Hanauer, A., Heilig, R., Levin, M., Moisan, J.P.,
Grzeschik, K.H. and Mandel, J.A. (1984). The
actin gene family in man: assignment of the gene
for skeletal muscle alpha-actin to chromosome 1,
an presence for actin sequences on autosomes 2 and
3, and on the X and Y chromosomes (Abstract).
Human Gene Mapping Conference VII, 487.

Harper, K., Winter, R., Pembrey, M., Hartley, D.,
Davies, K.E., and Tuddenham, E. (1984). 'A
clinically useful DNA probe closely linked to
Haemophilia A', Lancet, ii, 1.

Harper, M.E., and Saunders, G.F. (1981).
'Localisation of single copy DNA sequences of G-
banded human chromosomes by in situ
hybridisation', Chromosoma, 83, 431-439.

Harper, P.S., O'Brien, T., Murray, J.M., Davies,
K.E., Pearson, P.L., and Williamson, R. (1983).
'The use of linked DNA polymorphisms for genotype
prediction in families with Duchenne muscular
dystrophy', J. Med. Genet., 20, 252-254.

Hartley, D.A., Davies, K.E., Drayna, D., White, R.L.,
and Williamson, R. (1984). 'A cytological map of
the human X chromosome - localisation of genetic

markers by in situ hybridisation',
Nucleic Acids Res., 12, 5277-5285.

Hofker, M.H., Wapenaar, M.C., Goor, N., Bakker, B.,
Van Ommen, G.B., and Pearson, P.L. (1985).
'Isolation of probes detecting restriction
fragment length polymorphisms from X chromosome
specific libraries: potential use for diagnosis of
Duchenne muscular dystrophy', Hum. Genet, 70,
148-156.

Horwich, A.L., Fenton, W.A., Williams, K.R.,
Kalousek, F., Kraus, J.P., Doolittle, R.F.,
Konigsberg, W., and Rosenberg, L.E. (1984).
'Structure and expression of a complementary DNA
for the nuclear coded precursor of human
mitochondrial ornithine transcarbamylase',
Science, 224, 1068-1074.

Hulten, M. (1974). 'Chiasma distribution at
diakinesis in the normal human male', Hereditas,
76, 55-78.

Hulten, M.A., Palmer, R.W., and Laurie, D.A. (1982).
'Chiasma derived genetic map and recombination
fractions: chromosome 1', Ann. Hum. Genet., 46,
167-176.

Ingle, C., Williamson, R., De la Chapelle, A., Herva,
R.R., Haapala, K., Bates, G., Willard, H.F.,
Pearson, P., and Davies, K.E. (1985). 'Mapping DMD
sequences in a human X chromosome deletion which
extends across the region of the DMD mutation',
Am. J. Hum. Genet., 37, 451-462.

Jabs, E.W., Wolf, S.F., and Migeon, B.R. (1984).
    'Characterisation of a cloned DNA sequence that is
    present at centromeres of all human autosomes and
    the X chromosome and shows polymorphic variation',
    Proc. Natl. Acad. Sci. USA, 81, 4884-4888.

Jacobs, P.A., Mayer, M., Matsuura, J., Rhoads, F.,
    and Yee, S.C. (1983). 'A cytogenetic study of a
    population of mentally retarded males with special
    reference to the marker', Hum. Genet., 63, 139-
    148.

Jeffreys, A.J. (1979). 'DNA sequence variants in
    G gamma-, A gamma-, zeta-, and beta-globin genes
    of man', Cell, 18, 1-10.

Jeffreys, A.J., Wilson, V., and Thein, S.L. (1985).
    'Hypervariable "minisatellite" regions in human
    DNA', Nature, 314, 67-73.

Jolly, D.J., Esty, A.C., Bernard, H.U., and Friedman,
    T. (1982). 'Isolation of a genomic clone partially
    encoding human hypoxanthine
    phosphoribosyltransferase', Proc. Natl. Acad. Sci.
    USA, 79, 5038-5041.

Kan, Y.W., and Dozy, A.M. (1978). 'Polymorphism of
    DNA sequence adjacent to human beta-globin
    structural gene: Relationship to sickle mutation',
    Proc. Natl. Acad. Sci. USA, 75, 5631-5635.

Kingston, H.M., Harper, P.S., Pearson, P.L., Davies,
    K.E., Williamson, R., and Page, D. (1983).
    'Localisation of the gene for Becker dystrophy',
    Lancet, ii, 1200.

Kingston, H.M., Thomas, M.F.T., Pearson, P.L.,
    Sarfarazi, M., and Harper, P.S. (1983). 'Genetic
    linkage between Becker muscular dystrophy and a
    polymorphic DNA sequence on the short arm of the X
    chromosome', J. Med. Genet., 20, 255-258.

Kunkel, L.M., Smith, K.D., and Boyer, S.H. (1976).
    'Human Y chromosome-specific reiterated DNA',
    Science, 191, 1189-1190.

Kunkel, L.M., Smith, K.D., Boyer, S.H., Borgaonkar,
    D.S., Wachtel, S.S., Miller, O.J., Breg, W.R.,
    Jones, H.W., and Rary, J.M. (1977). 'Analysis of
    human Y-chromosome-specific reiterated DNA in
    chromosome variants', Proc. Natl. Acad. Sci. USA,
    74, 1245-1249.

Kunkel, L.M., Tantravahi, U., Kurnit, D.M.,
    Eisenhard, M., Bruns, G.P., and Latz, S.A. (1983).
    'Identification and isolation of transcribed human
    X chromosome DNA sequences', Nucleic Acids Res.,
    11, 7961-7979.

Laurie, D.A., Hulten, M., and Jones, G.H. (1981).
    'Chiasma frequency and distribution in a sample of
    human males', Cytogenet. Cell. Genet., 31, 153-166.

Leary, J.J., Brigati, D.J., and Ward, D.C. (1983).
    'Rapid and sensitive colorimetric method for
    visualising biotin-labelled DNA probes hybridised
    to DNA or RNA immobilised on nitrocellulose: bio-
    blots', Proc. Natl. Acad. Sci. USA, 80, 4045-4049.

Lebo, R.V., Carrano, A.V., Burkhardt-Schultz, K.,
    Dozy, A.M., Yu, L-C., and Kan, Y.W. (1979).

'Assignment of human beta-, gamma- and zeta-globin genes to the short arm of chromosome 11 by chromosome sorting and DNA restriction analysis', Proc. Natl. Acad. Sci. USA, 76, 5804-5808.

Lindgren, V., De Martinville, B., Horwich, A.L., Rosenberg, L.E., and Francke, U. (1984). 'Human ornithine transcarbamylase locus mapped to band Xp21.1 near the Duchenne muscular dystrophy locus', Science, 226, 698-700.

Littlefield, J.W. (1964). 'Selection for hybrids from matings of fibroblasts in vitro and their presumed recombinants', Science, 145, 709-710.

Malcolm, S., Barton, P., Murphy, C., and Ferguson-Smith, M.A. (1981). 'Chromosomal localisation of a single-copy gene by in situ hybridisation - human beta-globin genes on the short arm of chromosome 11', Ann. Hum. Genet., 45, 135-141.

Martin, J.P., and Bell, J. (1943). 'A pedigree of mental defect showing sex linkage', J. Neurol. Neurosurg. Psychol., 6, 154-157.

Mattei, M.G., Baeteman, M.A., Heilig, R., Oberle, I., Davies, K., Mandel, J.L., and Mattei, J.F. (1985). 'Three probe localisations by in situ hybridisation with respect to the fragile X breakpoint', submitted for publication.

McKusick, V.A. (1980). 'The anatomy of the human genome', J. Hered., 71, 370-391.

McKusick, V.A. (1983). Mendelian Inheritance in Man,

John Hopkins University Press Ltd, Baltimore, London.

McKusick, V.A., and Ruddle, F.H. (1977). 'The status of the gene map of the human chromosomes', Science, 196, 390-405.

Michelson, A.M., Markham, A., and Orkin, S.H. (1983). 'Isolation and DNA sequence of a full length cDNA clone for human X chromosome-encoded phosphoglycerate kinase', Proc. Natl. Acad. Sci. USA, 80, 472-476.

Mohandas, T., Sparkes, R.S., Hellkuhl, B., Grzeschik, K.H., and Shapiro, L.J. (1980). 'Expression of an X-linked gene from an inactive human X chromosome in mouse-human hybrid cells: further evidence for the noninactivation of the steroid sulfatase locus in man', Proc. Natl. Acad. Sci. USA, 77, 6759-6763.

Moser, H. (1984a). 'Duchenne muscular dystrophy: pathogenic aspects and genetic prevention', Hum. Genet., 66, 17-40.

Moser, H. (1984b). 'Review of studies on the proportion and origin of new mutants in Duchenne muscular dystrophy', in Research into the Origin and Treatment of Muscular Dystrophy (Eds. L.P. Ten Kate, P.L. Pearson and A.M. Stadhouders), pp. 41-52, Excerpta Medica, Amsterdam.

Moses, M.J., Counce, S.J., and Poulsen, D.F. (1975). 'Synaptoneal complex of man in spreads of

spermatocytes, with details of the sex chromosome
pair', Science, 187, 363.

Mueller, C.R., Davies, K.E., Cremer, C., Rappold, G.,
Gray, J.W., and Ropers, H-H. (1983a). 'Cloning of
genomic sequences from the human Y chromosome
after purification by dual beam flow sorting',
Hum. Genet., 64, 110-115.

Mueller, C.R., Davies, K.E., and Ropers, H.H.
(1983b). DNA sequence homologies on human X and Y
chromosomes (Abstract). Human Gene Mapping
Conference VII, 545.

Murray, J.M., Davies, K.E., Harper, P.S., Meredith,
L., Mueller, C., and Williamson, R. (1982).
'Linkage relationship of a cloned DNA sequence on
the short arm of the X chromosome to Duchenne
muscular dystrophy', Nature, 300, 69-71.

Nussbaum, R.L., Crowder, W.E., Nyhan, W.L., and
Caskey, C.T. (1983). 'A three allele restriction
fragment length polymorphism at the hypoxanthine
phosphoribosyltransferase locus in man',
Proc. Natl. Acad. Sci. USA, 80, 4035-4039.

Ohno, S. (1967). Sex Chromosomes and Sex Linked
Genes, Springer Verlag, New York.

Ohno, S., Becak, W., and Becak, M.L. (1964). 'X-
autosome ratio and the behavior pattern of
individual X chromosomes in placental animals',
Chromosoma, 15, 14-18.

Olsen, A.S., McBride, O.W., and Otey, M.C. (1980).

'Isolation of unique sequence human chromosomal deoxyribonucleic acid', Biochemistry, 19, 2419-2428.

Padgett, T.G., Stubbledield, E., and Varmus, H.E. (1977). 'Chicken macrochromosomes contain sequences related to the transforming gene of ASV', Cell, 10, 649-657.

Page, D., De Martinville, B., Barker, D., Wyman, A., White, R., Francke, U., and Botstein, D. (1982). 'Single-copy sequence hybridises to polymorphic and homologous loci on human X and Y chromosomes', Proc. Natl. Acad. Sci. USA, 79, 5352-5356.

Pai, G.S., Sprenkle, J.A., Do, T.T., Mareni, C.E., and Migeon, B.R. (1980). 'Localisation of loci for hypoxanthine phosphoribosyltransferase and glucose-6-phosphate dehydrogenase and biochemical evidence of nonrandom X chromosome expression from studies of a human X-autosome translocation', Proc. Natl. Acad. Sci. USA, 77, 2810-2813.

Pearson, P.L., and Van Ommen, G.B. (1984). 'Recent developments in DNA research of Duchenne muscular dystrophy', in Research into the Origin and Treatment of Muscular Dystrophy (Eds. L.P. Ten Kate,P.L. Pearson and A.M. Stadhouders), pp. 91-100, Excerpta Medica, Amsterdam.

Pembrey, M.E., Davies, K.E., Winter, R.M., Elles, R.G., Williamson, R., Fazzoni, T.A., and Walker, C. (1983). 'The clinical use of DNA markers linked to the gene for Duchenne muscular dystrophy',

Arch. Dis. Child., 59, 208-216.

Pembrey, M.E., Winter, R., and Davies, K.E. (1984).
   'A premutation that generates a defect at
   crossing-over explains the inheritance of fragile
   (X) mental retardation. J. Med. Genet., 21, 299.

Persico, M.G., Toniolo, C., Nobile, C., D'Urso, M.,
   and Luzzatto, L. (1981). 'cDNA sequences of human
   glucose 6-phosphate dehydrogenase cloned in
   pBR322', Nature, 294, 778-780.

Pinaev, G., Bardyopadhyay, D., Glekov, O., Shanbag,
   V., Johansson, G., and Albertsson, P.A. (1979).
   'Fractionation of chromosomes. I. A methodological
   study on the use of partition in aqueous two-phase
   systems and multiple sedimentation', Exp. Cell
   Res., 124, 191-203.

Polani, P.E. (1981). 'Abnormal sex development in
   man', in Mechanisms of Sex Differentiation in
   Animals and Man (Eds. C.R. Austin and R.G.
   Edwards), pp. 465, Academic Press, London.

Polani, P.E. (1982). 'Pairing of X and Y chromosomes,
   non-inactivation of X-linked genes, and the
   maleness factor', Hum. Genet., 60, 207-211.

Purrello, M., Alhadeff, B., Esposito, D., Szabo, P.,
   Rocchi, M., Truett, M., Masiarz, F., and
   Siniscalco, M. (1984). 'The human genes for
   Hemophilia A and B flank the X chromosome fragile
   site at Xq27.3', submitted for publication.

Race, R.R., and Sanger, R. (1975). Blood Groups in
    Man, 6th Edn. Blackwells, Oxford, London.

Rappold, G.A., Cremer, T., Cremer, A., Back, W.,
    Bogenberger, J. and Cooke, H.J. (1984).
    'Chromosome assignment of two cloned DNA probes
    hybridising predominantly to human sex
    chromosomes', Hum. Genet., 65, 257-261.

Renwick, J.H., and Schulze, J. (1965). 'Male and
    female recombination fractions for the nail-
    patella: ABO linkage in man', Ann. Hum. Genet.,
    28, 379-392.

Ropers, H-H., Wieacker, P., Wienker, T.F., Davies,
    K.E., and Williamson, R. (1983). 'On the genetic
    length of the short arm of the human X
    chromosome', Hum. Genet., 65, 53-55.

Sarfarazi, M., Harper, P.S., Kingston, H.M., Murray,
    J.M., O'Brien, T., Davies, K.E., Williamson, R.,
    Tippett, P., and Sanger, R. (1983). 'Genetic
    linkage relationships between the Xg blood system
    and two chromosome DNA polymorphisms in families
    with Duchenne and Becker muscular dystrophy',
    Hum. Genet., 65, 169-171.

Schmeckpeper, B.J., Smith, K.D., Dorman, B.P.,
    Ruddle, F.H., and Talbot, C.C. (1979). 'Partial
    purification and characterisation of DNA from the
    human X chromosome', Proc. Natl. Acad. Sci. USA,
    76, 6525-6528.

Sherman, S.L., Morton, N.E., Jacobs, P.A., and
    Turner, G. (1984). 'The marker (X) syndrome: a

cytogenetic and genetic analysis',
Ann. Hum. Genet., 48, 21-37.

Singer-Sam, J., Simmer, R.L., Keith, D.H., Shively,
    L., Teplitz, M., Itakura, K., Gartler, S.M., and
    Riggs, A.D. (1983). 'Isolation of a cDNA clone for
    human X-linked 3-phosphoglycerate kinase by use of
    mixture of synthetic oligodeoxyribonuceotides as a
    detection probe', Proc. Natl. Acad. Sci. USA, 80,
    802-806.

Singh, L., and Jones, J.W. (1982). 'Sex reversal in
    the mouse (Mus musculus) is caused by a recurrent
    non-reciprocal crossover involving the X and an
    aberrant Y chromosome', Cell, 28, 205-216.

Siniscalco, M. (1979). 'Approaches to human linkage',
    Prog. Med. Genet., 3, 221-307.

Solomon, E., and Bodmer, W.F. (1979). 'Evolution of
    sickle variant gene', Lancet, i, 923.

Sturtevant, A.H. (1913). 'The linear arrangement of
    six sex-linked factors in Drosophila as shown by
    their mode of association', J. Exp. Zool., 14,
    43-59.

Szabo, P., Purrello, M., Rocchi, M., Archidiacono,
    N., Alhadeff, B., Filippi, G., Toniolo, D.,
    Martini, G., Luzzatto, L., and Siniscalco, M.
    (1984). 'Cytological mapping of the human G6PD
    gene distally to the fragile X site suggests a
    high rate of meiotic recombination across this
    site', Proc. Natl. Acad. Sci. USA, in press.

Tchen, P., Fuchs, R.P.P., Sage, E., and Leng, M. (1984). 'Chemically modified nucleic acids as immunodetectable probes in hybridisation experiments', Proc. Natl. Acad. Sci. USA, 81, 3466-3470.

Tiepolo, L., Zuffardi, O., Fraccaro, M., Di Natale, D., Gargantini, L., Mueller, C.R., and Ropers, H-H. (1980). 'Assignment by deletion mapping of the steroid sulfatase X-linked icthyosis locus to Xp223', Hum. Genet., 54, 205-206.

Toole, J.J., Knopf, J.L., Wozney, J.M., Sultzman, L.A., Buecker, J.L., Pittman, D.D., Kaufman, R.J., Brown, E., Shoemaker, C., Orr, E.C., Amphlett, G.N., Foster, W.B., Coe, M.L., Knutson, G.J., Fass, D.N., and Hewick, R.M. (1984). 'Molecular cloning of a cDNA encoding human antihaemophilic factor', Nature, 312, 342-347.

Vehar, G.A., Keyt, B., Eaton, D., Rodriguez, H., O'Brien, D.P., Rotblat, F., Opperman, H., Keck, R., Wood, W.I., Harkins, R.N., Tuddenham, E.G.D., Lawn, R.M., and Capon, D.J. (1984). 'Structure of human factor VIII', Nature, 312, 337-342.

Webb, G.C., Rogers, J.G., Pitt, D.B., Halliday, J., and Theobald, T. (1981). 'Transmission of fragile (X)(q27) site from a male', Lancet, ii, 1231-1232.

Weissenbach J, Geldwerth D, Guellaen G, Fellous M and Bishop C (1984) Sequences homologous to the human Y chromosome detected in the female genome (Abstract). Human Gene Mapping Conference VII, 604.

Went, L.N., De Groot, W.P., Sanger, R., Tippett, P.,
    and Gavin, J. (1969). 'X-linked ichthyosis:
    linkage relationship with the Xg blood groups and
    other studies in a large Dutch kindred ', Ann.
    Hum. Genet., 32, 333-345.

Wieacker, P., Davies, K.E., Bevorah, B., and Ropers,
    H-H. (1983). 'Linkage studies in a family with X-
    linked recessive ichtyosis employing a cloned DNA
    sequence from the distal short arm of the X
    chromosome', Hum. Genet., 63, 113-116.

Wieacker, P., Davies, K.E., Cooke, H.J., Pearson,
    P.L., Williamson, R., Southern, E., Zimmer, J.,
    and Ropers, H-H. (1984). 'Towards a complete
    linkage map of the human X chromosome: regional
    assignment of 17 cloned single copy DNA sequences
    employing a panel of somatic cell hybrids',
    Am. J. Hum. Genet., 36, 265-276.

Wieacker, P., Wienker, T.F., Dallapiccola, B.,
    Bender, K., Davies, K.E., and Ropers, H-H. (1983).
    'Linkage relationship between retinoschisis, Xg
    and a cloned DNA sequence from the distal short
    arm of the X chromosome', Hum. Genet., 64, 143-
    145.

Willard, H.F., Smith, K.D., and Sutherland, J.
    (1983). 'Isolation and characterisation of a major
    tandem repeat family from the human X chromosome',
    Nucleic Acids Res., 11, 2017-2033.

Willard, H.F., and Smith, K.D. (1982).
    'Identification and characterisation of a repeated
    DNA fragment from the human X chromosome',

Cytogenet. Cell. Genet., 32, 327-335.

Wolf, G., Mueller, C.R., and Jobke, A. (1980).
'Linkage genes for chronic granulomatous disease
and Xg', Hum. Genet., 54, 269-271.

Wolff, G., Hameister, H., and Ropers, H-H. (1978).
'X-linked mental retardation: transmission of the
trait by an apparently unaffected male',
Am. J. Med. Genet., 2, 217-224.

Woo, S.L.C., Lidsky, A.S., Guttler, F., Chandra, T.,
and Robson, K.J.H. (1983). 'Cloned human
phenylalanine hydroxylase gene allows prenatal
diagnosis and carrier detection of classical
phenylketonuria', Nature, 306, 151-155.

Wood, W.I., Capon, D.J., Simonsen, C.C., Eaton, D.L.,
Gitschier, J., Keyt, B., Seeburg, P.H., Smith,
D.H., Hollingshead, P., Wion, K.L., Delwart, E.,
Tuddenham, E.G.D., Vehar, G.A., and Lawn, R.M.
(1984). 'Expression of active human factor VIII
from recombinant DNA clones', Nature, 312, 330-
337.

Worton, R.G., Duff, C., Sylvester, J.E., Schmickel,
R.D., and Willard, H.F. (1984). 'Duchenne muscular
dystrophy involving translocation of the dmd gene
next to ribosomal RNA genes', Science, 224, 1447-
1449.

Wyman, A.R., and White, R. (1980). 'A highly
polymorphic locus in human DNA',
Proc. Natl. Acad. Sci. USA, 77, 6754-6758.

Yang, T.P., Hansen, S.K., Oishi, K.K., Ryder, O.A.,
    and Hamkalo, B.A. (1982). 'Characterisation of a
    cloned repetitive DNA sequence concentrated on the
    human X chromosome', Proc. Natl. Acad. Sci. USA,
    79, 6593-6597.

Young, B.D., and Davies, K.E. (1983). 'Construction
    of a DNA library from the human X chromosome', in
    Cytogenetics of the Mammalian X Chromosome (Ed.
    A.A. Sandberg), pp. 479-491, A.R.Liss, New York.

FROM HEMOPHILIA B TO HEMOPHILIA A VIA THE FRAGILE X LOCUS :
GENES AND RECOMBINATION IN THE DISTAL REGION
OF THE HUMAN X CHROMOSOME LONG ARM

I. Oberlé and J.L. Mandel

INTRODUCTION

The region which includes bands q27 and q28, near the telomere of the long arm of the X chromosome, corresponds to about 0.5% of the human genome (one tenth of the X chromosome or $2 \ 10^4$ kilobases). Apart from the HLA region on chromosome 6 (which is ten times smaller), it is probably the part of the human genome where the density of defined genetic loci is the highest. It contains genes corresponding to important diseases : Hemophilia A and B, Glucose 6 phosphate dehydrogenase deficiency, Adreno-leukodystrophy, and the first three of them have been already cloned. The locus of the fragile X-mental retardation syndrome, characterized by some unique features among X-linked traits, is also located in this region.

A wealth of linkage data had been gathered over the years using the classic genetic markers : G6PD variants and Protan and Deutan color blindness. In fact, the linkage between hemophilia (A) and color blindness was, in 1937, the first to be established in man (1). More recently anonymous or gene-specific DNA probes which detect Restriction Fragment Length Polymorphisms (RFLPs) have been isolated from the q27-q28 region, several of them being of use for carrier detection and/or prenatal diagnosis of hemophilias and of the fragile X syndrome. These genetic markers have also allowed more detailed linkage studies which suggest that recombination in this region is not evenly distributed.

THE GENETIC AND PHYSICAL MAP OF THE q27-q28 REGION.

The various loci described in table I have been mapped using three main methodologies.

Initially thorough linkage studies were performed between G6PD, the Protan and Deutan color blindness and hemophilia A (HEMA), showing very close linkage between all the loci (2-4). Later adrenoleukodystrophy was joined to this cluster showing no recombination with the G6PD locus in a more limited number of meioses (5). In contrast, preliminary linkage analysis suggested that hemophilia B was not detectably linked to the G6PD-color blindness cluster (6, see also 7 for further references).

Using somatic cell hybrid lines containing a complex translocation with a breakpoint at the junction of the q27 and q28 bands Pai et al. mapped the G6PD gene in the q28 region (8). The cloning of the coagulation factor genes and of anonymous probes that detect RFLPs led to further refinement of the physical and genetic map. It came rather as a surprise that the factor IX gene was located in q27, as shown by blot hybridization to somatic hybrids (9-11), and appeared thus physically quite close to the G6PD cluster. In situ hybridization confirmed this localisation. In particular, the fragile site in Xq27.3 (associated to mental retardation, see below) served as a very useful cytogenetic marker. It was thus shown that the anonymous probe 52A (DXS51) and the factor IX gene were proximal to the fragile site, while the G6PD and factor VIII genes, and the St14 probe (DXS52), were all found in q28, distal to the fragile site (7, 12, 13).

Linkage studies could be reinitiated using RFPLs detected by anonymous probes (in particular, the St14 probe which is the most polymorphic marker known on the X chromosome (14)) and by the cloned coagulation factor genes (see Table II). This added two loci to the G6PD cluster :

the St14 and DX13 (DXS15) loci which are very tightly linked to the hemophilia A gene (15-17) while factor IX (HEMB) showed about 30% recombination with the loci in the G6PD cluster (14, 16, 18). The locus DXS51
defined by probe 52A is on the centromeric side of HEMB. The study of an
exceptional family showing recombination between Deutan and Protan color
blindness loci has suggested the order Hypoxanthine - Guanine Phosphoribosyl Transferase (HPRT) - Deutan - G6PD - Protan - Xqter (19).
However, this conclusion is only tentative since the HPRT locus (in q26)
and the G6PD cluster appear genetically unlinked and thus a double recombination event might have occurred.

The presence of a cluster of seven very tightly linked loci and the
large genetic distance between them and hemophilia B located in the
neighbouring band (see Table II) suggest a highly non random distribution
of recombination in this area (7, 14, 20). Crossing-overs would be favored in a region between HEMB (in q27) and the loci in q28 : the 30%
recombination figure corresponds to a frequency of a least 0.6 crossing-
overs per meiosis. It is interesting to note that such a recombination
hot spot would be close to a region that can show chromosome breakage in
certain individuals i.e., the fragile site at the Xq27-q28 interface (see
below).

THE HEMOPHILIAS.

Hemophilia A and B are bleeding disorder with similar clinical
features. Hemophilia A is due to the deficiency in coagulation factor
VIII : C, and has a prevalence of 1 in 7 000-10 000 newborn males (21).
Hemophilia B is caused by a deficiency in coagulation factor IX and is
about 4 times less frequent than hemophilia A (22, 23). An important
proportion of hemophilia cases is due to new or recent mutation events as
demonstrated first by Haldane (24). This is a feature shared by all

severe or lethal X-linked diseases which result in a diminished reproductive fitness of affected males. Replacement therapy with the appropriate
plasma concentrates or partially purified coagulation factor is effective
but the treatment is the cause of a high incidence of hepatitis (25), and
more recently of acquired immunodeficiency syndrome (26). Furthermore
some patient acquire inhibiting antibodies against the exogenously admi-
nistered coagulation factor. This occurs in approximately 6% of hemophilia A patients (27, 28) and in 2% of the patients affected with severe
forms of hemophilia B (29). Thus, families often request genetic counselling. Prior to the development of recombinant DNA approaches, detection
of heterozygous carrier females in hemophilia families was based on
pedigree analysis and on the measurement of the antigen levels and clotting activity corresponding to factor VIII : C and the autosomally coded
von Willebrand factor (sometimes called FVIII related antigen, FVIII :
RAg) for hemophilia A, or to factor IX for hemophilia B. However, the
accuracy of such assays was only 70 to 90%, the higher values necessi-
tating rigorous testing procedures and sophisticated statistical analysis
(21, 30-34). The broad range of values found in population of normal or
heterozygous females is due in part to biologic and methodologic variabi-
lities. More important, since these genes are X-linked, phenotypic
expression in carriers may vary according to the proportion of active X
chromosome carrying the mutations in the liver cells which synthesize the
coagulation factors. Prenatal diagnosis was possible after examination
of factor VIII (or IX) activity and antigen levels in foetal blood
obtained by foetoscopy at the 18 th to 20 th week of gestation, a
demanding procedure that can be performed by only a very small number of
well trained teams (35).

The economic and medical interest in the production of safer coagulation factors for therapy prompted several groups to clone the DNA sequences corresponding to the two coagulation factors. This was carried out first for factor IX (36-38), which is synthesized at a higher level, and for which information on amino acid sequence and site of synthesis was already available. The task was much more difficult for factor VIII : C which was very poorly defined biochemically, which is present at very low level in blood, and for which the site of synthesis was not known with certainty. The cloning of factor VIII sequences is probably one of the most striking accomplishment of genetic engineering technology (39, 40). The cloning of these genes has provided powerful tools for the analysis of hemophilia mutations, and for reliable prenatal and carrier diagnosis.

THE COAGULATION FACTOR VIII GENE.

Factor VIII : C is a large plasma glycoprotein that functions in the blood coagulation cascade as the cofactor for the factor IXa dependent activation of factor X. It can be activated proteolitically by various coagulation enzymes including thrombin. Purification of human or porcine factor VIII by affinity chromatography using monoclonal antibodies allowed the obtention of partial amino acid sequence. Oligonucleotide probes were synthesized and used to screen libraries of human (40) or porcine (39) genomic DNA. This unusual strategy was followed rather than the screening of cDNA libraries because of the uncertainty in the site of factor VIII synthesis. Genomic clones obtained were shown to be X-linked and allowed the demonstration that the corresponding mRNA is expressed in liver (39, 40). Complete cloning of the cDNA and genomic sequences was a demanding task, due to the very large size of the mRNA and gene. The

factor VIII gene is the longest human gene known at the present time and
covers 186 kb, i.e. 0.1% of the X chromosome length. It contains
26 exons, one of which (exon 14) is 3106 bp long, and is thus the largest
protein coding exon found in eukaryotic genes. The mRNA is 9 000 nucleotide long, coding for a protein of 2 332 amino acids. Derivation of the
protein sequence allowed several very interesting conclusions (41, 42).
Three types of structural domains are apparent : a triplicated region of
330 amino acids (A domains), a unique region of 980 amino acids (B
domain) and a duplicated region of 150 amino acids (C domain) arranged in
the order A1-A2-B-A3-C1-C2. A surprising observation was the very significant 35% homology of domain A with the copper binding plasma protein,
ceruloplasmin (39, 42). This suggests the possible involvment of copper
or other metal ions in the factor VIII activity. Some sequence homology
has also been found between factor VIII and coagulation factor V, which
can be related to common structural and biologic properties (39). The
domain structure is reflected in the intron exon organization with some
but not all intron boundaries conserved between the duplicated or triplicated region. Domain B poses an interesting problem. This 925 amino acids
region encompasses nearly all of the 3.1 kb long exon 14 where the end of
the A2 repeat and the start of the A3 repeat are also found. It has been
suggested by Gitschier et al. (41) that a processed gene might have been
incorporated into a primary short exon containing the A2-A3 boundary,
accounting for the very large size of exon 14.

The structural knowledge acquired through the cloning of factor VIII
sequences should be invaluable for the study of the function of
Factor VIII and its proteolytic cleavage products.

THE COAGULATION FACTOR IX GENE.

Coagulation factor IX is the precursor of a serine protease.
Cloning of its cDNA was based on the knowledge of the amino acid sequence
of the bovine protein, and on the knowledge of the site of synthesis :
the liver. Use of synthetic oligonucleotide probes in three different
laboratories allowed the cloning of the human cDNA, either directly (37,
38) or via isolation of bovine cDNA clones and human genomic clones
(36).

The mRNA is 2 800 nucleotides long with a protein coding sequence of
1 380 residues and a long 3' non coding sequence. Sequence analysis of
the cloned cDNA indicated that the precursor protein contains a N-
terminal signal sequence of 46 amino acids which is subsequently cleaved.
The human gene is 34 kb long, and consists of 8 exons and 7 introns. The
largest exon (2.2 kb) contains the whole 3' non coding region in addition
to protein coding sequence. The gene is thus 12 fold longer than the mRNA
and 24 fold longer than the protein coding sequence.

The relation between exons and protein domains has been analyzed by
Anson et al. (43). The protease catalytic region is coded by two exons
and the positions of introns differ from that found in other members of
the serine protease super gene family. Other exons show homology to the
epidermal growth factor, a feature also found in two other coagulation
proteins (factor X and protein C) as well as in tissue plasminogen
activator and in one domain of the receptor for low density lipoprotein
(44). These features suggest that the shuffling of exons that correspond
to domains of ancestral proteins has played an important role in
evolution of these genes (45).

APPLICATION OF DNA PROBES TO THE ANALYSIS AND DIAGNOSIS OF HEMOPHILIA
MUTATIONS.

A) Direct detection of mutations.

1. Deletions.

A first Southern blot analysis in hemophilia B patients suggested
that deletions might be very rare and prevalent only in the patients who
develope antibodies against factor IX in the course of the treatment
(46). In a more recent study, no deletions were detected in a series of
25 antibody free patients (47). However, at least one case of deletion in
the absence of factor IX inhibitors has been found (M. Goossens, personal
communication). A thorough analysis of hemophilia A patients showed only
two partial gene deletions in 92 patients, and indicated no firm correlation between antibody production and gross gene defects (48).

2. Point mutations.

Three distinct non sens mutations have been detected out of
102 hemophilia A cases analyzed, each of them resulting in the loss of a
TaqI restriction site (48, 49). Apart from such chance findings, it is
unlikely that Southern blot analysis will be of general use to reveal
point mutations in the factor VIII or IX genes. As for other severe X-
linked diseases it can be expected that the gene defects will be very
heterogeneous since the proportion of new or recent mutations is high.
Thus, each hemophilia family will probably carry a distinct mutation
(except if hot spots create recurrent mutations). Because of the very
large size of the genes, the sequence and functional analysis of each
mutation will represent a major undertaking and this knowledge is

unlikely to be useful for diagnosis in other families. The recent work of Rees et al. (50) illustrates this point : identification of a point mutation in an hemophilia B patient lacking detectable Factor IX antigen (CRM⁻) involved cosmid cloning, subcloning of exon containing fragments and sequencing of 5 200 nucleotides. This revealed a mutation in a donor splice site junction. The screening of 9 other CRM⁻ patients with an oligonucleotide specific for this mutation indicated that they all had a normal sequence at this position. The strategy used so efficiently in β thalassemia for detecting mutations prevalent in a population, based on analysis of RFLP haplotypes (51), is clearly irrelevant. Thus, it is unlikely that direct detection of hemophilia mutations will be possible for diagnostic purposes on a routine basis.

B) Indirect detection using polymorphic DNA markers.

The segregation analysis, in families at risk, of polymorphic DNA markers (RFLPs) located within or very close to the factor VIII or IX genes represent the method of choice for carrier detection (or exclusion) and prenatal diagnosis on trophoblast villi biopsy (15, 17, 49, 52-56). One drawback is the necessity of a minimal family analysis to identify the allele associated to the mutation in each case, and the limited informativenes of some of the markers used. (If a woman is homozygous for the marker, the latter cannot be used to trace the segregation of the mutation).

1. Hemophilia A.

a) Intragenic markers.

An extensive search for RFLPs in the factor VIII gene has revealed a single useful two allele polymorphism with the enzyme BclI which is

informative in about 40% of caucasian families (16). A BglI polymorphism
has been found in another study, but it appears that it is present at
high frequency only in American Blacks, which limits its usefulness for
diagnosis in other ethnic backgrounds (49). Thus, at present the intragenic polymorphic markers might allow diagnosis in only 50 to 60% of the
cases. In fact the factor VIII gene appear to show remarkable scarcity of
sequence variation since sequencing of the entire 9 000 bp cDNA in two
individuals detected only 2 bp differences (41).

b) Linked markers.

Two anonymous probes, DX13 (DXS15 in the Human Gene Mapping nomenclature) and St14 (DXS52) have been shown by linkage analysis to be
closely linked to Hemophilia A. DX13 detects a two allele BglII RFLP (50%
heterozygosity) and a first study showed no recombination between DX13
and Hemophilia A in 24 informative meioses (lod score of 5.4 at recombination fraction $[\theta]=0$) (15). A further linkage analysis between the DX13
RFLP and the BclI polymorphism present in factor VIII gene showed no
recombination in 27 meioses (16). However, in more recent studies (57)
recombination events have been detected between DX13 and hemophilia A
suggesting that the two loci might be separated by about 3 centimorgans
(cM).

The St14 probe is particularly interesting since it detects a very
polymorphic locus : 10 alleles are detected in TaqI digests, (80% heterozygosity) and other useful two allele RFLPs are present in MspI digests
(14). Linkage analysis to hemophilia A mutations (17) or to the BclI RFLP
in the factor VIII gene (16) showed no recombination in 105 meioses (the
cumulated lod score is 28.6 see table II). This indicates that the St14
locus is at less than 3 cM from HEMA at a 90% confidence level. (The 90%
confidence limit was calculated as the recombination value with a relative probability of 10% of the maximum likelihood).

It is not known at present whether St14 and DX13 are on the same side or on opposite side with respect to the hemophilia A locus. This would have important implications for the combined use of these probes for diagnosis purposes.

c) Use of the RFLP markers.

It is obvious that the intragenic factor VIII RFLPs should be used as a first choice, since the probability of recombination is very low (however given the large size of factor VIII gene, recombination cannot be totally excluded : a 200 kb distance might be equivalent to about 0.2% recombination). At present the available factor VIII RFLPs allow diagnosis only in about 50-60% of the cases. Therefore the St14 probe should be useful in the vast majority of the remaining cases because of its very high informativeness and because the confidence intervals of the recombination fraction with hemophilia A are smaller than for DX13. Care should be taken nevertheless in the interpretation with the linked probes since recombination events might occur. For carrier detection it is possible to combine biologic assays to the segregation analysis, which improves even further the accuracy of diagnosis (17). For prenatal diagnosis, the genetic counselor and the families should be aware of the possibility of diagnostic error due to rare recombination events, when the St14 or DX13 probes are used.

If DX13 and St14 would prove to bracket the hemophilia A locus, their combined use would result in almost 100% diagnostic accuracy in families informative for both probes, since errors would be possible only in the case of a double recombination in the DX13-St14 interval. The probability of such an event would be less than 10^{-3}. Analysis with the

St14 probe of those meioses showing recombination between DX13 and hemophilia A should give information on the relative position and distance of the two DNA markers with respect to hemophilia A.

2) Hemophilia B.

Four RFLP markers detected with factor IX probes have been described. All are two allele polymorphisms. The TaqI polymorphism has a 40% heterozygosity and is the most useful (10). The DdeI polymorphism described by Winship et al. (52) is a deletion-insertion polymorphism with a heterozygosity of about 35%. It is in linkage equilibrium with the TaqI RFLP so that the combined use of the two is informative in 60% of the families. (However detection of the DdeI polymorphism is slightly more delicate since it produces a 50 bp difference in a 1.7 kb fragment). The two other RFLPs are of little additional interest despite their relatively high heterozygosity, since they are in strong linkage disequilibrium with the TaqI RFLP. (i.e. the major allele of the TaqI RFLP show strong preferential association to the major alleles of both the XmnI (52) and MspI (58) RFLPs).

In the cases where the intragenic RFLPs are not informative, it might be possible to use linked markers detected by probe 52A (DXS51) and pX45h (DXS100). However, their linkage to hemophilia B is less well documented that the linkage of DX13 or St14 to hemophilia A, and recombination events have been detected between these probes and the HEMB locus. (The two probes are on the same side with respect to HEMB).

C) Detection of new mutations.

It has been first recognised by Haldane that in severe X-linked diseases an important proportion of cases (up to 1/3) will be due to new mutations (24). He postulated also that hemophilia mutations might occur more frequently in male than in female germ cells (due to the higher number of cell divisions in the male germ line than in the female one) (59). One would expect also that mutations will show a paternal age effect (i.e. will occur preferentially in older males). After some controversy, more recent statistical studies tend to support the validity of this hypothesis for hemophilia A (60). As a consequence most of the mothers of isolated cases of hemophilia A should be carriers. It is of great importance for genetic counselling purposes to determine, in a family of a sporadic case of hemophilia, in which person the new mutation has occurred since such a knowledge might allow to conclude that some family members will not transmit hemophilia.

This can be done when mutations can be detected directly (deletions or mutations which lead to a change in a restriction site). In some other cases the origin of the mutation can be traced by segregation analysis using intragenic RFLPs or linked markers. For hemophilia A, the large number of alleles detected with the St14 marker allows often the deduction of the genotype of family members who cannot be directly investigated as in the family presented in figure 1. This is of importance when one tries to trace an event which occurred in past generations (however the possibility of recombination has to be taken into account when the analysis is performed with a linked DNA marker).

Such studies should be helpful for determining the relative frequency of new mutations occurring in male or female germ cells, and investigate the effect of paternal age. In the three cases known to us, mutation occurred once in a female (48) and twice in a male (our unpublished results) (including one who was 60 at the time of conception, see figure 1).

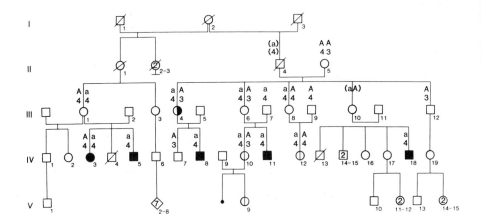

Figure 1 : Detection of a new mutation and diagnosis of carrier status in
an hemophilia A family by segregation analysis with the St14
probe.

In the sibship of the hemophiliac patients, the hemophilia
mutation segregates with maternal allele 3 in the three males.
The two sisters III5 and III6 had a probability of being
carrier of 74% and 3% respectively (based on FVIII coagulant
activity, see ref. 17). III5 has inherited from her mother
St14 allele 3, and III6 has inherited allele 2. Thus, their
final probability of being carrier is 98% and less than 0.5%
(assuming a very conservative value of 5% for the
recombination fraction between hemophilia A and the St14
locus). Analysis of sisters II6 and II8 of the obligatory
carrier II2 showed that allele 3 was inherited from the grand
father I1. Inheritance of hemophilia from the grand mother is
very unlikely since it would require at least 3 recombination
events in generation III (p < 3.10[-5]). It can be concluded
that a mutation occurred in the germ cells of the grand father
(who was 60 at the time of conception of II2).

ADRENOLEUKODYSTROPHY.

Adrenoleukodystrophy (ALD) is a generally severe disorder characterized by progressive cerebral demyelination and diminished function of the adrenal cortex. The basic defect is the impaired capacity to degrade very long chain fatty acids, a reaction which normally takes place in the peroxysome (61, 62). The increased levels of these fatty acids, particularly hexacosanoïc acid (C26) in cultured skin fibroblasts or amniotic fluid cells allow prenatal diagnosis and carrier identification of 90% of obligate heterozygotes (63, 64). More recently it has been shown that the biochemical diagnosis can be performed on chorionic villi biopsy (65).

Linkage analysis showed that the ALD locus was genetically close to G6PD : no recombination occurred in 18 meioses (4) (further analysis extended this number to 25 meioses without recombination, H. Moser, personal communication). However, the confidence interval for the recombination fraction is still fairly broad (about 0 to 10% at a 90% confidences interval). These data suggested that ALD was located in q28, within the G6PD-hemophilia A-color blindness cluster. Since the polymorphic locus detected by the St14 probe is very closely linked to hemophilia A (see above) this suggested that St14 might be useful in segregation studies of ALD families. In fact in a preliminary study we found no recombination in 9 meioses (65, and unpublished results).

Although further linkage studies are needed, segregation analysis with the St14 probe might be useful, in combination with measurements of fatty acid levels in plasma and fibroblasts, to improve identification of carriers. It might also complement the biochemical assays of prenatal diagnosis or replace them in cases where the biochemistry cannot be done due to logistical difficulties (in the latter case with careful consideration of the possibility of recombination). As for hemophilia, it should also be possible to trace, at least in some families, the origin of new mutations.

THE FRAGILE X MENTAL RETARDATION SYNDROME.

The fragile X mental retardation syndrome (Fra X) accounts for about one quarter to one third of families with X-linked mental retardation and it is present in approximately 1/2 000 newborn males. It may also account for 3-4% of all mental retardation in otherwise normal females. The Fra X syndrome is a pleiotropic trait consisting of 1) the presence of a fragile site on the X chromosome at the q27-q28 interface, induced in vitro by conditions which impair thymidylate synthesis (growth of periferal lymphocytes in media with low concentrations of thymidine and folic acid or in the presence of fluorodeoxyuridine (FUdR) or methotrexate), 2) a variable (moderate to severe) degree of mental retardation in hemizygous males, usually accompanied by characteristic physical features (macroorchidism and a typical facies), 3) a 35% risk of mental impairment in heterozygous females (for review, see 66-69).

The genetics of the Fra X syndrome departs from classic X-linked inheritance in several respects. The Fra X gene does not appear to be fully penetrant in males. Apparently normal males (cytogenetically and/or clinically) can transmit the disease, as first suggested from retrospective analysis of large pedigrees (reviewed in ref. 66 and 67). On the other hand the percentage of clinically expressing females is much higher than in other sex-linked diseases. Furthermore a segregation analysis suggested that the mutation rate at this locus is very high (7.2×10^{-4}) but that mutations occur only in sperm (70). The gene seems to be more penetrant in the offspring of daughters of transmitting males than in offspring of mothers of transmitting males (71).

The diagnosis of carrier females is difficult since only about one half of females who carry the Fra X mutation can be detected by their phenotype (mental retardation and/or fragile site expression (70). The

proportion of female showing the fragile site appears negatively correlated with both IQ and age (67). Prenatal diagnosis can be performed by
assaying for the presence of the fragile site in foetal cells obtained by
amniocentesis (72), foetal blood sampling (73) and more recently by
chorionic villi biopsy (74). The first two techniques can be performed in
only a few centers due to the difficulty of detection of the fragile site
in amniocytes or the greater obstetrical complexity of foetal blood
sampling compared to amniocentesis. The reliability of fragile site
detection in trophoblast cells is not yet documented. Thus, because of
the prevalence of the disease and of the problems encountered for genetic
counseling, it is important to improve the methods for carrier detection
and prenatal diagnosis.

ANALYSIS OF THE FRAGILE X SYNDROME WITH DNA MARKERS.

A) Linkage studies.

We have analyzed the segregation of TaqI RFLPs detected by the
coagulation factor IX (FIX) and St14 probes in 16 families with fragile X
mental retardation. Eleven families were informative for both FIX and
St14 probes. Several large families exhibited no recombination at
meiosis between the Fra X locus and either the FIX or the St14 probes
(Families 13 and 9, Fig. 2). In contrast, one family showed several
recombination events with the two probes (Family 5, Fig. 2).

In order to estimate the genetic distance between loci (expressed
as the fraction of meioses showing recombination between two loci) we
analyzed the segregation data using the Linkage computer program (75).
It is important in this case to take into account the problem of incom-
plete penetrance since a proportion of males or females who carry the

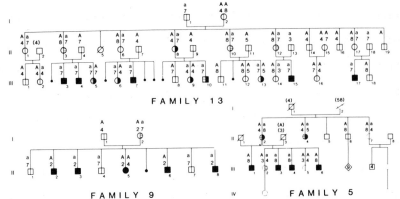

FAMILY 13

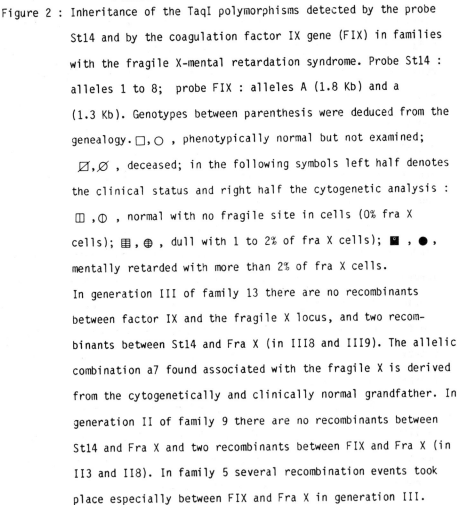

FAMILY 9   FAMILY 5

Figure 2 : Inheritance of the TaqI polymorphisms detected by the probe
           St14 and by the coagulation factor IX gene (FIX) in families
           with the fragile X-mental retardation syndrome. Probe St14 :
           alleles 1 to 8; probe FIX : alleles A (1.8 Kb) and a
           (1.3 Kb). Genotypes between parenthesis were deduced from the
           genealogy. □, ○ , phenotypically normal but not examined;
           ▨,⌀ , deceased; in the following symbols left half denotes
           the clinical status and right half the cytogenetic analysis :
           ⊡ ,⊕ , normal with no fragile site in cells (0% fra X
           cells); ▦,⊕ , dull with 1 to 2% of fra X cells); ▪ , ● ,
           mentally retarded with more than 2% of fra X cells.
           In generation III of family 13 there are no recombinants
           between factor IX and the fragile X locus, and two recom-
           binants between St14 and Fra X (in III8 and III9). The allelic
           combination a7 found associated with the fragile X is derived
           from the cytogenetically and clinically normal grandfather. In
           generation II of family 9 there are no recombinants between
           St14 and Fra X and two recombinants between FIX and Fra X (in
           II3 and II8). In family 5 several recombination events took
           place especially between FIX and Fra X in generation III.

mutation will not express it clinically or cytogenetically. Because only rough estimates of penetrances are available, we have considered the influence of varying penetrance values for males and females. Although the odds in favor of linkage vary with the various modes of calculation, the recombination fraction estimate remains quite stable (76). Part of the results obtained are shown in Fig. 3, where the relative probability of linkage is plotted as a function of the recombination fraction θ. The maximum lod score (i.e. the log of odds in favour of linkage) for the linkage between FIX and Fra X is 6.18 at a recombination fraction of 0.114 with 90% confidence limits of 0.044 to 0.225. Choo et al. (77) have published an analysis of 5 families showing 17% recombination between FIX and Fra X, which is compatible with our data. The St14 locus also shows about 10% recombination with Fra X (90% confidence limits of 0.040 to 0.185), with a maximum lod score of 9.5. The latter value is consistent with the linkage observed in other families between the G6PD locus and the fragile X syndrome (78, see also ref. 7).

The three point linkage analysis as well as simple examination of large nuclear pedigree (Fig. 2) places the Fra X locus between the two marker loci. Since a similar conclusion has been reached by in situ hybridization of the FIX and St14 probes to mitotic chromosome displaying the fragile site at Xq27 (13), this supports further the notion that the mutation is located in the same region as the cytogenetically demonstratable fragile site (78).

Linkage analysis does not provide evidence for heterogeneity among families as tested following Morton (79) suggesting that the mutations reside in the same region in the different families. However, positive evidence for heterogeneity is difficult to obtain, especially when rela-

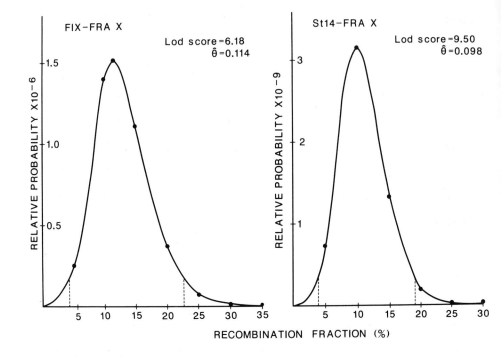

Figure 3 : Estimation of the recombination fraction between the fragile X
mental retardation syndrome (Fra X) and the polymorphic DNA
markers St14 and FIX. The relative probability of linkage
(RPL) is the ratio of the probability of obtaining the data if
the two loci are linked with a recombination fraction θ to the
probability that there is no linkage (θ = 0.5). The lod score
is the log of the maximum RPL obtained for the recombination
factor θ. The 90% limits for the recombination fractions
(dotted lines) are the values of θ for which the RPL is one
tenth of the maximum.

tively small pedigrees are analyzed. Recently higher recombination rates between FIX and Fragile X have been reported (80). Although it is unlikely that the relative position of the fragile X mutation with respect to the FIX and St14 loci is variable, it cannot be excluded that heterogeneity in recombination frequency occurs among families. It would thus be of interest to reanalyze data pooled from the various investigators, for instance by separating, for the calculations, families with clear transmission through normal males from the other families (the two families with normal male transmitters shown in figs. 2 and 4 yielded no recombination between FIX and Fra X in at least 20 meioses).

B) Diagnostic applications.

Our results establish the validity of the St14 and FIX probes as tools in the genetic analysis in Fra X families. These markers flank the disease locus and can thus be used in conjunction with cytogenetic tests for prenatal diagnosis and carrier detection in families informative at both loci. This is of interest since only about 50% of the carriers can be diagnosed by cytogenetic analysis and the reliability of detection of the fragile site on chorionic villi biopsy is not yet established. Given the recombination fraction between each test locus and the Fra X locus, double recombinants should occur in only about 1 to 2% of the meioses in the St14-FIX interval. The heterozygosity for the combined TaqI and DdeI RFLPs detected by the FIX probe is 60% in caucasians (52), while that of the TaqI and MspI RFLPs at the St14 locus is about 90% (14). Thus, 54% of the families would be doubly informative. However, in 25 to 30% of the cases a single recombination event will occur between the two marker loci, which will prevent diagnosis. Therefore, we can estimate that about 40% of the cases could benefit from a segregation analysis with the two

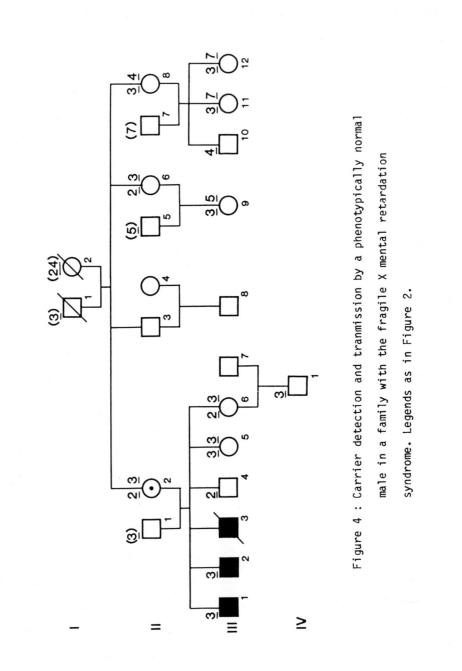

Figure 4 : Carrier detection and tranmission by a phenotypically normal
male in a family with the fragile X mental retardation
syndrome. Legends as in Figure 2.

probes. An exemple is shown in Fig. 4 where the sister (IV10) of an
affected male asked for genetic counselling on the occasion of a
pregnancy. She was clinically and cytogenetically normal, with thus a
30% chance only of being a carrier (assuming a penetrance of 0.56). She
has received from her doubly heterozygous mother FIX and St14 alleles
which are different from those of her affected brother or cousins. Risk
calculation shows that her final risk of being a carrier is only 1.3%.
The risks of diagnostic error (false positives and false negatives) have
been calculated for a few model families, showing that the percentage of
false negative results is very low when cytogenetic analysis is combined
to segregation study in families informative for the two flanking markers
(80a).

The marker study can also help to detect new mutations (as shown
for hemophilia A, see above) and/or those families in which the disease
was transmitted through normal males. Out of the 16 families analyzed, 2
were already known to have normal male transmitters, based on the pedigree alone (a conclusion only possible for large pedigrees). In family
13, the FIX segregation data were sufficient to establish male transmission beyond doubt (81), and the St14 data confirmed this conclusion. In
family 1 the FIX segregation data also suggested male transmission, which
was confirmed by further pedigree analysis (Fig. 4). Suggestive evidence
for male transmission (or for a new mutation arising in a male) has also
been obtained in two other families. These results would be in agreement
with the high percentage of phenotypically normal male carriers estimated
by Sherman et al. (71), although it is possible that an ascertainment
bias exists in favor of such families. It must be emphasized that detection of male carriers is of great importance in a genetic counselling
context. Finally, the segregation analysis with linked markers could help

in testing the hypothesis that penetrance might be different among sibships of normal male carriers and sibships of affected males (71) by
allowing us to infer the genotype of clinically and cytogenetically
normal males.

DOES THE FRA X MUTATION AFFECT RECOMBINATION ?

        The fragile site is visualised as a region which is not properly
packaged for mitosis, probably as a result of inadequate DNA replication
since it is induced by conditions which affect the synthesis of a DNA
precursor. One could wonder whether the presence of the mutation has an
effect on another chromosomal process, i.e. meiotic recombination. The
region which includes the fragile site seems to exhibit a high frequency
of recombination in normal families, since the two probes 52A and FIX
which map in q27 are at 0.30 recombination unit from the cluster of loci
in q28 which includes probes DX13 and St14, and the G6PD and hemophilia A
loci (7, 14, 18). The physical distance between the two groups of loci
can be estimated cytogenetically to correspond to about 5% of the X chromosome length and a recombination fraction of 0.12 would be expected if
the genetic distance was proportional to the physical distances. On the
other hand, Szabo et al. (7) have suggested that the recombination fraction between FIX and the G6PD cluster is much decreased in Fra X families. Our data do not support the hypothesis of Szabo et al. since the
recombination fraction between St14 and FIX in Fra X families, as estimated from three locus analysis, is about 0.19 (or 0.22 when estimated from
two point linkage data), while there was 28% recombination in normal
families. A χ^2 test showed that these differences are not significant.

X LINKED MENTAL RETARDATION WITHOUT THE FRAGILE SITE

The fragile X syndrome might account only for 1/3 of the families with "non-specific" X-linked mental retardation. In the study of Fishburn et al. (68) the fragile X was detected in 12 out of 45 pairs of brothers with non-specific mental retardation consistent with X linkage. Six had macroorchidism and no fragile X, and the others had none of the above signs. It would be of interest to determine whether such forms of mental retardation correspond to allelic variations at the fragile X locus or are due to mutations at different loci on the X chromosome. An approach to this problem is to perform segregation analysis with the same markers that show linkage to the fragile X. We have started such a study with a large family from Hawaï initially described by Proops et al. (82). With the St14 probe we found 5 recombinants in 15 meioses. The factor IX marker was uninformative, but two markers located in q27, and corresponding to probes 45h and 52A (with the order Fra X-FIX-52A-45h) show a minimum of 3 recombinants (B. Arveiler, unpublished results). A marker located in Xq13 (DXYS1) does not show any linkage. Thus, this preliminary results although compatible with a localisation of the putative mental retardation mutation in the region of the fragile site, do not give strong support for this hypothesis. On the other hand, results from Filippi et al. (78) suggested that mental retardation with macroorchidism but without fragile X shows more recombination with the G6PD locus (located in Xq28 close to St14) than the fragile X locus. One problem in such studies is that in the absence of a real pathognomonic sign, one might be dealing with heterogeneous entities, which would render a linkage study very difficult.

THE FRAGILE X : FUTURE PROSPECTS.

Although the presently available probes represent useful tools, it is desirable to find other markers closer to the Fra X locus and to increase the number of families informative for probes on the proximal side of the Fra X locus. The families already investigated are an extremely useful material to quickly map any new polymorphic probe with respect to the fragile site, St14 and FIX, since it is necessary to analyze only those meioses showing recombination between the Fra X and one of the test loci in order to know on which side of the fragile X locus is the new marker, and whether it is closer than the already available ones. This multipoint linkage approach is much more efficient in ordering loci than two point linkage data performed on different sets of families. By this way, we have recently mapped several polymorphic probes to the proximal side of the fragile site (Wrogeman K. et al., in preparation). However, if the Fra X region is a hot spot for recombination as has been suggested (7, 14), it might be difficult to find random probes genetically closer unless they are physically in the immediate vicinity of the mutation.

The fragile X mental retardation syndrome is still a very mysterious disease. We can list some questions which remain unanswered. What is the structure (normal and mutated) of the region which constitutes the fragile site. The high mutation rate suggests that this region might have a special organisation (tandem or inverted repeats for instance). Is the region a hot spot for meiotic recombination in normal individuals ? Does the mutation directly affect a specific gene (or genes) responsible for the phenotype ? An alternative hypothesis would be that induction of the fragile site occurs in vivo only in region or cells which have a limiting supply of folate or of DNA precursors such as thymidylate, and that as a result, expression of all the genes distal to the fragile site would be decreased. However, one should bear in mind that many other fragile sites exist on autosomes, with the same pattern of induction, but with no clinical consequences (83). Since the mutation appears to function in cis with respect to the fragile site, why is the expression generally so low in females compared to males (preferential X chromosome inactivation due to cell selection does not seem to be the explanation). Is the non pene-trance of the mutation in some males due to the action of modifying genes, or of environmental conditions, or due to a two step generation of the full mutation (the premutation hypothesis, 83a, see also ref. 71).

Ultimately, the cloning of the mutated region (and of its normal homologue) might give important clues to answer some of these questions. Even with the help of new cloning techniques which are being developed (jumping cloning vectors (84), microdissection of mitotic chromosomes, 85) this will be an extremely difficult task.

CONCLUSION : TOWARDS A DETAILED PHYSICAL AND GENETIC MAP OF THE q27-q28 REGION.

The tip of the long arm of the X chromosome might be a good model for testing new methods of analysis of large genomic regions. Well characterized cloned DNA segments (Factor VIII and G6PD genes, St14 and DX13 probes) already could provide a restriction map over more than 300 kb in the q28 band (approximatively 3% of the DNA in this band, if cytogenetic lengths are proportional to DNA content). Probably 600 or 700 kb could be mapped using in addition the many probes more recently isolated (see footnote to Table I). A similar length could be covered in the q27 band with already available probes.

TABLE I : loci in Xq27-q28

Loci	Disease or Phenotype	Cloned	in situ hybridization	somatic hybrids	linkage	Final Localisation
Genes						
Protein						
Coagulation Factor IX	Hemophilia B	yes	+	+	+	q27
Coagulation Factor VIII	Hemophilia A	yes	+	+	+	q28
G6PD	G6PD Deficiency (Favism)	yes	+		+	q28
Iduronate Sulfatase	Hunter Syndrome	no (partially)			±	q26-q28 [1]
Function						
Catabolism of very long chain fatty acids	Adrenoleukodystrophy	no			+	q28
Color Vision	Deutan Color Blindness	no			+	q28
	Protan Color Blindness	no			+	q28
	Torticollis, Keloids, Cryptorchidism and Renal Dysplasia	no				q28 [2]
Cytogenetic Marker						
Fragile Site	Mental Retardation	no			+	q27.3
Anonymous probes [3]						
52A (DXS51)		yes	+	+	+	q27
St14 (DXS52)		yes	+	+	+	q28
DX13 (DXS15)		yes	+	+	+	q28

(1) Based on the finding of a girl with typical features of Hunter's disease showing a X:5 translocation with a breakpoint in Xq26-27 (87). Preliminary linkage analysis suggests a localisation distal to Factor IX (88).

(2) Based on two females with partial manifestations of the syndrome who carry translocations with breakpoints in Xq28 (89).

(3) In addition, in the catalogue of cloned DNA fragments established at the 8th Human Gene Mapping Workshop (Helsinki, 1985), 44 anonymous probes have been assigned to the q26-q28 region, including 8 assigned to q28 and 17 assigned to q27-q28.

TABLE II : linkage analysis data between pairs of loci in the Xq27-q28 region

q27 to q28		
Loci	Recombinants(*)	scorable sibs
52A - St14	12/41	(14)
52A - DX13	12/40	(18)
52A - FVIII	6/27	(16)
FIX - DX13	6/21	(18)
FIX - St14	11/37	(14)
FIX(HEMB) - (CBD-CBP)	15/32	(7)

within q28		
Loci	Recombinants	scorable sibs
G6PD - CBD	3/238	(7)
G6PD - CBP	1/51	(7)
G6PD - HEM A	0/58	(3)
G6PD - ALD	0/18	(5)
(HEMA - FVIII) - St14	0/105	(16,17)
(HEMA - FVIII) - DX13	0/51	(16,17)
HEMA - (CBD, CBP)	1/46	(7)
St14 - DX13	0/41	(14)
CBP - CBD	6/106	(90)

* References are given in parenthesis.

Two features could be used to construct genomic libraries specific for this region. The HPRT locus (probably located in q26) provides a very useful marker to select somatic cell hybrids containing small portions of the human X chromosome around q26, obtained by X ray irradiation, and transferred into the rodent cell line by microcelling or mitotic chromosome transfer. A genomic library could then be constructed from such hybrids and screened with human specific repetitive DNA sequences. A second possibility would be to use the fragile site in Xq27.3 as a way of identification for microdissection of the Xq27-q28 region in mitotic chromosomes, followed by DNA microcloning. A similar experiment performed in mouse (using a robertsonian translocation for chromosome identification) allowed the obtention of several genomic DNA clones in the T locus (85).

The probes already available, and those that could be generated using region specific libraries could then be used as starting points for cosmid cloning with jumping (or hopping) vectors (84), or for restriction mapping of very large fragments in genomic DNA using pulse field electrophoresis combined to Southern blot analysis (86).

Regions of overlap between the restriction maps could be detected, and clusters of cloned segments identified. New RFLPs would also be generated, allowing the ordering of the clusters in a genetic map. This could be efficiently performed by analyzing those meioses that show recombination between the various polymorphic or disease markers already identified. Although such a scheme depends in part on new methodologies which might be more difficult to apply than expected, the establishment of a detailed physical and genetic map of this region appears as a reasonable goal.

ACKNOWLEDGEMENTS

We wish to thank C. Aron and A. Marrel for typing the manuscript. Research in the author's laboratory was supported by INSERM, CNRS and by a grant from the Fondation pour la Recherche Médicale Française.

REFERENCES

1. Bell, J. and Haldane, J.B.S. Proc. Roy. Soc. B. 123, 119-150, (1937).

2. Siniscalco, M., Filippi, G. and Latte, B. Nature (London) 204, 1062-1064, (1964).

3. Rinaldi, A., Velisavasakis, M., Latte, B., Filippi, G. and Siniscalco, M. Am. J. Hum. Genet. 30, 339-345, (1978).

4. Filippi, G., Mannucci, P.M., Coppola, R., Farris, A., Rinaldi, A. and Siniscalco, M. Am. J. Hum. Genet. 36, 44-71, (1984).

5. Migeon, B.R., Moser, H.W., Moser, A.B., Axelman, J., Sillence, D. and Norum, R.A. Proc. Natl. Acad. Sci. USA 78, 5066-5070, (1981).

6. Whittaker, D.L., Copeland, D.L. and Graham, J.B. Am. J. Hum. Genet. 14, 149-158, (1964).

7. Szabo, P., Purrello, M., Rocchi, M., Archidiacono, N., Alhadeff, B., Filippi, G., Tonido, D., Martini, G., Luzzatto, L. and Siniscalco, M. Proc. Natl. Acad. Sci. USA 81, 7855-7859, (1984).

8. Pai, G.S., Sprenkle, J.A., Do, T.T., Mareni, L.E. and Migeon, B.R. Proc. Natl. Acad. Sci. USA 77, 2810-2813, (1980).

9. Chance, P.F., Dyer, K.A., Kurachi, K., Yoshitake, S., Ropers, H.H., Wieacker, P. and Gartler, S.M. Hum. Genet. 65, 207-208, (1983).

10. Camerino, G., Grzeschik, K.H., Jaye, M., De La Salle, H., Tolstoshev, P., Lecocq, J.P., Heilig, R. and Mandel, J.L. Proc. Natl. Acad. Sci. USA 81, 498-502, (1984).

11. Boyd, Y., Buckle, V.J., Munro, E.A., Choo, K.H., Migeon, B.R. and Craig, I.W. Ann. Hum. Genet. 48, 145-152, (1984).

12. Purrello, M., Alhadeff, B., Esposito, D., Szabo, P., Rocchi, M., Truett, M., Masiarz, F. and Siniscalco, M. EMBO J. 4, 725-729, (1985).

13. Mattei, M.G., Baeteman, M.A., Heilig, R., Oberlé, I., Davies, K., Mandel, J.L. and Mattei, J.F. Hum. Genet. 69, 327-331, (1985).

14. Oberlé, I., Drayna, D., Camerino, G., White, R. and Mandel, J.L. Proc. Natl. Acad. Sci. USA 82, 2824-2828, (1985).

15. Harper, K., Winter, R.M., Pembrey, M.E., Hartley, D., Davies, K.E. and Tuddenham, E.G.D. Lancet II, 6-8, (1984).

16. Gitschier, J., Drayna, D., Tuddenham, E.G.D., White, R.L. and Lawn, R.M. Nature 314, 738-740, (1985).

17. Oberlé, I., Camerino, G., Heilig, R., Grunebaum, L., Cazenave, J.P., Crapanzano, C., Mannucci, P.M. and Mandel, J.L. N. Engl. J. Med. 312, 682-686, (1985).

18. Drayna, D., Davies, K., Hartley, D., Mandel, J.L., Camerino, G., Williamson, R. and White, R. Proc. Natl. Acad. Sci. USA 81, 2836-2839, (1984).

19. Purrello, M., Nussbaum, R., Rinaldi, A., Filippi, G., Traccis, S., Latte, B. and Siniscalco, M. Hum. Genet. 65, 295-299, (1984).

20. Hartley, D.A., Davies, K.E., Drayna, D., White, R.L. and Williamson, R. Nucl. Acids Res. 12, 5277-5285, (1984).

21. Levine, P.H. In "Hemostasis and Thrombosis", Colman, R.W., Hirsh, J., Marder, J.V. and Salzman, E.W. (eds.), Lippincott, Philadelphia, pp. 75-90, (1982).

22. McKee, P.A. In "The Metabolic Basis of Inherited Disease", 5th edn., Stanbury, J.B., Wyngaarden, J.B., Fredrickson, D.S., Goldstein, J.L. and Brown, M.S. (eds.), McGraw-Hill, pp. 1531-1560, (1983).

23. Hedner, U. and Davie, E.W. "Hemostasis and Thrombosis", Colman, R.W., Hirsh, J., Marder, J.V. and Salzman, E.W. (eds.), Lippincott, Philadelphia, pp. 29-38, (1982).

24. Haldane, J.B.S. J. Genet. 31, 317-326, (1935).

25. Ratnoff, O.D. J. Lab. Clin. Med. 103, 653-659, (1984).

26. Bloom, A.L. Lancet I, 1452-1454, (1984).

27. Gill, F.M. In "Factor VIII Inhibitors", Hoyer, L.W. (ed.), Alan. R. Liss, New York, pp. 19-29, (1984).

28. Feinstein, D.I. In "Hemostasis and Thrombosis", Colman, R.W., Hirsh, J., Marder, J.V. and Salzman, E.W. (eds.), Lippincott, Philadelphia, pp. 563-576, (1982).

29. Rizza, C.R. and Spooner, R.J.D. Br. med. J. 286, 929-933, (1983).

30. Graham, J.B., Barrow, E.S. and Elston, R.C. Ann. NY Acad. Sci. 240, 141-146, (1975).

31. World Health Organization Bull. WHO 55, 675-702, (1977).

32. Klein, H.G., Aledort, L.M., Bouma, B.N., Hoyer, L.W., Zimmerman, T.S. and De Metz, D.L. N. Engl. J. Med. 296, 959-962, (1977).

33. Mannucci, P.M. In "Recent Advances in Blood coagulation", Poller, L. (ed.), Churchill Livingstone, New York, pp. 193-210, (1981).

34. Duncan, B.M., Tunbridge, L.J., Duncan, E.M. and Lloyd, J.V. Brit. J. Haematol. 57, 113-121, (1984).

35. Misbahan, R.S., Rodeck, C.H. and Thumpston, J.K. In "The hemophilias. (Methods in Hematology, Vol. 5), Bloom, A.L. (ed.), Churchill Livingstone, New York, pp. 176-196, (1982).

36. Choo, K.H., Gould, K.G., Rees, D.J.G. and Brownlee, G.G. Nature 299, 178-180, (1982).

37. Jaye, M., De La Salle, H., Schamber, F., Balland, A., Kohli, V., Findeli, A., Tolstoshev, P. and Lecocq, J.P. Nucl. Acids Res. 11, 2325-2335, (1983).

38. Kurachi, K. and Davie, E.W. Proc. Natl. Acad. Sci. USA 79, 6461-6464, (1982).

39. Toole, J.J., Knopf, J.L, Wozney, J.M., Sultzman, L.A., Buecker, J.L.,
    Pittman, D.D., Kaufman, R.J., Brown, E., Shoemaker, C., Orr, E.C.,
    Amphlett, G.W., Foster, W.B., Coe, M.L., Knutson, G.J., Fass, D.N.
    and Hewick, R.M. Nature 312, 342-347, (1984).

40. Wood, W.I., Capon, D.J., Simonsen, C.C., Eaton, D.L., Gitschier, J.,
    Keyt, B., Seeburg, P.H., Smith, D.H., Hollingshead, P., Wion, K.L.,
    Delwart, E., Tuddenham, E.G.D., Vehar, G.A. and Lawn, R.M. Nature
    312, 330-337, (1984).

41. Gitschier, J., Wood, W.I., Goralka, T.M., Wion, K.L., Chen, E.Y.,
    Eaton, D.H., Vehar, G.A., Capon, D.J. and Lawn, R.M. Nature 312, 326-
    330, (1984).

42. Vehar, G.A., Keyt, B., Eaton, D., Rodriguez, H., O'Brien, D.P.,
    Rotblat, F., Oppermann, H., Keck, R., Wood, W.I., Harkins, R.N.,
    Tuddenham, E.G.D., Lawn, R.M. and Capon, D.J. Nature 312, 337-342,
    (1984).

43. Anson, D.S., Choo, K.H., Rees, D.J.G., Giannelli, F., Gould, K.,
    Huddleston, J.A. and Brownlee, G.G. EMBO J. 3, 1053-1060, (1984).

44. Südhof, T.C., Goldstein, J.L., Brown, M.S. and Russell, D.W. Science
    228, 815-822, (1985).

45. Gilbert, W. Science 228, 823-824, (1985).

46. Giannelli, F., Choo, K.H., Rees, D.J.G., Boyd, Y., Rizza, C.R. and
    Brownlee, G.G. Nature 303, 181-182, (1983).

47. Hassan, H.J., Orlando, M., Leonardi, A., Chelucci, C., Guerriero, R.,
    Mannucci, P.M., Mariani, G. and Peschle, C. Blood 65, 441-443,
    (1985).

48. Gitschier, J., Wood, W.I., Tuddenham, E.G.D., Shuman, M.A., Goralka,
    T.M., Chen, E.Y. and Lawn, R.M. Nature 315, 427-430, (1985).

49. Antonarakis, S.E., Waber, P.G., Kittur, S.D., Patel, A.S., Kazazian, H.H., Mellis, M.A., Counts, R.B., Stamatoyannopoulos, G., Bowie, E.J.W., Fass, D.N., Pittman, D.D., Wozney, J.M. and Toole, J.J. (1985) Submitted.

50. Rees, D.J.G., Rizza, C.R. and Brownlee, G.G. Nature 316, 643-645, (1985).

51. Antonarakis, S.E., Kazazian, H.H. and Orkin, S.H. Hum. Genet. 69, 1-14, (1985).

52. Winship, P.R., Anson, D.S., Rizza, C.R. and Brownlee, G.G. Nucl. Acids Res. 12, 8861-8872, (1984).

53. Grunebaum, L., Cazenave, J.P., Camerino, G., Kloepfer, C., Mandel, J.L., Tolstoshev, P., Jaye, M., De La Salle, H. and Lecocq, J.P. J. Clin. Invest. 73, 1491-1495, (1984).

54. Tønnesen, T., Søndergaard, F., Mikkelsen, M., Davies, K.E., Old, J., Winter, R.M. and Hauge, M. Lancet II, 1269-1270, (1984).

55. Tønnesen, T., Søndergaard, F., Güttler, F., Oberlé, I., Moisan, J.P., Mandel, J.M., Hauge, M. and Damsgard, E.M. Lancet II, 932, (1984).

56. Din, N., Schwartz, M., Kruse, T., Vestergaard, S.R., Ahrens, P., Caput, D., Hartog, K. and Quiroga, M. Lancet I, 1446-1447, (1985).

57. Winter, R.M., Harper, K., Davies, K.E., Hartley, D., Goldman, E. and Pembrey, M.E. Hum. Gene Map. 8, Cytogenet. Cell Genet., (1985), in press.

58. Camerino, G., Oberlé, I., Drayna, D. and Mandel, J.L. Hum. Genet, (1985), in press.

59. Haldane, J.B.S. Ann. Eugen. (London) 13, 262-271, (1947).

60. Vogel, F. and Motulsky, A.G. In Human Genetics : problems and approaches, Springer Verlag, Heidelberg, pp. 293-309, (1979).

61. Singh, I., Moser, A.B., Moser, H.W. and Kishimoto, Y. Pediatr. Research 18, 286-289, (1984).

62. Singh, I., Moser, A.E., Goldfischer, S. and Moser, H.W. Proc. Natl. Acad. Sci. USA 81, 4203-4207, (1984).

63. Moser, H.W., Moser, A.B., Trojack, J.E. and Supplee, S.W. J. of Pediatrics 103, 54-59, (1983).

64. Moser, H.W., Moser, A.B., Powers, J.M., Nitowsky, H.M., Schaumburg, H.H., Norum, R.A. and Migeon, B.R. Pediatr. Res. 16, 172-175, (1982).

65. Boué, J., Oberlé, I., Heilig, R., Mandel, J.L., Moser, A., Moser, H., Larsen, J.W., Dumez, Y. and Boué, A. Hum. Genet. 69, 272-274, (1985).

66. Mattéi, J.F., Mattéi, M.G., Auger, M. and Giraud, F. J. Génét. Hum. 32: 167-192, (1984).

67. Turner, G. and Jacobs, P.A. In: Advances in Human Genetics (Eds. Harris, H. and Hirschorn, K.), Plenum Press, pp. 83-112, (1983).

68. Fishburn, J., Turner, G., Daniel, A. and Brookwell, R. Am. J. Med. Genet. 14: 713-724, (1983).

69. Sutherland, G.R. Trends in Genet. 1: 108-112, (1985).

70. Sherman, S.L., Morton, N.E., Jacobs, P.A. and Turner, G. Ann. Hum. Genet. 48: 21-37, (1984).

71. Sherman, S.L., Jacobs, P.A., Morton, N.E., Froster-Iskenius, U., Howard-Peebles, P.N., Nielsen, K.B., Partington, M.W., Sutherland, G.R., Turner, G. and Watson, M. Hum. Genet. 69: 289-299, (1985).

72. Jenkins, E.C., Brown, W.T., Duncan, C.J., Brooks, J., Ben-Yishay, M., Giordano, F.M. and Nitowsky, H.M. Lancet II: 1292, (1981).

73. Webb, T., Butler, D., Insley, J., Weaver, J.B., Green, S. and Rodeck, C. Lancet II: 1423, (1981).

74. Tommerup, N., Søndergaard, F., Tønnesen, T., Kristensen, M., Arveiler, B. and Schinzel, A. Lancet I: 870, (1985).

75. Lathrop, G.M., Lalouel, J.M., Julier, C. and Ott, J. Proc. Natl. Acad. Sci. USA 81: 3443-3446, (1984).

76. Oberlé, I., Heilig, R., Moisan, J.P., Kloepfer, C., Mattéi, M.G., Mattéi, J.F., Boué, J., Froster-Iskenius, U., Jacobs, P.A., Lathrop, G.M., Lalouel, J.M. and Mandel, J.L. Proc. Natl. Acad. Sci. USA, submitted, (1985).

77. Choo, K.H., George, D., Filby, G., Halliday, J.L., Leversha, M., Webb, G. and Danks, D.M. Lancet II, 349, (1984).

78. Filippi, G., Rinaldi, A., Archidiacono, N., Ricchi, M., Balazs, I. and Siniscalco, M. Am. J. Med. Genet. 15: 113-119, (1983).

79. Morton, N.E. Cytogenet. Cell. Genet. 22: 15-36, (1978).

80. Connor, J.M., Colgan, J.M., Crossley, J.A., Imric, S.J., Shiach, C., Hann, I.M. and Forbes, C.D. (1985) Hum. Gene Map. 8, Cytogenet. Cell Genet., in press.

80a Oberlé, I., Mandel, J.L., Boué, J.,Mattei, M.G. and Mattei, J.F. Lancet I: 871, (1985).

81. Camerino, G., Mattéi, M.G., Mattéi, J.F., Jaye, M. and Mandel, J.L. Nature 306: 701-704, (1983).

82. Proops, R., Mayer, M. and Jacobs, P.A. Clinical Genet. 23: 81-96, (1983).

83. De la Chapelle, A. and Berger, R. Cytogenet. Cell. Genet. 37: 274, (1984).

83a Pembrey, M.E., Winter, R.M. and Davies, K.E. Amer. J. Med. Genet. 21, 709-719, (1985).

84. Collins, F.S. and Weissman S.M. Proc. Natl. Acad. Sci. USA 81: 6812-6816, (1984).

85. Rohme, D., Fox, H., Herrmann, B., Frischauf, A.M., Eström, J.E., Mains, P., Silver, L.M. and Lehrach, H. Cell 36: 783-788, (1984).

86. Schwartz, D.C. and Cantor, C.R. Cell <u>37</u>, 67-75, (1984).

87. Mossman, J., Blunt, S., Stephens, R., Jones, E.E. and Pembrey, M. Arch. Dis. Child. <u>58</u>, 911-915, (1983).

88. Upadhyaya, M., Bamforth, S., Harper, P.S., Sarfarazi, M., Thomas, N.S.T., Shaw, D.J., Meredith, A.L., Rees, D., Davies, K. and Young, I.D. Hum. Gene Map. 8, Cytogenet. Cell Genet., in press.

89. Zuffardi, O. and Fraccaro, M. Hum. Genet. <u>62</u>, 280-281, (1982).

90. Arias, S. and Quero, J. Cytogenet. Cell Genet. <u>25</u>, 132, (1979).

Human Genes and Diseases
Edited by F. Blasi
© 1986, John Wiley & Sons, Ltd.

MAPPING OF RARE X-LINKED GENES THROUGH DNA POLYMORPHISMS AND IDENTIFICATION OF CROSSOVER POINTS.

Giovanni Romeo

Traditional approach to human gene mapping.

Mapping of human genes on specific chromosomes (or regions of chromosomes) was essentially initiated at the beginning of the '70s when the approach of somatic cell genetics became widely used. The first autosomal gene mapped through somatic cell hybrids was thymidine kinase on chromosome 17 (Migeon and Miller, 1968) and since then more than 100 genes expressed in cultured human fibroblasts have been assigned to specific autosomes with a steady rate of increase until the late '70s, as shown in figure 1.

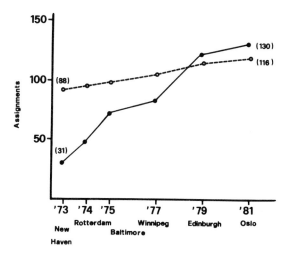

Fig. 1 - Cumulative number of assignments of autosomal(●——●) and X-linked genes (o--o) genes from successive Gene Mapping Workshops.

The apparent plateau reached by the number of genes mapped on autosomes at this time is indicative of the relative lack of further polymorphisms, and in particular of enzymatic markers, which were used
during the mapping "boom" of the preceeding years. While the gene
mapping of the '70s was quite important for the chromosomal assignment of autosomal loci, the rate of assignment of new genes to the
X chromosome did not change with regard to the preceeding years
(fig.1), because somatic cell hybrids could not provide more information in this respect than the traditional genetic analysis carried
out in human families. However major advances were obtained by somatic cell hybrids in the regional mapping of genes located on the X
chromosome, as well as on the autosomes. This progress resulted in
the construction of sintenic maps in which human genes could be assigned to specific regions of human chromosomes and therefore could be
ordered with respect to each other but did not give rise (with a few
limited exceptions) to true linkage maps with known genetic distances.

In order to transform the information coming from linear sequences
of genes (and from physical distances among these) in true genetic
maps it was necessary to calculate genetic distances among the same
genes by traditional linkage analysis. This goal could be partially
achieved for the human X chromosome map where linkage relationships
could be defined for two different gene clusters, the Xg and G-6-PD,
located on either end of the chromosome (McKusick, 1983).

The analysis of linkage for X-linked markers has always taken
advantage of the relative easiness of establishing phase relationships through the so called "grandfather's rule" (Stern, 1973).
This advantage together with the simplicity by which a gene can be
assigned to the X chromosome through its peculiar mode of transmission (diaginic), has made the mapping of X-linked genes (and in particular of rare disease loci) more feasible than for any autosomal

locus. In spite of these peculiar advantages the linkage analysis of
X-linked disorders has always been limited by the number of polymorphic markers available. The new class of polymorphisms already defined in the chapter by Williamson as RFLP's (Restriction Fragment
Length Polymorphisms) or more simply as DNA polymorphisms, has provided in the most recent years a wealth of markers that open new
possibilities to the analysis of multipoint linkage based on haplotype reconstruction.

This approach shall become increasingly important during the next few
years because it will yield information not only on the map position
of rare disorders (with obvious practical implications for genetic
counselling and prenatal diagnosis) but also on some basic biological
phenomena such as the unequal distribution of crossing-over and the
existence of hot-spots of recombination.

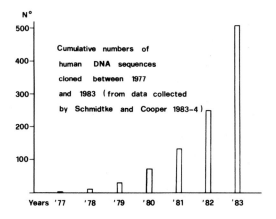

Fig. 2 - Exponential increase of the number of cloned
human DNA sequences.

Human recombinant DNA clones of the X chromosome.

The last seven years have witnessed an exponential growth of the molecular cloning of human DNA sequences, as documented by the number
of entries in the list prepared by Schmidtke and Cooper (1984) which
were 511 in January 1984 as compared to the 2 or 3 human recombinant
DNA clones existing in 1977. Figure 2 clearly shows that there has
been approximately a doubling of cloned human DNA sequences every
year during this period. These sequences are represented in part by
cDNA and genomic clones corresponding to genes with known function
(e.g. aldolase, $\alpha_1$antitrypsin, etc.) while the majority is represented by cloned DNA fragments of unknown function called arbitrary
DNA sequences. The nomenclature of arbitrary DNA sequences established by the Seventh International Workshop on Human Gene Mapping is
based on four pieces of information defined by letters or numbers in
the following way: Part. I = "D" for DNA
Part.II = "1, ..., 22, X, Y, XY, N," for the chromosomal assignment,
with XY for sequences homologous to the X and Y chromosomes, and "N"
for segments detected on numerous chromosomes other than X and Y;
Part.III = a symbol indicating the complexity of the genomic DNA
fragment(s) generated by restriction enzyme digestion and recognized
by that particular arbitrary DNA sequence: "S" stands for single or
unique DNA sequence, "Z" for repetitive sequence found clustered on
a single chromosome and "F" for small families of homologous sequences found on multiple chromosomes;
Part.IV: "1,2 ....." a sequential number to give uniqueness to the
above symbols.
DNA polymorphisms can be detected by clones of either type (genes
or arbitrary DNA sequences). The assignments of cloned arbitrary DNA
sequences and DNA polymorphisms to individual chromosomes, as reported by the proceedings of the Seventh International Workshop of Human
Gene Mapping (1983), are summarized in fig.3.

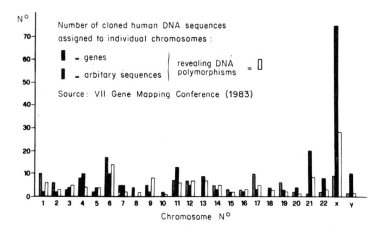

Fig. 3 - Number of cloned human DNA sequences and of DNA
polymorphisms mapped on specific chromosomes.

From this figure, which shows the distribution in the human genome
of the 115 cloned genes, 202 arbitrary DNA sequences and 127 RFLPs,
reported at the time of the workshop, it is apparent that the X chromosome contains 8% of the first type of clones (9 out of 115), 37%
of the second type of clones (75 out of 202) and 22% of the RFLPs
(28 out of 127). Since the X chromosome contains less than 5% of the
total human genome (haploid), this means that it holds a share of
cloned sequences and DNA polymorphisms greater than that expected on
the basis of its size. In particular the X chromosome ranks first
among the human chromosomes for its content in cloned arbitrary sequences and for the number of its DNA polymorphisms. These observations confirm that the interest of human geneticists for X-linked
markers, already established during the era of biochemical genetics
(1960s and '70s), has been revived in recent years by molecular genetics, and at the same time indicate that the map of the X chromosome
is potentially more complete than those of the remaining human chro-

TABLE 1 - REGIONALLY MAPPED X-LINKED RECOMBINANT CLONES REVEALING DNA POLYMORPHISMS

LOCUS/GENE NAME	CLONE NAME	REGIONAL ASSIGNMENT	RESTRICTION USED	FREQUENCY OF HETEROZYGOTES	REFERENCE
DXS18	pXQMGR1-22	p22.3-p21.1	Msp I	---	Hamerton et al. (1984
	782	p22.3 p21	Eco RI	.50	Pearson et al.(unpubl.)
DXS9	RC8	p22.3-p21	Taq I	.23	Murray et al. (1982)
DXS43	pD2	p22.3-p21	Pvu II	.40	Aldridge et al. (1984)
	pX23	p22.3-p21	Bgl II	.42	Willard et al. (unpubl.)
DXS47	83RId	p22.3-p21	Msp I	.04	Hamerton et al. (1984)
DXS67	pB24	p22.3-p21	Msp I	.15	Bruns et al. (1984)
	C7	p21	Eco RV	.30	Mandel et al. (unpubl.)
	754	p21	Pst I	.48	Pearson et al. (unpubl.)
OTC	---	p21	Msp I	---	Horwich et al. (1984)
DXS7	L 1.28	p11.3-p11	Taq I	.44	Pearson et al. (1984)
DXS41	p99-6	p21-cen.	Pst I	.49	Aldridge et al. (1984)
DXS14	58HA-I	p11-cen.	Msp I	.36	Aldridge et al. (1984)
DXS1	8EA	cen-q1	Taq I	.33	Bruns et al. (1984)
DXS62	MGU22	p11-q12	Hind III	.16	Wieacker et al. (1984)
DXS3	p19.2	q11-q22	Msp I	.28	Bruns et al. (1984)

Locus	Probe	Location	Enzyme	Value	Reference
DXYS1	pDP34	Xq13-22;Yp	Taq I	.36	Page et al. (1982)
DXS17	S21/S9	q21.3-q22	Taq I	.35	Drayna et al. (1984a)
DXS8	RB6	q24-q26	Msp I	.02	Hill et al. (1982)
DXS11	22-33	q24-qter	Taq I	.24	Aldridge et al. (1984)
DXS42	p43-15	q24-qter	Bgl II	.20	Aldridge et al. (1984)
DXS19	pXQMGRI-46	q25-qter	Taq I	.14	Hamerton et al. (1984)
DXS37	30RIb	q25-qter	Taq I	.11	Hamerton et al. (1984)
DXS49	78RIc	q25-qter	Msp I	.04	Hamerton et al. (1984)
HPRT	pHPT4	q26-q27	Bam HI	.34	Nussbaum et al. (1983)
F IX	---	q26-q28	Taq I	.33	Camerino et al. (1984)
DXS10	p6A-1	q26-qter	Taq I	.34	Nussbaum et al. (unpubl.)
DNF1	pAX-6	q26-qter	Msp I	.66	Balazs et al. (unpubl.)
DNF3	pAX-7	q26-qter	Sst I	.35	Balazs et al. (1984)
DXS51	52A	q27	Taq I	.38	Drayna et al. (1984 b)
PGK	---	q13	PstI+XbaI	---	Hutz et al. (1984)
DXS15	Dx13	q28-qter	Bgl II	.36	Harper K. et al. (1984)
---	St14	q28-qter	Bgl II	.90	Oberlé I. et al. (1985)

mosomes.

The 33 arbitrary sequences and genes reported in table 1 represent, to my knowledge, most of the X-linked recombinant clones available at the end of 1984 which detect DNA polymorphisms and which
have been regionally mapped on the X chromosome.

It is clear that the great number of polymorphic markers makes it
quite easy today to establish a linkage relationship between a DNA
polymorphism and any X-linked disorder, provided that one knows
which polymorphism is more likely to reveal such linkage. In practice any type of preliminary information pointing to a particular region of the X chromosome as the possible location of a rare X-linked gene will be very valuable because it will restrict the search
for linkage only to those polymorphisms mapping in that region and
therefore presumably in close linkage with the gene.

Physical and genetic mapping of rare X-linked genes.

In the past at least 17 X-linked disorders have been physically mapped to specific regions of the X chromosome using means other than
recombinant DNA techniques. This was accomplished as summarized in
table 2, by taking advantage of occasional observations of affected
females carrying X-autosomal translocations (as in the case of Duchenne, Hunter, Aicardi, etc.), or even of particular cytogenetic
markers (as in the case of Menkes linked to a centromeric chromosomal polymorphism or in the case of the mental retardation, macroorchidism syndrome associated by definition to a fragile site in q27).
Alternatively somatic cell hybrids have been utilized to map disease
loci when the corresponding missing gene product was known (as in
the case of X-linked ichthyosis, Fabry disease, PGK deficiency,
Lesch-Nyhan syndrome and G-6-PD deficiency).

TABLE 2 – X-LINKED DISORDERS ASSIGNED TO SPECIFIC REGIONS OF THE X CHROMOSOME BY TRADITIONAL MEANS

	SYMBOL	ASSIGNMENT	REFERENCE
STEROID SULFATASE DEFICIENCY	STS	p223-pter	VI HUMAN GENE MAPPING (1982)
AICARDI SYNDROME	AIC	p22	ROPERS ET AL. (1982)
OCULAR ALBINISM	OA	p22	McKUSICK (1983)
CHRONIC GRANULOMATOUS DISEASE	CGD	p21-pter	VI HUMAN GENE MAPPING (1982)
DUCHENNE	DMD	p21	VI HUMAN GENE MAPPING (1982)
MENKES	MNK	p11-q11	FRIEDREICH ET AL. (1983)
TESTICULAR FEMINIZATION	DHTR	p11-q11	MIGEON ET AL. (1981)
PGK DEFICIENCY	PGK	q13	McKUSICK (1983)
FABRY DISEASE	GALA	q21-q24	VI HUMAN GENE MAPPING (1982)
HUNTER DISEASE	MPS II	q26-q27	MOSSMAN ET AL. (1983)
LESCH-NYHAN SYNDROME	HPRT	q26-q27	PAI ET AL. (1980)
MR, MACROORCHIDISM,FRAGILE X	FRAXQ27	q27	McKUSICK (1983)
COLOR BLINDNESS, DEUTAN	CBD	q27-qter	McKUSICK (1983)
COLOR BLINDNESS, PROTAN	CBP	q27-qter	McKUSICK (1983)
G-6-PD DEFICIENCY	G6PD	q28	PAI ET AL. (1980)
TORTICOLLIS,KELOIDS,CRYPTORCHIDISM	TKC	q28-qter	ZUFFARDI ET AL. (1982)
COLOR BLINDNESS,BLUE-MONOCONE, MONO-CHROMATIC TYPE	CBBM	q	FLEISCHMAN ET AL. (1981)

More recently the gene coding for ornithine transcarbamylase (OTC) has been clóned (Horwich et al., 1984) and mapped in p21. It should be underlined that, in this as well as in previous cases, physical mapping of a gene coding for a given enzyme does not necessarily imply that a mutation causing a deficiency of that enzyme has the same map position. Although not common, the possibility exists that the enzymatic deficiency is caused by some type of regulatory mutation at a different locus. As an example, in mouse a regulatory locus for ß-glucuronidase, indipendent of the structural locus, has been demonstrated (Paigen, 1979). In order to exclude this possibility, linkage data have always to be collected from families in which the particular disorder segregates.

Linkage studies with DNA polymorphisms performed during the last two years in "normal" pedigrees or in pedigrees of patients affected with particular X-linked disorders, have also contributed the initial information which has been utilized for the construction of a tentative genetic map of the X chromosome. The sequence and reciprocal genetic distances of some polymorphic markers (some of which correspond closely to disease loci) on the long arm of the X chromosome have been estimated (Drayna et al.,1984b) as: DXYS1 - (.16)- DXS17 - - (.29)- HPRT - (0.5)- DXS51 - (.07)- F IX - ($<$ .01)- DXS15.

In the subterminal region of the chromosome the linear order of 3 disease loci has been estimated as: F IX, fragile site, F VIII (Purello et al., 1984). This sequence is consistent with the cytological mapping of G-6-PD on the distal side of the fragile X (Szabo et al., 1984). To account for the observation that F IX deficiency (hemophilia B) recombines freely with at least two loci of the G-6-PD cluster it has been proposed that the chromosomal region which includes the fragile X-site is normally a region of high meiotic recombination (Szabo et al., 1984).

As to the short arm, linkage data regarding several polymorphisms

listed in table 1 and Duchenne muscular dystrophy (DMD), X-linked
ichthyosis (STS), retinoschisis (RS), Menkes syndrome (MNK) and X-
linked retinitis pigmentosa (RP) reported at the VII International
Worshop on Human Gene Mapping are consistent with the following gene
order and map distances: Xg-(.15)-STS-(.10)-RS-(.15)-DMD-(.15)-DXS7-
-(.13)-DXS62, CENT. However the precision of the estimates (often
based on low lod scores) is insufficient to regard these sequences
as more than provisional. On the basis of the available data the assignment to the short arm of two other disease loci, namely MNK and
RP, for which a reliable linkage with DXS7 has been demonstrated
with a distance of 16 cM and 3 cM respectively from this locus, cannot be considered definitive. Even in the latter case (which shows
95% confidence limits at 0 and 15 cM) a single linkage cannot discriminate between a short arm and a long arm localization, always close
to the centromere. In principle this uncertainty can be overcome by
studying more informative families for other short arm polymophisms
but sometime this objective is not easily achieved because of the
rarity of the disorders being examined. This type of difficulty has
been encountered by our group in the mapping of a rare non-lethal
form of X-linked muscular dystrophy, called Becker muscular dystrophy
(BMD), and the strategy adopted is described in detail in the next
section.

Mapping by haplotype and crossover identification in single pedigrees.
Duchenne and Becker muscular dystrophies have been identified as
separate clinical entities (McKusick's catalog N.31020 and 31010)
but differentiation between them can be difficult in some cases due
to borderline clinical presentations. Becker (BMD) is more benign
than Duchenne (DMD), being characterized by later onset and by slower
progression, and is compatible with a normal reproductive function.
Heterozygotes for DMD, as well as BMD, can be diagnosed on the basis

of an indirect test consisting in the assay of an enzyme of muscle
origin, creatine-phosphokinase (CPK), in serum. Between 70 and 80%
of genetically certain heterozygotes show increased values of CPK in
their sera, which indicates that 30-20% of heterozygotes can go un-
dected on the basis of this test. The genetic mapping of these disorders, whose basic defects are still unknown, represents therefore
a prerequisite for the possible use of DNA polymorphisms in the diagnosis of heterozygotes. In fact it has been calculated that if an
X-linked locus such as the Duchenne is encompassed on each side by
4-8 polymorphic loci having a frequency of the minor allele between
0.2 and 0.4 then at least 90% of the potential heterozygotes can be
diagnosed in an unequivocal way on the basis of the information obtained by at least two polymorphisms located on either side of the
disease locus (Aldridge et al., 1984).

At present there is no proof contrasting the hypothesis that DMD and
BMD can be caused by allelic mutations of the same gene, but, since
the basic defects of these disorders are still unknown it is not
possible to produce evidence in favor of this hypothesis. In the
past this hypothesis was rejected because the DMD gene was thought
to reside in a region of the X chromosome different from that where
the DMD gene is located. The position of DMD on the short arm in
the sequence of genetic markers mentioned in the previous section is
well established because of the clear linkage relationships of DMD
with two short-arm polymorphic loci, DXS9 and DXS7 (with lod scores
>3 in both cases). On the contrary BMD was thought to be linked to
the G-6-PD cluster and therefore to reside in the long arm on the basis of a weak evidence of linkage (Zatz et al.1974).This assignment
has been revised because of the firm linkage established between BMD
and DXS7 with an estimated distance of 19 cM (Kingston et al.1983).
Since DXS7 maps at approximately 13 cM from the centromere, linkage
with this single marker could not be taken as a proof for the location

of Becker on the short arm of the X chromosome until another short
arm polymorphism was shown to be linked to Becker in an arrangement
similar to that described for Duchenne. Because of the relative rarity of Becker families, our group chose a different approach based
on the analysis of 10 different X-linked DNA polymorphisms in two
particularly informative pedigrees (Roncuzzi et al. 1985).
In the first pedigree (N.1001) the mother of the two brothers affected with BMD (III-3 of fig.4) was found to be heterozygous for the
polymorphisms revealed by clones D2, RC8, 58, pDP34 and 22-33 and
homozygous for L 1.28, 8, B24 and PGK. The phase of the informative
polymorphisms with respect to the BMD gene was ascertained in this
carrier mother because her father, who in turn was affected with
the disease, was available for study. Her two affected sons, who
inherited the same Becker gene from their maternal grandfather, were
found to be concordant for the allele 4 of polymorphism 58 but
discordant for all the other informative polymorphisms. It was therefore inferred that one of the two brothers   was a recombinant.
Since the relative position of most of these DNA polymorphisms was
established (table 1),it was possible to locate the Becker gene
with respect to their order on the X chromosome by taking advantage
of the information deriving from the identification of the points
of crossover in the recombinant affected sib. As schematized in
fig. 4 the X chromosome of individual IV-3 must have been the result
of at least two recombinational events, one on each side of the
Becker gene, to explain the concordance of the 2 sibs for the Becker
gene and for the DNA polymorphism 58 only. A greater number of recombinations, namely 3, was not accounted for by the observed disco-
rdance of the 2 sibs for the remaining polymorphisms.

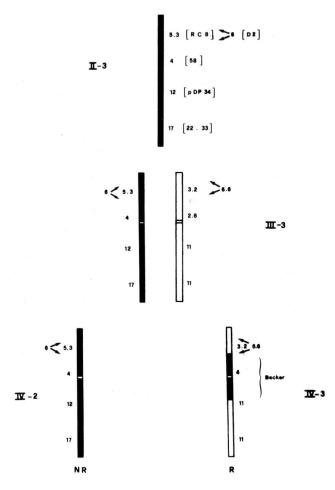

Fig. 4 - Haplotypes of the X chromosomes as defined by the informative DNA polymorphisms in 3 generations of pedigree 1001. The Becker gene must be located in that part of the X chromosome which is included between the 2 points of recombinations in individual IV-3 and which is also shared by the three Becker patients analyzed (II-3, IV-2, and IV-3) (From Roncuzzi et al., 1985).

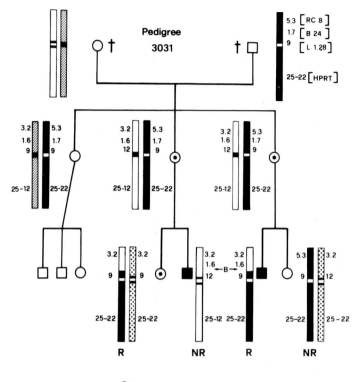

⊙ = positive CPK test

Fig. 5 - Sequence of informative DNA polymorphisms in pedigree 3031 and location of the Becker gene deduced from the hypothesis of a single crossover between DXS67 revealed by probe B24 and DXS7 revealed by probe L 1.28 in the recombinant X chromosomes (indicated by R) inherited by III-4 and III-6 (From Roncuzzi et al.1985). The phase of the polymorphic alleles transmitted as a single haplotype by the dead grandfather(I-2) to all his daughters has been deduced from somatic cell hybrids as shown in table 3.

This pedigree indicated therefore that BMD   had to be located
between the polymorphisms revealed by probes RC8 (locus DXS9) and
D2 (locus DXS43) on one side and the polymorphism revealed by probe
pDP34 (locus DXYS1)) on the other side.

The second pedigree (N.3031) was analyzed for the same DNA polymorphisms used with the first one and in addition for the polymorphism revealed by the HPRT gene (Nussbaum et al., 1983), characterized by 3 alleles (25-22, 22-18 and 25-12). A partial analysis of
DNA polymorphisms in this second pedigree had already been reported
by our group (Romeo et al., 1984) but at the time the phase of the
informative DNA polymorphisms in the two carrier mothers was not
known, because both maternal grandparents were already dead (fig.5).
Unequivocal information on this point was obtained by selecting hybrids deriving from the fusion of lymphocytes of the three sisters
of generation II with YH21 hamster fibroblasts. Those clones retaining a single human X chromosome (as documented by cytogenetic characterization), analyzed for several DNA polymorphisms, yielded the
necessary information about the alleles carried by at least a single
X-chromosome from each of the 3 sisters (table 3). In the case of
II-1 both X chromosome haplotypes were deduced directly from the
analysis of 3 informative DNA polymorphisms in hybrid clones. In the
case of the other two sisters (II-2 and 3) only one haplotype was
obtained directly from hybrid clones whereas the second was easily
inferred from the knowledge of the corresponding genotype (table 3).
The former sister who had two non-affected sons and a normal CPK
test (fig.5) shared with the latter two only the haplotype characterized by alleles 5.3 (locus DXS9), 1.7 (locus DXS67) and 9 (locus
DXS7), which must therefore have been contributed by the grandfather
of the patients, who was not affected with Becker.

TABLE 3 — IDENTIFICATION OF X CROMOSOME HAPLOTYPES OF THE 3 SISTERS OF PEDIGREE 3031 (from RONCUZZI ET AL. 1985)

CODE N. OF INDIVIDUALS	GENOTYPE			CODE N.OF SUBCLONES SELECTED IN HAT	N.(AND STATE) OF X CHROMOSOME RETAINED	HAPLOTYPES IDENTIFIED			ORIGIN **
	RC8	B24	L1.28			RC8	B24	L1.28	
II – 1	5.3 / 3.2	1.7 / 1.6	9 / 9	344	1 (active)	3.2	1.6	9	M
				352	1 (active)	5.3	1.7	9	P
II – 2	5.3 / 3.2	1.7 / 1.6	12 / 9	432	1 (active)	3.2	1.6	12	M
				452	1 (active)	3.2	1.6	12	M
II – 3	5.3 / 3.2	1.7 / 1.6	12 / 9	552–6TG *	1 (inactive)	5.3	1.7	9	P

* 6TG = clone counterselected in 6-thioguanine

** M = maternal or P = paternal haplotype deduced as described in the text. The 3 short-arm polymorphisms define the phase or arrangement of the corresponding alleles on each X chromosome isolated in somatic cell hybrids. The haplotypes thus identified together with the known genotypes of each of the 3 sisters lead to the reconstruction of the haplotype (P) contributed by the grandfather of the patients, who was not affected with BMD. This result establishes the phase of the BMD gene which must have been transmitted by the grandmother of the patients to two of her daughters (II – 2 and II – 3) together with alleles 3.2 – 1.6 – 12.

This conclusion is based on the assumption that the three sisters
have the same father, an assumption which is not contradicted by any
of the informative polymorphisms inherited by them. According to
this reconstruction of the paternal haplotype, individual II-1 received from her mother the haplotype characterized by alleles 3.2
(locus DXS9), 1.6 (locus DSX67) and 9 (locus DXS7), while her two
Becker carrier sisters (II-2 and 3) received the haplotype characterized by 3.2 (locus DXS9), 1.6 (locus DXS67) and 12 (locus DXS7),
which therefore identifies the chromosome carrying the Becker gene.
Based on these results it could be concluded that patient III-6 and
female III-4, who had a positive CPK test, carried recombinant X
chromosomes. In both X chromosomes at least one recombination must
have occurred between DXS67 and DXS7. If the Becker gene is localized in this region, the recombination occurred between the Becker
gene itself and DXS7 (fig.5). If instead the Becker gene is localized below DXS7, a greater number of recombinations, namely three,
occurred in each recombinant X chromosome of individuals III-4 and
III-6, which makes this alternative more unlikely. The support in
favor of the former alternative comes from recent linkage data collected by our group on a total of 17 Becker pedigrees (Fadda et al.
1985). These data show that loci DXS9 and DXS43 are closely linked
(with a recombination fraction of less than 0.01 and lod scores of
2.7) and both are located at approximately 15 cM from the Becker
locus. Since DXS9 and DXS43 are so closely linked that can be considered as parts of a single gene cluster, and since they are both
located distally to   DXS7, this implies that BMD maps between
DXS43 and DXS9 on one side and DXS7 on the other. This conclusion
is in agreement with the reported genetic distance between DXS9 and
DXS7 measured at 33 cM both in Duchenne and Becker families (Davies
et al., 1983, Fadda et al., 1985). The close similarity of genetic

distances which separate Becker or Duchenne genes from surrounding
DNA polymorphisms indicates that the two map in the same Xp region.
The approach we used in the mapping of the Becker gene allows to
locate in a given sequence of markers a gene whose map position on
the X chromosome is unknown. This might be particularly relevant for
other rare X-linked disorders, whose linkage analysis with DNA polymorphisms might be hampered by the lack of an adequate number of
informative families. Once the position of the rare gene has been
established in a given order of markers, it will be more convenient
to direct the search for linkage towards those polymorphic markers
which have already been mapped in that region.

Conditions for haplotype reconstruction through cell hybrids.

As just described, mapping of rare genes on the X chromosome can be
achieved using a strategy which is indipendent from fortuitous observations of linkage with X-linked DNA polymorphisms chosen at
random. This strategy can be exploited for mapping X-linked disorder
whose enzymatic (or biochemical) defects are either unknown or
have already been characterized. An example of the former case is
that just discussed of BMD, while the latter case can be exemplified
by Hunter disease (Mucopolysaccharidosis II) whose enzymatic defect
(iduronate-sulfate sulfatase) is well known. The map position of
the gene of Hunter disease, suspected to reside in q26-27 (table 2)
on the basis of a single X-autosomal translocation (Mossman et al.
1983), is far from being established.

Our group has been analyzing different DNA polymorphisms on eight
pedigrees with a family history of Hunter (Mochi et al., unpublished
data). The phase of the Hunter gene and of the informative polymorphisms is known only for 3 of the carrier mothers, whose fathers are
still alive. For the remaining carrier mothers the haplotype inherited from their fathers can be identified without ambiguity in hybrid

cells made by fusing their lymphocytes with a murine line. In this
case, in fact, it is possible to differentiate the hybrid cells
carrying the human X chromosome with the Hunter gene from the hybrid
cells carrying the human wild type gene on the basis of the enzymatic assay for iduronate-sulfate sulfatase. The method described in
table 3 has therefore been utilized to select hybrids with a single
active X chromosome derived from lymphocytes of this second group
of Hunter mothers, and haplotype reconstruction is now being carried
out.

If the basic defect of the disorder is not known, as in the case
of BMD (or if it cannot be assayed in the type of cells used for
somatic hybridization) then the reconstruction of the grandfather's
haplotype becomes more difficult. Even in such a situation it should
still be possible to identify the grandfather's haplotype if at
least two of his daughters have received a different X chromosome
from their mothers. Alternatively the 2 X chromosome haplotypes of
the grandmother might be deduced from the analysis of the polymorphic alleles carried by at least 2 of her sons, if again they have
received different X chromosomes. This type of information might
be blurred by recombinations occurred in the grandmother (which
might be difficult to identify if she is not available for study)
but the grandfather's haplotype should eventually become apparent
as the only one recurring unchanged in his daughters. Finally if a
fourth completely different haplotype is observed in one of the
grandmother's daughters, this constitutes a clear indication for
extramarital paternity.

In conclusion when the basic defect of the disorder cannot be
assayed in hybrid cells, it is not sufficient to prepare somatic
cell hybrids using lymphocytes of the carrier mother whose haplotypes should be studied, unless further information can be obtained

from her sibs. Since the amount of work required in the characterization of the grandfather's haplotypethrough somatic cell hybrids
might become quite heavy, it is often necessary to decide which pedigrees should be selected for this type of study. Several criteria
can be considered in taking this decision, the first obvious one
being the number of progeny born to a carrier mother whose haplotypes should be identified. It should be borne in mind that once the
haplotypes of this carrier are identified, not only the genotypes
of her sons can show the occurrence of recombinations but also those
of her daughters can yield equally useful information if their
father's X chromosome haplotype is known. Such a situation is exemplified by the information contributed by females III-4 and III-7
in the study of pedigree N.3031 summarized in fig.5.

An indication of the potential information a given pedigree may
contribute to the mapping of a given X-linked disease locus is given
by the discrepancy of affected males for one or more X-linked polymorphisms. This was the main criterion used by our group for selecting the pedigrees N.1001 and 3031 (fig.4 and 5) in the mapping
study of BMD.

In conclusion when the study of genotypes of carrier females and
affected males has been carried out on genomic DNA from perypheral
leukocytes or from lymphoblastoid lines, the decision should be
taken whether to use the approach of somatic cell hybrids for the
identification of haplotypes. Two different alternatives are sometime used in our laboratory   to delay this decision: the first
is represented by the storage of lymphocytes in liquid N_2 for several months before using them for fusion experiments, the second
consists in the freezing of whole cell cultures 48 hrs after fusion,
before starting the selection in HAT, which can be applied to thawed
hybrid cells kept for a period of several weeks or months in liquid
N_2.

Future research on the X chromosome through DNA polymorphisms.

The impact that molecular biology is having on human genetics is self evident and in the next few years an increasing number of genetic problems, both basic and applied, will be approached using molecular tools. DNA polymorphisms, in particular, provide the opportunity for the construction of a reasonable genetic map of each human chromosome (Botstein et al., 1980) and at the same time offer a practical indirect approach to the prenatal diagnosis of disorders whose basic defects are still unknown. This approach will become particularly important for well mapped X-linked disorders like Duchenne muscular dystrophy and mental retardation with fragile X syndrome or even for   disorders whose corresponding wild-type genes have been cloned, like hemophilia A and B. In fact because of the theoretical high frequency of new mutations one expects a great heterogeneity of molecular defects to be present in X-linked disorders.

The frequency of new mutations per se represents a problem which can be investigated using DNA polymorphisms closely linked to the gene causing a particular X-linked disease. As an example, this type of study can be performed in Duchenne pedigrees using DNA polymorphisms flanking this gene which are already available today.

If, using these polymorphisms, one identifies without ambiguity in a mother who has two or more affected sons the haplotype carrying the Duchenne mutation as derived from the maternal grandfather of the patients (who evidently was not affected) then the conclusion of a new mutation would be inescapable and would represent the first experimental evidence in favor of this type of phenomenon. The high frequency of new mutations postulated by Haldane (1935) to account for equilibrium of X-linked lethal genes has not received experimental support when it was possible to test it, as in the case of the Lesch-Nyhan syndrome (Francke et al., 1976). In order to approach the problem of new mutations in this way it will be necessary to identify a haplotype at risk characterized by closely linked polymorphisms located on either side of the gene. In fact the observation of one or more polymorphic alleles all inherited from the grandfather but located on only one side of the gene would leave open the possibility of a single crossing over occurring in a stem cell of a maternal gametic line between the polymorphic markers and the gene itself. This possible case of germinal mosaicism could be interpreted as tracing the inheritance of the mutation to the grandmother's and not to the grandfather's gamete.

It should be apparent from the above example that DNA polymorphisms can be used both for solving practical problems of diagnosis and for research. Especially for X-linked disorders these two aspects cannot be separated because research has to be carried out in heterozygotes at risk, who might present higher frequencies of recombination with respect to control women. This hypothesis has already been advanced for the fragile X locus which might represent a hot spot of recombination in females carrying the mutated gene (Szabo et al., 1984). This hypothesis should also be tested for the Duchenne locus to which a good number of DNA polymorphisms are closely linked on either side of the gene (table 1).

Unequal distribution of crossing-overs along the X chromosome seems already apparent from the comparison of existing linkage data with the physical map of the chromosome (Keats, 1983). In particular the frequency of recombination seems to be higher in the subterminal regions of the chromosome, where the Duchenne and fragile X loci are located, with respect to the more internal portions of the chromosome. The linkage data used for this comparison have not been separated according to their origin from control families or families at risk for particular disorders. This type of disaggregation

will become necessary, in the future, as just discussed.
Finally X-linked DNA polymorphisms make it possible today to devise
new experimental models for asking basic questions about fundamental
biological mechanisms such as X chromosome inactivation. One model
which can be utilized in this respect is represented by the 2 types
of hybrid cells obtained in experiments like those described in table 3: some of these hybrids contain the same X chromosome (identified by haplotype characterization)in 2 different functional states,
active and inactive as established by cytogenetic techniques (Roncuzzi et al., 1985).

Although DNA polymorphisms of the X chromosome are more numerous
than those located on any other single chromosome (fig.3) the discovery of a greater number of X-linked polymorphisms remains a prerequisite for answering the many questions of which only a few have
been mentioned in these pages. It would be highly useful to find new
ways of discovering DNA polymorphisms and in particular to devise
methods for directing the search of such polymorphisms to specific
regions of interest, like those flanking the fragile X site or the
Duchenne locus. These methods can originate only from work done by
molecular biologists but the questions which can be asked using
these tools will be directed to the solution of classical genetic
problems.

The new interest generated by the discovery of DNA polymorphisms
in linkage and in the mapping of disease loci is only the first step
toward a renewal of the methodology utilized so far by human geneticists. Linkage analysis itself will be utilized no longer for two
point-linkages only, but rather for multipoint linkage analysis as
already done for genes located on the short arm of chromosome 11
(Antonarakis et al., 1983).Computer programs for multipoint linkage
analysis have already been prepared (Lathrop et al., 1984).

The identification of single or multiple recombinations through
DNA polymorphisms will be equally feasible in the future for the
X chromosome and for the autosomes, but the basic biological questions peculiar to the former will probably make more interesting
the study of any X-linked problem.

REFERENCES

Aldridge,J., Kunkel,L., Bruns,G., Tantravahi,U., Laland,M., Brewster,
    T., Moreau,E., et al. (1984). A strategy to reveal high frequency
    RFLPs along the human X-chromosome. Am. J.Human Genet. 36,546-564.
Antonarakis,S.E., Phillips,J.A., Mallonee,R.L., Kazazian,H.H.,
    Fearon,E.R., Waber,P.G., Kronenberg,H.M., Ullrich,A., Meyers,D.A.
    (1983). ß-globin locus is linked to PTH locus and lies between
    the insulin and PTH loci in man. Proc.Nat.Acad.Sci. USA, 80,
    6615-6619.
Balazs,I., Purello,M., and Siniscalco,M. (1984). Properties of a
    common restriction-site polymorphism for the X-chromosome. Pro-
    ceedings of the Los Angeles conference (1983) Published by Cyto-
    genetics and Cell Genetics. 37, N.1-4.
Bhattacharya,S.S., Wright,A.F., Clayton,J.F., Price,W.H.,Phillips,C.I.
    McKeown, C.M.E., Jay,M., Bird,A.C., Pearson,P.L., Southern,E.M.,
    Evans,J.H. (1984).Close genetic linkage between X-linked retini-
    tis pigmentosa and a restriction fragment lenght polymorphism
    identified by recombinant DNA probe L 1.28. Nature 309, 253-258.
Botstein,D., White,R.L., Skolnick,M., and Davis,R.W. (1980). Con-
    struction of a genetic linkage map in man using restriction frag-
    ment lenght polymorphisms. Am.J.Hum.Genet. 32, 314-331.
Bruns,G., Aldridge,J., Kunkel,L., Tantravahi,U., Lalande,M., Dryja,T.
    and Latt,S.A. (1984). Molecular analysis of the human X-chromoso-
    me. Proceedings of the Los Angeles Conference (1983) Published by
    Cytogenetics and Cell Genetics. 37,N.1-4.

Camerino,G., Grzeschik,K.H., Jaye,M., De La Salle,H., Tolstoshev,P.,
    Lecocq,J.D., Heilig,R., Mandel,J.L. (1984). Regional localiza-
    tion on the human X chromosome and polymorphism of the coagula-
    tion factor IX gene (hemophilia B locus). Proc.Nat.Ac.Sci. U.S.A.
    81, 498-502.

Davies,K.E., Pearson,P.L., Harper,P.S., Murray,J.M., Brien,T.O.,
    Sarfarazi,M., Williamson,R. (1983). Linkage analysis of two clo-
    ned DNA sequences flanking the Duchenne muscular dystrophy locus
    on the short arm of the human X-chromosome. Nucleic Acids Resea-
    rch 11, 2303-2312.

Davies,K.E., Pearson,P.L., Harper,P.S., Murray,J.M., O'Brien,T.,
    Sarfarazi,M., and Williamson,R. (1984). Linkage analyses of Du-
    chenne muscular dystrophy using DNA sequences flanking the locus
    on the short arm of the human X chromosome. Proceeding of the
    Los Angeles Conference (1983) Published by Cytogenetics and Cell
    Genetics.37, N.1-4.

Drayna,D.T., Davies,K., Hartley,D., Williamson,R., and White,R.
    (1984 a). Genetic linkage on the human X chromsome. Proceedings
    of the Los Angeles Conference. Published by Cytogenetics and Cell
    Genetics. 37, N.1-4.

Drayna,D., Davies,K., Hartley,D., Mandel,J.L., Camerino,G., William-
    son,R., White,R. (1984 b). Genetic mapping of the human X chromo-
    some by using RFLPs. Proc.Nat.Ac.Sci. U.S.A. 81, 2836-2839.

Fadda,S., Mochi,M., Roncuzzi,L., Sangiorgi,S., Sbarra,D., Zatz,M.,
    Romeo,G. (1985). Definitive mapping of BMD in Xp by linkage to
    a DNA polimorphic cluster (DXS9 and DXS43). Submitted for publ.

Francke,U., Felsenstein,J., Gartler,S.M., Migeon,B.R., Dancis,J.,
    Seegmiller,J.E., Bakay,F., Nyhan,W.L. (1976). The occurrence of
    new mutants in the X-linked recessive Lesch-Nyhan disease. Am.J.
    Hum.Genet. 28, 123-137.

Friedrich,U., Horn,N., and Stene,J. (1984). Close linkage of the gene for Menkes disease to the centromere region of the human X-
chromosome. Ann.Hum.Genet. in press.

Haldane,J.B.S. (1935). The rate of spontaneous mutations of a human
gene. J.Genet. 31, 317-326.

Hamerton,J.L., Wang,H.S., Riddell,D.C., Beckett,J., Holden,J.J.A.,
Mulligan,L., Phillips,A., Simpson,N., White,B.N., and Wrogemann,
K. (1984). Assignment and regional localization of a series of
X chromosome specific DNA probes. Proceedings of the Los Angeles
Conference (1983) Published by Cytogenetics and Cell Genetics.
37, N.1-4

Harper,K., Pembrey,M.E., Davies,K.E., Winter,R.M., Hartley,D.,
Tuddenham,E.G.D.(1984). A clinically useful DNA probe closely
linked to hemophilia A. Lancet II, 6-8.

Hill,M.E.E., Davies,K.E., Harper,P., and Williamson,R. (1982). The
Mendelian inheritance of a human X chromosome-specific DNA sequence polymorphism and its use in linkage studies of genetic disease. Human Genetic 60, 222-226.

Horwich,A.L., Fenton,W.A., Williamson,K.R., Kalousek,F., Kraus,J.P.
Doolittle,R.F., Konisberg,W., Rosenberg,L.E. (1984) Structure
and expression of a cDNA for the nuclear coded precursor of human
mitochondrial OTC. Science 224, 1068-1074.

Hutz,M.H., Michelson,A.M., Antonarakis,S.E., Orkin,S.H., Kazazian,
N.H. (1984). Restriction site polymorphism in the phosphoglycerate kinase gene on the X chromosome. Human Genetics (in press).

Keats,B. (1983). Genetic mapping: X chromosome. Human Genetics 64,
28-32.

Kingston,H.M., Thomas,N.S.T., Pearson,P.L., Sarfarazi,M., Harper,P.S.
(1983) Genetic linkage between Becker muscular dystrophy and a
polymorphic DNA sequence on the short arm of the X chromosome.
J.Med.Genetics 20, 255-258.

Kingston,H.M., Sarfarazi,M., Thomas,N.S.T., and Harper,P.S. (1984). Localization of the Becker muscular dystrophy gene on the short arm of the X chromosome by linkage to cloned DNA sequences. Human Genetics 67, 6-17.

Lathrop,G.M., Lalouel,J.M., Julier,C., Ott,J. (1984). Strategies for multilocus linkage analysis in humans. Proc.Nat.Ac.Sc. U.S.A. 81, 3443-3446.

McKusick,V. (1983). Mendelian inheritance in man. Sixth edition. Johns Hopkins University Press.

Migeon,B.R. and Miller,C.S. (1968). Human-mouse somatic cell hybrids with single human chromosome (group E): link with thymidine kinase activity. Science, 162, 1005-1006.

Migeon,B.R., Brown,T.R., Axelman,J and Migeon,C.J. (1981). Studies of the locus for androgen receptor: localization on the human X and evidence for homology with the Tfm locus in the mouse. Proc. Nat.Acad.Sc. 78, 6339-6343.

Mossman,J., Blunt,S., Stephens,R., Jones,E.E., and Pembrey,M. (1983). Hunter's disease in a girl: association with X:5 chromosomal translocation disrupting the Hunter gene. Archives of Disease in Childhood. 58, 911-915.

Murray,J.M., Davies,K.E., Harper,P.S., Meredith,L., Mueller,C.R., and Williamson,R. (1982). Linkage relationship of a cloned DNA sequence on the short arm of the X-chromosome to Duchenne muscular dystrophy. Nature 300, 69-71.

Nussbaum,R.L., Crewder,W.E., Nyhan,W.L., and Caskey,C.T. (1983). A three allele restriction fragment-lenght polymorphism at the hypoxanine phosphoribosyltransferase locus in man. Proc.Natl. Acad.Sci. U.S.A. 80, 4035-4039.

Oberlé,I., Camerino,G., Heilig,R., Grunebaum,L.,. Cazenave,J.P., Crapanzano,C., Mannucci, P.M., Mandel J.L. (1985). A highy poly-

morphic DNA probe useful for genetic screening of hemophilia A.
N.Engl.J.Med. (in press).

Pai,G.S., Sprenkle,J.A., Do,T.T., Mareni,C.E., Migeon,B.R. (1980).
Localization of loci for HPRT and G6PD and biochemical evidence
of nonrandom X chromosome expression from studies of a human X-
autosome translocation. Proc. Nat.Ac.Sci. U.S.A. 77, 2810-2813.

Page,D., DeMartinville,B., Barker,D., Whyman,A., White,R, et al.
(1982). Single copy sequence hybridised to polymorphic and homologous loci on human X and Y-chromosome. Proc.Natl.Acad.Sci. USA
79, 5352-5356.

Paigen,K. (1979). Acid hydrolases as models of genetic control. Ann
Rev.Genet. 13, 417-466.

Pearson,P.L., Wieacker P., Bakker,B., and Prins,H.A. (1984). A
restriction site polymorphic marker for the short arm of the X-
chromosome. Clin.Genetics. In press.

Purrello,M., Alhadeff,B., Esposito,D., Szabo,P., Rocchi,M., Truett,
M., Masiarz,F., Siniscalco,M. (1984). The human genes for hemo-
philia A and B flank the X chromosome fragile site at Xq27.3.
EMBO Jour; (in press).

Romeo,G., Besana,D., Dworzak,F., Fadda,S., Mochi,M., Prosperi,L.,
Staffa,G., and Morandi,L. (1984). A single pedigree approach to
the genetic mapping of Becker muscular dystrophy using DNA polymorphisms of the X-chromosome. Muscular Dystrophy. Facts and
Perspectives. Supp.N.3, 29-34 of the Ital.J.Neurol.

Roncuzzi,L., Fadda;S., Mochi,M., Prosperi,L., Sangiorgi,S., Santamaria,R., Sbarra,D., Besana,D., Morandi,L., Rocchi,M., Romeo,G.
(1985). Am.J.Hum.Genet. (in press).

Ropers,H.H., Zuffardi,O., Bianchi,E., and Tiepolo,L. (1982).Agenesis
of the corpus callosum, ocular and skeletal anomalies (X-linked
dominant Aicardi's syndrome) in a girl with balanced X/3 transo-

cation. Hum.Genet. 61, 364-368.

Schmidtke,J., and Cooper,D.N. (1984). A list of cloned human DNA
sequences. Supplement. Hum.Genet. 67, 111-114.

Seventh International Workshop on Human Gene Mapping. (1984). Proceedings of the Los Angeles Conference (1983) Published by Cytogenetics and Cell Genetics. 37, N.1-4.

Stern,C. (1973). Principles of Human Genetics. 3rd Edition, Freeman
& Co., pp.337-340.

Szabo P., Purrello,M., Rocchi,M., Archidiacono,N., Alhadeff,B.,
Filippi,G., Toniolo,D., Martini,G., Luzzatto,L., Siniscalco,M.
Cytological mapping of the human G6PD gene distally to the fragile-X site suggests a high rate of meiotic recombination across
the site. Proc.Nat.Ac.Sci. USA (in press).

Wieacker,P., Davies,K., Cooke,H., Pearson,P.L., Williamson,R.,
Southern,E., Fraccaro,M., and Ropers,H.H. (1984 a). Mapping the
X chromosome by means of restriction fragment lenght polymorphisms: regional assignment of 17 cloned DNA sequences. Proceedings
of the Los Angeles Conference (1983) Published by Cytogenetics
and Cell Genetics 37, N.1-4.

Wieacker,P., Wienker,T.F., Mevorah,B., Dalla Piccola,B., Davies,K.,
and Ropers,H.H. (1984 b). Linkage relationships between Xg, steroid sulfatase (STS) and retinoschisis (RS), respectively, and
cloned DNA sequence from the distal short arm of the X chromosome. Proceeding of the Los Angeles Conference published by Cytogenetics and Cell Genetics. 37, N.1-4.

Wieacker,P., Horn,N., Pearson,P., Cooke,H., Davies,K., Wienker,T.F.
and Ropers,H.H.(1984 c). Linkage relationship between the Menkes
locus and cloned DNA sequences from the X-chromosome. Proceedings
of the Los Angeles Conference (1983) Published by Cytogenetics
and Cell Genetics. 37, N.1-4.

Zatz,M., Istkan,S.B., Sanger,R., Frotapessoa,O., Saldanha,P.H.(1974).
   New linkage data for the X-linked types of muscular dystrophy and
   G-6-PD variants, colour blindness and Xg blood group. J.Med.Genet.
   11,321-327.

Zuffardi,O., and Fraccaro,M. (1982). Gene mapping and serendipity.
   The locus for torticollis, keloids, cryptorschidsm and renal dy-
   splasia (31430, McKusick) is at X28, distal to the G6PD locus.
   Hum.Genet. 62, 280-281.

ACKNOWLEDGMENTS

The work carried out by our group and partly reviewed in this paper
was made possible by grants from the Italian C.N.R. (Progetto finalizzato Ingegneria Genetica e Basi Molecolari delle Malattie Ereditarie), from the Muscular Dystrophy Association (U.S.A.) and from
the Legato Dino Ferrari (Modena).

THE ROLE OF THE HPRT GENE IN HUMAN DISEASE

Douglas J. Jolly

A. INTRODUCTION

Until fairly recently, a picture of gout generally accepted
by most people except the victims themselves and probably their
doctors was that of a somewhat amusing and largely innocent disorder
typically seen in an overfed and over-indulgent old man, sitting by
the fire, sipping his brandy and being very careful not to move his
very painful, bandaged foot lest the slight movement set off a
torrent of the most agonizing pain. It is an ancient disease, with
evidence of gouty bone disease found to exist in ancient Egyptian
mummies and also recognized and described by the early Greek
physicians, although the Greeks were not always able to distinguish
it from other causes of inflammation and pain in the joints.
Hippocrates recognized that gout was associated with chalk-like
deposits in the bones or joints and possibly had a hereditary
nature. He also was among the first to understand something of the

nutritional causes of the disease and ascribed it to overindulgence and luxurious high living.

The English word gout, or precursors of it, was first used during the 13th century. It is derived apparently from the Latin gutta, or drop, and refers to the medieval principle of pathology that a toxic substance, such as the accumulated material found in the joints of gout patients, "drops" out of the blood into the tissue.

Until very recently, gout was of little general medical or scientific interest, but the explosion of biochemical, and now molecular, sophistication and their application to human disease has made it and its related diseases some of the most useful available "windows" on normal human physiology. Moreover, disorders of the purine synthetic pathways that lead to the development of gout and other defects in purine metabolism are producing several of the most interesting model systems for the development of the techniques of gene therapy in humans. Much of this is due to the discovery in the late 1960s and early 1970s of the role that the purine biosynthetic pathways, and especially the purine salvage pathway enzyme hypoxanthine guanine phosphoribosyl transferase (HPRT) play in gout and other clinical disorders of purine metabolism.

In addition, complete absence of the gene for HPRT in humans corresponds to a devastating genetic disease, Lesch-Nyhan syndrome, which has interesting neurological symptoms, as well as gouty arthritis symptoms.

This review is intended to summarize some of the recent developments in our understanding of the genetic and enzymic bases

for the disorders of purine metabolism that lead to urate deposits
in tissues and in the joints and thereby to the clinical symptoms of
gout. We shall also describe some results of studies on the
regulation of expression of the HPRT gene as a typical X-linked
gene, and some very recent studies on the development of models for
gene therapy in human disease. Other recent reviews of HPRT gene
mutations (Chinault and Caskey, 1984) and Lesch-Nyhan disease
(Wilson et al. 1983; Kelley and Wyngaarden, 1983) exist.

B. The Metabolic and Medical Features of the HPRT Gene

The enzyme hypoxanthine guanine phosphoribosyl transferase
(HPRT; EC 2.4.2.8) and its gene in humans and other mammalian
species are of interest for a number of reasons:

I. HPRT deficiency is associated with two important human
genetic diseases. One of these is Lesch-Nyhan disease (Kelley and
Wyngaarden, 1983; Nyhan, 1973), first identified in 1962, and
characterized by excessive production and accumulation of uric acid
and several extremely interesting and puzzling neurological
features. Children with this devastating disease show the
neurological symptoms of spasticity and choreoathetosis, an apparent
retardation, and a bizarre, aggressive self-mutilation behavior.
The children bite their tongues, the edges of their lips and
literally amputate their finger tips. The behavior is entirely
involuntary, compulsive and painful and can be prevented only by the
prolonged use of arm restraints and by removal of the teeth.
Affected children show a degree of apparent mental retardation,
although some experienced clinicians have felt that the retardation
may be exaggerated and overestimated because of the severe speech

defects associated with the motor dysfunction. Many of the affected children are communicative and have a good sense of humor. The disease is characterized by an almost complete lack of HPRT enzyme activity (less than 1-2% normal) and is sex-linked and has therefore been well documented almost exclusively in males although at least one female (Ogasawara et al. 1982) has now been described and is being characterized. The disease is rather rare, having been estimated to occur 1 in 100,000 live births.

The second major disorder resulting from HPRT deficiency is the classical disease gout and gouty arthritis (Kelley and Wyngaarden, 1983), which is the consequence of a partial deficiency of HPRT activity to a level of approximately 5-35% normal and is characterized mainly by episodes of arthritis and joint and bone tenderness so severe that even the weight of bed sheets or the vibrations caused by passing vehicles can cause excrutiatingly severe pain. While this form of adult gout is a particularly well characterized one, it accounts for a very small minority of clinical cases. Approximately 10% of all adults with gout have an overproduction of uric acid, and of those, only 10-30% are due to a partial deficiency of HPRT.

An interesting feature of these manifestations of HPRT deficiency is that the level of the measured HPRT deficiencies does not always seem to correlate closely with the severity of the disease (Bakay et al. 1979), although the 1-2% normal activity seems to represent a limit below which patients begin to show the severe neurological disorder characteristic of Lesch-Nyhan disease. However, a patient with the Lesch-Nyhan disease with 1-2% normal

HPRT might be much more severely affected than one with 0.1% normal HPRT. The same sort of dissociation between clinical severity and the degree of enzyme deficiency has been seen in gouty arthritis patients with partial enzyme defects, and the reasons for this effect are not understood.

The biochemical basis for the gouty arthritis found in patients with either partial or complete enzyme deficiency is well understood. HPRT is an enzyme of the purine salvage pathway and the scheme outlining the metabolic reactions in which it participates is shown in Fig. 1. Lack of HPRT activity in all the cells of the body leads to excessive production and accumulation of oxypurines which cannot be reutilized efficiently and are therefore excreted. The activity of the de novo purine biosynthetic pathway leading to the over-production of the major purine precursor, IMP, is enhanced in HPRT deficiency apparently because of the reduced pool sizes of GMP and AMP that normally regulate de novo purine biosynthesis by feed-back inhibition of the enzyme that regulates the first and rate limiting step of the purine biosynthetic pathway, glutamine phosphoribosyl amido transferase. Another possible mechanism of enhanced de novo purine biosynthesis involves the increased intracellular concentration of PRPP, normally utilized by HPRT but now available in higher concentrations to drive the de novo pathway. The increased rate of IMP production has the effect of increasing the rate of production and the intracellular and extracellular concentrations of nonreutilizable oxypurine hypoxanthine that is excreted into the extracellular fluid or into the tissue culture medium. In patients, this extracellular and soluble hypoxanthine is further metabolized to xanthine and uric acid by the enzyme xanthine

oxidase. These purines are both much less soluble than the

precursor hypoxanthine and exist in body fluids at levels very close

to their limits of solubility. Under conditions of inflammation,

acidosis, injury or even a small increase in concentration,

precipitation of sodium urate crystals takes place into joints and

into the kidney, causing gouty arthritis and kidney stones. The

problem of renal stones has been severe and life-threatening for

Lesch-Nyhan patients, and in fact, the first sign of Lesch-Nyhan

disease is sometimes the presence of uric acid crystals in the

urine. Fortunately, the gouty arthritis and the kidney stone

production can now be controlled to a large extent by the drug

allopurinol (Kelley and Wyngaarden, 1983) which blocks conversion of

hypoxanthine to xanthine and uric acid by inhibition of xanthine

oxidase.

Unlike the reasonably clear pathogenesis of the

hyperuricemia, gout and renal disease, the biochemical basis for

the neurological features of Lesch-Nyhan syndrome is not well

understood. The central nervous system dysfunction includes mental

retardation, spasticity and notably self-mutilation. The self-

mutilation is particularly bizarre since patients will ask to be

restrained in order to avoid hurting themselves. This behavior of

self-mutilation is therefore of much clinical and biochemical

interest since it is one of the very few, and perhaps, the only very

simple single gene defect available for the study of such

complicated aberrations of behavior and mentation. Lesch-Nyhan

disease therefore represents a very important and intriguing model

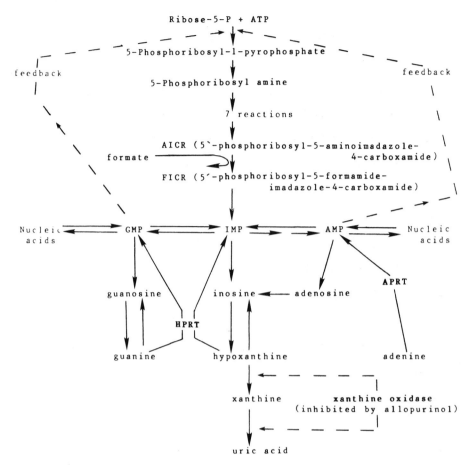

Figure 1. Purine biosynthetic and salvage pathways. The solid
arrows indicate enzymic reactions. The only enzymes shown are HPRT,
APRT (adenosine phosphoribosyl transferase) and xanthine oxidase.
Aminopterin blocks the biosynthetic pathway at the 1 carbon addition
step (addition of formate). In the HAT selection scheme for HPRT+
cells, hypoxanthine is then the sole source for synthesis of the
central precursor inosine monophosphate (IMP), through HPRT. In
selection for HPRT⁻ cells, thioguanine replaces hypoxanthine as a
substrate for HPRT, which leads to toxic products, further down the
metabolic pathways. GMP and AMP inhibit the first step in purine
biosynthesis by feedback, as shown. In Lesch-Nyhan disease cellular
levels of GMP and AMP are reduced, leading to increased biosynthetic
activity and IMP and hypoxanthine overproduction. The action of
allopurinol is as shown, inhibiting xanthine and uric acid
production by xanthine oxidase.

for understanding the mechanisms of normal and abnormal CNS function. Since the retardation, and compulsive and aggressive behavior are the consequence of simple genetic defects in a single gene, the Lesch-Nyhan disease is one of the best candidates currently available for studies of new forms of therapy through gene manipulation.

The specific activity of HPRT in normal brain tissue is approximately 10 to 50 fold higher than in most other organs of the body, and the highest level is found in the cells of the basal ganglia. In Lesch-Nyhan patients, the HPRT activity in the brain is, as elsewhere, greatly reduced compared to normal. No gross reproducible structural or histological abnormalities have been seen in the brains of Lesch-Nyhan patients, and it is therefore assumed that the CNS symptoms and signs are due to neurochemical dysfunction secondary to HPRT deficiency rather than structural brain damage or "wiring" abnormalities. Allopurinol, the xanthine oxidase inhibitor effective in reducing gouty arthritis and kidney stones, is completely ineffective in treating the neurological symptoms suggesting that other imbalances in purine metabolism deriving from aberrant levels of one or more purine derivatives in the brain must be responsible for the behavioral abnormalities.

Just how aberrant purine metabolism can lead to defects in neurological function is not at all clear, but it is likely to result from one of several possible mechanisms. Purine derivatives may act as neurotransmitters (Burnstock et al. 1975), they may interfere with normal neurotransmitter release (Fredholm and Hedqvist, 1980), or they may alter receptor sensitivity (Rodbell,

1980). For example, since these adenosine and guanine derivatives are putative modulators of neurotransmitter functions, they might interfere with the function of the adenosine receptors now known to be found in the brain (Daly et al. 1981). However, it has been shown that hypoxanthine and its known derivatives in the cerebrospinal fluid apparently do not compete with adenosine for the adenosine receptors, although they do interfere with the binding of diazepam (Skolnick et al. 1978), another neurotransmitter, to its receptors. The same effect can also be seen with the purine derivative caffeine and in fact long term administration of caffeine to underfed rats produces paw self-mutilation behavior reminsicent of Lesch-Nyhan patients (Nyhan, 1973). It has therefore been suggested that the bizarre Lesch-Nyhan syndrome behaviors are produced somehow by the blocking of diazepam receptors (Kopin, 1981).

Other kinds of neurotransmitters have also been implicated in the neurochemistry of the Lesch-Nyhan defect. Administration of the serotonin precursor 5 OH-tryptophan led to a transient correction of some of the aberrant biochemical CNS functions found in Lesch-Nyhan patients including altered amino acid levels and decreased homovanillic acid in the spinal fluid (Castells et al. 1979). However, although self-mutilation behavior was temporarily suppressed, no other neurological dysfunctions were affected and the mutilation behavior returned after a short time..

In the most detailed study to date, Lloyd et al. (1981) have compared neurotransmitter levels in Lesch-Nyhan brains to those in age-matched individuals. They found that dopamine and some enzymes

of the dopamine metabolism were depressed three to ten fold in some,
but not all, brain tissues of patients with the Lesch-Nyhan disease.
The levels of serotonin and choline acetyl-transferase in the brains
of such patients showed less impressive but significant differences
from normal in some parts of the brain, but no significant
differences were found in GABA (gamma amino butyric acid) or
norepinephrine levels. The findings of altered dopamine levels were
interpreted to mean that defects occur in terminal arborization of
dopamine neurons, since dopamine levels were normal in the
substantia nigra where the cell bodies of the dopamine neurons
occur, but low in the caudate nucleus and putamen where arborization
of these neurons occurs. How these neurotransmitter defects relate
to the excess purine produced by HPRT deficiency is unclear.

II. The locus is on the X chromosome. The HPRT gene was
first recognized to be on the X chromosome through human pedigree
analysis (Seegmiller et al. 1967). It has since been mapped to the
X q26-q27 region of the long arm of the X chromosome by cell
hybridization studies (Pai et al. 1980) and it is now known to be
physically close, in light microscropy terms, to the loci for a
number of other important genetic diseases such as Hemophilia A,
Hemophilia B, G6PD deficiency, color blindness and X-linked mental
retardation (McKusick, 1983). Thus, the HPRT gene is obviously a
useful starting point for physical characterization of a large
portion of a human chromosome by molecular chromosome walking
(Eickbush and Kafatos, 1982; Steinmetz et al. 1982) towards these
other loci.

It has recently been shown that genetic distances at the end of

the long arm of the X chromosome measured in centimorgans do not
correlate tightly with the apparent physical distances as measured
by in situ hybridization (Hartley et al. 1984). Thus, correlation of
molecular distances with recombinational and physical distances
should clarify mechanisms and roles of recombination "hot" or "cold"
spots in chromosomes and of the corresponding chromatin structure
and packaging. In addition, examination of the HPRT gene and its
activity in females allows investigation of the phenomenon of random
X chromosome inactivation or Lyonization (Martin, 1982; Lyon, 1972)
in mammalian cells. This is the process by which one of the two X
chromosomes in every female cell is randomly and stably inactivated.
For that matter, all but one X chromsome is inactivated in cells of
patients having more than two X chromosomes. The molecular basis of
this is not understood. It is known that in females heterozygous at
the HPRT locus with one X carrying the wild-type gene and the other
carrying a mutant gene, all bone marrow-derived cells are HPRT$^+$,
indicating that there has been either selection for the HPRT$^+$ cells
or nonrandom inactivation of the HPRT$^-$ X chromosome (Nyhan et al.
1970). We and others have recently used the isolated HPRT gene to
show that there are methylation differences between HPRT genes on
active and inactive X chromosomes (Wolf et al. 1984; Yen et al.
1984).

III. In tissue culture there are extremely strong
selections both for and against HPRT activity. The most useful
selection for HPRT$^+$ cells is that provided by HAT medium
(Hypoxanthine, Aminopterin, Thymidine) (Szybalski et al. 1962). De
novo purine and pyrimidine syntheses are blocked by aminopterin, a

dihydrofolate reductase inhibitor, and the synthesis of IMP from the
exogenous hypoxanthine (see Fig. 1) by HPRT activity is the sole
remaining and obligatory path to produce this essential central
precursor of purine components of nucleic acids. The thymidine is
converted through the action of thymidine kinase to TMP (thymidine
monophosphate) for the synthesis of pyrimidine nucleic acid
components. An alternative selection relies on the use of
hypoxanthine plus azaserine (Fujimoto et al. 1971), in which
azaserine blocks purine but not pyrimidine de novo biosynthesis. In
contrast, selection against the presence of HPRT is possible in
tissue culture through the addition of toxic analogues of
hypoxanthine such as 8-azaguanine or 6-thioguanine (Szybalski et al.
1962; Hochstadt, 1973) to the culture medium. These analogues are
efficient substrates for HPRT activity and exert their toxic effect
further down the purine biosynthetic pathways probably at the level
of ribonucleotide reductase, as non-competitive inhibitors.
Thioguanine seems more specific in this scheme, since azaguanine-
resistant, hypoxanthine-metabolizing HPRT activities have been found
in cells selected with 8-azaguanine (Hochstadt, 1973). In practice,
mutant cell lines that are azaguanine and thioguanine resistant are
usually found to be HPRT$^-$. The utility of these selection schemes
is enhanced by the fact that the HPRT gene is on the X chromosome and
that therefore all mammalian cells are effectively haploid for the
gene since XY or male cells contain only one X chromosome and female
cells have one X inactivated. Furthermore, the selections not only
work in all mammalian cells tested, but also have been used in other
eukaryotic cells such as yeast (Woods et al. 1983), to isolate HPRT$^-$

mutants from wild type cells.

C. HPRT Protein and Gene

The HPRT protein is a cytoplasmic enzyme, and in brain
tissue, where its activity is highest, it can represent up to
approximately 0.05% of the total soluble protein (Muensch and
Yoshida, 1977). The abundance of the message and, by implication,
the protein has been estimated approximately from the number of
clones in a cDNA library made from fibroblast RNA and screened for
HPRT positive clones (Jolly et al. 1983). This gives a message
frequency of 0.002% in fibroblasts. Since brain tissue has 10-50
times higher specific activity than other tissues (Kelley and
Wyngaarden, 1983), these numbers are in good agreement. Most recent
studies agree that the native form of the enzyme is a tetramer of
four identical subunits of 24,500 molecular weight (Arnold and
Kelley, 1971; Johnson et al. 1979). However, catalytically active
dimers of the human and mouse enzymes have been described (Johnson
et al. 1982). The protein, particularly that derived from human
erythrocytes, shows multiple forms after polyacrylamide gel
isoelectric focusing detection assays, and the number of apparent
isozymic forms increases apparently with aging of the sample
(Johnson et al. 1982), presumably due to post translational
modification.

The full amino acid sequence of the purified human HPRT
protein has been determined by classical amino acid sequencing
methods (Wilson et al. 1982), and the sequence has been compared
with that deduced from nucleic acid sequencing of the isolated full
length cDNA clones (Jolly et al. 1983). The amino acid and

nucleotide-deduced sequences for the human protein are in full
agreement, and indicate that the encoded N-terminal methionine
residue is cleaved from the full length polypeptide during protein
maturation.

Figure 2 illustrates the complete nucleotide sequences for
the human (Jolly et al. 1983) and mouse (Brennand et al. 1982;
Konecki et al. 1982; Melton et al. 1984) genes, and a comparison of
their encoded proteins is given in Figure 3. There are only seven
amino acid differences between the human and mouse proteins out of a
total of 218 amino acids, indicating that this gene is highly
conserved between these two species.

On the basis of comparison of amino acid sequence and
smoothed plots of amino acid physical characteristics with other

---

Figure 2. Nucleic acid sequence of the human and mouse cDNA genes
for HPRT. The sequence shown is the "sense" strand of the human
gene (Jolly et al. 1983). The first 15 G residues are from G tails
during cloning. The mouse sequence (Konecki et al. 1982; Melton et
al. 1984) is compared from base 75 to 953 in the line below the
human sequence; outside these bases at the 5' and 3' ends no
significant homology was found. A "+" indicates homology, the
altered bases are as shown. An "x" in either the human or mouse
line indicates that there is a base deletion with respect to the
other sequence. The methionine start codon is shown by "met" and
the stop codon by "***." Out of 654 bases total, there are 42 base
differences between the human and mouse genes (6.4%) in the actual
coding sequence. The positions of the exon-intron junctions are
identical in the mouse and human genes and are as indicated in the
line above the human sequence, with a space between the exons. The
introns sizes in kb in mouse are (Melton et al. 1984) 5' to 3':
10.8; 2.9; 6.5; 1.6; 3.9; 0.2; 0.4 and in human (C. Caskey, personal
communication): 13.8; 1.6; 7.5; 3.7; 4.5; 4.0; 0.17; 1.6.

The bases in brackets above the human sequence are deduced
nucleotide changes from identified amino acid changes in the HPRT
protein from three gouty arthritis and one Lesch-Nyhan patient (see
also Fig. 3 and Table 2).

```
      GGGGGGGGGG GGGGTCTTGC TGCGCCTCCG CCTCCTCCTC TGCTCCGCCAC CGGCTTCCT
             10         20         30         40          50         60
                                         m et
      CCTCCTGAGC AGTCAGCCCG CGCGCCGGCC GGCTCCGTTA TGGCGACCCG CAGCCCTGGC
      ---------- ----+++++++ +++A+++A++ ++TC+++++C+ ++C++++++++ +++T+++A++
             70         80         90        100         110        120
exon1 exon2
      GTC GTGATTA GTGATGATGA ACCAGGTTAT GACCTTGACC TATTTTGCAT ACCTAATCAT
      +++ +++++++ +C++++++++ ++++++++++ +++++A++++ +G++++T++ ++++++++++
             130        140        150        160        170        180
                                                              exon2 exon3
      TATGCTGAGG ATTTGGAAAG GGTGTTTATT CCTCATGGAC TAATTATGGA CAG GACTGAA
      +++++C++++ ++++++++++A A+++++++++ ++++++++++ +G++++++++ +++ +++++++
             190        200        210        220        230        240
        (G)
      CGTCTTGCTC GAGATGTGAT GAAGGAGATG GGAGGCCATC ACATTGTAGC CCTCTGTGTG
      A+A+++++++ +++++++C++ ++++++++++ ++++++++++ +++++++G++ ++++++++++
             250        260        270        280        290        300

      CTCAAGGGGG GCTATAAATT CTTTGCTGAC CTGCTGGATT ACATCAAAGC ACTGAATAGA
      ++++++++++ ++++++++G+ ++++++++++ ++++++++++ ++++T+++++ ++++++++++
             310        320        330        340        350        360
                                                         (G)exon3 exon4
      AATAGTGATA GATCCATTCC TATGACTGTA GATTTTATCA GACTGAAGAG CTATTGT AAT
      ++++++++++ ++++++++++ ++++++++++ ++++++++++ ++++++++++ +++C+++ +++
             370        380        390        400        410        420
        (T)
      GACCAGTCAA CAGGGGACAT AAAAGTAATT GGTGGAGATG ATCTCTCAAC TTTAACTGGA
      ++T+++++++ +G++++++++ ++++++T+++ ++++++++++ ++++++++++ ++++++++++
             430        440        450        460        470        480
exon4 exon5              exon5 exon6
      AAG AATGTCT TGATTGTGGA A GATATAATT GACACTGGCA AAACAATGCA GACTTTGCTT
      +++ +++++++ ++++++T++ + ++++++++++ ++++++++T+ ++++++++++ A+++++++++
             490        500        510        520        530        540
                                   exon6 exon 7
      TCCTTGGTCA GGCAGTATAA TCCAAAGATG GTCAAGGTCG CAAG CTTGCT GGTGAAAAGG
      +++C++++T+ A+++++C+G C++C++A+++ ++T+++++T+ ++++ ++++++ ++++++++++
             550        560        570        580        590        600
                                        exon7 exon8
      ACCCCACGAA GTGTTGGATA TAAGCCAGAC TT TGTTGGAT TTGAAATTCC AGACAAGTTT
      +++T+T++++ ++++++++++ C+G+++++++ ++ ++++++++ ++++++++++ ++++++++++
             610        620        630        640         650        660
                                        (A)     exon8 exon9
      GTTGTAGGAT ATGCCCTTGA CTATAATGAA TACTTCAGGG ATTTGAAT CA TGTTTGTGTC
      +++++T++++ ++++++++++ ++++++++G ++++++++A ++++++++ ++ C+++++++++
             670        680        690        700         710        720
                                             ***
      ATTAGTGAAA CTGGAAAAGC AAAATACAAA GCCTAAGATG AGAGTTCAAG TTGAGTTTGG
      ++++++++++ ++++++++++ C+++++++++ ++++++++++ ++C+xx++++ ++++A+C++C
             730        740        750        760        770        780

      AAACATCTGG AGTCCTATTG ACATCGCCAG TAAAATTATC AxATGTTCTAG TTCTGTGGCC
      +++T+CGA++ ++++++G+++ +TG+T+++++ ++++++++G+ +GG+++++++ +C++++++++
             790        800        810        820        830        840

      ATCTGCTTAG TAGAGCTTTT TGCATGTATC TTCTAAGAAT TTTATCTGTT TTGTACTTTA
      ++++++C+++ ++A+++++++ ++++++A+C+ +++++T++++ G+++x+++++ ++A+Tx++++
             850        860        870        880        890        900

      GAAATGTCAG TTGCTGCATT CCTAAACTGT TTxATTTGCAC TATGAGCCTA TAGACTATCA
      ++++++++++ +++++++G+C ++C+G++x++ ++G+++++++ ++++++++++ +++-------
             910        920        930        940        950        960

      GTTCCCTTTG GGCGGATTGT TGTTTAACTT GTAAATGAAA AAATTCTCTT AAACCACAGC
             970        980        990       1000       1010       1020

      ACTATTGAGT GAAACATTGA ACTCATATCT GTAAGAAATA AAGAGAAGAT ATATTAGTTT
            1030       1040       1050       1060       1070       1080

      TTTAATTGGT ATTTTAATTT TTATATATGC AGGAAAGAAT AGAAGTGATT GAATATTGTT
            1090       1100       1110       1120       1130       1140

      AATTATACCA CCGTGTGTTA GAAAAGTAAG AAGCAGTCAA TTTTCACATC AAAGACAGCA
            1150       1160       1170       1180       1190       1200

      TCTAAGAAGT TTTGTTCTGT CCTGGAATTA TTTTAGTAGT GTTTCAGTAA TGTTGACTGT
            1210       1220       1230       1240       1250       1260

      ATTTTCCAAC TTGTTCAAAT TATTACCAGT GAATCTTTGT CAGCAGTTCC CTTTTAAATG
            1270       1280       1290       1300       1310       1320

      CAAATCAATA AATTCCCAAA AATTTAAAAA AAAAA ---
            1330       1340       1350
```

sequenced phosphoribosyl transferases (Salmonella typhimurium ATP phosphoribosyl transferase and E. coli glutamine phosphoribosyl transferase), Argos et al. (1983) have proposed a common structure for the 120 N-terminal amino acids and have identified possible binding sites for hypoxanthine and PRPP. A separate search for homology of the HPRT sequence to all other known amino acid sequences has also shown the presence of a direct amino acid homology, consisting of a seven amino acid peptide homology to the E. coli enzyme glutamine phosphoribosyl transferase, (Friedmann et al. 1983). It is clear that the usefulness of such approaches toward the identification of functional domains in the HPRT molecule is limited. The correctness of such models remains to be defined by characterization of point mutations, monoclonal antibody studies and, ultimately, x-ray crystallography.

The genomic structure of the mouse gene has been determined (Fig. 2). This gene has 9 exons and 8 introns, and an extremely G-C rich region at the 5' end, but interestingly enough there are no recognizable transcriptional signals such as the prototypical TATAA box in the 5' untranslated region of the gene. Similar genetic

---

Figure 3. Amino acid sequence of the human and mouse HPRT proteins. The human (Jolly et al. 1983; Wilson et al. 1982) sequence is shown N terminus to C terminus with identical amino acids in mouse (Konecki et al. 1982; Melton et al. 1984) indicated by "+" and differences (seven in all) as shown. The corresponding boundaries of the exons in the coding sequence are as shown. The amino acids in parenthesis above the sequence are those corresponding to the mutations defined in three gouty arthritis and one Lesch-Nyhan patient (Wilson et al. 1984b).

```
                                      exon1  exon2
met  ala  thr  arg  ser  pro  gly  val  val  ile  ser  asp
 +   pro   +    +    +    +   ser   +    +    +    +    +
 1                        6                            12

asp  glu  pro  gly  tyr  asp  leu  asp  leu  phe  cys  ile
 +    +    +    +    +    +    +    +    +    +    +    +
13                       18                           24

pro  asn  his  tyr  ala  glu  asp  leu  glu  arg  val  phe
 +    +    +    +    +    +    +    +    +   lys   +    +
25                       30                           36
                                      exon2 exon3
ile  pro  his  gly  leu  ile  met  asp  ar g thr  glu  arg
 +    +    +    +    +    +    +    +    +    +    +    +
37                       42                           48
          (gly)
leu  ala  arg  asp  val  met  lys  glu  met  gly  gly  his
 +    +    +    +    +    +    +    +    +    +    +    +
49                       54                           60

his  ile  val  ala  leu  cys  val  leu  lys  gly  gly  tyr
 +    +    +    +    +    +    +    +    +    +    +    +
61                       66                           72

lys  phe  phe  ala  asp  leu  leu  asp  tyr  ile  lys  ala
 +    +    +    +    +    +    +    +    +    +    +    +
73                       78                           84

leu  asn  arg  asn  ser  asp  arg  ser  ile  pro  met  thr
 +    +    +    +    +    +    +    +    +    +    +    +
85                       90                           96
                              (arg)       exon3 exon4
val  asp  phe  ile  arg  leu  lys  ser  tyr  cys  asn  asp
 +    +    +    +    +    +    +    +    +    +    +    +
97                       102                          108
     (leu)
gln  ser  thr  gly  asp  ile  lys  val  ile  gly  gly  asp
 +    +    +    +    +    +    +    +    +    +    +    +
109                      114                          120
                                      exon4 exon5
asp  leu  ser  thr  leu  thr  gly  lys  asn  val  leu  ile
 +    +    +    +    +    +    +    +    +    +    +    +
121                      126                          132
     exon5 exon6
val  glu  asp  ile  ile  asp  thr  gly  lys  thr  met  gln
 +    +    +    +    +    +    +    +    +    +    +    +
133                      138                          144

thr  leu  leu  ser  leu  val  arg  gln  tyr  asn  pro  lys
 +    +    +    +    +    +   lys   +    +   ser   +    +
145                      150                          156
                         exon6 exon7
met  val  lys  val  ala  se r leu  leu  val  lys  arg  thr
 +    +    +    +    +    +    +    +    +    +    +    +
157                      162                          168
                                      exon7 exon8
pro  arg  ser  val  gly  tyr  lys  pro  asp  ph e val  gly
ser   +    +    +    +    +   arg   +    +    +    +    +
169                      174                          180

phe  glu  ile  pro  asp  lys  phe  val  val  gly  tyr  ala
 +    +    +    +    +    +    +    +    +    +    +    +
181                      186                          192
     (asn)                                  exon8 exon9
leu  asp  tyr  asn  glu  tyr  phe  arg  asp  leu  asn  his
 +    +    +    +    +    +    +    +    +    +    +    +
193                      198                          204

val  cys  val  ile  ser  glu  thr  gly  lys  ala  lys  tyr
 +    +    +    +    +    +    +    +    +    +    +    +
205                      210                          216

lys  ala
 +    +
217
```

organizations and sequence features have recently been found in
several other so-called "housekeeping" genes, and it may be that the
mechanisms of regulation of gene expression in these genes is not
the same as is found in genes representing highly differentiated
functions. The positions of the introns of the HPRT gene are shown
in Figs. 2 and 3. The human genomic sequences have proven somewhat
elusive, due in part to the existence of several pseudogenes (Patel
et al. 1984) on chromosomes 3, 5 and 11, and to difficulty in
cloning the extreme 5' end. However, it is now clear that the human
gene closely resembles the mouse gene, not only in coding sequence
but in the exact placement of the intron-exon boundaries (Kim,
Moores et al. unpublished data, Patel, Caskey et al. personal
communication). The 5' end of the human gene is also extremely G-C
rich and has no TATAA box, but there is no further homology of the
non-coding portion of the 5' end of the human gene to that of the
mouse 5' end. In addition, the 3' non-coding sequences and the
intron sizes are not apparently conserved in the human and the
mouse. Both the human and mouse genes are large and first estimates
of their size came from transfection experiments (Jolly et al. 1982)
with human DNA, putting it at >32 kilobases (kb). The mouse gene
is >33 kb (Melton et al. 1984) and the human gene now appears to
be about 36 kb in size. Because of the difficulties in cloning the
human HPRT gene, the gene has been described in a preliminary manner
in terms of the restriction fragments seen in Southern blots of
total DNA hybridized to cDNA and cDNA subfragment clones (Patel et
al. 1984; Jolly et al. unpublished results); this description is
given in Table 1.

TABLE 1
HUMAN HPRT GENE FRAGMENTS

X Linked (functional HPRT gene)

Exon (5'-3')	EcoR1	Hind III	BamH1
1	11	7.2	8.0
2,3	8.3	7	22 (12,polymorphic,16%)
4,5	10.5	4.9	3.7
6,7,8,9	8	17	25 (18,polymorphic,7%)

Autosomal ("pseudogenes")

	EcoR1	Hind III	BamH1
chrom 3	8.8	9.8	3.4
chrom 5	5.4	3.3	25
chrom 11	15.5	7.8	14.5
chrom 11	4.2	2.0	7.6

The sizes of the fragments are in kb. The data come mainly from Patel et al. (1984) with some also from unpublished data (Kim et al.). The assignments of exons 2-5 for the Hind III and Bam H1 fragments are tentative.

It has recently been possible to identify several normal variants of the human HPRT gene by the method of restriction fragment length polymorphism (RFLP) analysis. RFLP's are alleles of genes, revealed by digestion of chromosomal DNA with a particular restriction enzyme followed by Southern blotting and hybridization to a particular probe. They occur throughout the human (and other) genomes and usually represent seemingly innocuous point mutations. Their utility is that by following them through family pedigrees their genetic distance from each other and from phenotypic genetic

markers can be estimated by recombination analyses (Botstein et al.
1980). Thus by acquiring a large number of these throughout the
human genome, any new piece of DNA can be quickly mapped by linkage
to known markers and in addition correlation with phenotypic effects
such as susceptibility to coronary disease can be attempted. The
nature and frequency of the polymorphic variants in the HPRT gene
are also given in Table 1. The interpretation of these studies
requires a distinction between the fragments derived from the X-
linked HPRT sequences and those derived from the autosomal
pseudogenes. In addition to the Bam H1 pattern shown in Table 1,
the X-linked, authentic HPRT fragments show a pattern of Bam H1
polymorphisms indicated by the numbers in brackets. Table 1 shows
the relative frequencies seen in one extensive study of 83 X
chromosomes (Nussbaum et al. 1983) where a total of 23% deviate from
the major pattern. Another less detailed study of 46 X chromosomes
reveals a deviation from the normal pattern of Bam H1 fragments at a
frequency of 0.15 (Friedmann et al. 1983, Jolly et al. unpublished
data), in approximate agreement qualitatively and quantitatively
with the more extensive series. This polymorphism has no apparent
functional effect on HPRT activity and is present throughout the
normal and Lesch-Nyhan populations with apparently equal
frequencies.

The cloned human HPRT cDNA and genomic fragments have been
used to study the methylation patterns of the entire human HPRT gene
by restriction enzyme digests of active and inactive X chromosomes
in a variety of cell types (Wolf et al. 1984; Yen et al. 1984). It
is known that increasing methylation in autosomal genes generally

decreases their transcriptional activity, and it would be important
to determine if the mechanisms of gene inactivation during X-
chromosome inactivation are related to those thought to be involved
in the regulation of gene expression of non X-linked markers. There
are an extraordinarily large number of Hpa II (5'-CCGG-3') and Hha I
(5'-GCGC-3') sites at the 5' end of the HPRT gene. These
restriction enzyme recognition and cleavage sites contain the
nucleotide 5'CpG3' which is the major site for methylation in
mammalian genomes. The aforementioned enzymes are useful for the
detection of methyl C in the CpG doublets, since they cleave their
recognition sequence only if it is unmethylated. Thus, Southern
blots of DNA cut with these enzymes and hybridized to characterized
probes covering specific Hpa II and Hha I sites can reveal which of
these sites are methylated and which are unmethylated. The
existence of a large number of clustered sites susceptible to
methylation may serve as a kind of fail-safe mechanism for ensuring
and achieving properly regulated gene expression through methylation
at particularly crucial sites.

We have found that active human X chromosomes have a limited
set of characteristic patterns of methylation at the HPRT locus
which are invariant in various individual cell clones, cell types or
tissues. That is, there are obligatorily methylated sites, as well
as unmethylated ones. Inactive X chromosomes, on the other
hand, show a less consistent set of methylation patterns that have
not yet been correlated with the mechanism of X inactivation in
mammalian cells. Thus the specificity appears to lie in the

methylation of the active X rather than the inactive X chromosomes.
Using a subcloned single copy genomic intron fragment, pPB1.7 (Jolly
et al. 1982) derived from near the 5' end of the HPRT gene,
Vogelstein et al. (1985) have used the 22/12 kb polymorphism and the
active/inactive X HPRT gene methylation polymorphism to analyze
clonal origins of tumors. The advantage of this probe is that (a)
it reveals only this polymorphism and so simplifies interpretation
of blot results; (b) it gives different Hha I and Hpa II patterns
for active and inactive X chromosomes. Thus, in a tumor derived
from a female individual showing a 22 kb and 12 kb allele, a set of
simple digests and blots will reveal whether a tumor has only one
HPRT gene and hence one X chromosome inactivated (clonal origin) or
has a 50:50 mixture of the two different HPRT genes, and hence X
chromosomes, inactivated (polyclonal origin).

        We expect this probe, pPB1.7 will also be useful in linkage
studies with other X chromosome markers.

D. Patients with Lesch-Nyhan disease and with gouty arthritis-characterized HPRT mutations

        A genetic deficiency of any protein in the human can arise
in three ways: (a) through the synthesis of a defective protein
with altered properties resulting from a structural gene mutation,
as in sickle cell anemia; (b) decreased intracellular amounts of
protein due to a gene rearrangement, splicing defects or other
regulatory signals affecting synthesis at the RNA or protein levels,
as in some thalassemias; (c) a combination of these two effects
(i.e. a defective protein produced in reduced amounts). Type (a)
defects have been examined by characterizing the normal and abnormal

proteins ultimately at the amino acid level. However this is now more easily examined by cloning and sequencing the cDNA for the normal and abnormal proteins. Type (b) defects can be addressed by examination of the cloned genome itself or by looking for abnormal mRNA complementary to the gene. Mutant HPRT genes responsible for human clinical disease have been characterized in a number of ways. Initially, of course, all HPRT-deficient patients are characterized by reduced HPRT enzymic levels. Heterozygotes for mutant HPRT genes also can be identified, but the assay for the gene product is most usefully done with fibroblast enzyme preparations since there is an apparent strong selection for $HPRT^+$ cells in heterozygotes in the bone marrow-derived lymphoid and erythroid cells (Nyhan et al. 1970). Detection of heterozygotes in families of patients with gouty arthritis is much more difficult since the enzymic deficiency in patients is partial. Characterization of the HPRT mutations in individual families at the genetic level should make possible diagnosis through deficiency linked restriction fragment length polymorphisms (Wilson et al. 1983), or oligonucleotide hybridization as has been performed on thalassemia patients (Orkin et al. 1983).

Protein Studies

Protein material that can cross react with antibodies against the human HPRT protein (CRM) is usually not present in cells derived from patients with Lesch-Nyhan disease but is usually present in the serum of patients with gouty arthritis caused by HPRT deficiency (Wilson et al. 1983b; Upchurch et al. 1975; Bakay et al. 1976; Ghangas and Milman, 1975). Early reports of high frequency of cross reacting material (CRM^+) in cells of Lesch-Nyhan patients appears to have been due to impure antibodies. The lack of CRM in Lesch-Nyhan patients is not particularly surprising, since induced mutations of HPRT in animal cells are usually CRM negative, even though they can revert to give an altered protein and hence are candidates for point mutational events (Chinault and Caskey, 1984; Caskey and Kruh, 1979). That is, even mutations that are simply point mutations can give rise to CRM⁻ proteins and these CRM⁻ mutations are not necessarily candidates for regulatory gene defects or splicing mutations (see also RNA production data, below). Since HPRT-deficient gouty arthritis patients have HPRT proteins that retain some enzymic activity, those molecules are more likely to be recognized by antibodies to normal native HPRT.

Extending their studies on normal HPRT protein, Wilson et al. (1981) have purified HPRT protein from three patients with gouty arthritis showing a partial HPRT deficiency and from one patient with Lesch-Nyhan disease. They identified altered peptides, determined their amino acid sequences and identified 4 different single amino acid substitions (Wilson et al. 1983b) attributable to point mutations (see Table 2). In addition, they were then able to deduce the likely DNA mutations and in one case (ser to leu at amino acid 109) characterize the mutant DNA by restriction enzyme digestion and Southern blotting (Wilson et al. 1983a). This procedure permitted a diagnosis of a heterozygote sibling, not previously possible definitively by enzyme activity measurements.

Despite the success of the characterization of HPRT mutants for diagnostic purposes by protein characterization, it is likely that future characterization of Lesch-Nyhan mutations will be

TABLE 2

HPRT MUTATIONS CHARACTERIZED BY AMINO ACID ANALYSIS

Symptoms	Amino Acid (Posn.) Change	Deduced DNA (Posn.) Change
Gout	Arg to Gly (51)	CGA to GGA (250)
Gout	Ser to Leu (110)	TCA to TTA (428)
Gout	Ser to Arg (104)	AGC to AGG (411)
Lesch–Nyhan Disease	Asp to Asn (194)	GAC to AAC (679)

The data is taken from Wilson et al. (1983b). The amino acid positions refer to Fig. 3, the DNA positions refer to Fig. 2.

performed at the nucleic acid level. This is for several reasons: (i) there is not always a sufficient amount of protein available to purify; (ii) even if there is cross-reacting protein available, it may be unstable; (iii) modern techniques of molecular biology have made it technically less demanding, though not necessarily simple, to characterize mutations at the cloned cDNA or genomic DNA level.

Nucleic Acid Studies

Nucleic acid studies of HPRT deficiency diseases so far published have been limited to searches for aberrant gene patterns revealed by Southern blot hybridization of DNA from Lesch–Nyhan patients and their families. The information derived from these kinds of studies is limited, since the method will detect only gross rearrangements of the gene. Such large scale rearrangements have

been detected in two (Yang et al. 1984; Jolly et al. unpublished
results) Southern blot studies on the gene in Lesch-Nyhan patient
cells. The frequencies of rearranged genes were low (5/48 and 1/15
respectively). The current state of analyses of these six large
scale rearrangements is shown in Fig. 4. As with the protein
analyses all the mutations are different although in none of them
has the exact extent of the alteration been defined. Three are
deletions, one of them apparently complete, i.e. of the whole gene
(Fig. 4, line A), the other two, partial (lines B and C). Two
other mutations remain uncharacterized except for alterations in the
blot pattern covering the regions shown (lines D and F).

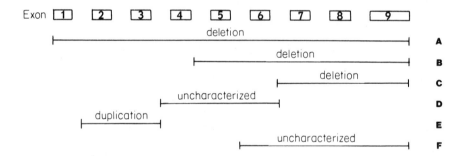

Figure 4. <u>Alterations in the HPRT gene in Lesch-Nyhan patients
detected by Southern blotting.</u> The HPRT gene is represented
schematically, and not to scale by 9 exons in the top line (exon 9
is by far the largest at approximately 640 b.p.). Rearrangements
covering the parts of the gene shown have been found. Lines A
through E have described by Young et al. (1984). Line F is
unpublished data (Jolly et al.).

The sixth mutation (Fig. 4, line E) is the best defined and
rather unusual. It involves the duplication of exons 2 and 3 in a
patient described by Gottlieb et al. (1982). Cells from this
patient produce a larger than normal RNA that can be expected to be

translated to a nonsense protein. The levels of HPRT activity in
this patient's cells are extremely low (0.01% normal) , but the
clinical symptoms are not severe. The patient is not mentally
retarded and has no self-mutilation behavior, but has the classic
Lesch-Nyhan symptoms of uric acid nephrolithiasis, spasticity and
choreoathetosis. In addition, it has been reported that cells in
culture from this patient revert to HAT resistance at high frequency
of 10^{-5} to 10^{-6} per generation (Aylsworth et al. 1984), resulting in
cells having HPRT enzyme levels 2-30% normal. It has been
conjectured that the reversion occurs in vivo and possibly at a
level sufficient to alleviate the symptoms resulting from HPRT
deficiency. One possible explanation for this effect is that
reversion of the duplication can occur if splicing in the
duplication is prevented or inhibited by a new point mutation, so
that the splice donor from the 3' end of the first exon 2 no longer
functions together with acceptor site in the adjacent exon 3, but
rather splices out the duplication and functions with the acceptor
site at the 5' end of the second exon 3. This would give a
functional, normal HPRT message. This scheme is of course totally
hypothetical at present, but gives an example of some of the kinds
of information on basic biological processes that may be obtained by
characterizing mutations in Lesch-Nyhan patients, as has been so
revealing and productive in studies of thalassemic patients.

Northern blot studies of RNA made in Lesch-Nyhan cells
having no apparent genomic rearrangements show that cells from 18/22
patients make apparently normal length mRNA while 3/22 have no
detectable RNA (Yang et al. 1984; C. Caskey personal communication).

Presumably, the genetic defect in the HPRT gene in the patients making normal length RNA are likely to be point mutations or small inserts or deletions. So far, no precise characterization of the defect has been made at the nucleic acid level for these predominant forms of HPRT mutation. It is likely, however, that these data will be acquired in the near future. These are the kind of mutations that originally were expected to give CRM^+ but inactive protein. However, as noted above, in general this is not the case. Apparently, most of these mutations severely disrupt the protein tertiary structure in general, so that most of the antigenic determinants are disrupted and antibody recognition does not occur.

There have been persistent but inconclusive reports of occurrence of regulatory loci for HPRT gene expression (Bakay et al. 1975; Watson et al. 1972). The defects in the Lesch-Nyhan patients demonstrating apparently normal gene pattern but lacking detectable message production might be considered candidates for such mutations, although it seems more likely that most of them will represent mutations in noncoding regions of the HPRT transcription unit itself or mutations in the promotor regions. We can expect that some of these Lesch-Nyhan mutations will over the next few years be characterized and that these may reveal a variety of different and interesting arrangements of non-functional HPRT genes.

It has been shown that for most instances of X-linked, fully penetrant and completely recessive genetic disorders, approximately 1/3 of all probands should demonstrate new mutations (Franke et al. 1976). However, enzymatic assays of presumed obligate carrier mothers for HPRT mutations of Lesch-Nyhan patients shown that the majority of such females (43 of 47 mothers) demonstrate the mutation. On the basis of this and similar studies, it has been suggested that the frequency of mutation at the HPRT locus, at least, and perhaps other loci, is greater in females than males. Since detection of HPRT mutations through the characterization of the genetic material by Southern blotting is much more reliable than through enzyme assays, especially for heterozygotes, the existence of this sex-based mutation frequency can be proved or disproved definitively and rather easily using cloned HPRT gene probes hybridized to blots of DNA from families with detectable rearrangements. So far, one pedigree has been investigated (Yang et al. 1984), and in this case the mutation in the carrier mother represents a new mutation, since the maternal grandmother carries no mutant allele.

E. Introduction of the HPRT gene into cells and animals and the prospects for gene therapy

One result of the isolation of the mammalian genes for HPRT has been the development of a powerful model to study the feasibility of introducing foreign genes in expressible forms into genetically marked cells. Using the calcium phosphate precipitation technique developed by Graham and Van de Eb for viral genes, the genomic HPRT gene was shown to be transferrable into tissue culture cells (fibroblasts) and this fact was used to isolate the human gene as a full length, cDNA clone (Jolly et al. 1982, 1983). Similarly, the cloned cDNA molecules can transfect HPRT⁻ cells at efficiencies close, or equal, to other selectable markers, when ligated to transcriptional signals including appropriate strong promotors from

SV40 or murine leukemia virus (MLV), long terminal repeats (LTRs)
(Jolly et al. 1982; Brennand et al. 1983; Miller et al. 1983) and
polyadenylation signals. These experiments have two long term
objectives: (1) to study and understand some of the normal genetic
regulatory control mechanisms of HPRT gene expression; and (2) to
examine the possibility and the feasibility of therapy (Anderson,
1984) for Lesch-Nyhan patients. For many of these kinds of studies,
the calcium phosphate precipitation techniques used for gene
transfer by transfection, and other commonly used methods of gene
introduction, are not ideal, for several reasons. The efficiency of
gene transfer is generally low, since at best approximately one cell
per 10^3-10^4 recipient cells takes up and expresses the foreign gene
in a stable and heritable fashion. Furthermore, many of the cells
with the most interesting expression control properties have
efficiencies of transfection very much less than 10^{-3} and often are
not susceptible to transfection at all. Gene rearrangements are
also quite common during gene uptake, and integration into the host
cell genome, making subsequent genetic characterization of such
markers of uncertain value. In many instances, inserted genes are
introduced in tandem arrays or in multiple copies, making it
difficult to define mechanisms of gene control in studies of gene
expression.

For these reasons, we and others have turned to efficient
and well characterized viral vectors, and in particular retroviral
vectors (Miller et al. 1983; Mulligan, 1983; Wei et al. 1981;
Shimotohno and Temin, 1981) to introduce foreign genes into
mammalian cells. Although it is likely that other viruses will be

used in the future (Smith et al. 1983; Van Doren and Gluzman, 1984;
Hermonat and Muzyczka, 1984), retroviral vectors at the
present time have associated with them a number of major advantages
that make them extremely useful. These advantages include: (i)
high efficiency of gene transfer—virtually 100% of a cell
population can be made to take up and express the foreign gene; (ii)
well characterized and cloned genomes that permit them to be
rearranged and recombined easily in vitro (Coffin, 1982); (iii) a
well characterized and understood life cycle that leads to
integration of a single copy   of the incoming gene(s) at defined
and recognized vector sequences into probably random sites in the
recipient cell genome (Varmus and Swanstrom, 1982; Groner and Hynes,
1982); (iv) relatively efficient gene expression compared with that
found with the calcium phosphate technique (Hwang and Gilboa, 1984).

Briefly, retroviruses are non lytic RNA viruses that pass
through a DNA stage after infection and integrate in the host genome
in a form called a provirus. The mechanisms of gene expression by
these viruses have been reviewed extensively (Varmus and Swanstrom,
1982; Groner and Hynes, 1982). The proviruses have been cloned in a
number of cases. The portions of the cloned retrovirus
corresponding to the viral genes have been excised and replaced with
a variety of non-viral genes, including the human HPRT gene (see
Fig. 5). Such a recombinant DNA molecule is then capable of being
inserted into cells by calcium phosphate-mediated transfection
precipitation. Of particular use is the fact that it can be rescued
as an infectious virus particle by supplying the missing viral
function of the gag, pol and env genes (see Fig. 5 and Fig. 6) in

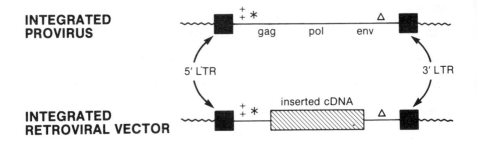

INTEGRATED
PROVIRUS

INTEGRATED
RETROVIRAL VECTOR

5'LTR   3'LTR

* Primer Binding Site, – Strand DNA Synthesis
⁺ Packaging Signal (ψ)
Δ + Strand DNA Synthesis

Figure 5. Structure of an integrated provirus HPRT retroviral
vector. The wavy lines denote flanking cellular DNA. In the
competent integrated provirus the 5' LTR has a polyadenylation site,
a promotor and enhancer sequences. Transcripts are read off this
promotor through the gag, pol and env genes and terminate in the
polyadenylation site in the 3' LTR (which is identical to the 5'
LTR). The gag gene codes for a viral structural protein that
confers some specificity for host cell infections (N,B tropism), the
pol gene specifies the reverse transcriptase and the env gene
specifies a protein that is embedded in the cellular membrane when a
virus particle buds off and which governs most of the host
specificity of the virus. Full length transcripts are read off this
operon, paired with a second identical RNA molecule and packaged
along with the reverse transcriptase enzyme inside the virus
particle which then buds off. The RNA pairing is a prerequisite for
packaging and is mediated through the packaging signal. The virus
then reinfects any cell with a receptor for its env gene product,
the RNA is uncoated, transcribed into cDNA by a complicated
mechanism involving the negative strand synthesis primed by a
cellular tRNA and subsequent positive strand syntheis started at the
site shown. This cDNA then integrates into the host genome,
apparently at random, to give the proviral structure once more.

    The gag, pol and env gene products can be supplied in trans,
and hence a transcript from a structure shown as the integrated
retroviral vector produces RNA molecules that can be packaged,
reinfect other cells and integrate once again. Unless the viral
genes are present in the reinfected cell, the integrated defective
HPRT cDNA provirus is not capable of producing viral particles and
hence remains where it has inserted,and produces RNA which is
translated into HPRT protein. DNA structures such as that of the
integrated cDNA vector can be generated in vitro and then
transfected into tissue culture cells.

trans. This can be done by simultaneous transfection with a cloned proviral form of a nondefective retrovirus, or subsequent infection with an infectious form of the nondefective, or helper virus. In either case, virus particles are produced that are competent to infect other cells and integrate their genes into the host cell DNA. Genes introduced by these methods into cells can then be expressed under the genetic control of the retroviral signals, but for all intents and purposes they become permanent and heritable new cellular genetic functions.

In addition it is possible to add a "helper" virus genome that is itself defective in packaging functions so that the net result is production of HPRT particles with no detectable competent helper virus (Mann et al, 1983; Watanabe and Temin, 1983; Miller et al. 1984a). Such preparations are called "helper free."

We have made several such HPRT virus preparations and examined their effect on mouse, rat and human Lesch-Nyhan HPRT⁻ fibroblasts, carcinoma and neuroblastoma cells, on immortalized lymphoblasts from Lesch-Nyhan patients, and on normal mouse bone marrow reimplanted into genetically matched mice. In summary, the results of these studies show that the HPRT gene is transferred to cells at efficiencies approaching 100%, producing cells having single integrated copies of the vector and its added gene per cell. In lymphoblast cells derived from Lesch-Nyhan patients, HPRT activities in HPRT infected cells were 4-23% normal wild-type values. In untreated lymphoblasts from patients with Lesch-Nyhan disease, many of the parameters of purine metabolism are grossly aberrant. These include high levels of extracellular, excreted

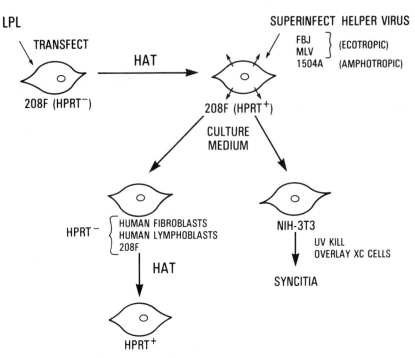

Figure 6. **Generation of infectious HPRT particles.** A structure
such as the cDNA retroviral vector described in Fig. 5 is generated
in vitro by molecular biological methods. In our case this vector
is called "LPL" and contains murine leukemia virus LTR's packaging
signals etc. plus the human HPRT cDNA. LPL is introduced into rat
HPRT⁻ fibroblasts (208F cells) by transfection and HPRT⁺ cells
selected with HAT. These cells are then superinfected with various
competent retroviruses. The 208F cells produce HPRT and
competent parent type retroviral particles which are assayed
by infection of HPRT⁻ cells and regular retrovirus
assays, as shown. Superinfection is only one way to add the gag,
pol and env helper functions. Other ways are to supertransfect
with a competent cloned proviral structure, or supertransfect with a
proviral structure deficient in the packaging signal (see Fig. 5).
This last gives rise, in principle, to virus preparations
consisting of only HPRT virus ("helper free").

hypoxanthine, large PRPP pools and others. Most of these aberrant

metabolic features are normalized roughly in direct proportion to

the percent of newly expressed HPRT activity attained in an HPRT

infected cell (Willis et al. 1984). That is, the introduction of a

functional, new HPRT activity even in low amounts is enough to

correct at least partially some of the major prime defects in HPRT-
deficient cells.

In a further series of experiments (Miller et al. 1984a) we
have asked if the vector can be used to introduce the gene into
somatic tissue of animals. In these experiments (see Fig. 7) mouse
bone marrow was infected with high titer "helper free" HPRT virus,
superinfected with MLV helper to allow spread of the virus and the
marrow reintroduced to irradiated genetically matched mice. These
mice survived only if the bone marrow graft replaced the endogenous
destroyed marrow. At various times between 31 to 133 days after the
marrow transplant, spleen foci and bone marrow were assayed for HPRT
virus production which was detected in all cases, thereby
demonstrating that the HPRT gene had entered stem cells in the
marrow and was being expressed. In addition in two out of ten of
the animals receiving marrow transplants, we were able to detect
highly significant amounts of human HPRT activity in spleen foci.

From these studies we know that it is possible to introduce
the HPRT gene efficiently into a variety of human and other
mammalian cells, and similar results have been obtained by numerous
other investigators using other gene markers (Sorge et al. 1984;
Joyner et al. 1983; Williams et al. 1984), in some cases with
apparent expression control mechanisms intact (Miller et al. 1984b;
Episkopou, 1984). In most cases, the levels of expression are low
compared to the native or resident wild type genes and the problem
of reduced and inappropriate expression of foreign genes remains a
characteristic and unexplained feature of all methods of gene
introduction. Methods of enhancing gene expression may become

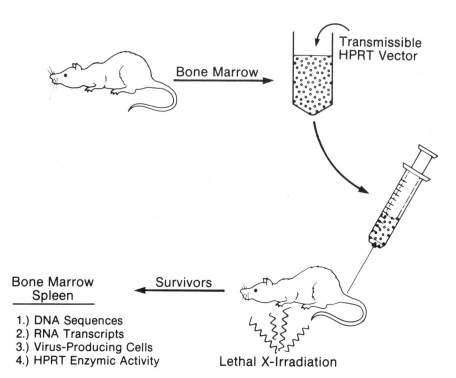

Figure 7. <u>Mouse bone marrow infection with HPRT</u>. The experiment is
illustrated diagrammatically. Bone marrow is removed from a mouse,
mixed with an HPRT virus preparation or with cells producing HPRT
virus. After a few hours, competent helper virus was also added, to
help spread the HPRT virus. The marrow is then reimplanted into
isogenic mice whose bone marrow has been destroyed by irradiation.
Surviving mice have marrow populations that are obligatorily
descended from the implant. At various times after reimplantation
the mice were sacrificed and their marrow and spleen loci examined
for: (1) DNA and RNA sequences homologous to the HPRT provirus; (2)
HPRT virus produced; (3) human HPRT activity. We were not able
directly to show the existence in RNA transcripts from the HPRT
provirus. However, in all cases HPRT virus was being produced, and
in two of ten cases human HPRT activity was clearly found in spleen
foci, by means of isoelectric focusing of cell extracts followed by
development to reveal HPRT activity.

available through the use of appropriate promotors or enhancers

fused to the gene, or by agents promoting integration in particular

or selected sites in the genome.

For studies of the feasibility of human gene therapy, the retrovirus vehicle itself may have to be modified further to some extent to decrease the likelihood of insertional mutagenesis and of inappropriate endogenous gene activation and/or recombination with genomic sequences to produce competent, potentially pathogenic viruses. In addition, it is not at all clear that methods of gene therapy currently envisioned will result in the alleviation of any of the neurological symptoms in Lesch-Nyhan patients. Nevertheless, it is likely that the HPRT deficiency model will continue to provide very useful information on the mechanisms of transfer of foreign genes into human cells via retroviral vectors, the fate and expression of such genes in vitro and in vivo, and the potential role of these kinds of manipulations in the management of otherwise untreatable and severe human disease. Models for human gene therapy will include genetic diseases such as Lesch-Nyhan disease or the rather more accessible disorders associated with adenosine deaminase (ADA) deficiency, and purine nucleoside phosphorylase (PNP) deficiency (Boss and Seegmiller, 1982), where the defective function is expressed predominantly in the cells of bone marrow derivation. It is in these disorders that rapidly turning over cells and large numbers of progenitor "stem" cells can be isolated from patients, modified genetically and returned to the patient by simple transfusion as an autologous marrow transplantation. Further advantages of these diseases include the fact that it is likely that only small amounts of activity (about 5%) are required to alleviate symptoms, that the consequences of the disease are severe and life-threatening, and that the normal genes have been isolated.

SUMMARY

1.   Human HPRT deficiency leads to two major forms of human disease.   Partial enzyme deficiency results in gouty arthritis, while an almost complete deficiency leads to the Lesch-Nyhan disease.   The latter is characterized by severe neurological dysfunction in addition to gouty arthritis, including retardation, choreoathetosis and aggressive and compulsive self-mutilation.   The biochemical basis for the neurological symptoms is not understood.

2.   The human and mouse cDNA (RNA copy) genes have been isolated and sequenced.   In addition, the amino acid sequence of the human protein has been directly determined.   The human and mouse proteins differ at 7 amino acids out of the total, (including the N terminal methionine, which is processed off during maturation) of 218.   There are 42 out of 654 nucleotide differences between the human and mouse genes in the amino acid coding region.

3.   The mouse genomic structure has been determined.   It has 9 exons and 8 introns with a total size of approximately 36kb.   The human gene is very similar with identical intron-exon junction points and approximately the same total gene size.   Both mouse and human presumed promotor region at the 5' end, lack a recognizable promotor in the form of a "TATAA" box and are very G-C rich, though not the same.   This may be a feature of most "housekeeping" genes.

4.   HPRT gene point mutations in three gouty arthritis and one Lesch-Nyhan patient have been identified by peptide sequencing.   Six gross gene rearrangements have been identified in Lesch-Nyhan HPRT genes.   However it is likely that most mutations are point mutations or small deletions.   So far all gene mutations identified are

different from all others.

5. The gene has been engineered into retrovirus vehicles which allows its efficient introduction into a wide variety of cells, including mouse marrow stem cells. This may allow treatment of Lesch-Nyhan patients as a model of gene therapy.

Acknowledgements

This work was supported in part by grants from NIH, The Kroc Foundation, and The Gould Foundation to T. Friedmann. The author thanks Dr. C.T. Caskey and his coworkers for generously allowing access to their unpublished results; Dr. B. Volgenstein and A. Feinberg for permission to quote their manuscript in press, Dr. T. Friedmann for his continuing support and encouragement, and Mrs. Y. Cabrera for her careful help in preparing the manuscript.

REFERENCES

Anderson, W.F. (1984). 'Prospects for human gene therapy', Science 226, 401–409.

Argos, P., Hanei, M., Wilson, J.M., and Kelley, W.M. (1983). 'A possible nucleotide–binding domain in the tertiary fold of phosphoribosyltransferases', J. Biol. Chem. 258, 6450–6457.

Arnold, W.J., and Kelley, W.N. (1971). 'Human hypoxanthine–guanine phosphoribosyltransferase: purification and subunit structure', J. Biol. Chem. 246, 7398–7404.

Aylsworth, A.S., Kredich, N.M., Jackson, L.G. and Rao, K.W. (1984). 'Spontaneous reversion of a human hypoxanthine phosphoribosyltransferase (HPRT) deficient mutant', Amer. J. Hum. Genet. 36, 201S.

Bakay, B., Becker, M.A., and Nyhan, W.L. (1976). 'Reaction of antibody to normal human hypoxanthine phosphoribosyltransferase with products of mutant genes', Arch. Biochem. Biophy. 177, 415–426.

Bakay, B., Nissinen, E., Sweetman, L., Franke, U., and Nyhan, W.L. (1979). 'Utilization of purines by an HPRT variant in an intelligent, non-mutilative patient with features of the Lesch-Nyhan syndrome', Pediatr. Res. 13, 1365–1370.

Bakay, R., Nyhan, W.L., Croce, C.M., and Koprowski, H. (1975). 'Reversion in expression of hypoxanthine–guanine phosphoribosyl transferase following cell hybridization', J. Cell Sci. 17, 567–578.

Boss, G.R. and Seegmiller, J.E. (1982). 'Genetic defects in human purine and pyrimidine metabolism', Ann. Rev. Genet. 16, 297–328.

Botstein, D., White, R.L., Skolnick, M., and Davis, R.W. (1980). 'Construction of a genetic linkage map in man using restriction fragment length polymorphisms', Am. J. Hum. Genet. 32, 314–331.

Brennand, J., Chinault, A.C., Konecki, D.S., Melton, D.W., and Caskey, C.T. (1982). 'Cloned cDNA sequences of the hypoxanthine/guanine phosphoribosyltransferase gene from a mouse neuroblastoma cell line found to have amplified genomic sequences', Proc. Nat. Acad. Sci. USA 79, 1950–1954.

Brennand, J., Konecki, D.S., and Caskey, C.T. (1983). 'Expression of human and chinese hamster hypoxanthine–guanine phosphoribosyltransferase cDNA recombinants in cultured Lesch-Nyhan and chinese hamster fibroblasts', J. Biol. Chem. 258, 9593–9596.

Burnstock, G. (1975). 'Purinergic transmission,' in Handbook of Psychopharmacology (Eds. L.L. Iversen, S.D. Iversen, and S.H. Snyder), pp. 131–194, Plenum Press, New York.

Caskey, C.T. and Kruh, G.D. (1979). 'The HPRT locus', Cell 16, 1–9.

Castells, S. and Chakrabarti, C., Winsberg, B.G., Hurwic, M., Perel, J.M., and Nyhan, W.L. (1979). 'Effects of L-5-hydroxytryptophan on monoamine and amino acids turnover in the Lesch-Nyhan syndrome,' J. Aut. Dev. Dis. 9, 95–103.

Chinault, A.C. and Caskey, C.T. (1984). 'The hypoxanthine phosphoribosyltransferase gene: a model for the study of mutation in mammalian cells', Prog. Nucl. Acid Res. Mol. Biol. 31, 295–313.

Coffin, G. (1982). 'Structure of the retroviral genome', in RNA Tumor Viruses, 2nd Edition, (Eds. R. Weiss, N. Teich, H. Varmus, and J. Coffin), pp. 261–368, Cold Spring Harbor Laboratory, New York.

Daly, J.W., Bruns, R.F., and Snyder S.H. (1981). 'Adenosine receptors in the central nervous system relationship to the central actions of methylxanthines', Life Sci. 28, 2083-2097.

Eickbush, T.H. and Kafatos, F.C. (1982). 'A walk in the chorion locus of bombyx mori', Cell 29, 633-643.

Episkopou, V., Murphy, A.J.M., and Efstratiadis, A. (1984). 'Cell-specified expression of a selectable hybrid gene', Proc. Natl. Acad. Sci. USA 81, 4657-4661.

Francke, U., Felsenstein, J., Gartler, S.M., Migeon, B.R., Dancis, J., Seegmiller, J.E., Bakay, B., and Nyhan, W.L. (1976). 'The occurence of new mutants in the X-linked recessive Lesch-Nyhan disease', Am. J. Hum. Genet. 28, 123-137.

Fredholm, B.B., and Hedqvist, P. (1980). 'Modulation of neurotransmission by purine nucleotides and nucleosides', Biochem. Pharmacol. 29, 1635-1643.

Friedmann, T., Esty, A., Filpula, D., Jolly, D., and Doolittle, R. (1983). 'Characterization of an expressible human hypoxanthine phosphoribosyltransferase cDNA', in Banbury Report 14: Recombinant DNA Applications to Human Disease, pp. 91-96, Cold Spring Harbor Laboratory, New York.

Fujimoto, W.Y., Subak-Sharpe, J.H., and Seegmiller, J.E. (1971). 'Hypoxanthine-guanine phosphoribosyltransferase deficiency: Chemical agents selective for mutant or normal cultured fibroblasts in mixed heterozygote cultures', Proc. Natl. Acad. Sci. USA 68, 1516-1519.

Ghangas, G.S. and Milman, G. (1975). 'Radioimmune determination of hypoxanthine phosphoribosyltransferase crossreacting material in erythrocytes of Lesch-Nyhan patients', Proc. Natl. Acad. Sci. USA 72, 4147-4150.

Gottlieb, R.P., Koppel, M.M., Nyhan, W.L., Bakay, B., Nissinen, E., Borden, M., and Page, T. (1982). 'Hyperuricaemia and choreoathetosis in a child without mental retardation or self-mutilation—a new HPRT variant', J. Inher. Metab. Dis. 5, 183-186.

Graham, F.L., and van der Eb, A.J. (1973). 'A new technique for the assay of infectivity of human adenovirus 5 DNA', Virology 52, 456-467.

Groner, B. and Hynes, N.E. (1982). 'Long terminal repeats provide regulatory signals at the ends of retroviral genes', Trends Biochem. Sci. 7, 400-403.

Hartley, D.A., Davies, K.E., Drayna, D., White, R.L., and Williamson, R. (1984). 'A cytological map of the human X chromosome—evidence for non-random recombination', Nucl. Acids Res. 12, 5277-5285.

Hermonat, P.L. and Muzyczka, N. (1984). 'Use of adeno-associated virus as a mammalian DNA cloning vector: transduction of neomycin resistance into mammalian tissue culture cells', Proc. Natl. Acad. Sci. USA 81, 6466-6470.

Hochstadt, J. (1973). 'The role of the membrane in the utilization of nucleic acid precursors', Crit. Rev. Biochem. 2, 259-310.

Hwang, L.-H.S. and Gilboa, E. (1984). 'Expression of genes introduced into cells by retroviral infection is more efficient than that of genes introduced into cells by DNA transfection', J. Virol. 50, 417-424.

Johnson, G.G., Eisenberg, L.R., and Migeon, B.R. (1979). 'Human and mouse hypoxanthine-guanine phosphoribosyltransferase: dimers and tetramers', Science 203, 174-176.

Johnson, G.G., Ramage, A.L., Littlefield, J.W., Kazazian, H.H. Jr. (1982). 'Hypoxanthine-guanine phosphoribosyltransferase in human erythroid cells: posttranslational modification', Biochemistry 21, 960-966.

Jolly, D.J., Esty, A.C., Bernard, H.U., and Friedmann, T. (1982). 'Isolation of a genomic clone partially encoding human hypoxanthine phosphoribosyltransferase', Proc. Natl. Acad. Sci. USA 79, 5038-5041.

Jolly, D.J., Okayama, H., Berg, P., Esty, A.C., Filpula, D., Bohlen, P., Johnson, G.G., Shively, J.E., Hunkapillar, T., and Friedmann, T. (1983). 'Isolation and characterization of a full-length expressible cDNA for human hypoxanthine phosphoribosyl-transferase', Proc. Natl. Acad. Sci. USA 80, 477-481.

Johner, A., Keller, G., Phillips, R.A., and Bernstein, A. (1983). 'Retrovirus tranfer of a bacterial gene into mouse haematopoietic progenitor cells', Nature 305, 556-558.

Kelley, W.N. and Wyngaarden, J.B. (1983). 'Clinical syndromes associated with hypoxanthine-guanine phosphoribosyltransferase deficiency', in The Metabolic Basis of Inherited Disease (Eds. J.B. Stanbury, J.B. Wyngaarden, D.S. Fredrickson, J.L. Goldstein, and M.S. Brown), pp. 1115-1143, McGraw-Hill Book Company, New York.

Konecki, D.S., Brennand, J., Fuscoe, J.C., Caskey, C.T., and Chinault, A.C. (1982). 'Hypoxanthine-guanine phosphoribosyltransferase genes of mouse and chinese hamster: construction and sequence analysis of cDNA recombinants', Nucl. Acids Res. 10, 6763-6775.

Kopin, I.J. (1981). 'Neurotransmitters and the Lesch-Nyhan syndrome', N. Eng. J. Med. 305, 1148-1149.

Lloyd, K.G., Hornykiewicz, O., Davidson, L., Shannak, K., Farley, I., Goldstein, M., Shibuya, M., Kelley, W.N., and Fox, I.H. (1981). 'Biochemical evidence of dysfunction of brain neurotransmitters in the Lesch-Nyhan syndrome', N. Eng. J. Med. 305, 1106-1111.

Mann, R., Mulligan, R.C., and Baltimore, D. (1983). 'Construction of a retrovirus packaging mutant and its use to produce helper-free defective retrovirus', Cell 33, 153-159.

Martin, G.R. (1982). 'X-chromosome inactivation in mammals', Cell 29, 721-724.

McKusick, V.A. (1983). Mendelian Inheritance in Man. Catalogs of Autosomal Dominant, Autosomal Recessive, and X-linked phenotypes, Sixth Edition, The Johns Hopkins University Press, Baltimore and London.

Melton, D.W., Konecki, D.S., Brennand, J., and Caskey. C.T. (1984). 'Structure, expression, and mutation of the hypoxanthine phosphoribosyltransferase gene', Proc. Natl. Acad. Sci. USA 81, 2147-2151.

Miller, A.D., Eckner, R.J., Jolly, D.J., Friedmann, T., and Verma, I.M. (1984a). 'Expression of a retrovirus encoding human HPRT in mice', Science 225, 630-632.

Miller, A.D., Jolly, D.J., Friedmann, T., and Verma, I.M. (1983). 'A transmissible retrovirus expressing human hypoxanthine phosphoribosyltransferase (HPRT): gene transfer into cells obtained from humans deficient in HPRT', Proc. Natl. Acad. Sci. USA 80, 4709-4713.

Miller, A.D., Ong, E.S., Rosenfeld, M.G., Verma, I.M., and Evans, R.M. (1984b). 'Infectious and selectable retrovirus containing an inducible rat growth hormone minigene', Science 225, 993-998.

Muensch, H. and Yoshida, A. (1977). 'Purification and characterization of human hypoxanthine/guanine phosphoribosyltransferase', Eur. J. Biochem. 76, 107-112.

Mulligan, R.C. (1983). 'Construction of highly transmissable mammalian cloning vehicles derived from murine retroviruses' in Experimental Manipulation of Gene Expression (Ed. M. Inouye), pp. 155-173, Academic Press, New York.

Nussbaum, R.L., Crowder, W.E., Nyhan, W.L., and Caskey, C.T. (1983). 'A three-allle restriction-fragment-length polymorphism at the hypoxanthine phosphoribosyltransferase locus in man', Proc. Natl. Acad. Sci. USA 80, 4035-4039.

Nyhan, W.L., Bakay, R., Connor, J.D., Marks, J.F., and Keele, D.K. (1970). 'Hemizygous expression of glucose-6-phosphate dehydrogenase in erythrocytes of heterozygotes for the Lesch-Nyhan syndrome', Proc. Natl. Acad. Sci. USA 65, 214-218.

Nyhan, W.L. (1973). 'The Lesch-Nyhan syndrome', Ann. Rev. Genet. 7, 41-60.

Ogasawara, N., Kashiwamata, S., Oishi, H., Hara, Kl, Watanabe, K., Miyazaki, S., Kumagi, T., and Hakamada, S. (1982). 'Hypoxanthine-guanine phosphoribosyl transferase (HGPRT) deficiency in a girl'. Adv. in Exp. Med. and Biol. 165A, 13-18.

Orkin, S.H., Markham, A.F., and Kazazian, H.H. Jr. (1983). 'Direct detection of the common Mediterranean-thalassemia gene with synthetic DNA probes. An alternative approach for prenatal diagnosis', J. Clin. Invest. 71, 775-779.

Pai, G.S., Sprenkle, J.A., Do, T.T., Mareni, C.E., and Migeon, B.R. (1980). 'Localization of loci for hypoxanthine phosphoribosyltranferase and glucose-6-phosphate hydrogenase and biochemical evidence of nonrandom X chromosome expression from studies of a human X-autosome translocation', Proc. Natl. Acad. Sci. USA 77, 2810-2813.

Patel, P.I., Nussbaum, R.L., Framson, P.E., Ledbetter, D.H., Caskey, C.T., and Chinault, A.C. (1984). 'Organization of the HPRT gene and related sequences in the human genome', Som. Cell Mol. Genet. 10, 483-493.

Rodbell, M. (1980). 'The role of hormone receptors and GTP-regulatory proteins in membrane transduction', Nature 284, 17-22.

Seegmiller, J.E. (1967). 'Enzyme defect associated with a sex-linked human neurological disorder and excessive purine synthesis', Science 155, 1682-1684.

Shimotohno, K. and Temin, H.W. (1981). 'Formation of infectious progeny virus after insertion of herpes simplex thymidine kinase gene into DNA of an avian retrovirus', Cell 26, 67-77.

Skolnick, P., Marangos, P.J., Goodwin, F.K., Edwards, M., and Paul, P. (1978). 'Identification of inosine and hypoxanthine as endogenous inhibitors of [^{3}H]diazepam binding in the central nervous system. Life Sci. 23, 1473-1480.

Smith, G.L., Mackett, M., and Moss, B. (1983). 'Infectious vaccinia virus recombinants that express hepatitis B virus surface antigen', Nature 302, 490-495.

Sorge, J., Wright, D., Erdman, V., and Cutting, A.E. (1984). 'Amphotropic retrovirus vector system for human cell gene transfer', Mol. Cell. Biol. 4, 1730-1737.

Steinmetz, M., Winoto, A., Minard, K., and Hood, L. (1982). 'Clusters of genes encoding mouse transplantation antigens', Cell 28, 489–498.

Szybalski, W., Szylbalska, E.H., and Ragni, B. (1962). 'Genetic studies with human cell lines', Natl. Cancer Inst. Monograph 7, 75–89.

Van Doren, K. and Gluzman, Y. (1984). 'Efficient transformation of human fibroblasts by adenovirus-simian virus 40 recombinants', Mol. Cell Biol. 4, 1653–1656.

Upchurch, K.S., Leyva, A., Arnold, W.J., Holmes, E.W., and Kelley, W.N. (1975). 'Hypoxanthine phosphoribosyltransferase deficiency: association of reduced catalytic activity with reduced levels of immunologically detectable enzyme protein', Proc. Natl. Acad. Sci. USA 72, 4142–4146.

Varmus, H., and Swanstrom, R. (1982) 'Replication of retroviruses', in RNA Tumor Viruses, 2nd Edition (Eds. R. Weiss, N. Teich, H. Varmus, and J. Coffin), pp. 369–512, Cold Spring Harbor Laboratory, New York.

Vogelstein, B., Fearon, E.R., and Feinberg, A.P. (1985). 'Determining the clonality of human tumors using restriction fragment length polymorphisms', Science, in press.

Watanabe, S. and Temin, H.M. (1983). 'Construction of a helper cell line for avian reticuloendotheliosis virus cloning vectors', Mol. Cell. Biol. 3, 2241–2249.

Watson, B., Gormley, I.P., Gardiner, S.E., Evans, H.J., and Harris, H. (1972). 'Reappearance of murine hypoxanthine guanine phosphoribosyl transferase activity in mouse A9 cells after attempted hybridisation with human cell lines', Exp. Cell Res. 75, 401–409.

Wei, C.M., Gibson, M., Spear, P.G., and Scolnick, E.M. (1981). 'Construction and isolation of a transmissible retrovirus containing the src gene of harvey murine sarcoma virus and the thymidine kinase gene of herpes simplex virus type 1', J. Virology 39, 935–944.

Williams, D.A., Lemischka, I.R., Nathan, D.G., and Mulligan, R.C. (1984). 'Introduction of new genetic material into pluripotent haematopoietic stem cells of the mouse', Nature 310, 476–480.

Willis, R.C., Jolly, D.J., Miller, A.D., Plent, M.M., Esty, A.C., Anderson, P.J., Chang, H.-C., Jones, O.W., Seegmiller, J.E., and Friedmann, T. (1984). 'Partial phenotypic correction of human Lesch-Nyhan (hypoxanthine-guanine phosphoribosyltransferase-deficient) lymphoblasts with a transmissible retroviral vector', J. Biol. Chem. 259, 7842–7849.

Wilson, J.M., Baugher, B.W., Landa, L., and Kelley, W.N. (1981). 'Human hypoxanthine-guanine phosphoribosyltransferase. Purification and characterization of mutant forms of the enzyme', J. Biol. Chem. 266, 10306–10312.

Wilson, J.M., Frossard, P., Nussbaum, R.L., Caskey, C.T., and Kelley, W.N. (1983a). 'Human hypoxanthine-guanine phosphoribosyltransferase. Detection of a mutant allele by restriction endonuclease analysis', J. Clin. Invest. 72, 767–772.

Wilson, J.M., Tarr, G.E., Mahoney, W.C., Kelley, W.N. (1982). 'Human hypoxanthine-guanine phosphoribosyltransferase; complete amino acid sequence of the erythrocyte enzyme', J. Biol. Chem. 257, 10978–10985.

Wilson, J.M., Young, A.B., and Kelley, W.N. (1983b). 'Hypoxanthine-guanine phosphoribosyltransferase deficiency. The molecular basis of the clinical syndromes', N. Eng. J. Med. 309, 900–910.

Wolf, S.F., Jolly, D.J., Lunnen, K.D., Friedmann, T., and Migeon, B.R. (1984). 'Methylation of the hypoxanthine phosphoribosyltransferase locus on the human X chromosome: Implications for X-chromosome inactivation', Proc. Natl. Acad. Sci. USA 81, 2806–2810.

Woods, R.A., Roberts, D.G., Friedmann, T., Jolly, D., and Filpula, D. (1983). 'Hypoxanthine: guanine phosphoribosyltransferase mutants in Saccharomyces Cerevisiae', Mol. Gen. Genet. 191, 407–412.

Yang, T.P., Patel, P.I., Chinault, A.C., Stout, J.T., Jackson, L.G., Hildebrand, B.M., and Caskey, C.T. (1984). 'Molecular evidence for new mutation at the HPRT locus in Lesch-Nyhan patients', Nature 310, 412–414.

Yen, P.H., Patel, P., Chinault, A.C., Mohandas, T., and Shapiro, L.J. (1984). 'Differential methylation of hypoxathine phosphoribosyltransferase genes on active and inactive human X chromosomes', Proc. Natl. Acad. Sci. USA 81, 1759–1763.

Human Genes and Diseases
Edited by F. Blasi
© 1986, John Wiley & Sons, Ltd.

THE UTILIZATION OF THE HUMAN PHOSPHOGLYCERATE KINASE GENE IN THE

INVESTIGATION OF X-CHROMOSOME INACTIVATION

Michael A. Goldman, Stanley M. Gartler, Elisabeth A. Keitges, and
Donald E. Riley

INTRODUCTION

Normal development in mammals and most other animals requires
chromosomal balance. Significant autosomal dosage changes usually
result in serious abnormalities, the so-called aneuploidy effect
(1-4). Chromosomal balance requires not only one pair of each autosome, but the same effective autosome to X chromosome ratio in the
two sexes. The process that brings about effective equality of X
chromosome expression between XX and XY individuals is called dosage
compensation. The effects of X chromosome aneuploidy are usually
less severe than those of autosomal imbalance because of dosage compensation. There are several ways in which dosage compensation is
achieved in forms with XY systems of sex determination, and in
mammals, a widespread mechanism is X-chromosome inactivation
(reviewed in ref. 5). In this article we will be concerned with the
utilization of the X-linked phosphoglycerate kinase (Pgk) gene in
the molecular analysis of the X-chromosome inactivation process.

169

X-chromosome inactivation first occurs in the early blastocyst when one of the two X chromosomes in female cells becomes inactivated, bringing about an effective X chromosome to autosome ratio equal to that in the male. At the cytological level, inactivation appears as the formation of sex chromatin. The process follows sequential steps in embryonic differentiation (see ref. 5 for review). It takes place first in trophectoderm, next in extra-embryonic endoderm, and last in the embryonic ectoderm. In the extra-embryonic lineages, X inactivation occurs preferentially in the paternal X chromosome, while in the embryonic lineage the inactivation process is random with respect to paternal and maternal X chromosomes. Once X inactivation occurs, this cytologically recognized differentiative event becomes fixed in all subsequent cell generations. The inactive X chromosome is heteropycnotic throughout the cell cycle and replicates asynchronously as do other forms of heterochromatin. The one exception to the permanence of X inactivation occurs in the germ line where the inactive X chromosome is reactivated around the time of entry into meiosis (5).

X-chromosome inactivation not only brings about dosage compensation between 2X females and 1X males; it also compensates for additional X chromosomes generated through meiotic error in that all but one of the X chromosomes in a cell are inactivated.

At the molecular level three sequential, and possibly related, steps must be considered in the X inactivation process—initiation of inactivation at a single site or control center (Xce), spreading of inactivation along the chromosome, and maintenance of the inactivated state throughout subsequent cell generations. In addition, a full explanation of the molecular genetics of X inactivation must take into account the nonrandom pattern of inactivation characteristic of the extra-embryonic lineage, the reactivation of individual genes on the inactive X induced by 5-azacytidine (6-8), the spreading of inactivation into autosomal regions in X-autosome translocations (9), and the maintenance of X-linked loci which are not inactivated (10).

There is good evidence that X inactivation is initiated at a single site on the X chromosome (5). Any one of a number of steps, such as inversion, membrane attachment, or methylation could be the initiating event (5). Difficulties arise in trying to explain the presence of a single active X chromosome per autosomal set, and consequently some rather complicated models have been proposed to explain the patterns of inactivation observed in various aneuploid and polyploid conditions (5). While the initiating event of X inactivation may be the most important developmental step in the process, it will also be the most difficult to study. Xce may represent only a small fraction of X chromosome DNA, and efficient selection or screening systems to identify recombinant DNA clones

with Xce are not immediately apparent. The spreading feature of X
inactivation will also be difficult to solve. One must consider the
spread of inactivation from a single site to more than 100 million
base pairs of DNA, a phenomenon which suggests the need for signal
amplification stations along the X chromosome (5, 22).

In view of these conceptual difficulties with the initiation
and spreading aspects of X inactivation, the major effort of most
groups has been to focus on the end result of X inactivation, the
inactivated X-linked gene. The hope is that an understanding of the
molecular difference between an active and inactive gene will
provide clues to unraveling the molecular basis of other features of
X inactivation.

The work described here focuses on the use of the Pgk gene to
probe the molecular basis of X inactivation. (We designate the X-
linked, functional Pgk gene "Pgk-1" and the autosomal, functional
Pgk gene "Pgk-2." We frequently refer to "Pgk" for brevity when the
meaning is clear from the context.) The Pgk locus is ideal for
studies of X-chromosome inactivation for several reasons. (1) The
X-linked Pgk gene is known to be inactivated and subject to reactivation with 5-azacytidine when present in interspecific somatic cell
hybrids. (2) The gene exhibits nonrandom paternal X inactivation in
extra-embryonic cell lineages. (3) In the mouse, Pgk maps close to
the X inactivation center or Xce locus (11-13). The pattern of X-
chromosome inactivation in X-chromosome abnormalities supports
mapping of the X inactivation center in man to cytological band Xq13
which is also the position of the Pgk gene (reviewed in ref. 5).
(4) An autosomal gene (Pgk-2) encodes a protein having the same

enzymatic function as the PGK encoded by the X-linked gene. The autosomal gene is active only in testis, in which the single X chromosome is inactive (for review, see ref. 5). A comparative study of the regulation of these two genes would be of great interest. (5) The family of Pgk-related sequences includes at least one autosomal and one X-linked pseudogene (67). Comparison of pseudogenes on the active and inactive X chromosomes may lead to an understanding of differences between active and inactive X chromatin apart from those directly related to gene transcription.

CLONING THE HUMAN X-LINKED Pgk GENE

In order to carry out molecular studies of X inactivation at the Pgk locus, it was necessary to clone the Pgk gene. As the amino acid sequence of the human PGK protein encoded on the X chromosome was known, mixed oligonucleotide probes were prepared corresponding to selected amino acid sequences in the protein (14). These mixed probes were used to screen cDNA libraries; positive colonies were picked, purified, and eventually sequenced to check for correspondence to the known amino acid sequence of the X-linked PGK protein. A similar approach was used independently by Singer-Sam et al. (14) and by Michelson et al. (15) to obtain cDNAs corresponding to the entire amino acid sequence of the protein. The sequence of pHPGK-7e, which contains 82 bp of 5' untranslated sequence and 437 bp of 3' untranslated sequence, including the poly(A) tail, was reported by Michelson et al. (15). Singer-Sam et al. (16) have

recently isolated a similar clone, pGK825, containing 30 bp of 5'
and 400 bp of 3' untranslated sequence.

MOLECULAR MAPPING OF THE HUMAN Pgk GENE FAMILY

Classical genetic studies have demonstrated X linkage of
somatic cell PGK in all mammalian species where genetic variants
have been found. The mapping of the human X-linked Pgk gene has
been further defined by interspecific somatic cell hybrid studies
which placed it in cytological band Xq13 (on the long arm of the
chromosome near the centromere). The tools of molecular biology, in
addition to providing independent confirmation of the classical
mapping data, permit several new levels of genetic mapping. These
levels include the recognition of related but nonfunctional
sequences (pseudogenes) and the study of gene fine structure by
mapping of restriction sites and by nucleotide sequencing.

Southern blots of human genomic DNA from cells varying in the
number of X chromosomes were probed with labeled Pgk cDNAs,
revealing a number of X-linked and non-X-linked fragments (14, 15,
17, 67). The fragment sizes produced by various restriction digests
of genomic DNA are summarized in Table 1. This complexity suggests
that the Pgk cDNA probes bear homology not only with the X-linked,
functional Pgk-1 gene, but with pseudogenes, the testis-specific
Pgk-2 gene, or genes coding for related proteins. Further work has
assigned those fragments which hybridize to the cDNA in Southern
blots to specific chromosomes, and determined their role in encoding
the PGK protein (67; Gartler & T. Shows, unpublished).

TABLE 1.   Human genomic digest patterns observed with various Pgk
cDNA probes.

EcoR I			BamH I	
Fragment	Linkage & Function		Fragment	Linkage & Function
16.0 Kb	X ψ 5'		> 10 Kb	A?
11.4 Kb	X F 5'		9.3 Kb	X
9.6 Kb	AF 5 'M 3'		6.7 Kb	X F
8.1 Kb	A(6) ψ 5' M		4.9 Kb	X
6.5 Kb	X F M 3'		3.8 Kb	X F 5'
4.0-4.2 Kb	X F M		2.1 Kb	A
3.6-3.8 Kb	X ψ 5' M 3'			
2.8-3.1 Kb	X F 3'			

Hind III			Pst I	
Fragment	Linkage & Function		Fragment	Linkage & Function
19.0 Kb	X F 5'			
>7.6 - 8.6 Kb	X M		14.0 Kb	X
<7.6 - 8.6 Kb	A 5' M 3'		7.2 Kb	X
5.6 Kb	X F 3'		6.0 Kb	A
4.0 - 5.0 Kb	A(6) ψ 5' M 3'		4.3 Kb	A
2.7 - 3.0 Kb	X ψ 3'		4.0 Kb	X M
0.8 Kb	A(19 or 6) M		2.0 Kb	X 5'
0.6 Kb	ψ M			
0.5 Kb	X ψ 5'			

Data compiled from unpublished results in our laboratory and
references 7, 14, 15, 16, 18, 19, 22, 30, 64 and 67. Molecular
weights are approximate and may differ among laboratories; a range
is given where these differences may lead to confusion. Symbols for
"Linkage & Function" are as follows: X = X linked, A = autosomal,
A(n) = mapped to autosome n, F = part of functional gene, ψ =
pseudogene, M = hybridized with middle Hind III fragment of cDNA
probe pGK825, 5' = hybridized with 5' fragment, 3' = hybridized with
3' fragment.

The human <u>Pgk</u> cDNA probe was first used to confirm the Xq13 assignment of the gene. These studies also revealed an X-linked pseudogene proximal to the functional gene. Figure 1 is a photograph of a Southern blot of <u>EcoR</u> I-cleaved DNA from human-rodent somatic cell hybrids varying in human X chromosome content, with the <u>Pgk</u> cDNA as probe. Only three human bands (4.2, 3.8 and 3.1 Kb; all X linked) are resolvable from rodent sequences in this blot. As expected, no <u>Pgk</u>-like sequences are detected in hybrids containing only the distal part of Xq (lanes J → M). The first detectable human <u>Pgk</u>-like sequence is the 3.1 Kb band in lane I. This hybrid expresses human PGK activity and, therefore, we can definitively state that the missing 3.8 Kb human band is not essential for PGK activity. This sequence must represent a pseudogene and map more proximal to the centromere than does the functional gene. These findings are in agreement with a recent report of Michelson et al. (67).

The <u>Pgk</u> cDNA probe also hybridized with 4.5 and 0.8 Kb <u>Hind</u> III fragments which were identified as autosomal by dosage studies (XY:4X comparisons). The 4.5 Kb band has been mapped to chromosome 6 by interspecific somatic cell hybrid studies (19, 67). Genetic studies have shown that the mouse autosomal <u>Pgk</u> gene is on chromosome 17. Since human chromosome 6 is homologous to mouse chromosome 17 (ref. 13), it was assumed that the 4.5 Kb <u>Hind</u> III fragment represented the functional human autosomal <u>Pgk</u> gene (19). However, our data strongly suggest that this fragment does not contain introns (the band hybridizes equally with the 3', 5' and middle

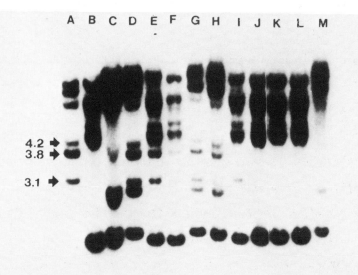

Fig. 1.   Hybridization of ^{32}P-labeled Pgk cDNA to DNA from various
          human, rodent, and hybrid cell lines with the hybrids
          carrying varying portions of the human X chromosome. Lane
          A, human; B, mouse; C, hamster; D, hamster-human, intact X;
          E, mouse-human, Xqter→p223; F, mouse-human, Xqter→p21; G,
          hamster-human, Xqter→p113; H, hamster-human, Xqter→p11; I,
          mouse-human, Xqter→q12; J, mouse-human, Xqter→q213; L,
          mouse-human, Xqter→q26, M, hamster-human, Xqter→q28.
          Arrows indicate distinguishable human bands. Lane I lacks
          the 3.8 Kb fragment, but expresses human PGK activity;
          therefore, the fragment, which is X linked, cannot be part
          of the X-linked functional Pgk gene (64). (Unpublished
          data of P. Chance, H. Ropers, and S. Gartler.)

regions of the Pgk cDNA). Furthermore, a recent study (18) has

reported the sequence of part of the 4.5 Kb fragment. It contains a

region with 85% homology to the X-linked Pgk cDNA (see below and

ref. 18). In view of known electrophoretic and immunological

differences between the X-linked and autosomal PGK proteins, this

sequence similarity would be unexpected if the clone did contain the

authentic autosomal gene. For these reasons, it seems likely that the 4.5 Kb sequence on chromosome 6 is not the functional autosomal gene.

Another autosomal sequence, a 0.8 Kb Hind III fragment, has been mapped by interspecific somatic cell hybrid studies to chromosome 6 by Michelson et al. (67) and in this laboratory to chromosome 19 (Gartler and T. Shows, in prep.). The 0.8 Kb Hind III fragment hybridizes with the middle fragment of the X-linked Pgk cDNA but not with the 5' or 3' segments. Michelson et al. (67) reported that the 4.5 Kb Hind III fragment did not hybridize with the 3' cDNA probe. These authors concluded that there are 2 autosomal Pgk genes -- a functional gene, represented by the < 7.6 Kb Hind III fragment, and a pseudogene truncated at its 3' end, presumably represented by the 4.8 and 0.8 Kb fragments. According to our data, however, the 3' cDNA does hybridize with the 4.5 Kb Hind III fragment, and the sequence of that fragment reported by Tani et al. (18) includes 3' sequences. This suggests to us that the 4.8 Kb and 0.8 Kb fragments are independent (non-overlapping). As our own somatic cell hybrid mapping data (Gartler & T. Shows, unpublished) favor a chromosome 19 locus for the 0.8 Kb Hind III fragment, we suggest that the 0.8 Kb Hind III fragment may be a part of the functional testis-specific Pgk gene, and that it maps to chromosome 19. However, a complete understanding of the situation will not be possible until all Pgk-related sequences have been cloned in their entirety.

STRUCTURE OF THE Pgk GENE AND RELATED SEQUENCES

The Pgk cDNA probe recognizes the functional, X-linked Pgk gene, an X-linked pseudogene, an autosomal pseudogene, and possibly the functional autosomal gene encoding testis-specific PGK (17-19,

67). In order to allow detailed analysis of the structure of the

functional, X-linked Pgk gene, it is necessary to identify DNA

probes which recognize only this region. Potentially important

regulatory regions may be 5' or 3' to the gene, and may not be

recognized by cDNA probes. Therefore, in order to study the struc-

ture and regulation of the Pgk-1 gene in detail, with special

reference to regions potentially important in the regulation of its

expression, it is necessary to isolate recombinant DNA clones con-

taining fragments of the Pgk gene itself.

GENOMIC CLONING AND SEQUENCE ANALYSIS

The large size of eukaryotic genes--from <6 Kb (20) to >150 Kb

(21)--presents special problems for cloning, as most vectors carry

less than 20 Kb of inserted DNA. The X-linked Pgk gene is at least

23 Kb in length (67). The problem can be solved by cloning large,

overlapping fragments of the genome until the region of interest has

been completely covered. Several clones have already been obtained

from different recombinant DNA libraries and analyzed in detail by

our group and by others (16, 18, 19, 67).

Tani et al. (18) prepared a recombinant DNA library in the

vector λEMBL3, containing inserted DNA from a partial Mbo I digest

of DNA from a human male. They obtained several clones homologous

to the human cDNA. Singer-Sam et al. (16) showed that a 25-mer

oligonucleotide probe, synthesized to correspond in sequence to the

extreme 5' end of the pHPGK-7e cDNA, revealed a 3.9 Kb BamH I fragment on hybridization with human genomic DNA. A 3.8 Kb BamH I

fragment was also present in clone λEMBL-3PGK-I identified by Tani,

and it hybridized with the 25-mer probe (16). Singer-Sam et al.

(16), therefore, sequenced the region of the genomic clone which was

homologous to the 5' end of the cDNA. They successfully identified

the first exon, approximately 440 bp of 5' flanking sequence, and

the first intron-exon splice junction (see also ref. 22).

The sequence of the promoter region of the Pgk gene revealed

several interesting features. (1) There were three transcription

start sites which appeared to be used with equal frequency in vivo.

(2) The 5' flanking region was extremely GC-rich (70%). (3) No

TATA or CAAT homologies, common to many promoter regions, were

found. (4) There was an 8 bp direct repeat, including a 6 bp

sequence also found in the SV40 21 bp repeats which are involved in

transcription initiation. Interestingly, the X-linked Hprt gene in

mouse (23) also lacks a TATA homology, is highly GC-rich in the 5'

flanking region, and contains a 10 bp direct repeat bearing homology

to the SV40 21 bp repeat. Further, there is a 30 bp region of the

mouse Hprt 5' flanking sequence which bears 80% homology to the

human Pgk sequence (16). However, these interesting features are

not common to all X-linked genes. The Factor VIII (21) and Factor

IX (24) genes both have apparent TATA homologies, and the promoter

regions are not GC-rich. Several features of the Pgk and Hprt

promoters are also found in two autosomal housekeeping genes, HMG

CoA Reductase (25) and Aprt (26). Most prominent among these

features is the presence of the sequence CCGCCC, which is repeated

six times in the SV40 promoter in a region that binds a transcription factor (27).

We have screened the Maniatis library (28) using the 5' portion of the Pgk cDNA as probe (unpublished). We identified clone λPGK-G3 which, on the basis of restriction mapping data, may contain the X-linked pseudogene of Michelson et al. (67).

J. Singer-Sam (City of Hope, Duarte, California) obtained a clone from the Lawn and Maniatis library (29), designated λPGK-Gl, which hybridizes with the 3' cDNA probe. We have analyzed this clone in some detail (see Fig. 2) because we believe it to contain a region of DNA which is differentially sensitive to nuclease digestion in active-X chromatin (see below and ref. 30). Sequence data suggest that the 5.8 Kb EcoR I fragment of λPGK-Gl contains an intron with a splice junction at position 836 of the pHPGK-7e cDNA sequence (15), and at least positions 836 through 965 of uninterrupted coding region (J. Singer-Sam, unpublished). We have inferred the 3'-5' orientation shown on the map, and believe that the clone extends about 9 Kb 3' to the coding sequence. This conclusion is consistent with the data of Michelson et al. (67) concerning the intron organization of the gene.

One clone from the λEMBL3 library appears to represent an autosomal pseudogene. This clone, designated λEMBL-3PGK-M, apparently overlaps one previously described by Szabo et al. (19) and mapped to the HLA region of chromosome 6. Tani et al. (18) have shown that a particular region of the clone contains no introns, bears roughly 85% homology to the cDNA, contains a poly(A) tract and polyadenylation signal, and is flanked by direct repeats. These features are hallmarks of pseudogenes dispersed by a retrovirus-like mechanism.

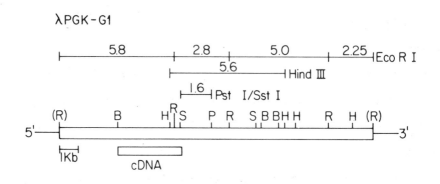

Fig. 2. Partial restriction map of λPGK-G1, a genomic clone
containing the 3' end of the human Pgk gene. Top line
indicates the size in Kb of EcoR I fragments; second line
indicates the 5.6 Kb Hind III fragment containing a DNase
I-sensitive site (30); third line indicates a 1.6 Kb Pst
I/Sst I fragment used as probe for mapping the sensitive
site and in methylation studies. S = Sst I, P = Pst I, B =
BamH I, R = EcoR I, Bg = Bgl II. Rectangle marked cDNA
indicates maximum region of homology to pGK825 probe.

The sequence contains a termination codon at amino acid 306, pre-
cluding the production of a functional protein.

Michelson et al. (67) have described in detail a clone isolated
from a flow-sorted X chromosome library. The clone contains an X-
linked Pgk pseudogene, and accounts for the 16.0 Kb and 3.6 Kb EcoR
I fragments seen in genomic digests. The pseudogene lacks introns,
contains mutations, and is flanked by 16 bp direct repeats. Unlike
other pseudogenes, however, the X-linked Pgk pseudogene has no
poly(A) tract and contains homology to the gene extending both 3'
and 5' to the transcribed region.

DISPERSION OF THE Pgk GENE FAMILY

A Pgk cDNA probe recognizes several sequences in the human
genome in addition to the functional, X-linked Pgk gene (14-19, 30,
67). This suggests a family of genes having a common evolutionary
origin. Members of a gene family may include functional genes or
nonfunctional genes (pseudogenes). Duplicate genes may diverge in
function or regulation from the parent copies, e.g., the α- and β-
globin genes. Pgk-1 and Pgk-2 produce proteins having similar
catalytic properties, but differing substantially in amino acid
sequence. Moreover, the regulation of the two genes is quite
different; Pgk-1 is active in all cells except testis, while Pgk-2
is active only in testis (see ref. 5). As two Pgk loci differing
similarly in their tissue specificity are found in mouse as well as
man, it appears that duplication of this gene preceded in time the
divergence of the rodent and primate lineages. In both cases, the
constitutively-expressed Pgk locus is found on the X chromosome,
while the other is found on an autosome. It is tempting to speculate that duplication of this X-linked gene to an autosome represents an evolutionary adaptation to inactivation of the X chromosome
during spermatogenesis.

Two classes of pseudogenes are widely recognized. One
represents a simple case of duplication in which the gene structure
remains intact. The duplicated copy may be found closely linked
with the parent gene, or inserted onto another chromosome. The
accumulation of mutations in the coding and regulatory regions of

the pseudogene may result in its producing no transcript, a non-translatable transcript, or a nonfunctional protein. The other major class includes the "processed" pseudogenes in which a mature message produced by a gene has been reverse-transcribed, and the reverse transcript (or cDNA) has been inserted into the genome by a retrovirus-like mechanism. This type of pseudogene evolution generally leads to dispersion of the copy from the region of the parent gene. As far as is known, the site of integraton is random. Several typical characteristics of dispersed, processed pseudogenes have emerged. (1) Their location in the genome is random. (2) They lack introns. (3) Their homology to the parent gene is confined to the transcribed region of the gene. (4) They retain a poly(A) tail reminiscent of the poly(A) tail of the message. (5) They are flanked by direct repeats of the genomic sequence into which they were inserted.

Two human Pgk pseudogenes have been cloned and sequenced (18, 67). The X-linked pseudogene, which we designate Pgk-ψ1, maps close to the Pgk-1 gene, though its exact distance is not known. Pgk-ψ1 has most of the characteristics of a typical dispersed, processed, pseudogene (67), but its nonrandom location (near the parent gene) is a puzzle (64, 67). Pgk-ψ1 lacks introns, but homology to the parent gene extends several base pairs 3' and 5' to the transcribed region, and does not include a poly(A) tail (67). Szabo et al. (19) and Tani et al. (18) independently cloned an autosomal pseudogene which we here designate Pgk-ψ2. Several workers have concluded that this pseudogene maps to chromosome 6 (19, 67 and Gartler and T.

Shows, unpublished). Michelson et al. (67) also suggest that the functional autosomal gene for Pgk (Pgk-2) also lies on chromosome 6, suggesting again tight linkage between a pseudogene and the parent gene. These authors invoke a mechanism in which pairing between a gene and a reverse transcript is followed by a gene conversion event, correcting the original gene to the spliced form (see ref. 68, proposing this mechanism for the origin of a globin pseudogene). In evaluating these hypotheses, we must keep in mind that the exact distance between gene and pseudogene remains unknown, and the position of insertion may still be random. Secondly, our mapping data (Gartler and T. Shows, unpublished) conflict with that of Michelson et al. (67) in that we feel that the functional Pgk-2 gene could be on chromosome 19. Chromosome 6 might then contain two pseudogenes. The chromosome 6 pseudogene which has not yet been cloned may represent an unprocessed pseudogene which was transcribed and processed, giving rise to Pgk-Ψ2 by a conversion event. The existence of homology beyond the usual limits of transcription might relate to the existence of multiple transcription start sites, including promoters several hundred base pairs upstream, which have been reported in several housekeeping genes (see, e.g., ref. 16, 69).

APPLICATION OF Pgk MOLECULAR GENETICS TO X INACTIVATION

The aim of the present work is to compare the Pgk genes on the active and inactive X chromosomes with respect to chromatin structure, DNA modification, and nucleotide sequence. From these

comparisons, differences should be found which might explain the
molecular basis for the regulation of Pgk gene expression. Analysis
of sequences near the Pgk gene may lead us to the X inactivation
center (Xce), which is closely linked to the Pgk locus (11-13).

CHROMATIN CONFORMATION AND X INACTIVATION

Recently, a great deal of information has accumulated concerning the differences between active or potentially-active vs.
inactive chromatin (for review, see refs. 31-33). Most of these
studies have involved autosomal genes which are transcribed only in
certain tissues and at relatively high levels. However, recent
results from this laboratory (30) and that of Migeon (66) have
demonstrated differences in nuclease sensitivity between housekeeping genes (Pgk, Gd and Hprt) on the active and inactive X chromosomes. Among the many differences between active and inactive
chromatin is an enhanced sensitivity of active chromatin to nuclease
digestion (32). Differing sensitivity to nuclease treatment is
generally regarded as an indication of differing chromatin conformation.

At least two levels of nuclease sensitivity have been
described. Broad or intermediate nuclease sensitivity involves
large (~100 Kb) domains which are preferentially sensitive to DNase
I in comparison to surrounding chromatin or the corresponding
chromatin domains in cells in which the genes are not active. Such
domains of sensitivity have been described in detail for the chick

β-globin gene (34) and for the chick ovalbumin gene (35, 36). A
second level of DNase I sensitivity is hypersensitivity, in which
short segments of DNA (~100 bp) are preferentially digested by low
levels of nuclease (31). Hypersensitive sites may be 3' (37–39) or
5' (40–43) to genes, or within the genes themselves (44). Hypersensitive sites are often indicative of active transcription
(reviewed in ref. 33). Most hypersensitive sites include or are
very close to regions sensitive to other nucleases (45, 46).

Since the inactive X chromosome is heterochromatic, we antici-
pated that gross differences would be found between any inactive and
active X-linked gene pair analyzed for molecular features of
chromatin structure. Initial studies were carried out on male cell
lines to determine the degree of DNase I sensitivity of the active
Pgk gene (30). Nuclei were prepared and treated with various
concentrations of DNase I. DNA was then purified, cleaved with
restriction enzymes, electrophoresed, blotted, and then hybridized
with labeled Pgk probe. The gene appeared to be relatively insensi-
tive to DNase I digestion, but at a high DNase I concentration (near
that at which bulk chromatin is digested) a characteristic subband
was detected in preparations secondarily cleaved with Hind III. By
comparing the prominence of the subband in cell lines varying in X
chromosome composition (one to four X chromosomes and none to three
inactive X chromosomes), we determined that the subband was derived
from the active X chromosome. Recent work in our laboratory (Riley
et al., unpublished) has revealed a site at the 5' end of the Pgk
gene which appears to be hypersensitive to DNase I. Similar results

have been obtained in our laboratory using probes for the X-linked
Factor IX and Hprt genes (Riley et al., unpublished), and by Wolf
and Migeon (66) studying the Hprt and Gd genes. We are presently
mapping and studying in greater detail the DNase I-sensitive sites
we have identified in the Pgk gene, and we expect to identify
additional sites using a genomic clone as a probe for the 5'
flanking regions of the gene (see ref. 22).

In view of the heterochromatic nature of the inactive X, we
were surprised that the molecular differences revealed by our
nuclease sensitivity studies (30) were so subtle. Because the cDNA
probe for the Pgk gene reveals both autosomal and X-linked pseudogenes as well as the functional, X-linked and autosomal genes (see
ref. 67), we were able to compare the kinetics of digestion of the
X-linked pseudogene on the active X chromosome with that of the
corresponding sequence on the inactive X chromosome. As the X-
linked pseudogene appears to be a processed pseudogene which lacks a
promoter (67), we may presume that it is transcriptionally inactive.
The comparison is, therefore, between inactive regions on the
"active" and inactive X chromosomes. We can detect no difference in
the kinetics of digestion of these two chromatin regions using DNase
I. Therefore, the non-transcribed pseudogene on both the active and
the highly condensed inactive X chromosome are in the same chromatin
conformation, at least as assayed by DNase I sensitivity studies.
These results, however, do not rule out differences in chromatin
conformation which are not detected with this technique.

Our studies also seemed to contradict nick translation studies
on fixed metaphase preparations (49), which suggested global

differences in DNase I sensitivity between active and inactive X
chromosomes. In the nick translation studies, fixed mammalian cells
in metaphase are incubated in a nick translation system using
labeled nucleotides; after autoradiography, one of the two X
chromosomes in a female cell was not labeled, while the autosomes
and the other X chromosome were labeled. The interpretation of the
authors is that the inactive X chromosome is resistant to DNase I
attack and, therefore, does not become labeled with the same efficiency as the active, nuclease-sensitive X chromosome.

Since detection of nuclease-sensitive sites requires double
strand breaks and nick translation requires only single strand
nicking, it was possible that the difference between our DNase I
results for the _Pgk_ gene and the results of the nick translation
studies was due to the resistance of the inactive X chromosome to
single strand nicking. We, therefore, carried out a combined DNase
I and S1 digestion followed by restriction enzyme treatment,
blotting, and hybridization with the labeled _Pgk_ probe (Riley et
al., unpublished). S1 treatment cleaves single stranded segments of
DNA and converts single strand nicks and gaps to double strand
breaks which are detectable by the usual methods used to identify
nuclease sensitive sites. Without S1 digestion, significant
degradation of bands hybridizing with the probe is not evident until
incubation with DNase I levels of 4.5 μg/ml. However, if DNase I
treatment is followed by S1 digestion, degradation of DNA is
observed at 0.45 μg/ml DNase I. Thus, as expected from the known
preference of DNase I for making single strand nicks in pure DNA

(48, 49) and single stranded gaps in chromatin (50), there are many

single strand nicks and gaps produced in chromatin by DNase I treat-

ment that go undetected when one looks only for sensitive and hyper-

sensitive sites. These combined DNase I and S1 digestion studies

show that the undetected single strand nicks were produced equally

in both the active and inactive Pgk genes. Apparently, the reason

for the resistance to nick translation of inactive X chromatin is

the inaccessibility of this chromatin to polymerase, a much larger

molecule than DNase I. Further evidence for equal susceptibility of

inactive X chromatin and active chromatin to general DNase I attack

is provided by experiments in which fixed interphase cells are

incubated with DNase I and the sex chromatin and dispersed chromatin

are monitored. Results of these studies indicate that both sex

chromatin and dispersed chromatin are digested with similar kinetics

(51).

GENOMIC CLONING OF THE 3' DNase I-SENSITIVE SITE

In order to sequence and to map precisely the nuclease-

sensitive site found at the 3' end of the Pgk gene (30), it was

necessary to clone this region of the genome. The insert of λPGK-G1

(J. Singer-Sam, unpublished; see Fig. 2) contained a 5.6 Kb Hind III

fragment with homology to the cDNA. The 5.6 Kb Hind III fragment is

believed to contain the nuclease-sensitive region. We identified a

1.6 Kb fragment within the 5.6 Kb Hind III fragment which appeared

to be free of repetitive sequences and to contain no homology to the cDNA. When used as a probe, the fragment hybridizes to a 5.6 Kb region of genomic DNA, and to a subband produced by DNase I digestion of active-X chromatin.

Preliminary restriction analysis of the genomic region containing the nuclease-sensitive site suggests that this site is 3' to the transcribed region of the Pgk gene, rather than within an intron or exon. More detailed mapping of this nuclease-sensitive region is in progress.

DNA MODIFICATION STUDIES

The role of DNA methylation in the regulation of gene expression is unclear. A wealth of evidence has been presented that DNA hypomethylation correlates with transcriptional activity (52), but there are several reports to the contrary (53-55). Circumstantial evidence points to an involvement of DNA methylation in X inactivation (5). For example, inactive X-chromosome DNA functions poorly in DNA-mediated gene transfer before but not after 5-azacytidine inhibition of methylation (6, 56). Yen et al. (57) have studied the human Hprt gene and have pointed out a general trend toward hypomethylation of the active gene, though the pattern is by no means simple (see also ref. 58). Yen et al. (57) even reported a site near the third exon which was hypomethylated on the inactive X. Wolf et al. (58) reported hypomethylation on the inactive X of several Hha I sites in the first intron of the human Hprt gene. Methylation may, in fact, play a pivotal role in "marking" the X chromosome for inactivation by binding of factors to, or release from, specific regulatory sites. Toniolo et al. (61) have studied the 3' end of the human gene for glucose-6-phosphate dehydrogenase (Gd). They found that most of the CpG dinucleotides were methylated in male DNA. Of five sites which were uniformly unmethylated in male DNA, two were 50% methylated in female DNA. Study of other X-linked genes with respect to methylation is of obvious interest.

Riggs and colleagues (22) have found a remarkably clear difference in methylation between the Pgk promoter regions of active and inactive X chromosomes. The 812 bp BamH I/EcoR I fragment which includes the Pgk transcription start site and about 400 bp of 5' flanking sequence (see refs. 16, 22) contains eight Hpa II sites. All eight appear to be methylated on the inactive X chromosome, while none appear to be methylated on the active X (22). This is the clearest correlation between X inactivation and DNA methylation reported to date.

Because detection of the nuclease-sensitive regions allows us to focus on regions of potential regulatory significance, we feel that we are likely to find significant methylation differences around the nuclease-sensitive sites on the 3' and 5' ends of the gene. Tissue-specific methylation patterns have indeed been reported in 5' nuclease-sensitive sites in several genes (31, 59, 60) including X-linked genes (22, 66; see below).

Our preliminary experiments to detect methylation differences between the active and inactive X chromosomes using the Pgk cDNA revealed no consistent differences. However, when we used a single-copy probe from the 3' flanking region of the gene (Fig. 2) we found

several bands present in a Bcl I/Hpa II digest of female DNA which

were not present in male DNA (Fig. 3). Although further experiments

are needed to determine the nature of these differences, the

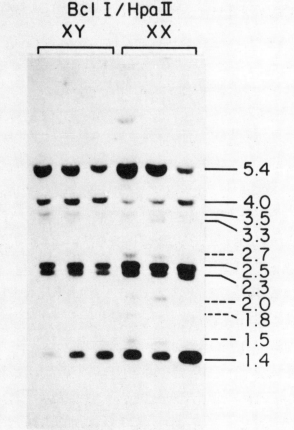

Fig. 3.   Methylation differences between active and inactive X
          chromosomes.   DNA from three females and three males was
          digested with Bcl I and Hpa II, fractionated on an 0.8%
          agarose gel, and blotted to nitrocellulose.   The blot was
          probed with a 1.6 Kb Pst I/Sst I fragment from clone λPGK-
          G1.   Female-DNA-specific bands are seen at 2.7, 2.0, 1.8,
          and 1.5 Kb (broken lines at right).

presence of these additional bands suggest hypomethylation of the

inactive X chromosome in the 3' region of the Pgk gene. Hypo-

methylation of regulatory sequences on an inactive gene is

unexpected. If confirmed, however, these observations might reflect

the interaction of X-chromosome DNA in these hypomethylated regions

with a repressor protein or with a protein involved in chromatin

condensation.

SENSITIVITY OF THE HUMAN Pgk GENE TO RESTRICTION ENDONUCLEASES

An enhanced sensitivity of chromatin to restriction endo-

nuclease digestion may reflect an active chromatin conformation.

Restriction endonuclease treatments which have been shown (or

suspected) to reflect active chromatin conformation include Msp I

(45; see also ref. 66), EcoR I, Ava II, Hae III, Taq I (62), BstN I

(44), Xba I (63) and BamH I (M. Groudine, pers. comm).

It is possible that regions which are not DNase I sensitive are

sensitive to restriction endonuclease digestion. For example,

Parslow and Granner (44) found that the immunoglobulin kappa gene

contained a DNase I hypersensitive region in lymphoid cells which

were actively transcribing immunoglobulin, while this site was

absent in non-transcribing cells. However, sensitivity to BstN I

was found in both transcribing and non-transcribing lymphoid cells,

but not in non-lymphoid tissues. These and similar data on the

chick β-globin locus (45) point to some degree of independence

between various kinds of nuclease sensitivity, and suggest that some

forms of nuclease hypersensitivity may reflect a potential for gene transcription.

We have preliminary data suggesting an Msp I-sensitive site
present on inactive-X chromatin. Nuclei were prepared from cell
lines having one or four X chromosomes and treated with increasing
concentrations of Msp I. DNA was purified, secondarily cleaved with
EcoR I, and blotted to nitrocellulose. Probing with the pGK825 cDNA
revealed three subbands (at 2.7, 5.2 and 5.4 Kb) resulting from Msp
I treatment of nuclei from the 4X cell line, but not from those of
the single X cell line (Fig. 4). Though we stress that these

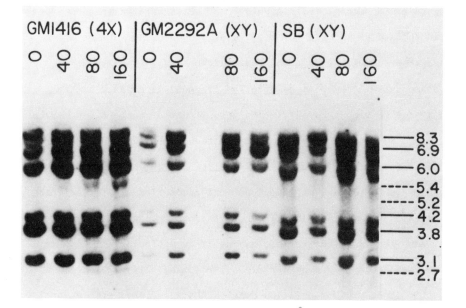

Fig. 4. Msp I sensitivity of inactive X chromosome DNA. Nuclei
were prepared from three lymphoblastoid cell lines. GM1416
contains 4 X chromosomes, GM2292A is from a 46,XY male
with Lesch-Nyhan syndrome, and SB is from a 46,XY male
(ref. 65). Nuclei were treated for 30 min at 37°C with
varying amounts of Msp I in 300 µl RSB containing approximately 120 µg DNA and 1 mM dithiothreitol. 0, 40, 80, or
160 units of enzyme (BRL) were added to each aliquot.
These are equivalent to 0, 0.33, 0.67 and 1.33 units per
µg. Subbands are visible at 2.7, 5.2 and 5.4 Kb (broken
lines at right) in 4X DNA treated with enzyme.

results are preliminary, they are exciting in that the existence of
a nuclease-sensitive site on the <u>inactive</u> X chromosome could
indicate binding of a protein which might be involved in repressing
gene expression in X inactivation.

DNA POLYMORPHISM AT THE <u>Pgk</u> LOCUS

One of the most important contributions of recombinant DNA
technology to medicine has been the discovery that apparently
neutral DNA sequence polymorphisms may occur at high frequency, and
can be used as markers for genetic disorders. One form of DNA
sequence polymorphism is the restriction fragment length
polymorphism (RFLP). A restriction fragment is defined by the
distance between two restriction sites. One of these sites may be
abolished, or another site may be introduced between the two sites,
by mutation. This change in a restriction site would result in a
change in the molecular phenotype--the production of a restriction
fragment of a new length. If the mutational change causes no loss
of gene function, the variant may drift to appreciable frequency;
the variant is then considered a polymorphism. In combination with
family studies and population data, RFLPs associated with a particular locus may be useful in prenatal detection of mutant alleles,
in detecting heterozygous carriers for a particular mutant allele,
or in detecting linkage with other loci. It is not surprising,
therefore, that many human DNA probes have been screened for the
existence of restriction fragment length variants.

Hutz et al. (17) have described a RFLP apparently associated with the human X-linked Pgk locus. DNA samples from 163 individuals were digested with Pst I, fractionated on agarose gels, and blotted to nitrocellulose. Probing with the Pgk cDNA revealed a 4.0 and a 2.0 Kb fragment present in 60% of all X chromosomes, but a 6.0 Kb fragment in about 40% of X chromosomes, suggesting a high frequency of absence of the internal Pst I site in the populations studied. The Pst I RFLP should prove very useful as a genetic marker for the Xq13 chromosome region in both clinical and basic research applications.

SUMMARY, CONCLUSION AND OUTLOOK

The X-linked human Pgk gene has been cloned and partially
characterized, and some preliminary results have been obtained
regarding active vs. inactive gene comparisons of chromatin structure and methylation patterns. As yet we can say nothing definitive
about what role, if any, these differences may play in X inactivation. The studies showing that DNA from the inactive X chromosome
in mature somatic cells does not function in transformation of the
Hprt gene strongly imply modification of the inactive X chromosome
at the DNA level. However, methylation studies with the Hprt, Gd
Pgk genes have revealed a complexity of methylation patterns
including hypermethylation of parts of the active X gene. Resolution of just what difference is critical in expression, differen-
tiating between cause and effect, and extrapolating to the spreading
and initiation aspects of X inactivation are still, unfortunately,
long-range goals. The Pgk system may be of special value in
unraveling some of these difficult questions. A unique autosomal
Pgk locus exists and should allow an informative comparison between
an X-linked housekeeping gene and an autosomal, tissue-specific gene
encoding proteins of identical enzymatic function. The proximity of
Pgk to the X-inactivation control center may be useful in
identifying the starting point of this very important event in early
mammalian development.

ACKNOWLEDGEMENTS

This study was supported by NIH grants HD16659 and GM15253 to S.M.G. and by Institutional Cancer Grant IN-26Y from the American Cancer Society to M.A.G. We are grateful to A. Riggs and J. Singer-Sam for sharing unpublished data, for supplying DNA probes, and for their comments on the manuscript; K. Dyer, N. Ellis and J. Graves for comments on the manuscript; and P. Green for typing.

REFERENCES

1. Muller, H. J., League, B. B., and Offermann, C. A. 1931.
    Effects of dosage changes of sex-linked genes, and the compen-
    satory effects of the gene differences between male and female.
    Anat. Rec. (Abstr.) 51, 110.

2. Sandler, L., and Hecht, F. 1973. Genetic effects of
    aneuploidy. Am. J. Hum. Genet. 25, 332–339.

3. Gorlin, R. J. 1977. Classical chromosome disorders. In New
    Chromosomal Syndromes, ed. J. J. Yunis, pp. 59–117. New York:
    Academic Press.

4. Lewandowski, R. C., Jr., and Yunis, J. J. 1977. Phenotypic
    mapping in man. In New Chromosomal Syndromes, ed. J. J. Yunis,
    pp. 369–394. New York: Academic Press.

5. Gartler, S. M., and Riggs, A. D. 1983. Mammalian X-chromosome
    inactivation. Ann. Rev. Genet. 17, 155–190.

6. Mohandas, T., Sparkes, R. S., and Shapiro, L. J. 1981.
    Reactivation of an inactive human X chromosome: evidence for X
    inactivation by DNA methylation. Science 211, 393–396.

7. Graves, J. A. M. 1982. 5-azacytidine-induced re-expression of
    alleles on the inactive X chromosome in a hybrid mouse cell
    line. Exp. Cell Res. 141, 95–105.

8. Lester, S. C., Korn, N. J., and DeMars, R. 1982. Derepression
    of genes on the human X chromosome: evidence for differences
    in locus-specific rates of derepression and rates of transfer
    of active and inactive genes after DNA-mediated transformation.
    Somatic Cell Genet. 8, 265–284.

9. Mattei, M. G., Mattei, J. F., Ayme, S., and Giraud, F. 1982.
    X-autosome translocations: cytogenetic characteristics and
    their consequences. Hum. Genet. 61, 295–309.

10. Shapiro, L. J., Mohandas, T., Weiss, R., and Romeo, G. 1979.
    Noninactivation of an X chromosome locus in man. Science 204,
    1224–1226.

11. Cattanach, B. M., and Papworth, D. 1981. Controlling elements
    in the mouse. V. Linkage tests with X-linked genes. Genet.
    Res. 38, 57–70.

12. Cattanach, B. M., Bucher, T., and Andrews, S. J. 1982.
    Location of Xce in the mouse X chromosome and effects of PGK-1
    expression. Genet. Res. 40, 103–104.

13. Buckle, V. J., et al. 1984. Chromosome maps of man and mouse
    II. Clinical Genet. 26, 1-11.

14. Singer-Sam, J., Simmer, R. L., Keith, D. H., Shively, L.,
    Teplitz, M., Itakura, K., Gartler, S. M., and Riggs, A. D.
    1983. Isolation of a cDNA clone for human X-linked
    3-phosphoglycerate kinase by use of a mixture of synthetic
    oligodeoxyribonucleotides as a detection probe. Proc. Natl.
    Acad. Sci. USA 80, 802-806.

15. Michelson, A. M., Markman, A. F., and Orkin, S. H. 1983.
    Isolation and DNA sequence of a full-length cDNA clone for
    human X chromosome-encoded phosphoglycerate kinase. Proc.
    Natl. Acad. Sci. USA 80, 472-476.

16. Singer-Sam J., Keith, D. H., Tani, K., Simmer, R. L., Shively,
    L., Lindsay, S., Yoshida, A., and Riggs, A. D. 1984. Sequence
    of the promoter region of the gene for human X-linked 3-
    phosphoglycerate kinase. Gene 32, 409-417.

17. Hutz, M. H., Michelson, A. M., Antonarakis, S. E., Orkin, S.
    H., and Kazazian, H. H., Jr. 1984. Restriction site
    polymorphism in the phosphoglycerate kinase gene on the X
    chromosome. Hum. Genet. 66, 217-219.

18. Tani, K., Singer-Sam, J., Munns, M., and Yoshida, A. 1985.
    Molecular cloning and structure of an autosomal processed gene
    for human phosphoglycerate kinase. Gene, in press.

19. Szabo, P., Grzeschik, K., and Siniscalco, M. 1984. A human
    autosomal phosphoglycerate kinase locus maps near the HLA
    cluster. Proc. Natl. Acad. Sci. USA 81, 3167-3169.

20. Stambrook, P. J., Dush, M. K., Trill, J. J., and Tischfield, J.
    A. 1984. Cloning of a functional human adenine phosphori-
    bosyltransferase (APRT) gene: identification of a restriction
    fragment length polymorphism and preliminary analysis of DNAs
    from APRT-deficient families and cell mutants. Somatic Cell
    Molec. Genet. 10, 359-367.

21. Gitschier, J., et al. 1984. Characterization of the human
    factor VIII gene. Nature 312, 326-330.

22. Riggs, A. D., Singer-Sam, J., and Keith, D. H. 1985.
    Methylation of the Pgk promoter region and an enhancer way-
    station model for X-chromosome inactivation. In The
    Chemistry, Biochemistry and Biology of DNA Methylation, eds. A.
    Razin and G. L. Cantoni. New York, A. R. Liss, in press.

23. Melton, D. W., Konecki, D. S., Brennand, J., and Caskey, C. T.
    1984. Structure, expression and mutation of the hypoxanthine
    phosphoribosyltransferase gene. Proc. Natl. Acad. Sci. USA 81,
    2147-2151.

24. Anson, D. S., et al. 1984. HMG CoA Reductase: a negatively regulated gene with unusual promoter and 5' untranslated regions. Cell 38, 275-285.

25. Reynolds, G. A., et al. 1984. HMG CoA Reductase: a negatively regulated gene with unusual promoter and 5' untranslated regions. Cell 38, 275-285.

26. Dush, M. K., Sikela, J. M., Khan, S. A., Tischfield, J. A., and Stambrook, P. J. 1985. Nucleotide sequence and organization of the mouse adenine phosphoribosyltransferase gene: presence of a coding region common to animal and bacterial phosphoribosyltransferases that has a variable intron/exon arrangement. Proc. Natl. Acad. Sci. USA 82, 2731-2735.

27. Dynan, W. S., and Tjian, R. 1983. The promoter-specific transcription factor Sp1 binds to upstream sequences in the SV40 early promoter. Cell 35, 79-87.

28. Maniatis, T., et al. 1978. The isolation of structural genes from libraries of eucaryotic DNA. Cell 15, 687-701.

29. Lawn, R. M., Fritsch, E. F., Parker, R. C., Blake, G., and Maniatis, T. 1978. The isolation and characterization of linked δ-and β-globin genes from a cloned library of human DNA. Cell 15, 1157-1174.

30. Riley, D. E., Canfield, T. K., and Gartler, S. M. 1983. Chromatin structure of active and inactive human X chromosomes. Nucl. Acids Res. 12, 1829-1845.

31. Elgin, S. C. R. 1981. DNase I-hypersensitive sites of chromatin. Cell 27, 413-415.

32. Weisbrod, S. 1982. Active chromatin. Nature 208, 289-295.

33. Lowenhaupt, K., Keene, M. A., Cartwright, I. L., and Elgin, S. C. R. 1982. Chromatin structure of eukaryotic genes: DNase I hypersensitive sites. Stadler Symp. 14, 69-85.

34. Stalder, J., Larsen, A., Engel, J. D., Dolan, M., Groudine, M., and Weintraub, H. 1980. Tissue-specific DNA cleavages in the globin chromatin domain introduced by DNase I. Cell 20, 451-460.

35. Lawson, G. M., Knoll, B. J., March, C. J., Woo, S. L. C., Tsai, M., and O'Malley, B. W. 1982. Definition of 5' and 3' structural boundaries of the chromatin domain containing the ovalbumin multigene family. J. Biol. Chem. 257, 1501-1507.

36. Stumph, W. E., Baez, M., Lawson, G. M., Tsai, M., and O'Malley,
     B. W. 1983. Chromatin structure of the ovalbumin gene domain.
     In UCLA Symposia on Molecular and Cellular Biology, vol. 26,
     "Gene Regulation," ed. B. W. O'Malley. New York: Academic
     Press.

37. Kuo, M. T., Mandel, J. L., and Chambon, P. 1979. DNA
     methylation: correlation with DNase I sensitivity of chicken
     ovalbumin and conalbumin chromatin. Nucl. Acids Res. 7,
     2105-2113.

38. Weischet, W. O., Glotov, B. O., Schnell, H., and Zachau, H. G.
     1982. Differences in the nuclease sensitivity between the two
     alleles of the immunoglobulin kappa light chain genes in mouse
     liver and myeloma nuclei. Nucl. Acids Res. 10, 3627-3645.

39. Fritton, H. P., Sippel, A. E., and Igo-Kemenes, T. 1983.
     Nuclease hypersensitive sites in the chromatin domain of the
     chicken lysozyme gene. Nucl. Acids Res. 11, 3467-3485.

40. Wu, C. 1980. The 5' ends of Drosophila heat shock genes in
     chromatin are hypersensitive to DNase I. Nature 286, 854-860.

41. Groudine, M., Kohwi-Shigematsu, T., Gelinas, R.,
     Stamatoyannopoulos, G., and Papayannopoulou, T. 1983. Human
     fetal to adult hemoglobin switching: changes in chromatin
     structure of the β-globin gene locus. Proc. Natl. Acad. Sci.
     USA 80, 7551-7555.

42. Kaye, J. S., Bellard, M., Dretzen, G., Bellard, F., and
     Chambon, P. 1984. A close association between sites of DNase
     I hypersensitivity and sites of enhanced cleavage by micro-
     coccal nuclease in the 5'-flanking region of the actively
     transcribed ovalbumin gene. EMBO J. 3, 1137-1144.

43. Tuan, D., and London, I. 1984. Mapping of DNase I -
     hypersensitive sites in the upstream DNA of human embryonic
     ε-globin gene in K562 leukemia cells. Proc. Natl. Acad.
     Sci. USA 81, 2718-2722.

44. Parslow, T. G., and Granner, D. K. 1983. Structure of a
     nuclease-sensitive region inside the immunoglobin kappa gene:
     evidence for a role in gene regulation. Nucl. Acids Res.
     11, 4775-4792.

45. McGhee, J. D., Wood, W. I., Dolan, M., Engel, J. D., and
     Felsenfeld, G. 1981. A 200 base pair region at the 5' end of
     the chicken adult β-globin gene is accessible to nuclease
     digestion. Cell 27, 45-55.

46. Larsen, A., and Weintraub, H. 1982. An altered DNA conformation detected by S1 nuclease occurs at specific regions in active chick globin chromatin. Cell 29, 609–622.

47. Kerem, B. S., Goitein, R., Richler, C., Marcus, M., and Cedar, H. 1983. In situ nick-translation distinguishes between active and inactive X chromosomes. Nature 304, 88–90.

48. Thomas, C. A. 1956. The enzymatic degradation of desoxyribose nucleic acid. J. Am. Chem. Soc. 78, 1861–1868.

49. Young, E. T., and Sinsheimer, R. L. 1965. A comparison of the initial actions of spleen deoxyribonuclease and pancreatic deoxyribonuclease. J. Biol. Chem. 240, 1274–1280.

50. Riley, D. E. 1980. Deoxyribonuclease I generates single-stranded gaps in chromatin deoxyribonucleic acid. Biochemistry 19, 2977–2992.

51. Dyer, K. A., Riley, D. E., and Gartler, S. M. 1985. Analysis of inactive X chromosome structure by in situ nick translation. Chromosoma 92, 209–213.

52. Riggs, A. D., and Jones, P. A. 1983. 5–methylcytosine, gene regulation, and cancer. Adv. Cancer Res. 40, 1–30.

53. Gerber-Huber, S., et al. 1983. In contrast to other Xenopus genes the estrogen-inducible vitellogenin genes are expressed when totally methylated. Cell 3, 43–51.

54. Vedel, M., Gomez-Garcia, M., Sala, M., and Sala-Trepat, J. M. 1983. Changes in methylation pattern of albumin and α-fetoprotein genes in developing rat liver and neoplasia. Nucl. Acids Res. 11, 4335–4354.

55. Ott, M., Sperling, L., and Weiss, M. C. 1984. Albumin extinction without methylation of its gene. Proc. Natl. Acad. Sci. USA 81, 1738–1741.

56. Venolia, L., Gartler, S. M., Wassman, E. R., Yen, P., Mohandas, T., and Shapiro, L. J. 1982. Transformation with DNA from 5–azacytidine-reactivated X chromosomes. Proc. Natl. Acad. Sci. USA 78, 2352–2354.

57. Yen, P. H., Patel, P. I., Chinault, A. C., Mohandas, T., and Shapiro, L. J. 1984. Differential methylation of hypoxanthine phosphoribosyltransferase genes on active and inactive human X chromosomes. Proc. Natl. Acad. Sci. USA 81, 1759–1763.

58. Wolf, S. F., Jolly, D. J., Lunnen, K. D., Friedmann, T., and
     Migeon, B. R. 1983. Methylation of the hypoxanthine phos-
     phoribosyltransferase locus on the human X chromosome:
     implications for X-chromosome inactivation. Proc. Natl. Acad.
     Sci. USA 81, 2806–2810.

59. Weintraub, H., Larsen, A., and Groudine, M. 1981. α-Globin-
     gene switching during the development of chicken embryos:
     expression and chromosome structure. Cell 24, 333–344.

60. Ginder, G. D., and McGhee, J. D. 1981. DNA methylation in the
     chicken adult β-globin genes: a relationship with gene
     expression. In Organization and Expression of Globin Genes,
     eds. G. Stamatoyannopoulos & A. W. Nienhuis, pp. 191–201. New
     York: A. R. Liss.

61. Toniolo, D., et al. 1984. Specific methylation pattern at the
     3' end of the human housekeeping gene for glucose-6-phosphate
     dehydrogenase. EMBO J. 3, 1987–1995.

62. Sweet, R. W., Chao, M. V., and Axel, R. 1982. The structure
     of the thymidine kinase gene promoter: nuclease hypersen-
     sitivity correlates with expression. Cell 31, 347–353.

63. Widmer, R. M. 1984. Chromatin structure of a hyperactive sec-
     retory protein gene (in Balbiani ring 2) of Chironomus. EMBO
     J. 3, 1635–1641.

64. Chance, P. F., Kurachi, K., Ropers, H., Weiacker, P., and
     Gartler, S. M. 1983. Regional assignment of the human Factor
     IX and 3-phosphoglycerate kinase by molecular hybridization.
     Am. J. Hum. Genet., 35, 187A.

65. Royston, I., Smith, R. W., Buell, D. N., Huang, E., and Pagano,
     J. S. 1974. Autologous human B and T lymphoblastoid cell
     lines. Nature 251, 745–746.

66. Wolf, S. F., and Migeon, B. R. 1985. Clusters of CpG
     dinucleotides implicated by nuclease hypersensitivity as
     control elements of housekeeping genes. Nature 314, 467–469.

67. Michelson, A. M., Bruns, G. A., Morton, C. C., and Orkin, S. H.
     1985. The human phosphoglycerate kinase multigene family. J.
     Biol. Chem. 260, 6982–6992.

68. Nishioka, Y., Leder, A., and Leder, P. 1980. Unusual α-
     globin-like gene that has cleanly lost both globin intervening
     sequences. Proc. Natl. Acad. Sci. USA 77, 2806–2809.

69. McGrogan, M., Simonsen, C. C., Smouse, D. T., Farnham, P. J.,
     and Schimke, R. T. 1985. Heterogeneity at the 5' termini of
     mouse dihydrofolate reductase mRNAs. J. Biol. Chem. 260, 2307–
     2314.

Human Genes and Diseases
Edited by F. Blasi
© 1986, John Wiley & Sons, Ltd.

METALLOTHIONEIN GENE REGULATION IN MENKES' DISEASE

Arturo Leone

INTRODUCTION

The history of the discovery of metallothioneins is
intimately connected with the studies on the distribution and
toxicity of heavy metals. The first report of natural accumulation
of cadmium in human tissues was published in 1941 by the Russian
biochemist, D.P. Malyuga, who determined by colometrical absorption
the content of this metal in human kidney, spleen and lung
(Malyuga, 1941). Subsequently, in 1957 and 1960 two different
reports from B.L. Vallee's laboratory showed that a protein isolated from equine kidney cortex was capable of binding cadmium and
other transition elements (Kagi and Vallee, 1960; Margoshes and
Vallee, 1957). The protein was given the name metallothionein, and
in the following years reports from several laboratories demonstrated the ubiquitous distribution of this protein in higher and
lower eukaryotes, its response to induction by heavy metals and
glucocorticoids and its highly conserved amino acid and nucleotide
sequence. At the First International Meeting on Metallothionein
and Other Low Molecular Weight Metal Binding Proteins in 1976,
metallothionein characteristics were outlined as: a) small molecu-

lar weight (6,000-7,000 daltons); b) high metal content; c) distinctive amino acid composition (high cysteine content, no aromatic
amino acids; d) unique distribution of cysteinyl residues in the
amino acid sequence; and e) optical features typical of metal
thiolates (Nordberg and Kojima, 1979).

In the following sections we will examine first the structure and function of metallothionein and then its abnormal regulation in a human genetic disorder, Menkes' disease. This X-linked
trait is characterized by a very high level of copper in certain
tissues. The metal is bound to metallothionein and is not available as a cofactor to copper enzymes, giving the typical clinical
picture of a general copper deficiency.

We have found that the metallothionein gene is induced in
Menkes' cells by low levels of copper via a trans-acting mechanism
and that other events appear at the molecular level in these cells
in response to copper poisoning. These data suggest that the basic
defect of this disease could be in one or more steps of the copper
transport pathway.

A. METALLOTHIONEIN Protein Chemistry

Metallothioneins are polymorphic proteins consisting of two
major classes, MTI and MTII (Kissling and Kagi, 1979; Kagi et al.,
1984). The number of isoforms can vary from two in rodents to four
in horse and rabbit (Kojima et al., 1979; Huang et al., 1979;
Klensen et al., 1983). In humans, the pattern is more complicated
since Kagi and colleagues have identified one MTII and four
different MTI isoforms in human liver by high liquid pressure
chromatography (Kagi et al., 1984). The amino acid sequence

analysis of the different mammalian isoforms shows some interesting
common features: a very high cysteine content, one third of the
total amino acid composition; absence of aromatic amino acids;
particular positioning of cysteine residues along the molecule; and
repetitive occurrence of a tripeptide sequence Cys-x-Cys, where x
stands for a residue other than cysteine. This tripeptide is
present seven times along the molecule and it is suggested to be
the primary chelation site for the seven group-2B metal ions which
usually bind to the protein (Winge and Miklossy, 1982). MTs are
cytoplasmatic proteins and 90% of the MT mRNA is associated with
free ribosomes. The difference between MTII and MTI isoforms are
single amino acid replacements which presumably occur in sites not
essential for metal binding (Kagi and Nordberg, 1979; Griffith et
al., 1983). The comparison of the amino acid sequence of metallothioneins isolated from mammals and from lower eukaryotic organisms
such as mollusk and fungi, gives a high degree of homology. The
average of identity in the primary sequence varies in the different
phyla; in mammals it ranges between 75% and 85%; 42% identity is
found between the mammalian and the arthropode forms and 32%
between Neurospora crassa metallothionein and the corresponding
N-terminal segment of the other sequenced MTs (Kagi et al., 1984;
Lerch, 1979). Several group Ib and IIb metals are known to bind
metallothioneins (Nielson et al., 1985). Cu, Zn, Cd and Hg MTs
have been described in different tissues in human and other
eukaryotes (Kagi et al., 1984). The ^{113}Cd-NMR studies suggested a
model in which the metal ions bind to two separate polynuclear
thiolate clusters (Otvos and Armitage, 1980). In Cd-MT the two

domains coordinate respectively 4 and 3 Cd (II) ions exclusively

through thiolate bonds (Winge and Miklossy, 1982; Boulanger et al.,

1982). As shown in Fig. 1, 11 cysteine residues form the carboxy-

terminal A cluster and the other 9 residues are disposed in tetra-

hedral geometry at the amino-terminal B domain. The formation of

one cluster occurs independently of influences from the other

domain: the saturation of the B domain with Cu (I) or Ag (I) does

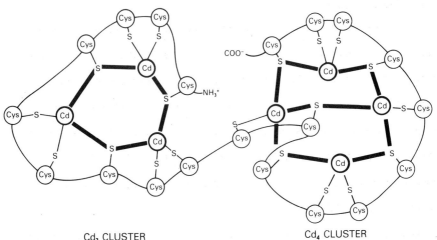

Cd₃ CLUSTER   Cd₄ CLUSTER

Fig. 1: Schematic representation of the two metal-cysteine

clusters in Cd-MT (Otvos and Armitage, 1980; Winge and

Miklossy, 1982).

not prevent the binding of either Cd (II) or Ag (I) to the other

with the usual stoichiometry (Nielson and Winge, personal communi-

cation). The strong interaction with the metals has been suggested

to protect the native protein from proteolysis since the metal free

apoprotein is degraded by peptidases (Webb, 1972; Winge and

Miklossy, 1982). The metal content of MT is highly variable and depends on species, organ and history of heavy metal exposure. For example, MT isolated from human liver autopsy samples contains almost exclusively zinc, whereas MT from kidney contains substantial levels of cadmium and copper. Therefore, the occurrence of Cd, Cu and Zn-MT could possibly reflect both the natural heavy metal exposure of the organs and the expression of different MT isoforms.

Inducibility of MT

Several agents are known to induce metallothionein synthesis in vivo. Administration of heavy metals (Cd, Zn or Cu), glucocorticoid hormones, interferon or various stresses, including heat and cold exposure, stress exercise, carbon tetrachloride injection, alkylating agents or bacterial endotoxin have been shown to induce metallothionein synthesis (Kagi and Nordberg, 1979; Karin and Herschman, 1979; Oh et al., 1978; Durnam and Palmiter, 1984; Durnam et al., 1984; Friedman and Stark, 1985). The regulation of MT gene expression has been more carefully studied in in vitro cell lines. Karin and collaborators have demonstrated a primary transcriptional induction of MT mRNA in response to Zn and dexamethasone in HeLa cells and in rat hepatocytes (Karin et al., 1980a; Karin et al., 1980b). The kinetics of both inducers appear the same for the first eight or nine hours, when synthesis reaches a maximum level, then Zn induction drops while dexamethasone continues its effect at the same level for several hours (Karin et al., 1981). The same transient induction appears with Cd in CHO cells (Enger et al., 1979). Heavy metals and glucocorticoids induce MT synthesis by independent mechanisms. The biphasic curve of induction by Zn is

not due to a toxic effect of the metal itself nor to its disappearance following the induction of MT. Dexamethasone, on the other
hand, does not act indirectly through Zn to induce MT, since MT mRNA
induction by dexamethasone occurs after removal of extracellular Zn
and in the presence of cycloheximide, which blocks dexamethasone-
induced Zn uptake (Karin et al., 1980b). Moreover, two different
regulatory regions responsive for the glucocorticoid and metal
induction have been characterized in the 5' promoter region of the
human HMTII$_A$ gene (see next Section). A third inducer, a bacterial
endotoxin or lipopolysaccharide, LPS, isolated from Salmonella
typhosa, also regulates MT synthesis (Durnam et al., 1984). In
mouse liver the response to LPS is strong, equivalent to Zn induction. The action of LPS is not mediated by metals and/or glucocorticoids and the region responsive for LPS induction has been
suggested to lie between nucleotides -85 and -350 from the mouse MTI
transcriptional initiation site. This finding could explain the
high level of MT produced after exposure to different stresses,
heat, cold, overexercise, etc. and suggests that elevated MT synthesis could be involved in a more general pathway of response to
inflammation. MT expression is also differentially regulated in
various tissues. In the mouse, Cd-induced MTI mRNA synthesis is
higher in the liver and in the kidney then in the heart, muscle,
brain and intestine. The testis is the only tissue not induced, but
shows a high basal level compared with the others (Durnam and
Palmiter, 1981). In addition, a differential transcriptional
response to the various metals has been observed in the kidney and
liver of the mouse: Cd and high doses of Zn are the best inducers in

both tissues, while Cu exhibits a high induction level in the liver, but not in the kidney (Durnam and Palmiter, 1981).

MT gene organization

The rapid increase of translatable mRNA in response to heavy metals and the relatively high representation of MT mRNA after induction, approximately 2% of the total RNA, has allowed the isolation of MT cDNA clones. Palmiter and colleagues first identified a mouse MTI cDNA clone which has been successively used by several laboratories to isolate by cross-hybridization MT clones from other species (Durnam et al., 1980). Karin and Richards screened a Cd-induced HeLa cell cDNA library and found a metal induced cDNA clone of human metallothionein II, $HMTII_A$ (Karin and Richards, 1982a).

In humans, the MTs are encoded by a complex multigene family with at least 12-14 members (Karin and Richards, 1982b; Richards et al., 1984; Varshray and Glutman, 1984; Schmidt et al., 1985). Up to now, five pseudogenes and four functional MT genes have been cloned and identified: two MTII class genomic clones, called respectively $HMTII_A$ and $\psi HMTII_B$, and seven MTI class genomic clones called $HMTI_A$, $\psi HMTI_C$, $\psi HMTI_D$, $HMTI_E$, $HMTI_F$, $\psi HMTI_G$ and $\psi HMTI_H$. These genes were isolated from λ genomic libraries using as probe human or monkey cDNA clones. To test their functional role, an efficient in vitro assay based on the properties of bovine papilloma virus (BPV) has been developed. BPV and BPV-derived recombinants are able to replicate as stable nuclear episomes in transformed cells without lysis of the cells, the average number of copies ranging between 50 and 100 per cell. The insertion of metal-

lothionein genes in a BPV derived vector and transformation of
rodent fibroblasts through DNA-calcium phosphate transfection
confers metal resistance to the cells (Karin et al., 1983; Schmidt
et al., 1985). The expression of the functional MT gene is regulated by the concentration of Cd in the growth medium, therefore
cells harboring multiple copies of MT-BPV plasmids became resistant
to Cd and are able to replicate at a concentration of 50 μM $CdCl_2$
while untransformed and BPV-transformed cells die at
10 μM $CdCl_2$. BPV plasmids containing the $HMTII_A$, $HMTI_A$, $HMTI_E$ and
$HMTI_F$ genes all confer metal resistance to transfected cells,
proving their functional role; the other genomic isolates do not
render transfected cells Cd-resistant and their nucleotide
sequences show mutations characteristic of pseudogenes.

The structure of the HMT genes has been determined through
DNA restriction analysis and nucleotide sequencing. The MT genes
are small, an average of I kilobase, and contain three exons and
two introns, obeying the GT-AG splicing rule. The splicing signals
are located at the same positions within the coding region of the
four functional HMT genes and in the mouse MTI gene and a typical
polyadenylation site is present in the 3' untranslated region. All
functional genes have a TATA or TATA-like box in the 5' flanking
region. These sequences are present in most of the known eukaryotic coding genes isolated and are assumed to represent a binding
region for RNA polymerase II or other common transcriptional
factors.

The comparison of the entire sequences of the $HMTII_A$, $HMTI_E$
and $HMTI_F$ genes and their 5' flanking regions shows 70% homology

among the MTI class genes but only 35% homology between MTI and

MTIIa genes. This result suggests that during evolution the diver-

gence between MTI and MTII isoforms arose before the amplification

of the multiple forms of MTI. The actual function of the different

MT genes in human has been widely discussed and different mech-

anisms have been proposed: first, different isoforms might bind

specific metals or have a specific role in their metabolism;

second, a functional rather than structural difference in their

control sequence could reflect the necessity for the cells to

respond to the various inducers, metals, glucocorticoids or

bacterial endotoxin; third, there might be random genetic drift,

without any real functional purpose, following the amplification

step. The availability of the different probes of the MT genes

will constitute an interesting tool for the understanding of this

mechanism. Recently, a tissue specific response to Cd has been

shown for the MHTI$_E$ and HMTI$_F$ genes in in vitro cell lines and in

human liver; in the same experiments HMTII$_A$ does not appear to be

tissue specific but rather has a general, ubiquitous response (C.J.

Schmidt, personal communication).

The characterized MT five pseudogenes have a variety of

mutations and rearrangements. ψHMTI$_C$ and ψHMTI$_D$ have in-phase

termination codons and/or single base pair deletions affecting the

reading frame (Richards et al., 1984). HMTI$_G$ has a stop codon

which terminates translation at amino acid 40, other amino acid

substitutions and no TATA or TATA-like box at the 5' region

(Schmidt et al., 1985). ψMT-II$_B$ and ψHMTI$_H$ are different from the

other pseudogenes in that they completely lack introns; both carry

several amino acid mutations, contain a polyA stretch at the 3' end
and are flanked by two slightly imperfect direct repeats which
might have been generated by insertion-site duplication (Karin and
Richards, 1982b; Varshney and Gedemn, 1984). The absence of a
promoter-like region at the 5' flanking region and the unaltered
coding capacity of the ψHMTII$_B$ and of the ψHMTI$_H$ sequences impli-
cates the generation of these genes through a cDNA intermediate. A
restriction site polymorphism linked to the HMTII$_B$ gene has also
been described (Karin and Richards, 1982b).

The chromosomal location of the functional human metallothionein gene family has been assigned to chromosome 16. Using
human-mouse cell hybrids and probes from MTI and MTII functional
genes, HMTII$_A$, HMTI$_A$, HMTI$_E$, HMTI$_F$, and ψHMTI$_G$ have been assigned
on chromosome 16. The other genes are dispersed: two are on the
long and short arms of chromosome 1; two others are localized on
chromosome 18 and 20 and ψHMTII$_B$ is located on chromosome 4
(Schmidt et al., 1984; Karin et al., 1984a).

The overall organization of the HMT gene family resemble
other multigene families such as the globins (Jahn et al., 1980;
Leder et al., 1980). The cluster of functional genes found on a
single chromosome, most probably in the proximal region of chromosome 16, 16cen/16q21, is of particular interest for the understanding of Menkes' disease, an X linked disorder of copper metabolism
(see next Section). The localization of the MT genes on chromosome
16 and other autosomes and the negative finding of MT-related
sequences linked to the X chromosome definitely rule against a mutation in the metallothionein gene itself and implicate a trans-
acting defect in MT gene regulation.

MT gene expression

The isolation of mouse and human MT-genomic clones, the dual response to metals and glucocorticoids and the "housekeeping" function of MT sparked interest in studies of MT transcriptional regulation in transfected cells.

The entire mouse MTI gene has been introduced in human or monkey cells and has been shown to retain its inducibility by Cd, either after integration into the genome (Mayo et al., 1982) or by acute transfection as an extrachromosomal element on an SV40 derived plasmid vector (Hamer and Walling, 1982). Both groups found that the transfected MT gene does not respond to glucocorticoids; this lack of response has also been described in Cd-resistant mouse sarcoma cell lines, where amplification of the HMTI gene occurs in response to metal poisoning (Mayo and Palmiter, 1982) and in transgenic mice carrying a mouse MT promoter fused to the structural gene of the Herpes virus timidine kinase (Palmiter et al., 1982). The 5 to 10-fold induction of MT by dexamethasone in normal mouse sarcoma cell lines and the presence of glucocorticoid receptors in the cell lines used and in the transgenic mice exclude a trivial deficiency in the glucocorticoid regulatory pathway. Furthermore, the transfected human MTII gene does not respond to glucocorticoids as an episomal BPV-derived vector, while it retains dexamethasone inducibility after integration into the genome (Karin et al., 1983; Karin et al., 1984b; Karin et al., 1984c). Together, these observations suggest that a modification of some cis-acting sequence occurs when the MMTI gene is transfected or amplified or when the HMTII gene is in a episomal state.

The introduction into rat TK⁻ cells or human cell lines of a fusion gene containing the coding region of viral timidine kinase ligated to the 770 bp of the 5' flanking sequence of the $HMTII_A$ gene, or to 860 bp of the promoter region of the MTI_A gene, provided interesting information on the differential regulation of these two genes (Karin et al., 1984c; Richards et al., 1984). $HMTI_A$ and $HMTII_A$-TK hybrids were readily detected by S1 mapping in transfected cells after induction with Cd; surprisingly, only the $HMTII_A$-TK fusion gene responded to Zn and dexamethasone while $HMTI_A$-TK gene did not.

The glucocorticoid responsive elements have been located between nucleotides -237 and -268 for the human $MT-II_A$ gene; this region is the binding site of purified glucocorticoid receptor and is protected against DNase I digestion in the presence of the receptor (Karin et al., 1984c). The deletion of this site does not affect the basal level of transcription of the $HMTI_A$ gene nor its ability to be induced by heavy metals. Furthermore, the comparison of the $HMTI_A$ nucleotide sequence -267 to -243 with another glucocorticoid regulated promoter, the mouse mammary tumor virus LTR, revealed strong homology with two regions in position -191 to -170 and -132 to -113. In MMTV these same regions are protected against DNase I in the presence of purified receptor (Scheidereit et al., 1983). A common receptor binding site consensus sequence TGGT.ACAAA.TGTTCT has been proposed by Karin and coworkers (Karin et al., 1984c). A related sequence is found in the first intron of the human growth hormone gene, another glucocorticoid regulated gene (DeNoto et al., 1981; Robins et al., 1982).

The different inducibility of the two HMT genes to Cd, Zn and dexamethasone is particularly intriguing. Since the regions responsive for metal regulation are highly homologous in the MT gene family (see next paragraph) the response of the $HMTI_A$-TK fusion gene to Cd and only to very high doses of Zn could mimic the real functional role of the $HMTI_A$ gene in human. The wider response of $HMTII_A$ to Zn and dexamethasone in gene transfer experiment and its ubiquitous induction in in vitro cell lines with different tissue origins raise the hypothesis of a more general involvement of this gene in zinc metabolism. Also the basal level of expression of the two genes differs, $HMTII_A$ being five times higher than $HMTI_A$, further suggesting a more specific role of the $HMTI_A$ gene in response to Cd or to metal poisoning (Richards et al., 1984). The lower transcriptional activity would be compensated by the higher copy number of HMTI genes, which are responsible for the synthesis of 50% of the total human MT proteins (Karin and Herschmann, 1979). Interestingly, the human interferon gene family has an analogous organization, the single β-interferon gene being expressed at higher levels than the multiple α-interferon genes (Hayes, 1979; Goeddel et al., 1981).

How do metals regulate MT gene expression? Up to now it has not been possible to give a complete answer to this question even though gene transfer experiments and an in vivo competition assay have made it possible to map the cis-acting sequence required for metal regulation and to detect at least one class of cellular factors involved in the regulation of the MMTI promoter.

The region essential for transcription and metal regulation

lies in the first 150 bp of the 5' flanking region of the human and
mouse MT genes (Brinster et al., 1982; Mayo et al., 1982; Carter et
al., 1984; Karin et al., 1984c; Stuart et al., 1984). As shown in
Fig. 2, comparison of the 5' flanking region of the mouse MTI gene
and the human HMTI$_E$, HMTI$_F$ and HMTII$_A$ genes reveals a striking
homology in specific regions (Carter et al., 1984). Two duplicated
regions of twelve nucleotides in the human genes and of nine nucleotides in the mouse gene show a very high sequence homology (91%), a
possible conserved consensus sequence being 5'   -TGCGCCCGC;C/T;C-
3'. The position of the proximal repeat is centered 14-17 bp
upstream of the TATA box and is retained at the same place in the
different promoters. The distance of the second repeat is more

Fig. 2: Comparison of mouse and human MT gene 5' flanking
        sequences (Glanville et al., 1981; Karin and Richards,
        1982; Schmidt et al., 1985). The sequences were aligned
        at the conserved TATA sequences. The duplicated heavy
        metal control sequences are indicated by boxes, the G-rich
        sequences by broken underlines and the polyndromes by
        arrows.

variable, between 95 and 115 bp upstream of the TATA box. Other

partial homologies of the repeat are found elsewhere in the MMTI

promoter, but their precise role in transcription has not been

determined. A GC rich sequence is also present in all the

mentioned functional MT genes in the region between the two

repeats, and sequence homologies are also present in the region

upstream of the repeats (Stuart et al., 1984). These purine-pyrimi-

dine stretches are potentially able to form Z DNA structures and a

functional transcriptional role has been proposed, since the dele-

tion of both elements reduces the basal activity of the $HMTII_A$

promoter (Karin et al., 1984c).

The analysis of the transcription products of several dele-

tions and mutations in the 5' region of the MMTI promoter have

shown that the 151 bp upstream of the TATA box are needed to retain

the metal inducibility and 30% of the expression of the complete

promoter (Carter et al., 1984). From a functional point of view,

the promoter region can be divided in two distinct regions, prox-

imal and distal. The proximal region, between -84 and -15, gives

high induction ratios but low transcription levels when present by

itself. The distal region, between -84 and -151, gives less signif-

icant induction but higher transcription rates. Therefore, both

the proximal and the distal repeats are needed for efficient tran-

scription and metal regulation. The insertion of two synthetic

repeats of the heavy metal control element in the 5' promoter

region of viral timidine kinase, at approximately the same position

as in the MMT promoter, confers a metal response to it (Stuart et

al., 1984).

An *in vivo* competition assay has recently shown that a class of cellular factors positively regulate MMTI transcription (Seguin et al., 1984). The distal control region, between nucleotides -151 and -78, is essential for both efficient transcription and competition, while the proximal region, between -78 and -15, fails to compete. The synergistic effect of the two regions has been confirmed in these experiments: the rate of transcription and the competition are enhanced when both repeats are present in conjunction. The identification and the purification of the cellular factors regulating MT gene expression will be required to clarify the actual mechanism of metal induction.

B. MENKES' DISEASE   Introduction

In 1962 Menkes reported the finding of a new X-linked trait affecting five boys in the same family. The disease was lethal, characterized by progressive psychomotoric retardation leading to death before three years of age (Menkes et al., 1962). A more detailed clinical picture was obtained in the following years (Aguilar et al., 1966), but only in 1972 did Danks and collaborators observe a direct link between copper deficiency and Menkes' disease (Danks et al., 1972a; 1972b). The Australian group noticed a high similarity between the peculiar shape of the hair of Menkes' patients and the steely wool of sheep grazing on copper deficient soil. The subsequent analysis of the copper content in several organs revealed that Menkes' patients were accumulating the metal in several tissues but suffered from a severe decrease in its serum levels (Danks et al., 1972a). It was later shown that the intracellular metal was not exchangeable in this form and the lack in the function of many copper enzymes reflected the inability of the cell to release the stored metal.

The basic defect of Menkes' disease is still unknown, but progress has recently been made using cultured cells from Menkes' patients. These cells represent an interesting model for molecular characterization of the disease since they accumulate copper in the same manner as the tissue of origin (Horn, 1976). The intracellular copper has been shown to be bound to metallothionein, and cultured cells synthesize increased levels of this protein in response to low concentration of copper in the growth medium (Riorden and Joelicoeur-Paquet, 1982; Leone et al., 1985).

DNA transfer experiments have shown that the metallothionein promoter is abnormally induced in Menkes' fibroblasts by copper via a trans-acting mechanism (Leone et al., 1985).

Clinical picture

A normal gestation with premature birth has been observed in several Menkes' babies. The characteristic alterations in the hair have been reported in the first week, but often develop as late as three or four months of age. The hair is sparse, brittle, breaks off easily and accounts for the name Menkes' steely or "kinky hair" disease. Other minor symptoms such as feeding difficulties, depigmentation and relaxation of the skin have been reported. Birth weight is usually normal. Hypothermia and hypotonia can develop in the first days of life. The routine neonatal examinations appear normal and it is difficult to diagnose the disease at this stage. Development proceeds normally for some weeks or months, but after that a quite distinct pattern of

symptoms appears. Convulsions are commonly observed, accompanied
by vomiting, diarrhea and feeding problems. Severe mental retardation develops with somnolence, anorexia and hypothermia. The
skeletal x-ray shows generalized osteoporosis with metaphyseal
fractures. Arteriography reveals irregularity and elongation in
the lumen of the major arteria accompanied by bleeding and subdural
hemorrhagia. Death usually occurs between six months and three
years of age, but longer survival of a few patients has been
reported (Baerlocker et al., 1983; Lucky and Hsia, 1979). The
autoptic examination reveals the most critical alteration in the
connective and in the nervous tissues; the elastic layers of the
major arteria are fragmented and modifications of the Purkinje
cells of the cerebellum are present together with massive
hemorrhagia.

Genetics

Menkes' disease has an incidence of 1 out of 100,000 births.
X chromosome linkage has been shown by pedigree analysis and lyonization of clonal cell lines from heterozygous females (Horn, 1980;
Horn et al., 1980). Analysis of hybrid cell lines obtained by
fusing Menkes' fibroblasts with normal cells indicates that the
trait is mostly recessive (Brown et al., 1984).

The localization of the Menkes' gene has been proposed
close to the centromere, on the long arm of the X chromosome, by
C-banding polymorphism in five Danish families segregating the
Menkes' gene (Horn et al., 1984) and by comparison with the mottled
(Mo) mutation in the mouse, an X-linked recessive trait which
strongly resembles Menkes' disease (Hunt, 1974; Danks, 1977). The

Mo locus has been located in the region between phosphoglycerate
kinase (Pgk-1) and alpha galactosidase (Ags), closely linked to the
centromere (McKusick, 1978). Since in human these two loci have
the same distance from the centromere it has been suggested that
there is a close linkage between this region and the Menkes' locus
(Horn, 1983). Attempts have been made to find a specific restriction fragment length polymorphism with X chromosome specific probes
mapping near the centromere (Wieacker et al., 1983; Wienker et al.,
1983).

Animal model

In 1974 Hunt described a mutation in a strain of mouse very
similar to Menkes' disease in human (Hunt, 1974). Five alleles
have been characterized: the male Mottled mouse (Mo), which dies in
utero; the Dappled mouse (Dp), with death at birth; the Brindled
mouse (Br), whose life span is fourteen days; the sterile viable
Brindled mouse (Mo^{vbr}) and the Blotchy mouse (Mo^{blo}) which is
viable, fertile and die approximately at 150 days of age. All these
strains show X linked genetic transmission, poor pigmentation of
the hair, copper accumulation in several tissues and pathological
alterations in the connective and nervous system.

The mottled mouse mutants appear to be biochemically homologous to human Menkes' disease patients. The Cu levels are elevated in the gut mucosa, testis and kidney and deficient in the
liver, brain and plasma (Camakaris et al., 1979). The excess intracellular copper is bound to metallothionein in the kidney of Br
mutant (Port and Hunt, 1979; Prins and Van der Hamer, 1980; Hunt
and Clarke, 1983) and elevated levels of MTI mRNA has been found in

brain of this mutant mouse, despite having only one fourth the copper concentration of that in normal brains (Camakaris et al., 1983). The levels of several copper enzymes are reduced (Rowe et al., 1974; Hunt, 1977; Holstein, 1979) but can be increased by treatment of the animals with copper (Mann et al., 1979). In vitro cultured cells from Br mutants are sensitive to high levels of copper and have impaired efflux, as the human Menkes' fibroblasts (Camakaris et al., 1980).

The mouse Blotchy (Mo^{blo}) allele is dominant in somatic cell hybrids containing a normal expressed human X chromosome (Wienker et al., 1981), while the Brindled mouse allele behaves as a recessive trait in mouse-mouse cell hybrids (Camakaris et al., 1983). Similar experiments fusing normal mouse and Menkes' fibroblasts exclude dominance of the latter (Brown et al., 1984).

Further studies are needed to prove the existence of different complementation groups among the Mo alleles. Interestingly, in human, two variants of Menkes' disease have been described (Procopis et al., 1981; Williamse et al., 1982) and more recently, an X linked Ehler-Danlos syndrome type IX has been shown to resemble Menkes' disease in several aspects of its abnormal copper metabolism (Kuivaniemi et al., 1982; Peltonen et al., 1983). These studies could reveal if the mutations in the mouse Mo locus are homologous to the one(s) in Menkes' and if one or more genes clustered on the X chromosome are responsible for the alteration in copper metabolism.

Copper metabolism

The homeostasis of copper in the body is balanced between intestinal absorption and biliary excretion. The absorption of ingested copper takes place in the duodenum through the cells of the epithelial mucosa. The process is rather quick; high peaks of label radiotracers have been detected in the blood of patients 90-150 minutes after oral administration of ^{64}Cu or ^{67}Cu. 90% of the copper in the plasma is bound to ceruloplasmin, a liver glyco-protein, which tightly incorporates the metal at synthesis. This protein distributes the metal to the various organs. Injection of radioactive copper is followed by rapid appearance of ceruloplasmin in the blood and the isotope accumulates in the extrahepatic tissues only after the ^{64}Cu-ceruloplasmin complex is formed (Owen, 1965; Sarkar, 1981). Decreased levels of this protein are an index of severe copper deficiency, as is found in Menkes' disease.

The rest of the metal in the plasma is bound to albumin and, in small amounts, amino acids, i.e. histidine, threonine and glutamine, and small peptides. Copper is loosely bound to these complexes, in order to be rapidly exchanged to other tissues.

Albumin binds the metal in the early hours after administration and transport it to the liver. The liver is a key organ in copper homeostasis; the metal accumulates here more than in other tissues. Distribution of the metal is regulated through the synthesis of ceruloplasmin and excretion takes place in the biliary tract, where copper is found complexed to proteins of different molecular weights (Evans, 1973; Sternlieb, 1980; Hrgovcic and Schullenberg, 1984).

The molecular basis of intracellular copper transport are at present largely unknown. The physiochemical properties of the metal to form stable complex with different ligands, due to its very polarizable configuration, tends to exclude a passive diffusion through the cellular membrane, in support of a model of active or facilitated cellular transport.

Some copper proteins have been biochemically characterized and their gene mapped. The greatest part of them are autosomally inherited; therefore only the X linked genes should be considered as possible candidates for Menkes' disease.

Copper proteins

Mammalian cuproenzymes and proteins can be classified by function: oxidases, hydrolases, superoxide dismutase, alpha fetoprotein, albumin and metallothionein (O'Dell, 1976; Hsie and Hsu, 1980; Hrgovcic and Schullenberger, 1984).

Oxidases. This is the largest class and includes ceruloplasmin, cytochrome c oxidase and amino oxidases. All these enzymes reduce oxygen to either water or hydrogen peroxide.

Ceruloplasmin (E.C.1.16.3.1) is a multifunctional serum copper protein found in many mammals. It it a glycoprotein with a molecular weight of 132,000 daltons and contains 6 atoms of copper per molecule. After the transport of copper, its most important function resides in iron mobilization and oxidation of Fe(II) to Fe(III), allowing the release of iron from ferritin and its attachment to transferrin. Severe deficiencies of copper-ceruloplasmin have been shown in Wilson's disease, a single gene autosomal recessive inherited disorder (Sternlieb and Schienberg, 1979;

McKusick, 1978; Danks, 1983). Plasma copper is abnormally low in
Menkes' babies after the first two weeks of life (Grover et al.,
1979; Danks et al., 1972a and 1972b).

   Cytochrome c oxidase (E.C.1.9.3.1.) is the terminal oxidase
in the mitochondrial electron transport system catalyzing the
transfer of electrons from cytochrome c to oxygen. The synthesis
of ATP necessary for all the energy requiring processes is linked
to the mitochondrial electron chain, therefore this enzyme has a
key importance in cellular aerobic metabolism. Cytochrome c
oxidase is localized in the inner mitochondrial membrane. This
heme-protein is composed of four redox centers, two containing
copper and two containing cytochrome a and a_3, respectively.
Between seven and eight subunits contribute to form this protein
with a molecular weight of 270,000, of which 20% is estimated to be
lipid. Since the structural genes are localized on the mitochondrial genome, inheritance does not follow the Mendelian law.

   Cytochrome oxidase activity is reduced in leukocytes and -
fibroblasts from Menkes' patients (Willemse et al., 1982; Noojen et
al., 1981; Maehara et al., 1983) but copper therapy increases mitochondrial respiration (French, 1977). Deficient activity of cytochrome c oxidase in the brain has been associated with neonatal
ataxia.

   Amino oxidases catalyze oxidative deamination of amines by
molecular oxygen to give aldehyde and hydrogen peroxide. Monoamino
oxidase (E.C.1.4.34) contains copper and pyridoxal and is found in
the human plasma. Mitochondrial MAO A belongs to a different class
of amino oxidases and its gene has been localized on the X chromo-

some, close to PGK (Shows et al., 1982). It is still not clear if this enzyme contains copper.

Diaminooxidase (E.C.1.4.3.6) is a copper enzyme and it is found in several tissue as intestinal mucosa, lungs, kidney and placenta. The serum levels of this oxidase are severely reduced in Menkes' patients (Garnica et al., 1977) and disorders in polyamine metabolism have been observed (Rennart et al., 1980). Deficiencies in mono and di-aminooxidases affect histamine turnover, causing rashes of urticaria.

Lysyl oxidase (E.C.?) initiates the cross linking of collagen and elastin, catalyzing the oxidative deamination of the epsilon amino group in lysyl and hydroxylysyl residues. The enzyme is secreted extracellularly; its molecular weight is 30,000 or a multiple of it. The activity of lysyl oxidase is copper dependent; therefore the gross alteration in the elastic lamina of the arteries and in the bone found in Menkes' patients has been attributed to the severely decreased activity of this enzyme (Royce et al., 1980; Peltonen et al., 1983). The chromosomal location of this enzyme has not been established.

Hydrolases. Tyrosinase (E.C.1.10.3.1.) catalyzes the oxidation of tyrosine to dopa and from dopa to quinone, which is used in melanin biosynthesis. Its activity is reduced in copper deficiency status and the failure of pigmentation in the hair reflects the decreased level of melanin synthesized. The gene has been tentatively located on an autosome (McKusick, 1978).

Dopamine beta hydroxylases (E.C.1.14.17.1) catalyzes the conversion of dopamine into adrenaline. The enzyme has a molecular weight of 70,000 daltons and is located in the adrenal medulla; a soluble form is also present in the blood. Autosomal inheritance has been proposed (Dunnette and Weinshilboum, 1982). Lower activities of the enzyme are present in the serum and in the brain of Menkes' patients (Henkin and Grover, 1978; Rohmer et al., 1977).

Superoxide Dismutase (E.C.1.15.1.1)

The molecule weight of this cytosolic enzyme is approximately 33,000 daltons. A manganese and a copper-zinc form have been identified. The SOD copper-zinc form gene has been mapped on chromosome 21 (Shows et al., 1982). The enzymes play an important role in the conversion of toxic superoxide radicals to peroxide and oxygen. The deficient detoxifying action of this enzyme could be responsible for neuronal death in Menkes' disease, since these cells are very sensitive to the toxic action of free radicals. In erythrocyte from Menkes' patients its activity is either normal or highly increased (Williams et al., 1977). Only in one patient has a reduced SOD activity been found (Rohmer et al., 1977).

α-Fetoprotein. α-Fetoprotein shows an extensive nucleotide sequence homology with albumin and is one of the major plasma proteins in early fetal life. It contains one cupric chelate site. Its function during fetal life is still poorly understood. The high functional similarity with albumin raises the possibility of its involvement in metal transport during early fetal life.

Albumin. The role of this protein in the transport of copper in the serum has already been discussed. Copper binds at a specific site located at the amino acid terminal end of the protein. The

gene has been mapped in humans to chromosome 4 (Shows et al.,
1972).

Copper distribution in Menkes' patients

A variety of studies have shown the abnormal and characteristic distribution of copper in tissues of Menkes' patients
(Heydorn et al., 1975; Danks et al., 1972b; Horn, 1984). Several
fold increases of the metal are present in the kidney, spleen,
pancreas, gut and placenta; moderate increases are present in other
tissues such as skin, lung and muscle, and very low and moderate
levels, respectively are found in the liver and in the brain.
Serum copper and ceruloplasmin are low.

Attempts to restore a normal distribution of copper in the
body have not been successful. Treatment of patients with paren-
teral copper-EDTA therapy increased the extrahepatic storage,
especially in the kidney and normalized the serum and hepatic
copper levels (Williams and Atkin, 1981; Nojen et al., 1981; Akima
et al., 1978; Yazaki et al., 1983) but no remission of the clinical
symptoms was observed. The abnormal copper distribution in the
tissues of Menkes' fetuses and the presence of increased copper
levels in the placenta suggest that the damage of the metabolic
pathway occurs in utero before birth (Nojen et al., 1981; Horn,
1984). Interestingly, studies of copper therapy on brindled mice
have shown striking differences with the results obtained in humans.
Intramuscular administration of a single dose of copper at seven
days of age restores to normal the life expectancy; a parallel
increase in the levels of some cuproenzymes has been observed,

followed by a return to low levels after some weeks (Mann et al.,
1979; Royce et al., 1982; Danks et al., 1983).

Copper metabolism and metallothionein expression in cultured cells

In vitro cultured Menkes' skin fibroblasts and lymphoblasts
are able to accumulate copper to much higher extents than normal
control lines (Horn, 1976; Goka et al., 1976; Beratis, et al.,
1978). The differences reported range from three to more than ten
fold (Beratis et al., 1978; Riorden and Jolicoeur-Paquet, 1982;
Camakaris et al., 1980). The role of uptake is influenced by the
amount of copper in the growth medium and by the growth status of
the cells. This characteristic of the Menkes' cells provides the
only prenatal diagnosis currently available. Horn has shown that
amniocytes from patients accumulate isotopes of copper at higher
levels than cells from normal individuals (Horn, 1976; Horn, 1981)
and moreover, that two cell populations are present in skin
biopsies from obligate female heterozygous carriers, one behaving
normally, the other overaccumulating ^{64}Cu. This data confirms the
predicted mosaicism for an X-linked disease (Horn, 1980; Horn et
al., 1980).

The initial rates of uptake of copper are very similar in
normal and Menkes' cells, and impaired efflux of the metal is
responsible for its accumulation in the mutant cell (Camakaris, et
al., 1980; Goka and Howell, 1978). In the cell copper is bound to
metallothionein (Beratis et al., 1978; Onishi et al., 1980; Onishi
et al., 1981; Riorden and Jolicoeur-Paquet, 1982). The protein
purified from Menkes' cells shows the same copper-binding capacity
as the one from normal cells; no difference has been found in the

copper binding sites of proteins from both sources (LaBadie et
al., 1981a; LaBadie et al., 1981b).

In our lab we have compared the effect of copper on normal
and Menkes' fibroblasts, focusing on the expression of metallothionein genes (Leone et al., 1985). The most striking difference
between Menkes' and normal cells is the effect of low concentrations of copper on MT synthesis. As shown in Fig. 3, in the
Menkes' line concentrations of $CuSO_4$ between 50 and 200μM strongly
induce MT synthesis. In contrast, in the normal line 50 and 100μM
$CuSO_4$ have no effect while 200μM give only a small increase. The
effect is specific for copper since Zn and dexamethasone induce MT
synthesis in both lines. Moreover, the MT induction by low doses
of copper appears to be a general characteristic of Menkes' cells.
The analysis of cell lines from three different families also show
increased MT synthesis of concentrations of 100μM $CuSO_4$. Interestingly, we obtained more variable results from heterozygous cell
lines, presumably due to lyonization of the X-linked mutation
followed by selective outgrowth of the normal population (Horn,
1980).

The amount of copper in the nutrient media and the growth
status of the cells can greatly influence the extent of induction
and the basal level of expression of metallothionein. Hence, a
more detailed picture of the effect of copper on MT synthesis is
obtained by growing and labeling the cells in the presence of
metal-chelated fetal calf serum. In the Menkes' cells, MT synthesis increased sharply between 50 and 600μM $CuSO_4$, then plateaued
at 700 to 1000μM. In normal cells, strong induction occurred only

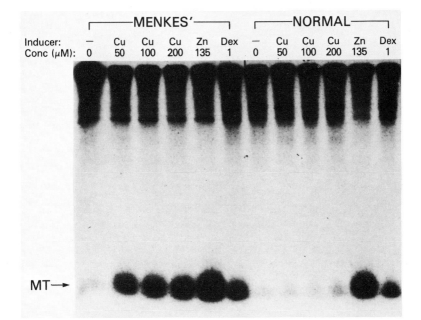

Fig. 3: MT Synthesis: Menkes' cells (GM220) or normal cells
(GM323) were preincubated for 24 h in EMEM plus 20%
dialyzed fetal calf serum then treated with the indicated
concentration of $CuSO_4$, $ZnCl_2$, or dexamethasone for 9 hr.
The cells were labeled with ^{35}S-cysteine in medium
containing 0.5% dialyzed serum during the final hour of
induction. Cell extracts were prepared,
carboxymethylated, and electrophoresed through a
nondenaturing 20% polyacrylamide gel. Reproduced by
permission of Cell, MIT.

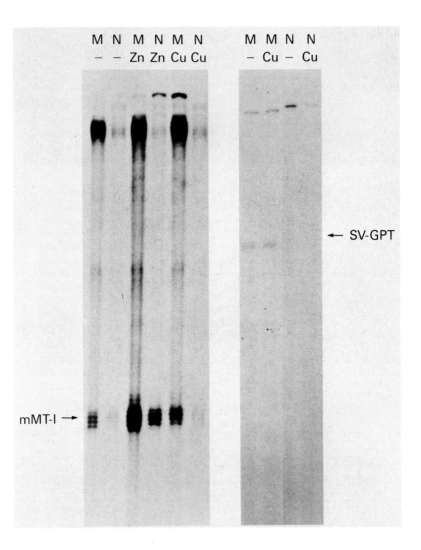

at $CuSO_4$ concentrations greater than 700µM and expression

plateaued at a level approximately one-half that observed in the

mutant line. Therefore, the main difference between the Menkes'

and normal cells is the concentration of copper needed for induc-

tion rather than the absolute level of MT synthesis. By in vitro

translation and S_1 mapping experiments we attributed the increased

MT synthesis to increased mRNAs levels. Two genes, $HMTII_A$ and

$HMTI_E$, are affected in parallel (Karin and Richards, 1982a;

Schmidt et al., 1985).

To distinguish between a cis or a trans activation of MT

genes in Menkes' cells, we used a transient transfection assay. A

plasmid containing 2000 bp of 5' flanking sequences and 68 bp of

---

Fig. 4: Transfection with an MT Fusion Gene: Transformed Menkes'

cells (AdSVGM 220, "M") or normal cells (AdSVGM 323, "N")

were transfected with a plasmid containing a mouse

MMTI-CAT fusion gene by a modification of the calcium

phosphate precipitation method with sodium butyrate

enhancement. The cells were treated with no metal, 200µM

$CuSO_4$, or 135µM $ZnCl_2$ in medium with 20% untreated fetal

calf serum during the final 12 hr of the experiment. RNA

was extracted and analyzed by nuclease S1 mapping with

single-stranded probes specific for the 5' end of the

mouse MT-1 gene (left) or the 3' end of the SV-GPT

transcription unit (right). RNA from mock transfected

cells gave no specific protected fragments. Reproduced by

permission of Cell, MIT.

5' untranslated sequence of the mouse MT-I gene fused to the
coding sequences of the E. coli Chloramphenicol Acetyl Transferase
was introduced into Menkes' and normal transformed fibroblasts
through calcium-phosphate precipitation technique. The transfection efficiency of the two lines was monitored through an SV_{40}-
guanine phosphoribosyl transferase (SV-GPT) transcription unit
present on the same plasmid. The use of adenovirus transformed
cell lines was a necessary step in order to obtain efficient
uptake of the DNA (Van Doren and Gluzman, 1984).

As shown in Fig. 4, $200\mu M$ $CuSO_4$ induced transcription of
the mouse MT_I promoter only in Menkes' line while Zn had similar
effects on both lines. A three fold induction by copper is detected by densitometer scanning in the Menkes' line; zinc induction
ratios were eight fold in normal and mutant cell lines. The
expression of the SV_{40}-gpt unit was not altered by copper,
although a difference of three times in transfection efficiency
existed between the two lines.

The effect of copper on Menkes' cells is not limited to
the induction of MT synthesis. A broader influence of this metal
on the metabolism of these cells is suggested by two different
observations. First, several groups (Chan et al., 1978; Camakaris
et al., 1980; Beratis et al., 1978; Leone et al., 1985) have shown
that Menkes' cells are more sensitive to copper than normal cells.
As shown in Fig. 5, after two days in culture in the presence of
$600\mu M$ $CuSO_4$, Menkes' cells show morphological alterations, inhibition of growth and cell death while the normal control cells do
not. In contrast, Zn and Cd are equally toxic to both cell lines.

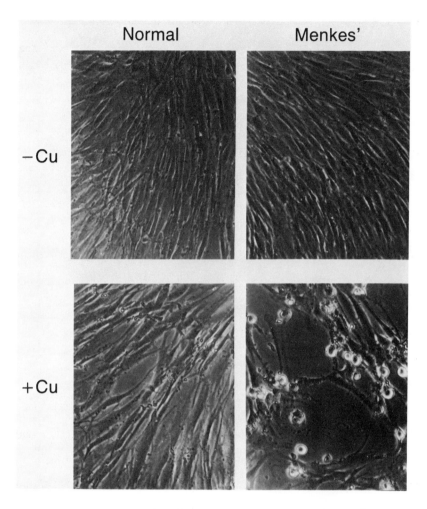

Fig. 5: Copper Toxicity: Menkes' cells (GM 220) or normal cells
(GM 323) were grown to 70% confluency in EMEM plus 20%
untreated fetal calf serum then incubated in the absence
or presence of 600μM CuSO₄ for an additional 48 hr. The
cells were photographed under phase contrast. Reproduced
by permission of Cell, MIT.

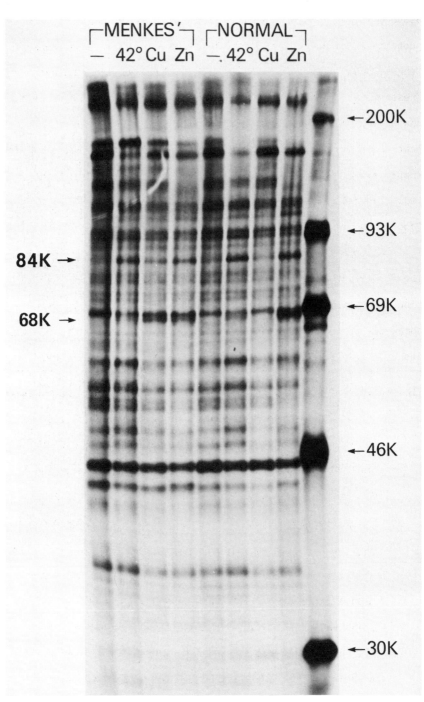

Second, as shown in Fig. 6, copper induces the synthesis of two proteins of molecular weight 83,000 daltons and 68,000 daltons in Menkes' cells but not in normal cells. The same proteins are present in both cell lines after induction with Zn or exposure for one hour to 42°C. We think that these proteins are part of the heat-shock family, since it has been shown that toxic metal concentrations induce the synthesis of such proteins (Levinson et al., 1980; Courgeon et al., 1984).

Conclusions

Copper shows a severe effect on Menkes' cell metabolism. The strong induction of MT synthesis via a transcriptional activation of its promoter in trans, the appearance of other two heat-shock like proteins following exposure to high doses of copper and its cytotoxic effects suggest that the basic defect of Menkes'

Fig. 6: Induction of 84,000 daltons and 68,000 daltons proteins: Menkes' cells (GM 220) or normal cells (GM 323) were preincubated for 24 hr in EMEM plus 20% dialyzed fetal calf serum. One plate of cells was shifted to 42° for 9 hr. The other plates were treated with no metal, 200μM CuSO4, or 135μM ZnCl2 for 9 hr at 37°. The cells were labeled with ^{35}S-cysteine in medium containing 0.5% dialyzed serum during the final hour of induction. Total cell lysates were electrophoresed through a 9% polyacrylamide-SDS gel. The markers (right lane) were ^{14}C-methylated myosin, phosphorylase-b, bovine serum albumin, ovalbumin, and carbonic anhydrase (Amersham). Reproduced by permission of Cell, MIT.

mutation affects a factor which acts at the early stage of the
metal regulatory pathway.

A defect in a molecule that either binds or metabolizes a
significant fraction of the intracellular copper under normal
physiological conditions would lead to an increase in the concentration of free or available copper inside the cell. The increase
of transcription of the MT gene would represent the response of
the cell to the increased intracellular concentrations of the
metal. Clearly the defective factor is not MT itself, since the
localization of MT genes on chromosome 16 excludes its X-linkage
(Karin et al., 1984a; Schmidt et al., 1984). This model would
also explain the increased sensitivity of Menkes' cells to copper
and the abnormal synthesis of the heat-shock proteins as a result
of the increased intracellular levels of the metal. Finally, we
have to postulate a role of this factor in the activation of
copper enzyme in order to explain the decreased activities of
these enzymes even in cells containing elevated levels of copper.

A variation of this model is that a factor with higher
affinity for copper is synthesized in Menkes' cells. We have to
postulate that this factor is toxic to account for the increased
copper sensitivity and the induction of the 68,000 daltons and the
84,000 daltons proteins.

We tend to exclude a second model of a mutation in a
factor which directly activates MT gene transcription by binding
to the control DNA sequences. Cells overproducing MT due to gene
amplification are more resistant, not sensitive, to copper (Durnam
and Palmiter, 1984). Moreover, the induction of the two 68,000

daltons and 84,000 daltons proteins would not be explained by this model since no obvious primary homology has been found between sequenced heat shock gene control regions and these of MT genes (Pelham, 1982).

In humans another genetic disease affects copper metabolism. Wilson's disease is an autosomal recessive trait characterized by abnormal accumulation of copper in the liver and in the brain, together with low levels of ceruloplasmin in the serum, leading to hepatic damage and neurological defects (Danks, 1983). The excess copper in the liver is bound to MT or MT aggregates. The minor abnormalities in the accumulation of MT protein and MT mRNAs found in skin fibroblasts from Wilson's patients (Leone and Hamer, unpublished results) tends to exclude a more general primary alteration in MT gene regulation in favor of a defect in a liver-specific copper liganding factor.

Therefore, in Wilson's disease and in Menkes' disease the related mottled mutation in the mouse, MT gene regulation seems to be affected indirectly via a primary alteration(s) in copper metabolism, rather than directly through mutation(s) affecting MT gene structure or DNA binding regulatory factors. Our present knowledge of the molecular mechanism controlling copper transport, liganding and enzyme activation is unfortunately at a primitive state. The recent development of recombinant DNA techniques could play a key role in association with traditional biochemical approaches in the understanding of such metabolic pathway and its alterations.

Summary

    Metallothioneins are a family of ubiquitous, cysteine rich
proteins, whose amino acidic and genomic sequences have been
highly conserved during evolution. MT synthesis is induced by
heavy metals, glucocorticoids and a bacterial lipopolysaccharide
in vivo and in vitro. MT forms stable complexes with heavy
metals.

    One MTII$_A$ gene, four MTI class genes and five pseudogenes
have been isolated in humans. The cluster of MT genes is located
on chromosome 16. The cloned, transfected genes retain metal
inducibility. The first 150 bp of the 5' flanking region of mouse
and human MT genes are essential for transcription and metal
regulation. Two control regions have been identified. The distal
region, between -151 and -78 is essential for efficient
transcription and binding of cellular factor(s) which regulates MT
gene expression.

    In Menkes' disease, a lethal X-linked recessive disorder,
copper accumulates intracellularly bound to MT. Low doses of
copper induce MT synthesis in Menkes' fibroblasts, but not in
normal controls. Transfection experiments using the mouse MTI
promoter fused to CAT show that the effect of copper in MT
transcription is in trans.

    Menkes' cells are more sensitive to copper than normal
controls and respond to copper poisoning by synthesizing two
heat-shock like proteins. A mutation affecting copper transport
or metabolism is discussed.

Acknowledgements

I thank D. Hamer for his suggestions and critical reading of the manuscript and G. Gray and C. Ray for excellent typing.

REFERENCES

Aguilar, M.J., Chadwick, D.L., Okuyama, K., and Kamoshita, S. (1966). 'Kinky hair disease. I. Clinical and pathological features', J. Neuropathol. Exp. Neur. 25, 507-522.

Akima, M., Nonaka, H., Senzui, M., Kitazawa, Y., Ichimori, S., Aoki, T., and Kanekawa, H. (1978). 'Two autopsy cases of Menkes' kinky hair disease', Adv. Neurol. Sci. (Jpn.) 22, 427.

Baerlocher, K., Steinmann, B., Rao, V.H., Gitselmann, R., and Horn, N. (1983). 'Menkes' disease (Md): clinical, therapeutical and biochemical studies', J. Inher. Metab. Dis. 6, 87-88.

Beratis, N.G., Price, P., LaBadie, G., and Hirschhorn, K. (1978). ^{64}Cu metabolism in Menkes' and normal-cultured skin fibroblasts', Pediatr. Res. 12, 699-702.

Brown, R.M., Camakaris, J., and Danks, D. (1984). 'Observation on the Menkes' and brindled mouse phenotypes in cell hybrids', Som. Cell. Mol. Genet. 10, 321-330.

Boulanger, Y., Armitage, I.M., Miklossy, K.A., and Winge, D.R. (1982). '^{113}Cd NMR study of a metallothionein fragment', J. Biol. Chem. 257, 13717-13719.

Brinster, R.L., Chen, H.Y., Warren, R., Sarthy, A., and Palmiter, R.D. (1982) 'Regulation of metallothionein-thymidine kinase fusion plasmids injected into mouse eggs', Nature 296, 39-42.

Camakaris, J., Mann, J.R., and Danks, D.M. (1979). 'Copper metabolism in mottled mouse mutants. Copper concentrations in tissues during development', Biochem. J. 180, 597-604.

Camakaris, J., Danks, D.M., Ackland, L., Cartwright, E., Borger, P., and Cotton, R.G.H. (1980). 'Altered copper metabolism in cultured cells from Menkes' syndrome and mottled mouse mutants', Biochem. Genet. 18, 117-131.

Camakaris, J., Phillips, M., Danks, D.M., Brown, R., and Stevenson, T. (1983) 'Mutations in humans and animals which affect copper metabolism', J. Inher. Metab. Dis. 6, 44-50.

Carter, A.D., Felber, B.K., Walling, M.-J., Jubier, M.F., Schmidt, C.J., and Hamer, D.H. (1984). 'Duplicated heavy metal control sequences of the mouse metallothionein-I gene', Proc. Natl. Acad. Sci. USA 81, 7392-7396.

Chan, W.Y., Garnica, A.D., and Rennert, O.M. (1978). 'Cell culture studies of Menkes' kinky hair disease', Clin. Chim. Acta 88, 495-507.

Courgeon, A.M., Maisonhaute, C., and Best-Belpomme, M. (1984). 'Heat shock proteins are induced by cadmium in Drosophila cells', Exp. Cell Res. 153, 515-521.

Danks, D.M. (1977). 'Copper transport and utilization in Menkes' syndrome and in mottled mice', Inorg. Perspect. Biol. Med. 1, 73-100.

Danks, D.M. (1983). 'Hereditary disorders of copper metabolism in Wilson's disease and in Menkes' disease', In: The Metabolic Basis of Inherited Disease. Standbury, J.B., Wyngaarden, J.B., Frederickson, D.S., Goldstein, J.L., and Brown, M.S., Eds. McGrow Hill, pp. 1251-1268.

Danks, D.M., Campbell, P.E., Stevens, B.J., Mayne, V., and Cartwright, E. (1972a). 'Menkes' kinky hair syndrome. An inherited defect in copper absorption with widespread effects', Pediatrics 50, 188-201.

Danks, D.M., Campbell, P.E., Walker-Smith, J., Stevens, B.J., Gillespie, J.M., Blomfield, J., and Turner, B. (1972b). 'Menkes' kinky hair syndrome', Lancet 1, 1100-1103.

Danks, D.M., Camakaris, J., Herd, S., Mann, J.R., and Phillips, M. (1983). 'Copper transport in Menkes' disease,' In: Biological Aspects of Metals and Metal Related Diseases. B. Sarkar, Ed., Raven Press, New York, NY USA, 133-146.

DeNoto, F.M., Moore, D.D., and Goodman, H.M. (1981). 'Human growth hormone DNA sequence and mRNA structure: Possible alternativesplicing', Nucl. Acids Res. 9, 3719-3730.

Durnam, D., Perrin, F., Gannon, F., and Palmiter, R.D. (1980). 'Isolation and characterization of the mouse metallothionein-I gene', Proc. Natl. Acad. Sci. USA 77, 6511-6515.

Durnam, D.M., and Palmiter, R.D. (1981). 'Transcriptional regulation of the mouse metallothionein-I gene by heavy metals', J. Biol. Chem. 256, 5712-5716.

Durnam, D.M., Hoffman, J.S., Quaife, C.J., Benditt, E.P., Chen, H.Y., Brinster, R.L., and Palmiter, R.D. (1984). 'Induction of mouse metallothionein-I mRNA by bacterial endotoxin is independent of metals and glucocorticoid hormones', Proc. Natl. Acad. Sci. USA 81, 1053-1056.

Durnam, D.M., and Palmiter, R.D. (1984). 'Induction of metallotionein-I mRNA in cultured cells by heavy metals and iodoacetate: Evidence for gratuitous inducers', Mol. Cell. Biol. 4, 484-491.

Dunette, J., and Weinshilboum, R. (1982). 'Family studies of plasma dopamine-β-hydroxylase thermal stability', Am. J. Hum. Genet. 34, 84-99.

Enger, M.D., Rall, L.B., and Hildebrand, C.E. (1979). 'Thionein gene expression in Cd^{++} variants of the CHO cell: correlation of thionein synthesis rates with translatable mRNA levels during induction, deinduction, and superinduction', Nucl. Acids Res. 7, 271-288.

Evans, G.W. (1973). 'Copper homeostasis in the mammalian system', Physiol. Rev. 53, 535-570.

French, J.H. (1977). 'X-chromosome linked copper malabsorption (X-cLCM)', In: Handbook of Clinical Neurology, Vol. 29, Part 3, Vinken, P.J. and Bruyn, G.W., Eds., Elsevier, New York, NY USA, 279.

Friedman, R.L., and Stark, G.R. (1985). 'α-Interferon-induced transcription of HLA and metallothionein genes containing homologous upstream sequences', Nature 314, 637-639.

Garnica, A.D., Frias, J.L., and Rennert, O.M. (1977). 'Menkes' kinky hair syndrome: is it a treatable disorder?', Clin. Genet. 11, 154-161.

Glanville, N., Durnam, D.M., and Palmiter, R.D. (1981). 'Structure of mouse metallothionein-I gene and its mRNA', Nature 292, 267-269.

Goeddel, D.V., Leung, D.W., Dull, T.J., Gross, M., Lawn, R.M., McCandliss, R., Seeburg, P.H., Ullrich, A., Yelverton, E., and Gray, P.W. (1981). 'The structure of eight distinct cloned human leukocyte interferon cDNAs', Nature 290, 20-26.

Goka, T.J., Stevenson, R.E., Hefferan, P.M., and Howell, R.R. (1976). 'Menkes' disease: a biochemical abnormality in cultured human fibroblasts', Proc. Natl. Acad. Sci. USA 73, 604-606.

Goka, T.M., and Howell, R.R. (1978). 'Copper metabolism in Menkes' disease', Monogr. Hum. Genet. 10, 148-155.

Griffith B.B., Walters, R.A., Enger, M.D., Hildebrand, C.E., and Griffith, J.K. (1983). 'cDNA cloning and nucleotide sequence comparison of Chinese hamster metallothionein I and II mRNAs', Nucl. Acids Res. 11, 901-910.

Grover, W.D., Johnson, W.C., and Henkin, R.I. (1979). 'Clinical and biochemical aspects of trichopolidystrophy', Ann. Neurol. 5, 65-71.

Hamer, D.H., and Walling, M.J. (1982). 'Regulation in vivo of a cloned mammalian gene: cadmium induces the transcription of a mouse metallothionein gene in SV40 vectors', J. Mol. Appl. Genet. 1, 273-288.

Henkin, R.I., and Grover, W.D. (1978). 'Trichopolidystrophy (TDP): new aspects of pathology and treatment in trace elements' In Metabolism in Men and Animals, Kirchgessner, M., Ed., Freisins Weihenstephan, W. Germany, 405-418.

Heydorn, K., Damsgaard, E., Horn, N., Mikkelsen, M., Tygstrup, I., Vestermark, S., and Weber, J. (1975). 'Extra-hepatic storage of copper. A male foetus suspected of Menkes' disease', Hum. Genet. 29, 171-175.

Holstein, T.J., Fung, R.Q., Quevedo, W.C., and Bienicki, T.C. (1979). 'Effect of altered copper metabolism induced by mottled alleles end diet on mouse tyrosimase (40662)' Proc. Soc. Exp. Biol. Med. 162, 264-268.

Horn, N. (1976). 'Copper incorporation studies on cultured cells for prenatal diagnosis of Menkes' disease', Lancet 1, 1156-1158.

Horn, N. (1980). 'Menkes' X linked disease: heterozygous phenotypes in uncloned fibroblast cultures', J. Med. Genet. 17, 257-261.

Horn, N. (1981). 'Menkes' X-linked disease: prenatal diagnosis of hemizygous males and heterozygous females', Prenatal Diagn. 1, 107-120.

Horn, N. (1983). 'Menkes' X-linked disease: prenatal diagnosis and carrier detection', J. Inher. Metabl. Dis. 6, 59-62.

Horn, N. (1984). 'Copper metabolism in Menkes' disease', In: Metabolism of trace metals in man', Chan, W.-Y., Ed. CRC Press, Boca Raton, FL, USA vvII, 25-52.

Horn, N., Mooy, P., and McGuire, V.M. (1980). 'Menkes' X-linked disease: two clonal cell populations in heterozygotes', J. Med. Genet. 17, 262-266.

Horn, N., Stene, J., Mollekaer, A.-M., and Friedrich, U. (1984). 'Linkage studies in Menkes' disease. The Xg blood group system and C-banding of the X chromosome', Ann. Hum. Genet. 48, 161-172.

Hrgovcic, M.J., and Shullenberger, C.C. (1984). 'Copper and lymphomas', CRC Press, Boca Raton, FL USA.

Hsieh, S.H., and Hsu, J.M. (1980). 'Biochemistry and metabolism of copper' In zinc and copper in medicine', Karcioglu, Z.A. and Sarper, R.M., Eds., Charles C. Thomas, Springfield, IL, Ch 4.

Huang, I.-Y., Tsunoo, H., Kimura, M., Nakashima, H., and Yoshida, A. (1979). 'Primary structure of mouse liver metallothionein-I and -II', In: Metallothionein, Kagi, J.H.R. and Nordberg, M., Eds., Birkhauser Verlag, Basel, pp. 169-172.

Hunt, D.M. (1974). 'Primary defect in copper transport underlies mottled mutants in the mouse', Nature 249, 852-854.

Hunt, D.M. (1977). 'Catecholamine biosynthesis and the activity of a number of copper-dependent enzymes in the copper deficient mottled mouse mutants', Comp. Biochem. Physiol. (C). 57, 79-83.

Hunt, D.M., and Clarke, R. (1983). 'Metallothionein and the development of the mottled disorder in the mouse', Biochem. Genet. 21, 1175-1194.

Jahn, C.L., Hutchison, C.A., Phillips, S.J., Weaver, S., Haigwood, N.L., Voliva, C.F., and Edgell, M.H. (1980). 'DNA sequence organization of the β-globin complex in the BALB/c mouse', Cell 21, 159-168.

Kagi, J.H.R., and Vallee, B.L. (1960). In Metallothionein, Birkhauser Verlage, Basel.

Kagi, J.H.R., and Vallee, B.L. (1960). 'Metallothionein: a cadmium- and zinc-containing protein from equine and renal cortex', J. Biol. Chem. 235, 3460-3465.

Kagi, J.H.R., Vasak, M., Lerch, K., Gilg, D.E.O., Hunziker, P., Bernhard, W.R., and Good, M. (1984). 'Structure of mammalian metallothionein', Environ. Health Persp. 54, 93-103.

Karin, M., and Herschman, H.R. (1979). 'Dexamethasone stimulation of metallothionein synthesis in HeLa cell cultures', Science 204, 176-177.

Karin, M., Herschman, H.R., and Weinstein, D. (1980a). 'Primary induction of metallothionein by dexamethasone in cultured rat hepatocytes', Biochem. Biophys. Res. Commun. 92, 1052-1059.

Karin, M., Andersen, R.D., Slater, E., Smith, K., and Herschman, H.R. (1980b). 'Metallothionein mRNA induction in Hela cells in response to zinc or dexamethasone is a primary induction response', Nature 286, 295-297.

Karin, M., Slater, E.P., and Herschman, H.R. (1981). 'Regulation of metallothionein synthesis in HeLa cells by heavy metals and glucocorticoids', J. Cell. Physiol. 106, 63-74.

Karin, M., and Richards, R. (1982a). 'Human metallothionein genes: molecular cloning and sequence analysis of the mRNA', Nucleic Acids Res. 10, 3165-3173.

Karin, M., and Richards, R. (1982b). 'Human metallothionein genes: primary structure of the MT-II gene and a related processed gene',Nature 299, 797-802.

Karin, M., Cathala, G., and Nguyen-Huu, M.C. (1983). 'Expressiono and regulation of a human metallothionein gene carried on an autonomously replicating shuttle vector', Proc. Natl. Acad. Sci. USA 80, 4040-4044.

Karin, M., Eddy, R.L., Henry, W.M., Haley, L.L., Byers, M.G., and Shows, T.M. (1984a). 'Human metallothionein genes are clustered on chromosome 16', Proc. Natl. Acad. Sci. USA 81, 5494-5498.

Karin, M., Haslinger, A., Holtgreve, H., Cathala, G., Slater, E., and Baxter, J.D. (1984b) 'Activation of a heterologous promoter in response to dexamethasone and cadmium by metallothionein gene 5'-flanking DNA', Cell 36, 371-379.

Karin, M., Haslinger, A., Holtgreve, H., Richards, R.I., Krauter, P., Westphal, H.M., and Beato, M. (1984c). 'Characterization of DNA sequences through which cadmium and glucocorticoid hormones induce human metallothionein-II$_A$ gene', Nature 308, 513-519.

Kissling, M.M., and Kagi, J.H.R. (1977). 'Primary structure of human hepatic metallothioneins', FEBS Lett. 82, 247-250.

Kissling, M.M., and Kagi, J.H.R. (1979). 'Amino acid sequence of human hepatic metallothioneins', In Metallothionein J.H.R. Kagi and M. Nordberg, Eds., pp. 145-151, Birkhauser Verlag, Basel.

Klauser, S., Kagi. J.H.R., and Wilson, K.J. (1983). 'Characterization of isoprotein patterns in tissue extracts and isolated samples of metallothioneins by reverse-phase high-pressure liquid chromatography', Biochem. J. 209, 71-80.

Kojima, Y., Berger, C., and Kagi. J.H.R. (1979). 'The amino acid sequence of equine metallothioneins', In Metallothionein J.H.R. Kagi and M. Nordberg, Eds., pp. 153-161, Birkhauser Verlag, Basel.

Kuivaniemi, H., Peltonen, L., Palotie, A., Kaitila, I., and Kivirikko, K.I. (1982). 'Abnormal copper metabolism and deficient lysyl oxidase in a heritable connective tissue disorder', J. Clin. Invest. 69, 730-733.

LaBadie, G.U., Beratis, N.G., Price, P.M., and Hirschhorn, K. (1981a). 'Studies of the copper-binding proteins in Menkes' and normal cultured skin fibroblast lysates', J. Cell. Physiol. 106, 173-178.

LaBadie, G.U., Hirschhorn, K., Katz, S., and Beratis, N.G. (1981b). 'Increased copper metallothionein in Menkes' cultured skin fibroblasts', Pediatr. Res. 15, 257-261.

Leder, P., Hansen, J.N., Kondel, D., Leder, A., Nishioka, Y, and Talkington, C. (1980). 'Mouse globin system: a functional and evolutionarily analysis', Science 209, 1336-1342.

Leone, A., Pavlakis, G.N., and Hamer, D.H. (1985). 'Menkes' disease: abnormal metallothionein gene regulation in response to copper', Cell 40, 301-309.

Lerch, K. (1979). 'Amino-acid sequence of copper-metallothionein from Neurospora crassa', In Metallothionein J.H.R. Kagi and M. Nordberg), Eds., pp. 173-179, Birkhauser Verlag, Basel.

Levinson, W., Oppermann, H., and Jakson, J. (1980). 'Transition series metals and sulfhydryl reagents induce the synthesis of four proteins in eukaryotic cells', Biochim. Biophys. Acta 606, 170-180.

Lucky, A.W., and Hsia, Y.E. (1979). 'Distribution of ingested and injected radiocopper in two patiens with Menkes' kinky hair disease', Pediatr. Res. 13, 1280-1284.

Maehara, M., Ogasawara, M., Mizutani, N., Watanabe, K., and Suzuki, S. (1983). 'Cytochrome c oxidase deficiency in Menkes' kinky hair disease', Brain Dev. 6, 533-540.

Malyuga, D.P. (1941). 'On cadmium in organisms', Dokl Akad. Nauk SSSR 31, 145-147.

Mayo, K.E., and Palmiter, R.D. (1982). 'Glucocorticoid regulation of the mouse metallothionein-I gene is selectively lost following amplification of the gene', J. Biol. Chem. 257, 3061-3067.

Mayo, K.E., Warren, R., and Palmiter, R.D. (1982). 'The mouse metallothionein-I gene is transcriptionally regulated by cadmium following transfection into human or mouse cells', Cell 29, 99-108.

Mann, J.R., Camakaris, J., Danks, D.M., and Walliczek, E.G. (1979). 'Copper metabolism in mottled mouse mutants: copper therapy of brindled (Mo) mice', Biochem. J. 180, 605-612.

Margoshes, M., and Vallee, B.L. (1957). 'A cadmium protein from equine kidney cortex', J. Am. Chem. Soc. 79, 4813-4814.

McKusick, V. (1978). 'Medelian inheritance of man', 5th ed., Johns Hopkins, Baltimore.

Menkes, J.H., Alter, M., Steigleden, G.K., Weakley, D.R., and Sung, J.H. (1962). 'A sex linked recessive disorder with retardation of growth, peculiar hair, and focal cerebellar degeneration', Pediatrics 29, 764-779.

Nielson, K.B., Atkin, C.L., and Winge, D.R. (1985). 'Distinct metal-binding configurations in metallotionein', J. Biol. Chem. 260, 5342-5350.

Nooijen, J.L., de Groot, C.J., and van den Hamer, C.J.A., Monnens, L.A.H., Willemse, J., and Niermeijer, M.F. (1981). 'Trace element studies in three patients and a fetus with Menkes' disease. Effect of copper therapy', Pediatr. Res. 15, 284-289.

Nordberg, M., and Kojima, Y. (1979). 'Metallothionein and other low molecular weight metalbinding proteins', In Metallothionein J.H.R. Kagi and M. Nordberg, Eds., pp. 41-135, Birkhauser Verlag, Basel.

O'Dell, B.L. (1976). 'Biochemistry and physiology of copper in vertebrates', In Trace Elements in Human Health and Disease, vv 1 Zinc and Copper, A.S. Presad, Eds., chapter 24, Acad. Press, New York.

Oh, S.H., Deagen, J.T., Whanger, P.D., and Weswig, P.H. (1978). 'Biological function of metallothionein. V. Its induction in rats by various stresses', Am. J. Physiol. 234, E282-E285.

Onishi, T., Inubushi, H., Tokugawa, S., Muramatsu, M., Nishikawa, K., Suzuki, Y., and Miyao, M. (1980). 'Abnormal copper metabolism in Menkes' cultured fibroblasts', Eur. J. Pediat. 134, 205-210.

Onishi, T., Suzue, J., Nishikawa, K., Muramatsu, M., and Miyao, M. (1981). 'Nature of copper and zinc compounds in tissues from a patient with Menkes' kinky hair syndrome', Eur. J. Pediatr. 137, 17-21.

Otvos, J.D., and Armitage, I.M. (1980). 'Structure of the metal clusters in rabbit liver metallothionein', Proc. Natl. Acad. Sci. USA 77, 7094-7098.

Owen, C.A. (1965). 'Metabolism of radiocopper (^{64}Cu) in the rat', Am. J. Physiol. 209, 900-904.

Palmiter, R.D., Chen, H.Y., and Brinster, R.L. (1982). 'Differential regulation of metallothionein thymidine kinase fusion genes in transgenic mice and their offspring', Cell 29, 701-710.

Pelham, H.R.B. (1982). 'A regulatory upstream promoter element in the Drosophila hsp70 heat-shock gene', Cell 30, 517-528.

Peltonen, L., Kuivaniemi, H., Palotie, A., Horn, N., Kaitila, I., and Kivirikko (1983). 'Alterationts in copper and collagen metabolism in the Menkes' syndrome', J. Biol. Chem. 22, 6156-6163.

Port, A.E., and Hunt, D.M. (1979). 'A study of the copper-binding proteins in liver and kidney tissue of neonatal and mottled mutant mice', Biochem. J. 183, 721-730.

Prins, H.W., and van der Hamer, C.J.A. (1980). 'Abnormal copper-thionein synthesis and impaired copper utilization in mutated brindled mice: Model for Menkes' disease', J. Nutr. 110, 151-157.

Procopis, P., Camakaris, J., and Danks, D.M. (1981). 'A mild form
of Menkes' steely hair syndrome', J. Pediatr. 98, 97-99.

Rennart, D.M., Chan, W.Y., Cushing, W., and Griesmann, G. (1980).
'Polyamine metabolism in Menkes' kinky hair disease', Clin.
Chem. Acta 103, 375-380.

Richards, R.I., Heguy, A., and Karin, M. (1984). 'Structural and
functional analysis of the human metallothionein-I_A gene:
differential induction by metal ions and glucocorticoids', Cell
37, 263-272.

Riordan, J.R., and Joelicoeur-Paquet, L. (1982). 'Metallothionein
accumulation may account for intracellular copper retention in
Menkes' disease', J. Biol. Chem. 257, 4639-4645.

Robins, D.M., Paek, I., Seeburg, P.H., Axel, R. (1982). 'Regulated
expression of human growth hormone genes in mouse cells' Cell
29, 623-631.

Rohne, A., Krub, J.P., Mennesson, M., Mandel, P., Mack, G., and
Zawislak, R. (1977). 'Maladie de Menkes. Etudes the 2 enzymes
cupro dependentes', Pediatrie 32, 447.

Royce, P.M., Camarakis, J., and Danks, D.M. (1980). 'Reduced lysyl
oxidase activity in skin fibroblasts from patients with Menkes'
syndrome', Biochem. J. 192, 579-586.

Royce, P.M., Camakaris, J., Mann, J.R., and Danks, D.M. (1982).
'Copper metabolism in mottled mouse mutants: effects of copper
therapy upon lysyl oxidase activity in brindled (Mo^{br}) mice',
Biochem. J. 202, 369-371.

Rowe, D.W., McGoodwin, E.B., Martin, G.R., Sussman, M.D., Grahn,
D., Faris, B., and Franzblau, C. (1974). 'A sex linked defect in
the cross-linking of collagen and elastin associated with the
mottled locus in mice', J. Exp. Med. 139, 180-192.

Sarkar, B. (1981). 'Transport of copper', in Metal Ions in
Biological Systems, H. Sigel, Ed., chapter 6, Marcel Dekker, New
York.

Schmidt, C.J., Hamer, D.H., McBride, O.W. (1984). 'Chromosomal
location of human metallothionein genes: implications for
Menkes' disease', Science 224, 1104-1106.

Schmidt, C.J., Jubier, M.F, and Hamer, D.H. (1985). 'Structure and
expression of two human metallothionein-I isoform genes and a
related pseudogene', J. Biol. Chem. 260, 7731-7737.

Scheidereit, C., Geisse, S., Westphal, H.M., Beato, M. (1983).
'The glucocorticoid receptor binds to defined nucleotide
sequences near the promoter of mouse mammary tumour virus',
Nature 304, 749-752.

Seguin, C., Felber, B.K., Carter, A.D., Hamer, D.H. (1984). 'Competition for cellular factors that activate metallothionein gene transcription', Nature 312, 781-785.

Shows, T.B., Sakaguchi, A.Y., Naylor, S.L. (1982). 'Mapping of the human genome cloned genes, DNA polymorphism and inherited disease', in Advances in human genetics, H. Harris and K. Hirschhorn), Eds., chapter 5, Plenum Press, New York.

Sternlieb, I. (1980). 'Copper and the liver', Gastroenterology 78, 1615-1628.

Sternlieb, I., and Schienberg, H. (1979). 'The role of radiocopper in the diagnosis of Wilson's disease', Gastroenterology 77, 138-142.

Stuart, G., Searle, P.F., Chen, H.Y., Brinster, R.L., and Palmiter, R.D. (1984). 'A 12-base-pair DNA motif that is repeated several times in metallothionein gene promoters confers metal regulation to a heterologous gene', 81, 7318-7322.

Van Doren, N.K., and Gluzman, T. (1984). 'Efficient transformation of human fibroblasts by adenovirus-Simian virus 40 recombinants', Mol. Cell. Biol. 4, 1653-1656.

Varshney, U., and Gedamu, L. (1984). 'Human metallothionein MT-I and MT-II processed genes', Gene 31, 135-145.

Webb, M. (1972). 'Binding of cadmium by rat liver and kidney', Biochem. Pharmacol. 21, 2751-2765.

Wieacker, P., Horn, N., Pearson, P., Wienker, T.F., McKay, E., and Ropers, H.H. (1983). 'Menkes kinky hair disease: a search for closely linked restriction fragment length polymorphism', Hum. Genet. 64, 139-142.

Wieacker, P., Horn, N. and Ropers, H.H. (1981). 'Abnormal copper metabolism in cultured cells of mottled mice: no correction by hybridization with human fibroblasts', Mouse Newsletter 65, 14-15.

Wienker, Wieacker, P., Cooke, H.J., Horn, N., and Ropers. H.H. (1983). 'Evidence that the Menkes locus maps on proximal Xp', Hum. Genet. 65, 72-73.

Willemse, J., Van Den Hamer, C.J.A., and Prins, H.W., and Jonker, P.L. (1982) 'Menkes' kinky hair disease I. Comparison of classical and unusual clinical and biochemical features in two patients', Brain Dev. 4, 105-114.

Williams, D.M., Atkin, C.L., Frens, D.B., and Bray, P.F. (1977). 'Menkes' kinky hair syndrome: studies of copper metabolism and long term copper therapy', Pediatr. Res. 11, 823-826.

Williams, D.M., Atkin. C.L. (1981). 'Tissue copper concentrations
of patients with Menkes' kinky hair disease', Am. J. Dis. Child.
135, 375-376.

Winge, D.R., and Miklossy, K.A. (1982). 'Domain nature of
metallothionein', J. Biol. Chem. 257, 3471-3476.

Yazaki, M., Wada, Y., Kojima, Y., Tanaka, A., Issiki, G.,
Kawamura, N., and Hirose, M. (1983). 'Copper-binding proteins in
the liver and kidney from the patients with Menkes' kinky hair
disease', Tohoku J. Exp. Med. 139, 97.

β-GLOBIN GENE DISORDERS IN ITALY AND THE MEDITERRANEAN AREA

S.Ottolenghi and C.Carestia

1. INTRODUCTION

The natural history of β-globin gene disorders and of the mutations by which they are caused begins very far in the past and to understand when they were produced and by what mechanisms they were diffused is so interesting that it has been extensively studied. Certainly, the possibility of identifying the mutations and of following a mutation through a linkage to specific haplotypes will open vaster horizons in this field.

The scientific history of β-globin gene disorders begins in 1925 with the communication of Thomas Cooley to the "Transaction of the American Pediatric Society" on the first description of the clinical phenoptype of "Cooley's anemia". From then on, three periods can be easily distinguished, corresponding to three different aspects of this subject. The first, lasting to about 1955, dealt with the definition at the clinical level of Cooley's anemia, and of the heterozygous condition, named in Italy by E. Silvestroni "microcytic anemia". The second, from about 1955 to about 1975, dealt with the characterization of hemoglobin variants by fingerprint and other biochemical techniques, as well as the study of the biosynthetic patterns of globin chains and of the mRNA metabolism aspects. The third period deals with the description of mutations causing different clinical syndromes and their molecular mechanisms; with questions about mechanisms regulating the differential expression of β-globin genes; with some particular aspects of the adult erythropoiesis, which seems to be probably regulated by biochemical factors inducing a cellular selection. All these aspects interact tightly, because the definition of a disorder needs several different investigations.

In this article we will first discuss mutations causing β-globin gene disorders and the molecular mechanisms by which they act with particular reference to conditions usually found in Italy and the Mediterranean area. A number of excellent review articles dealing with several aspects of β-thalassaemia have recently appeared (Spritz 1983, Collins 1984a, Nienhuis 1984). Second, we will discuss the methodological approach to diagnosis and prevention today. Topics on organization, structure and expression of β-globin genes and on classification of the disorders will introduce these subjects.

1.1 ORGANIZATION AND STRUCTURE OF THE β –GLOBIN GENES

β-globin genes are a family of five genes (Esfradiatis 1980),coding for very similar polypeptide chains, named ε , ^{G}γ , ^{A}γ , δ and β ; they are all of 146 aminoacids (Dayhoff 1972, Gale 1979) and differ from each other at a maximum of 39 positions (γ from β) and at a minimum of 10 positions (δ from β)(Dayhoff 1972). They have been mapped on the short arm of chromosome 11 (Gusella 1979, Lebo 1979, Scott 1979) and are arranged on a segment of 60 kb, most of which represent intergenic regions. The genes are arranged in the order ε , ^{G}γ , ^{A}γ , δ , β 5' to 3' accor-ding to the direction of the transcription, as illustrated in Fig. 1. The duplicated ^{G}γ – and ^{A}γ –genes (Slightom 1980) code for polypeptide chains which differ by a single amino-acid at position 136 (Gly instead of Ala). A polymorphic ^{A}γ^{Sardinia}-globin gene at ^{A}γ locus has been described as having threonine instead of isoleucine at position 75 (Saglio 1979). A pseudogene ψβ , which does not code for a functional chain, has been found between ^{A}γ – and δ –globin genes (Fritsch 1980).

Polymorphic restriction enzyme sites have been described within coding regions, introns and flanking sequences, the most important of which are reported in Fig. 1. Specific association patterns of these polymorphisms (haplotypes) have been shown to have an unexpectedly high frequency in normal populations (see section 2.3); their use in the molecular cloning strategy of β –thalassaemia genes and diagnosis will be referred to later.

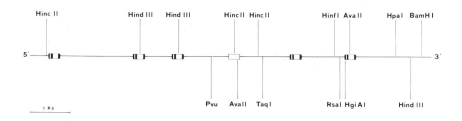

Fig.1 Non- α globin gene cluster. Some of the most important polymorphic restriction enzyme sites are represented; in the order Hinc II (Antonarakis 1982a), Hind III (Jeffreys 1979), Pvu II (Old 1983), Hind III (Jeffreys 1979), Ava II (Old 1984), Taq I (Maeda 1983), Hinf I (Moschonas 1982, Kohen 1982), Rsa I (Semenza 1984b), HgiAI (Orkin 1982a), Ava II (Antonorakis 1982a), Hpa I (Kan 1978), Bam HI (Kan 1980), Hind III (Tuan 1983).

     The structure of these genes is schematically represented
in Fig. 2. Each gene consists of three exons separated by two
introns; transcription proceeds from the 5' end of the first exon
to the 3' end of the third exon. At least three sequences exist
upstream of the first exon, that are considered to be essential in
regulating transcriptional activity: the so called TATA, CCAT and
CACC boxes (Dierks 1983). In addition, conserved sequences exist
at the 5' and 3' borders of the introns, presumably necessary for
the correct splicing of the primary transcript to yield the
processed mature mRNA (Mount 1982). Finally, a polyadenilation
signal (AATAAA) responsible for post-transcriptional addition of
poly-A tracts to the pre-mRNA exists in the third exon, as well
as possibly a more distant termination site (Proudfoot 1976,
Salditt-Georgieff 1983, Gil 1984).

                    1.2 GLOBIN GENE EXPRESSION

     The first globin genes to be expressed are the ζ- (an α
type gene) and the ε-gene (a non-α globin gene), whose chains
form Hb Gower I ($\zeta2\varepsilon2$), the main hemoglobin in the first few weeks
of gestation (Wheatherall 1983a, Peschle 1985a). Later on, (at
5-7 weeks of gestation) α- and γ-globin synthesis is activated,
while ζ- and ε-globin synthesis is repressed and compleletely
disappears by the 18th week. At this intermediate stage, Hb
Gower II ($\alpha2\,\varepsilon\,2$), Hb Portland ($\zeta2\,\gamma\,2$) and Hb F ($\alpha2\,\gamma\,2$) appear.
     Hb A ($\alpha2\,\beta\,2$) is first detected at the 7-8th week, when it
represents 2-3% of the total hemoglobin;its proportion steadily
increases up to 20% until birth, when the switch to the final
hemoglobin synthetic pattern rapidly occurs, and within a few
weeks, Hb F declines almost to the adult level of less than 1%,
with a major change in the relative representation of $^G\gamma$- and
$^A\gamma$-globin chains (fetal $^G\gamma/^A\gamma$ ratio 3:1;adult ratio 2:3); Hb F is
thus almost completely replaced by adult hemoglobins Hb A :97% and
HB A$_2$ ($\alpha2\delta2$):2,5%. The residual Hb F appears to be confined to a
small proportion (0,2-7%) of erythroid cells (so called F cells);
no Hb F can be detected by present methods in the remaining cells
(Boyer 1975, Wood 1975).
     Throughout all these changes, the synthesis of α-type
globin chains ($_A\zeta$ and α) is always equal to that of non-α globin
chains (ε,$^G\gamma$,$^A\gamma$, δ and β); the mechanisms regulating the relative activities of the unlinked α-like and non-α globin clusters
remain unknown. Similarly, little is known of the precise molecular mechanisms regulating the embryonic fetal→ adult switches.
     A number of studies have been carried out using non-dif-
ferentiated erythroid precursors stimulated to divide and undergo
maturation in culture; these experiments indicate that the hemoglobin switch occurs in a single population of cells (from the
embryonic to the fetal and adult periods) which is progressively
reprogrammed during development to express the appropriate
genes (Papayannopoulou 1977, Comi 1980, Peschle 1980). These
experiments also show that the relative activities of all
non-α globin genes are almost precisely coordinated during

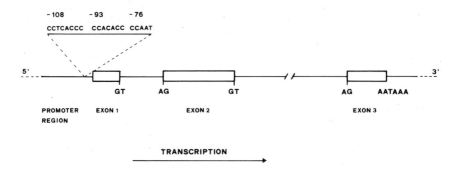

Fig.2   Schematic structure of the β-globin gene. Boxes represent the three exons; straight lines the introns. At the 5' side of the gene three conserved sequences, considered to be necessary for promoter functions. At the exon-intron junctions GT and AG are essential part of a splice signal; AATAAA is the poly-A addition signal.

these transitions so that the $^G\gamma /^G\gamma + ^A\gamma$ ratios are directly correlated with $\gamma /\gamma + \beta$ ratios (Comi 1980). Interestingly both "in vivo" and "in culture" studies indicate that γ-globin synthesis tends to precede β-globin synthesis during cell maturation(Papayannoulou 1979, Gianni 1980, Dover 1983). It has been suggested that a transient phase of γ-globin synthesis may be a property of all early erythroblasts, not only during the fetal but also during the adult period (Peschle 1981, Peschle 1985b); why the synthesis of significant amounts of Hb F remains then confined to few adult erythroid cells is still a matter of speculation (Peschle 1985b).

1.3   INHERITED DISORDERS OF β-GLOBIN GENES

Mutations of the β-globin gene cluster have been reported to produce alterations of the expression. These can be divided in two classes.

In the first, point mutations or small deletions affect the expression of a single gene, at the quantitative or qualitative level, without affecting the γ / β globin switch. The following belong to this class:

1) mutations that modify the characteristics of the final product of a gene. This is the case of the <u>variant hemoglobins</u>. A gene codes for a globin with one or more substituted,deleted or added aminoacids. Several mutations of this kind have been de-

scribed until now, the most commonly known being the Hb S-variant, that is present in Africa and much less so in Southern Italy and Greece. The clinical importance of variant hemoglobins depends on the physical and chemical characteristics of the new globin chain produced. The molecular mechanisms at the genetic level are clear and we do not consider it necessary to treat this extensive subject here.

2) mutations that modify the level of the expression of one gene, from its decrease down to its disappearance; these disorders, named "thalassaemia", can affect any globin gene. The known mutations of this type may alter the RNA transcription, splicing or translation process. The β-thalassaemia mutations are the most important in this group because of their clinical relevance. So far, up to 30 different mutations causing β-thalassaemia have been described; they can lead to β°-(absent β-globin gene expression), or β^{+}-thalassaemia (low expression) of a variable degree, from severe to mild forms.

3) mutations that modify the final product of a gene and at the same time the level of the expression of the same gene. This is the case of the variant hemoglobins causing also thalassaemia mutations, such as Hb E and Hb Knossos.

The second class of defects is represented by mutations altering the activities of several (fetal and adult) genes. These conditions lead to persistent high levels of Hb F in adults.

Inherited disorders of the expression of the β-globin genes not linked to the β-globin gene cluster have been reported. They can involve either the expression of single genes(β-thalassaemia) or the abnormal synthesis of γ-globin chains. The described cases will be discussed in the next section.

2. β-THALASSAEMIA

β-thalassaemia is the most frequent cause of genetic diseases in the Mediterranean area; its frequency varies from 2-3% to 10-15% -even 20%- in regions along the Mediterranean coast. The regions with a higher frequency of β-thalassaemia mutations are: Sardinia, Sicily, Southern Italy, the Po river Delta, Greece, Turkey, Crete, Cyprus, Algeria, Southern Spain and Portugal. This distribution resembles that of malaria to which the high frequency of β-thalassaemia mutations has been ascribed.

2.1   HEMATOLOGICAL AND BIOCHEMICAL PHENOTYPES CAUSED BY β-THALASSAEMIA DISORDERS

A mutation can be defined "β-thalassaemic" if it produces a decrease in the synthesis of the β-globin chain synthesis. Hence by definition, a heterozygote for a β-thalassaemia mutation is characterized by an imbalance of the biosynthetic ratio between β- and α-globin chains. This ratio, normally equal to 1, decreases to values around 0,5 with some variability, due principally to the type of mutation. The excess α-chains in thalassaemics may be destroyed in part by activation of a proteolytic process; the remainig α-chains, by precipitation within the cell, lead to the decreased life span of the erythrocytes and to anemia. However, compensatory mechanisms contribute in decreasing the excess of α-chains, thus improving the survival of red cells. The major mechanism consists in an increased synthesis of β-globin chains directed by the normal β-globin gene in trans. In addition an increased synthesis of δ-globin chains is observed, detectable as an increase of HbA$_2$ measured in % of the total Hb (>3.5%) or as total quantity per red cell (>0.75 pg per red cell). It is not clear whether this effect depends on an increased rate of transcription of the β- and δ-globin genes or whether it is a post-transcriptional event. These two mechanisms ensure a quantity of Hb per cell of about 20 pg, instead of the 15 pg expected because of the decreased activity of the β-globin gene. From the medical point of view this is a decrease of MCH (mean cell hemoglobin); the physical decrease of Hb per red cell leads to a decrease of MCV (mean cell volume) to about 60-70 fl (μ^3) as well as to morphological modifications of the cells (microcytosis, hypochromia, anisocytosis). An additional compensatory mechanism, at the cellular proliferation level, is the increased number of red cells; because of these processes, the heterozygous subjects may reach almost normal levels of Hb.

The described hematological findings (the decrease of MCH and MCV and the increase of Hb A$_2$) are present in most heterozygotes. However in a small percentage of cases these abnormalities are not clearly detectable.

The absence of red cell alterations but modifications with high value of Hb A$_2$ have been observed in patients who are heterozygotes also for α-thalassaemia mutations (Melis 1983a, Kanavakis 1982). The mechanism of this effect is unknown, but it is suggested that the lack of an excess of α-globin chains and therefore of cellular damage, may be allowing the maturing erythroblasts to better achieve compensation by either improved expression of the normal β-globin gene or changes in the cell cycle kynetics.

The presence of red cell alterations without increase of Hb A$_2$ has also been reported (Kattamis 1979) and probably reflects a double heterozygosis due to β- and δ-thalassaemia mutations.

Certain β-thalassaemia mutations whithout hematological alterations have been named "silent" One type is due to variant hemoglobins, not easily detectable by usual laboratory methods, such as Hb Knossos (Fessas 1982) which also causes β-thalassaemia (in cis). Another type is the only one type of β-thalassaemia inherited not in linkage with the β-globin gene cluster (Schwartz 1969). Determination of the biosynthetic ratio between β- and α-globin chains is an important diagnostic step to define the thalassaemic phenotype in these cases.

At the clinical level β-thalassaemia heterozygotes are essentially asymptomatic because of the compensatory mechanisms previously described; most carriers can be detected, however, by simple and relatively inexpensive laboratory methods, which constitute the basis for populations screening.

On the contrary the presence of two β -thalassaemia mutations in the same genome causes severe anemia. The biosynthetic
ratio between β- and α-globin chains ranges from 0 to 0.20.
This severe deficit of β -chain synthesis causes massive α -chain
precipitation, severe loss of maturing erythroblasts in the bone
marrow (ineffective erythropoiesis) and vastly decreased red cell
survival, and hence anemia. This in turn stimulates (via erythro-
poietin) bone marrow expansion, leading to active and diffuse
erytropoiesis. The peripheral Hb may be as low as 3-4 g/dl and is
mostly represented by HbF (80-98%).

It is important to note that the percentage of Hb F is not
a reliable indicator of the gravity of the disease; rather,
clinical conditions depend on the level of the hemoglobin that can
be maintained without transfusions. Thus, patients with 100% HbF
and 15 g Hb/dl (as in homozygous HPFH) are clinically well, while
patients with 98% Hb F and 5 g Hb/dl (as in homozygous β°-tha-
lassaemia) are obviously in a very poor condition.

The mechanisms for the increased Hb F levels are different
in these two conditions; in the former case a genuine hyper-
activity of γ -globin genes is observed; in the latter, γ -globin
gene activity remains low as normally, but only those cells
(F-cells) that already produce γ -globin chains are able to buffer
the excess of α -chains and to survive in the circulation.
Because of the great expansion of the bone marrow, even such a
small proportion of cells is able to reach a sizeable mass.

Patients with two β -thalassaemia mutations in general
present the symptoms of a severe anemia around the age of six
months, and they subsequently require red cell transfusions
throughout all their lives. However, because of the wide heterogeneity of the molecular alterations produced by the different
mutations, the symptoms of the disease can appear later and in
milder forms. In some cases it can be difficult to decide if and
when to begin transfusional therapy, which is directed to
maintain the level of 10-11 g/dl of Hb so as to prevent excess
erythopoiesis and its consequent alterations. This means a
hyper-transfusional regimen (about 1.5 blood transfusion per
month) which has in part abolished the clinical differences
between patients.

In a sizeable number of cases (5-10%), patients with
two β -thalassaemia mutations show mild clinical disorders, and
can survive without red cell transfusions. This condition is one
of those giving "Thalassaemia Intermedia", which will be treated
separately, as it represents a different clinical situation.

2.2   THE MOLECULAR MECHANISMS OF β -THALASSAEMIA

Mutations causing β-thalassaemia may affect transcription, processing or translation.

A. Mutations affecting transcription (promoter mutations)

A mutation at position -87 relative to the transcription initiation site (Cap site) has been found in Italian and Cypriot patients (Orkin 1982a, Treisman 1983) and is responsible for highly depressed transcription of the cloned gene when examined upon reintroduction in the cell for a transient expression assay. The position of this mutation corresponds to the CACC box (independently defined by studies with the rabbit β-globin gene as a conserved sequence of functional significance) (Dierks 1983). These studies, therefore, strongly indicate that this mutation may be responsible for the in vivo phenotype. This is the only example of the transcriptional defect in Mediterraneans although mutations of the TATA box have been found in other ethnic groups (Poncz 1982, Orkin 1983, Antonarakis 1984a).

B. Mutations affecting RNA-processing

These mutations generally decrease the frequency of normal splicing, generating abnormal mRNAs that cannot be translated into normal proteins and decay rapidly; accumulation of an unspliced precursor may also occur.
The precise removal of RNA corresponding to introns requires identification of the invariant GT and AG dinucleotides at the 5' and 3' of introns (Breathnach 1978); these sequences, which are part of a less precisely conserved consensus sequence (Mount 1982), may participate in the pairing with small nuclear RNA, named U1RNA (Rogers 1980). Short regions of partial homology with the consensus sequences are scattered along the β-globin gene, but are seldom (or never) used during normal processing.

B.1 Mutations that alter splice sites

These mutations may either hit the invariant GT dinucleotide or the consensus sequence surrounding it (Orkin 1982b, Treisman 1983). While in the former case, the result is the complete inhibition of normal splicing (hence β°-thalassaemia), in the latter case normal mRNA can still be generated albeit at a lower frequency (hence β⁺-thalassaemia).   In either case, abnormal (cryptic) splice sites become prominent (Treisman 1983, Antonorakis 1984a), indicating that the infrequent utilization of these sites during normal processing may be due to competition with the normal sites. Delayed processing of intron 1 RNA may also be caused by a mutation at the exon-2-intron-2 junction (Treisman 1982).

B.2 Mutations that create an alternative splice site,
either in exons or in introns

These mutations may create either a new 5' (donor) or a
new 3' (acceptor) site. A GT dinucleotide within a partially
conserved consensus sequence is present on codon 25 (first exon)
(Fig. 3); this is the splice site found to be activated in the
preceding example (Treisman 1983). Three different mutations
alter this splice site, strongly activating it and leading to
concurrent decrease of the normal splicing process at the nearby
normal GT site; of these mutations, one (found in Black
β-thalassaemics)(Goldsmith 1983) is silent at the aminoacid level,
while two (Arous 1982, Orkin 1982b, Orkin 1984a) generate the
mutant $\beta^{E26glu\to lys}$, and the $\beta^{Knossos\ 27ala\to ser}$ -globin chains.
Because the normal splice sites continue to be used at low
frequency, these mutations cause β^{+}-thalassaemia of modest
gravity; in particular, Hb Knossos, which is not distinguishable
from Hb A by standard electrophoretic methods (Fessas 1982),

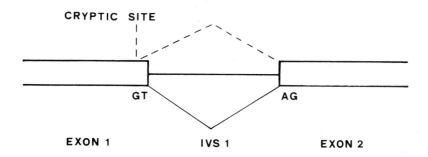

		(β^{+})	(β^{E})	$(\beta^{KNOSSOS})$
		A	A	T
CRYPTIC SITE		G G T G G T	G A G	G C C
CODONS		24 25	26	27
POSITION N T		125	129	132
NORMAL IVS1		C A G G T	T G G	T
CONSENSUS		$^{C}_{A}$ A G G T	$^{A}_{G}$ A G	T

Fig.3 β-thalassaemia mutations creating alternative splice sites
in the exon 1. The solid lines indicates the normal splice, dashed
lines the abnormal splice. The similarities between the cryptic
site in the exon 1 and consensus splice signals in normal intron 1
(IVS I) are shown at the top; three types of mutations (see text)
are reported.

mimics a "silent" β-thalassaemia gene (see section on "Thalassaemia Intermedia"). A GT (5') site may also be created within an
intron, again leading to competition with the corresponding normal
site (Orkin 1982a, Treisman 1983)).

Two mutations (Spence 1982, Dobkin 1983) generating new GT
sites (in IVS II nt 745 and 705, respectively) appear to be
particulary interesting because the new sites also lead to the
activation of an AG site 5'to the mutation, used in conjunction
with the normal GT site (Fig.4). At least one of these mutations
(IVS II nt 745) still allows a substantial utilization of the
normal GT-AG signals, hence generating β[+]-thalassaemia.

Finally, an AG site is created within the intron 1 by a
mutation at nt 110, leading to the common β-thalassaemia
originally found in Cyprus (Busslinger 1981, Spritz 1981,
Westaway 1981, Fukumaki 1982); this mutation causes 90% of the
splicing events to become abnormal, hence leading to severe
β[+]-thalassaemia. It is of interest that the defect in the
intron 1 also delays processing of intron 2 sequences; moreover
the abnormal mRNA sequence generated by the use of the new splice
site appears to be very unstable, at least in reticulocytes,
possibly because it cannot be fully translated due to a shift in
the reading frame that generates an abnormal stop codon within the
exon 2.

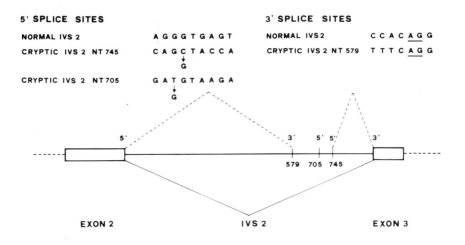

Fig.4   β-thalassaemia mutations creating alternative splice
sites in the intron 2. The solid lines indicate the normal splice,
the dashed lines the abnormal splice. The similarities between
cryptic and normal splice sites are indicated. Two mutations
(nt 705 and nt 745) are shown.

C.   Mutations affecting mRNA translation

One of the most frequent β -thalassaemias in the Mediterranean area is caused by a nucleotide substitution converting
codon 39 from a glutamine (CAG) to a non-sense signal (TAG),
thereby leading to premature termination (Moschonas 1981, Orkin
1981, Pergolizzi 1981, Trecartin 1981, Gorski 1982).   The
β-globin mRNA level in these patients is low, suggesting that the
untraslated mRNA may be degraded at an abnormally high rate.
However, when a cloned β-globin gene containing this mutation is
introduced into monkey cells, the level of accumulated β-globin
mRNA is lower by a factor of 8 than that of a normal β-globin
gene (Humphries 1984). This result is apparently not dependent
on altered processing or instability in the cytoplasm, and might
result from defective nuclear-cytoplasmic transport. This is in
agreement with data suggesting an intranuclear defect in the
"accumulation" of mutant β^{39}-mRNA in erythroblasts (Takeshita
1984).

A number of additional defects at the translational level
are caused either by some of the splicing mutations discussed
above (Spritz 1981, Chang 1983) or by deletions of nucleotides
in the exon sequence (Orkin 1981, Takeshita 1984) leading to
altered reading frames, in turn generating in-phase termination
codons. As in the case of the codon 39 non-sense mutation,
these abnormal mRNAs may be present at a low level.

D.   Mutations unlinked to the β -globin gene

The previously described mutations affect only the
β-globin gene in cis. Several years ago Schwartz (1969) reported
two children in a family of Albanian descent with apparently
homozygous   β -thalassaemia of intermediate gravity.   Their
Hb level was 8 g/dl, with 80-90% Hb A, and substantially
decreased β / α globin synthetic ratios (0.3). Surprisingly, the
father was hematologically "normal" although his β / α globin
synthetic ratio was decreased.

Recent analyses of the family indicate that the mater-
nally-derived β-thalassaemia gene is of the β -thalassaemic
type, while the father transmitted either one or the other
β -globin gene to his two "homozygous" children (Semenza 1984a).
The paternal cloned β -globin genes were completely normal as to
nucleotide sequences and activity when introduced into cells.
This is the first evidence that a β-thalassaemia determinant
unlinked to the β -globin gene can exist; the defective gene might
be coding for a transacting factor necessary for one of the many
steps required for the globin chain synthesis.

2.3 LINKAGE TO HAPLOTYPES OF β-THALASSAEMIA MUTATIONS

Restriction fragment polymorphisms have been detected in the non-α globin gene cluster (Fig.1) (Kan 1978, Jeffreys 1979, Kan 1980, Driscoll 1981, Antonarakis 1982a, Kohen 1982, Moschonas 1982, Maeda 1983, Old 1983, Tuan 1983, Semenza 1984b, Wainscoat 1984) and extensive evidence for the existence of linkage disequilibrium between these sites has been provided mainly by the works of Orkin's and Kazazian's groups (Orkin 1982a, Antonarakis 1982a).

Two clusters exist within which linkage disequilibrium can be demonstrated, namely a 5' 32 kb region from the 5' end of the ε-globin gene to the 3' end of the δ-globin gene and a 3' region extending from the 5' end of the β-globin gene to approximately 16 kb beyond it. These two clusters are separated by an approximately 9 kb long region, within which a high recombination rate occurs which is responsible for the absence of linkage disequilibrium between sites contained in the 5' and

HINC II	HIND III	HIND III	HINC II	HINC II	AVA II	BAM HI		
+	−	−	−	−	+	+	I	
−	+	+	−	+	+	+	II	FRAMEWORK 1
−	+	−	+	+	+	+	IX	
−	+	−	+	+	+	−	III	
+	−	−	−	−	+	−	Va	FRAMEWORK 2
+	−	−	−	−	+	−	Vb	
−	+	−	+	−	+	−	VIII	
−	+	−	+	+	−	+	IV	
−	+	+	−	−	−	+	VI	FRAMEWORK 3
+	−	−	−	−	−	+	VII	
−	+	−	−	−	−	+	X	

Fig.5   Haplotypes as described by Orkin et al (1982a), grouped on the basis of the β-gene framework. Haplotypes Va and Vb differ by the presence or the absence respectively of a polymorphic Hinf I site 5' to the β-globin gene. Haplotype I can be further distinguished into two haplotypes by the presence or the absence of a polymorphic Ava II site 3' to the ψβ-gene. This site is present on the cluster carrying the IVS I nt 110 mutation, but absent on most normal Haplotypes I in the Cypriot population (Wainscoat 1984).

3' clusters. This 9 kb region is defined by a polymorphic Taq I site 5' to the δ-globin gene and the polymorphic HgiA I site in the β-globin gene (Orkin 1982a, Maeda 1983, Chakravarti 1984).

Additional polymorphic changes are detectable within the β-globin gene by DNA sequencing (Fig.6) (Orkin 1982a), which contribute to define three different normal β-globin gene structures, called "frameworks".

Three different frameworks are known (Fig.6): Framework 2 differs from Framework 1 by a single nucleotide substitution (G-T) in the large intron, while Framework 3 is distinguished from Framework 2 by four additional polymorphic mutations (three in the large intron and one in the first exon); two of these mutations are detectable by the disappearance of a restriction enzyme site (for HgiA I and Ava II, respectively).

Each type of β-globin gene (framework) can be associated with different 5' cluster haplotypes (Fig. 5). But out of all the possible associations only 10 major haplotypes are detectable in the Mediterranean population (Fig 5), with some variability of the respective frequencies in different areas (Table I).

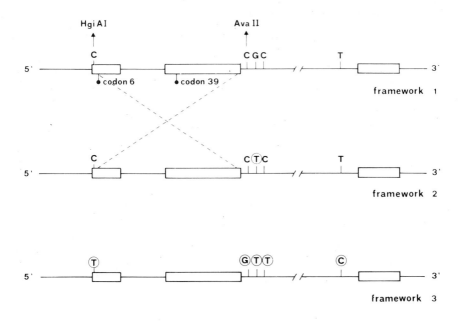

Fig.6 The figure shows the position of two different β-thalassaemia mutations (in codon 6 and 39 respectively), relative to polymorphic sites in the β-globin gene defining three frameworks. Polymorphic nucleotides are indicated. The scheme shows how a mutation at codon 6 can spread from Framework 1 to Framework 2 by a simple crossing over event; on the other hand the mutation at codon 39 cannot spread by the same mechanism from Framework 1 to Framework 3.

To each of these haplotypes specific β -thalassaemia
mutations in the Mediterranean area had originally been found to
be linked (Orkin 1982a) (Fig.7), based on sequence analysis
of cloned β -thalassaemia genes; as each haplotype appeared to
correspond to a different β -thalassaemia gene, identification of
the "new" β -thalassaemia mutations in each population had been
made easier by the strategy of cloning and sequencing only
those β-thalassaemia genes appearing to be linked to "new" haplotypes. Although this strategy has the theoretical disadvantage
(Robertson 1983) that, when two different mutations occur in
β -globin genes linked to the same haplotype, the rarer mutation
may probably be overlooked, it has in practice worked remarkably
well. In fact, of 156 Mediterranean β-thalassaemia chromosomes
studied Kazazian (Kazazian 1984a) had found only six to be
unassignable (by hybridization with oligonucleotides or direct
digestion with restriction enzymes) to any of the previously
known mutations. (Two mutations, Hb Knossos and IVS II nt 705 had
not been investigated).

However, the association between a specific β-thalassaemia
mutation and a specific haplotype is not absolute (Kazazian
1984a); for example of 53 chromosomes carrying the IVS I nt 110
mutation, 51 are characterized by Haplotype I, the two remaining
ones by Haplotypes II and IX respectively. These data suggest
that certain mutations "spread" from one chromosome background to
another, the most striking example being that of nonsense codon
39 β -thalassaemia which spread from its common Haplotype II
association to at least eight new chromosome environments
(Pirastu 1983a).

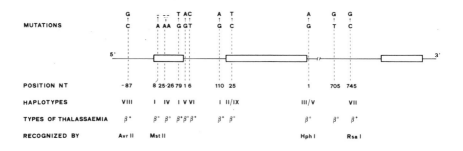

Fig.7 β-thalassaemia mutations, as described in the text, are
represented on a globin gene scheme. On the first line the type
of mutation is shown; - indicate deletion; position in nucleotide
order, starting from nucleotide 1 of the intron and exon respectively; haplotypes associated most commonly to the mutations are
indicated as well as the type of thalassaemia (β ° or β [+]).
Specific mutations can be recognized by the disappearance of
Avr II, Mst II or Hph I sites and the appearance of Rsa I site.

Similarly, the frameshift codon 6 mutation can be found in Haplotypes I, Va and IX (Kazazian 1984a). This phenomenon probably occurs through recombination between the two clusters discussed above. As Haplotype I is very common (34%) among normal chromosomes, while haplotype II is relatively rare (7%) (Antonarakis 1982a), it is not surprising that the IVS I nt 110 mutation has apparently spread much less than the codon 39 non-sense mutation, since recombination of the former type of β-thalassaemia gene would occur in many cases with a chromosome with similar haplotype; obviously, however, additional factors, such as for example the antiquity of the original mutation, affect this phenomenon.

In most instances of a mutation spreading from one haplotype to another, the mutation appears to remain in the same context of polymorphism of the β-globin gene (called framework) (Kazazian 1984a). Thus, the simple recombination event between the 5' and 3' clusters mentioned above may generate 8 out of 10 of the observed cases of spreading-mutations. A crossing-over event occurring at the level of the β-globin gene itself may also justify the appearance of the same mutation (codon 6 frameshift) in haplotypes I and IX (framework 1) and the haplotype Va (framework 2).

In all these instances the mutation probably occurred as a single rare event, fixed by selection in malarial areas. In only one case (presence of the codon 39 non-sense mutation in a haplotype VII chromosome) this event cannot be accounted for by any of these mechanisms, since haplotype VII is associated with framework 3 that cannot be generated from framework 1 by simple crossing over events .

Thus, it is possible that this mutation occurred twice independently in the same area, as also reported in Africa and Asia for β^E- and β^S-genes (Antonarakis 1982b, Antonarakis 1984b, Pagnier 1984).

The observation that individual mutant β-globin genes can be joined by recombination to different β-globin gene clusters has important implications for antenatal diagnosis of these disorders (Orkin 1984b); moreover as outlined in section 5, these recombination events may explain some differences in the phenotype of patients carrying the same β-thalassaemia mutation.

2.4   β-THALASSAEMIC MUTATIONS IN THE MEDITERRANEAN AREA: CHARACTERIZATION OF MUTATIONS AND HAPLOTYPES

Orkin's and Kazazian's original studies (Orkin 1982a, Antonarakis 1982a, Kazazian 1984a) have been carried out on mixed samples of patients of Mediterranean descent immigrated to the United States, and the relationship between haplotypes and β-thalassaemia mutations remains to be determined.

Preliminary studies have been carried out in Italy and other countries to define the main type of the β-thalassaemia.

This is of considerable importance both for the purpose of rendering the antenatal diagnosis of β-thalassaemia easier, and for tracing the origin and diffusion of the different mutations in different areas.

Data on the frequencies of different haplotypes in chromosomes carrying β-thalassaemia mutations in different samples of Italian origin are reported in Table II. Data on identification by oligonucleotides of specific mutations found in some of the described haplotypes are also given in Table III.

In this survey of some hundred chromosomes carrying β-thalassaemia mutations, seven only out of the ten haplotypes described in the preceding section have been found (Haplotype IV, VIII and X being the missing ones). Some additional haplotypes

β-Globin Gene Framework	Haplotypes	1	2	3	4	5
1	I	34%	40%	24%	50%	50%
	II	7%	10%	5%	–	–
	IX	12%	10%	5%	17%	7%
	- + - - + + +	–	–	5%	–	–
	+ - - + + + +	–	3%	–	–	–
	- - - - - + +	–	3%	–	–	–
2	III	7%	8%	–	3%	–
	V	18%	10%	18%	10%	–
	- + + - + + -	–	5%	5%	5%	–
	- + - - + + -	–	–	5%	–	–
	- + + - - + -	–	–	5%	–	–
3	VII	10%	3%	18%	3%	13%
	IV	3%	8%	–	7%	6%
	- + + - + - +	–	–	10%	7%	–
	- - - - - - +	–	–	–	–	6%
	Others	9%				19%
Tot.Chromosomes		122	39	21	30	16

TABLE I. Frequencies of restriction fragment lenght polymorphism haplotypes in the β-globin gene cluster in normal chromosomes from Italians (lines 1,2,3,4) and Algerians (line 5). Data have been obtained on a mixed sample in the case 1 (Kazazian 1984a), on a Campanian sample in the case 2 (Carestia, manuscript in preparation), and on Sardinian samples in the cases 3 and 4 (Wainscoat 1980, Cao, manuscript in preparation); in Algerians in the last case (Beljord 1983).

have been found: - - - - + + +, described in two individuals from the Southern Italy region of Puglia, had previously been reported in Africans (Antonarakis 1984a); - + + - - + +, described in Sardinia and in Campania, might derive from crossing-over between a 5' cluster typical of Haplotype VI and a 3' cluster common to Haplotypes I, II and IX; - + - - + + +, found in Sardinia, might derive from a crossing over between a rare 5' and the same common 3'cluster as above; - + + - + - +, described in Campania, also reported in Chinese and Indians (Cheng 1984, Kazazian 1984b), probably represents a crossing-over between the 5' cluster of Haplotytpe II and the β$^+$ IVS II nt 745 thalassaemia gene (usually found in Haplotype VII) as demonstrated by Rsa I digestion.

β-Globin Gene Framework	Haplotypes	1	2	3	4	5	6	7
1	I	31%	39%	41%	28%	39%	17%	6%
	II	24%	20%	24%	23%	15%	65%	75%
	IX	5%	4%	8%	10%	14%	13%	13%
	- + + - - + +	-	-	-	3%	-	3%	3%
	- + - - + + +	-	-	-	-	-	2%	3%
	- - - - + + +	-	-	0.5%	-	-	-	-
	- - - - - + +	-	-	-	-	6%	-	-
2	III	7%	7%	2%	3%	13%	-	-
	V	9%	13%	6%	11%	9%	-	-
	VIII	2%	1%	-	-	-	-	-
	- + + - + + -	-	-	-	-	1%	-	-
	- - - - + + -	-	-	-	-	2%	-	-
3	VI	11%	9%	13%	12%	-	-	-
	VII	11%	7%	5%	10%	-	-	-
	IV	-	1%	-	-	-	-	-
	X	-	1%	-	-	-	-	-
	- + + - + - +	-	-	-	-	-	-	-
	- - - - - - +	-	-	-	-	1%	-	-
Tot.Chromosomes		45	162	182	40	80	126	32

TABLE II.   Frequencies of restriction fragment lenght polymorphism haplotypes in the β-globin gene cluster in β-thalassaemia patients of Italian (lines 1,3,3,4,6,7) and Algerian origin (line 5). Data have been obtained on mixed samples in cases 1,2 and 3 (Orkin 1982a, Kazazian 1984a, Saglio manuscript in preparation); on a Campanian sample in the case 4 (Carestia manuscript in preparation); in Sardinian samples in the cases 6 and 7 (Wainscoat 1983a, Cao manuscript in preparation); in Algerians in case 5 (Beljord 1983).

It is most interesting that in Sardinia a single mutation (non-sense condon 39) is highly predominant, although linked to different haplotypes, but still having the same Framework 1, a fact suggesting a relatively ancient introduction of this mutation in the island followed by crossing-over (see discussion above). Only in the Nothern area of Sardinia (the $_+$ province of Sassari), however, an uncharacterized type of β -thalassaemia appear to be present at a relatively high frequency : 10-15% of β -thalassaemia genes (Longinotti 1982).

In the Po river Delta area three different haplotypes (I, II and IX) are found (Del Senno 1984, Pirastu 1984a); these correspond, as shown by hybridization to oligonucleotides, to IVS I nt 110 and codon 39 non-sense mutations. This limited heterogeneity of β -thalassaemia mutations and haplotypes contrasts with the variety found in other regions, and points to a relatively recent introduction of β -thalassaemia genes in this previously malarial area.

In other regions, as indicated by data collected on a mixed sample (Saglio, manuscript in preparation), and on a Campanian one (Carestia, manuscript in preparation) there is a wide heterogeneity of haplotypes, which underlines the hetero-geneity of mutations. Differences between data collected by Orkin and Kazazian and those collected on patients living in Italy seem to be very small, and refer to certain specific mutations, such as

Origin of samples	Oligonucleotides	HAPLOTYPES						
		I	II	IX	V	VI	altri	Tot.
Mixed	β^{39}	4	17					21
	β IVS1 nt110	21	1					22
	β IVS1 nt1				10			10
	β IVS1 nt6					6		6
Sardinia	β^{39}	21	82	16			6	125
Ferrara	β^{39}	1	4	17				22
	β IVS1 nt110	8						8

TABLE III.   Relationship between specific mutations (as detected byoligonucleotides) and haplotypes in Italian population. Data have been obtained on a mixed sample (Saglio manuscript in preparation); on a Sardinian sample (Cao manuscript in prepara-tion); on the Ferrara sample (Del Senno 1984, Pirastu 1984a).

IVS II nt 745 that is unusually common in Sicily (Maggio, personal communication) and in Campania, where it is present in Haplotype VII and a rare haplotype (Carestia, manuscript in preparation). The β^{39}-mutation is almost exclusive in Sardinia, while IVS I nt 110 is highly predominant in Cyprus.

These data, though preliminary, may have some practical influence on the antenatal diagnosis of β-thalassaemia; while in Sardinia this is feasible in most cases simply by hybridization of oligonucleotides specific for the codon 39 non-sense mutation, in other regions more extensive studies are necessary. However, the haplotype heterogeneity detected insures by itself that diagnosis by indirect restriction analysis of polymorphisms will almost always be feasible in families at risk, even if these originate, as commonly happens, from the same region. It should be pointed out that haplotype frequencies are not necessarily similar in the normal and thalassaemic populations in the same area (compare Table I and Table II), usually allowing observation of at least one polymorphic difference between the normal chromosome and thalassaemic one. For example, the recently detected Ava II polymorphism close to $\psi\beta$-gene greatly increases the feasibility of antenatal diagnosis in the Cypriot and Italian populations (Wainscoat 1984).

3. THALASSAEMIA INTERMEDIA

Thalassaemia intermedia is a rather common clinical entity, which -with the more sophisticated analyses at present available- is increasingly referred to the hematologist and to the pediatrician. Its frequency is about 5-10% as compared to thalassaemia major in Italy.

Patients with thalassaemia intermedia get along without transfusion at 6-9 g/dl or more of hemoglobin, thus resulting in a milder clinical course than is usual in Cooley's anemia patients. However, some of these patients may have to be transfused anyway to avoid skeletal deformities and the effects of severe anemia (Wheatherall 1983a). It is necessary to point out that different medical groups often adopt different operational definitions and treatments for these patients.

Hematological and biochemical alterations are extremely heterogeneous. A general pattern is represented by the deficit of the β-chain and increase of the γ-chain synthesis with a somewhat variable (but usually around 0.4-0.5) non-α / α biosynthesic ratio. β-globin synthesis may vary from "absent" to "high level" according to the different mutations. Similarly the Hb A_2 level is usually increased in β-thalassaemias, but may be reduced when other syndromes are responsible ($\delta\beta$-thalassaemia).

The difficulty to define clear hematological and biochemical patterns for thalassaemia intermedia is due to the fact that this syndrome may be caused by interaction of different genes,some of which have been defined by analysis at the DNA level. The relative data are reported below, but there are several situations where other unknown genetic and biochemical factors affect the clinical course of the disease.

Three genetic types of alteration causing thalassaemia intermedia have been clearly defined. The mechanisms by which they act are a quantitatively high level of synthesis of non- α globin chains or a decreased production of α -globin chains.

3.1 MILD DEFECTS IN β -GLOBIN SYNTHESIS

Mild defects in β -globin chain synthesis have been reported in thalassaemia intermedia and are of two types; silent and mild.

Silent β-thalassaemia gene is a condition where hematological parameters do not show any alteration in carriers who, detected as obligate heterozygotes, have only a decreased β / α globin biosynthetic ratio. Several such cases have been reported in Italy, most notably in Sicily (Musumeci 1983), and Campania (Mastrobuoni, personal communication) where the frequency can be 2-3% of thalassaemia carriers.

The silent β-thalassaemia gene, when interacting with a β ° or a β⁺ mutation, causes typical cases of thalassaemia intermedia, as reported in Sicily (Musumeci 1982), Greece (Fessas 1982), Campania (Mastrobuoni, personal communication). The mean value of Hb is between 6 and 9 g/dl. Two of these silent genes have been described in more detail.

Silent β -thalassaemia gene not in linkage with the β -globin gene cluster. It was first described by Schwartz (1969) in a family of Albanian descent, and has recently been demonstrated to be inherited indipendently of the β -globin gene cluster by different children of the same family (Semenza 1984a) (see Section 2.1).

The original silent β -thalassaemia of Albanian descent might be present in Southern Italy, where immigration from the opposite coast occurred and where Albanian villages are still reported.

Silent β-chain variant linked to silent β -thalassaemia. The only cases of this type have been described in Greece (Fessas 1982) and Algeria (Rouabhi 1984); it is due to a G→T substitution at codon 26 in the first exon of the β -gene. This mutation causes a variant β-globin chain and at the same time creates an alternative splicing site causing β-thalassaemia. The variant β-globin chain, named Hb Knossos, is not detected by conventional laboratory methods, and it is hence named "silent". It has been revealed by electrophoresis in urea-triton on polyacrilammide gels (Fessas 1982, Arous 1982).

Heterozygotes show no abnormality; Hb A_2 is normal because of an associated δ-thalassaemia mutation in cis to the β-Knossos gene (Mamalaki 1983); the β /α biosynthetic ratio is around 0.4-0.5. In genetic compounds with common β -thalassaemia, Hb is around 6g/dl, Hb F is about 10%, and the β /α ratio is 0.3.

The typical family is one with a child affected by thalassaemia intermedia, in spite of the apparent normality of one of the parents.

B) Mild β-thalassaemia gene: Certain β-thalassaemia genes have been named "mild" , because they cause thalassaemia intermedia in homozygous conditions or in genetic compounds with common β-thalassaemia mutations. In carriers, they produce the same hematological alterations as other mutations; the imbalance in the β / α biosynthetic ratio can be lower.

In some of these cases, the mutations have been characterized at the DNA level.

The IVS I nt 6 mutation associated to Haplotype VI (see Section 2.4) has been found in several regions. In Cyprus (Wainscoat 1983a) this mutation has not been found in thalassaemia major, but is present in homozygotes or compound heterozygotes (with IVS I nt 110 mutation) with thalassaemia intermedia. The same gene is probably responsible for the mild cases observed in Portugal (Tamagnini 1983), Sicily (Troungos 1983) and other Italian regions (Sampietro, personal communication; Saglio personal communication; Mastrobuoni, personal communication). Homozygotes for this mutation are characterized by unusually low levels of Hb F (10-20%), increased Hb A_2 and relatively high (7-8g/dl) total hemoglobin. In compound heterozygotes, Hb F levels may be slightly higher. It is likely that this gene is one of the main causes of thalassaemia intermedia in the Mediterranean area.

Another mild β-thalassaemia gene, observed in Cyprus (Wainscoat 1983a),may be identical (based on haplotype analysis) to a gene observed in an Italian and known as -87 mutation (Orkin 1982a, Kazazian 1984a).

3.2 CONCURRENT α-THALASSAEMIA

α-thalassaemia concurrent with homozygous β-thalassaemia decreases the imbalance in the β/α biosynthetic ratio and should therefore, increase the life span of erythroid cells. Based on this hypothesis, several studies have been carried out to detect α-thalassaemia genes in β-thalassaemics and to evaluate the effect of these genes on the hematological parameters and clinical conditions.

Indeed, in β-thalassaemia heterozygotes, the presence of α-thalassaemia, due to deletion of one or two α-globin genes, improves the hematological parameters and the β /α biosynthetic ratio, as reported in Greek (Bate 1977, Weatherall 1981, Wainscoat 1983b) and Sardinian families (Furbetta 1979, Melis 1983b). However, in β-thalassaemia homozygotes, the effect of concurrent α-thalassaemia is less clearly defined.

In Cypriots, α-thalassaemia (mainly due to loss of one, and even of two-three α-globin genes) has been detected in patients with thalassaemia intermedia (mostly of the IVS I nt 110 type) with a high frequency (14/27 patients), while only 4/30 patients with classical Cooley's anemia showed this defect (loss of one α-globin gene) (Wainscoat 1983c). This study suggests that

in this population α-thalassaemia may improve the hematological
and clinical conditions by decreasing the non-α / α biosynthetic
imbalance.

However in the Sardinian ̀β°-thalassaemia homozygotes,
29/50 patients with thalassaemia intermedia are reported to have
also α-thalassaemia, while 44/109 patients with Cooley's anemia
to have α-thalassaemia (Cao manuscript in preparation). The
authors suggest that the loss of two α-globin genes but not that
of only one gene may have an ameliorating effect in this population; additional factors may play a role in determining the mild
phenotype of thalassaemia intermedia in this population.

Differences in the type of β-thalassaemia genes involved,
the extent of the α-globin chain synthesis decrease, and the
presence of additional modifying factors may on the one hand
explain some of the contrasting views in these studies; on the
other hand, the criteria adopted in defining thalassaemia intermedia may also have affected the conclusions reached by the
different groups.

3.3 INCREASED LEVEL OF γ-GLOBIN SYNTHESIS

Many cases of thalassaemia intermedia can be characterized
by an abnormally high capacity to synthesize γ-globin chains
and non-α / α globin synthesic ratios substantially higher than
those usually observed in β-thalassaemia; thus a relatively large
proportion of erythroid cells may escape premature destruction in
the bone marrow and in the circulation, and a modest degree of
anemia may result. Different types of genetic disorders may be
involved, which will be discussed in the next section.

4.   HEREDITARY PERSISTENCE OF FETAL HEMOGLOBIN (HPFH) AND
δβ-THALASSAEMIA

These syndromes are characterized by a common factor, the
increased γ-globin chain synthesis, but their molecular basis
and hematological phenotypes are most heterogenous. Although some
of these syndromes are rare, the effect on γ-globin synthesis
may be so pronounced, that the interaction with a β-thalassaemia
gene usually results in thalassaemia intermedia, rather than
Cooley's anemia; therefore, a sizeable proportion of all
thalassaemia intermedia cases may be due to the presence
of HPFH or δβ-thalassaemia mutations.

Traditionally, a distinction is made between Hereditary
Persistence of Fetal Hemoglobin (HPFH) and δβ-thalassaemia; the
former is a very mild disorder usually characterized (in the
heterozygous state) by an essentially balanced non-α / α globin
ratio, the latter is a more severe form with unbalanced (though
less than in β-thalassaemia) synthetic ratios.

These differences generally depend on the higher level of γ-globin synthesis observed in many types of HPFH (typically of African origin) than in $\delta\beta$-thalassaemia; thus, Hb F levels never exceed 15% in heterozygous $\delta\beta$-thalassaemia, while they may be as high as 30% or more in heterozygous HPFH.

Another important factor is the level of residual β- and δ-globin chain synthesis; while in $\delta\beta$-thalassaemia the β- and often the δ-globin genes are inactive, in at least some types of HPFH the activity of these genes may be undetectably (Swiss HPFH) or only 20/30% decreased (Greek HPFH) .

Another traditional distinction is between pancellular and heterocellular types; in the former each red cell (in a hetero-zygote) can be shown to contain at least some Hb F, while in the latter only a variable proportion of red cells is positive at the diagnostic test. The distinction is, at least to some extent, artificial since the heterocellular rather than pancellular distribution is likely to depend on the relative levels of hemoglobin F and on the threshold for the detection with different techniques. Thus, conditions characterized by a small increase in Hb F synthesis (Swiss Type HPFH) appear to be heterocellular, while conditions, like Negro HPFH, where Hb F levels are high, appear to be pancellular. Though useful as an initial diagnostic marker,the Hb F distribution is no longer a particularly precise criterion for classification, because different molecular mecha-nisms now appear to underlie similar phenotypic conditions.

The following description of these disorders is essential-ly based on the type of mutations (deletional or not), and on the presence of residual expression of the β-globin genes in non deletional forms. The first paragraph (Heterocellular HPFH) treats a number of non deletional disorders, not well defined at the phenotypic and molecular level, but with continuously increasing evidence about their involvement in some cases of thalassaemia intermedia.

4.1 HETEROCELLULAR HPFH

Up to 5 % of normal individuals have slightly raised levels of Hb F (1-2% or more) with normal $^G\gamma/^A\gamma$ ratio (2:3) and increased numbers of Hb F containing cells (F cells) (Weatherall 1983a, Cappellini 1982). These individuals are considered to carry a variety of HPFH genes (known according to the area of origin of the affected individuals as "Swiss type", "British type" and so on). These genes considerably raise the Hb levels when in association with a β-thalassaemia gene (Wood 1976, Cappellini 1981), and being relatively common may be a rather frequent cause for mild syndromes in homozygous β-thalassaemics.

The definition of this gene is somewhat arbitrary as there is no precise cutoff point between the normal and HPFH levels of Hb F (Cappellini 1981). Haplotype analysis shows that, while some of these genes (for example, Indian and British HPFH) are closely linked to the non-α globin cluster(Old 1982a), others are unlinked and therefore presumably acting in trans on the $^G\gamma$-, $^A\gamma$-globin clusters of both chromosomes (Gianni 1983). Some families with the latter condition have recently been studied, and in one of these evidence of a high expression of a $^A\gamma$-globin gene from one chromosome and of the polymorphic variant $^{Sardinia}\gamma$-gene (Saglio 1979) from the other chromosome, has been reported (Giampaolo 1984).

A "Silent" heterocellular HPFH gene has been shown to exist in non-Mediterranean populations ; this gene linked to a specific haplotype and found in conjunction with the β^S-globin gene may not increase the Hb F level in heterozygotes, but greatly stimulates Hb F production in Hb S homozygotes (carrying this HPFH gene on one only of the two β^S chromosomes), presumably because of stress erythropoiesis (Wainscoat 1985).

This type of "silent HPFH gene", if present in the Mediterranean area, might represent the underlying mechanism for the high Hb F production observed in certain cases of thalassaemia intermedia, for which no obvious explanation is otherwise apparent.

Further investigation of haplotypes linked to γ-globin genes in β-thalassaemia homozygotes, with thalassaemia intermedia (unexplained on the basis of previously defined genetic factors) might reveal γ-globin gene clusters with a higher than normal capacity to compensate for β-globin chain synthesis deficiency.

4.2   NON-DELETIONAL FORMS WITH β-GLOBIN SYNTHESIS IN CIS

Non deletional syndromes have been described in Greece (Greek HPFH)(Sofroniadou 1975, Clegg 1979) and in Southern Italy (Camaschella 1979) (two families from Puglia with a disorder similar to Greek HPFH). These conditions, found only in the heterozygous state, are characterized by the production of Hb F, mostly (95%) of the $^A\gamma$-type; in at least the Italian family (Camaschella 1979), the β-globin gene on the HPFH chromosome is active, though at a slightly decreased level (70% of the normal). Compound heterozygotes for Greek HPFH and thalassaemia are also known (Sofroniadou 1975, Camaschella 1979, Clegg 1979); these patients have a very mild syndrome, characterized by a relatively high Hb level (11-12 g/dl) and approximately 40% Hb F.

The high expression of the $^A\gamma$-globin gene has been tentatively ascribed to single point mutations in one of two regions upstream to this gene (Fig.8): the CCAAT box (Collins 1985, Gelinas 1985) and a GC rich region (Giglioni 1984) centered around position -200, where a mutation is found also in an overexpressed $^G\gamma$-globin gene from a Negro type of $^G\gamma$ - HPFH (Collins 1984).

There is as yet no explanation for the low activity of the β - and δ-globin genes adjacent to the $^A\gamma$-globin gene. Analysis of the nucleotide sequence of the β-globin gene in the Italian case of HPFH shows that a functionally defective -globin gene does not need to be structurally abnormal; it is suggested (Giglioni 1984) that the abnormal transcriptional activation of the $^A\gamma$-globin gene region inhibits or delays the ordered progression (from γ- to β-globin region) of chromatin structural changes normally occurring during erythroblast maturation, leading to impaired activity of the β-globin gene, and possibly of the δ-globin gene.

4.3 NON DELETIONAL FORMS WITHOUT β-GLOBIN SYNTHESIS IN CIS

A non deletional syndrome (Sardinian δβ°-thalassaemia) with some similarity to Greek HPFH is found in Italy, particularly in Sardinia, where several independent families have been described (Cao 1982, David 1982)). This condition is clinically more severe than Greek HPFH, although the level (15%) and type (^{A}γ) of Hb F are similar; the difference is due to the fact that in Sardinian δβ°-thalassaemia the β-globin gene is inactive (Cao 1982), carrying the codon 39 non-sense mutation (Guida 1984, Pirastu 1984b). Compound δβ°-thalassaemia / β°-thalassaemia have a severe type of thalassaemia intermedia (Cao 1982) that may require, at least in some cases, splenectomy and blood transfusion. Association of the codon 39 non-sense mutation with several different 5' clusters has been discussed in Section 2.4; Sardinian δβ°-thalassaemia may represent a case of recombination between a rather common β°-thalassaemia gene and a mutant ^{A}γ-globin gene of the type described in Greek HPFH.

Indeed, recent experiments in our laboratory indicate that the ^{A}γ-globin gene of Sardinian δβ°-thalassaemia patients carries the -196 mutation observed in one case of Greek type HPFH (see section 4.2).

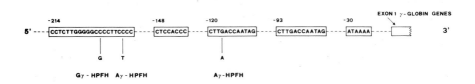

Fig.8   Mutations in the promoter regions of some HPFH genes. A scheme of the TATA, duplicated CCAAT boxes, CACC box and -200 region is represented from right to left with single nucleotide substitution detected in a ^{G}γ-HPFH (Collins 1985), a ^{A}γ-HPFH (Giglioni 1984) and two ^{A}γ-HPFH (Collins 1985, Gelinas 1985).

4.4   DELETIONAL FORMS

Four different types of deletions in the non-α globin gene cluster are observed in the Mediterranean region as reported in Fig. 9.

Hb Lepore is probably the commonest being found particularly in the Italian regions of Campania and Abruzzo, but also in other areas, including Iugoslavia and Macedonia.

Hb Lepore is a typical β-thalassaemia condition characterized by the absence of normal δ- and β-globin chains, and by the presence of an abnormal globin chain, shown to be the product of a fusion between normal δ- and β-globin chains (Baglioni 1962).

Three different Lepore globin chains are known, due to different sites of fusion: in Hb Lepore Hollandia the site of fusion is between codon 22 and 50; in Hb Lepore Baltimore between codon 50 and 86; in Hb Lepore Boston between codons 87 and 116 (Wheatherall 1983). The large homology between δ- and β-globin chains, which differ each other only at ten amino acid positions, renders impossible a more precise identification of the fusion point at protein level.

At the heterozygous state it is associated with low levels of Hb F; the variant chain is produced at the rate of 10%. The homozygous patient are affected by a syndrome similar to Cooley's anemia.

At the DNA level Hb Lepore is due to a deletion of about 6 kb, whose breakpoints are in regions of intense homology in the δ- and β-globin genes, thus resulting in a fusion 5'δ - 3' β globin gene (Flavell 1978, Ottolenghi 1979). The precise site of recombination in Lepore Boston globin gene has been restricted to a region of 59 nucleotides between codon 88 and nt 12 of IVS II, which is, therefore, of β-origin (Bank 1981, Mavilio 1983).

These data about the structure of β-globin gene Lepore have not clarified the molecular mechanism of the low expression of this gene; the presence of regulatory sequences in IVS I and 5' regions originating from δ-globin gene could play a role in the quantitative regulation of this gene expression.

Recent studies show that the Hb Lepore Boston is the only type of Lepore present in Italy, and that, on the basis of haplotype characterization, all patients carry a unique mutation, which thus probably occurred only once in the past (Mavilio 1983).

Of the other three deletions, Sicilian δβ-thalassaemia is, by far, the commonest, being found also in Calabria, Basilicata and Puglia. Heterozygous patients have, on average, 10% of Hb (60-70% of which is of $^A\gamma$-type) (Ottolenghi 1982a); several homozygous and doubly heterozygous (δβ-thalassaemia/ β -thalassaemia) patients are known, all characterized by a relatively mild condition that does not require, in general, regular blood transfusion (Gallo 1983, Musumeci 1983, Pirrone 1983).

The deletion, shown in Fig.9, has its 5' end point in the δ-globin gene and its 3' end point few kb beyond β-globin gene. (Ottolenghi 1979, Bernards 1979, Fritsh 1979, Ottolenghi

1982a,Tuan 1983). As shown by the characterization of the deletion and of the polymorphic sites Hind III in $^A\gamma$ - and $^G\gamma$ -
globin genes in 15 unrelated families, Sicilian δβ-thalassaemia
probably originated from a single mutational event (Carè 1984).

      In Spanish δβ-thalassaemia , a more extensive deletion is
found (Ottolenghi 1982a, Ottolenghi 1982b, Ottolenghi 1985), going
from   3.5 kb 5' to the δ-globin gene to more than 70 kb beyond
this β-globin gene. It is characterized by similar hematological
conditions to those described in the Sicilian type; several mildly
affected homozygous patients from three unrelated families are
known (M.Baiget, personal communication).

      Finally, three unrelated families with HPFH have been
observed recently in Italy (Saglio, manuscript in preparation);
no case of association with β-thalassaemia is known.   The hematological phenotype indicates a very mild condition (20-30% Hb F
in heterozygotes, almost normal hematological indices).

      The deletion has its 5' endpoint in an Alu I region of
repetitive DNA between the 5' endpoint of a Negro HPFH (Jagader-
seen 1983) and that of a Spanish δβ-thalassaemia (Ottolenghi
1982b). It was suggested that this Alu I region may include
sequences   of   regulatory   significance,   whose   differential
deletions in various syndromes may be responsible for different
levels of Hb production.

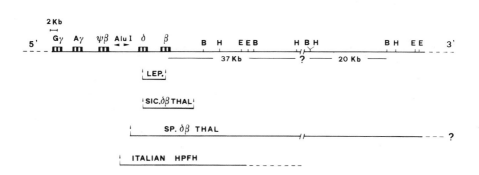

Fig.9   Deletions found in Hb Lepore, Sicilian and Spanish
δβ-thalassaemia and an Italian HPFH are reported. Horizontal
bars show the extent of the deletion; dashed lines indicate DNA
segments whose deletion is possible, but not proved.

5. DIAGNOSIS AND PREVENTION OF THE β–GLOBIN GENE DISORDERS

The extensive knowledge of the mutations that cause
β -thalassaemia disorders, the availability of restriction enzymes
and of specific DNA probes have brought about such a development
in the diagnosis of β -thalassaemia, especially at prenatal level,
that it has become a reference model for all monogenic disorders;
this is because DNA analysis allows us to identify the primary
mutations also in the absence of the phenotypic expression.

However techniques allowing the analysis of β-thalassaemia
at the phenotypic level retain importance at several steps of the
diagnosis. In particular, it should be pointed out that more than
95% of carriers can be detected already by the simple traditional
methods to measure red cell characteristics (MCV, MCH, etc) and
Hb A_2; more sophisticated techniques, like ion exchange chromatography on CM 52 cellulose (Clegg 1966) of radioactively labelled
globin chains from erythroid cells are now required only for
the diagnosis of atypical cases (silent β -thalassaemia gene,
presence of an α -thalassaemia gene, thalassaemia intermedia).
In addition, this technique is still of invaluable importance
in antenatal diagnosis of β -thalassaemia because of its precision
and of its simplicity relative to DNA techniques, as well as in
those cases (see below) that cannot be diagnosed by DNA analysis.
It was first introduced for this purpose in June 1974 (Kan 1975)
and afterwards extensively used all over the world; 3,959 antenatal diagnoses up to december 1982 have been reported (Alter
1984). The CM technique is able to identify the small ammounts
(approximately 10% relative to γ-globin chains) of the β -globin
chains synthesized by fetal reticulocytes at about the 18th week
of pregnancy, and to discriminate homozygotes from both normals
and heterozygotes (although problems may occasionally arise in
discriminating homozygous β -thalassaemia from heterozygotes).
The possibility of obtaining essentially pure fetal blood by
phetoscopy (Hobbins 1974) (that has replaced the earlier placen-
tocentesis technique) has now made the diagnosis easier and
allowed the use of analytical methods for non radioactive hemoglobin, as isoelectrophocusing (Ferrari 1984) . Unfortunately
these techniques can be used only at a very advanced stage of
pregnancy (18th week), implying a long wait for the mother and,
in case of homozygosis of the fetus, a very late abortion. There
are other disadvantages: this technique requires a specialized
obstetrical skill; it is associated to some risk of fetal loss
(around 5% in Centers with large experience), of maternal
complications (usually leakage of amniotic fluid)(about 1.6%)
and of premature birth (about 5%) (Alter 1984).

For these reasons, direct analysis at DNA level is now
preferred by several Centers, when possible. In fact, fetal cells
can be easily and safely obtained from the amniotic fluid at 16th
week of pregnancy; more recently, placental villi, obtained at
the 8th week of pregnancy have proven to be a good source of DNA
for antenatal diagnosis (Williamson 1981, Old 1983b).

DNA analysis allows to diagnose specific genetic alterations (direct analysis) or to establish the inheritance of a
mutation by a chromosome segregation study of the family
(indirect analysis). Table IV summarises essential elements of
the two different strategies and some of the clinical diagnostic
aspects (Weatherall 1983b, Orkin 1984, Weatherall 1984).

The direct analysis strategy is going to become most relevant at diagnostic level; it is based on different methods
allowing to detect the mutation responsible for the functional
defect of the gene. Thus it does not require extensive family
studies, but only the knowledge of the alteration at molecular
level.

Three different "direct" approaches are feasible:

1) detection of abnormal restriction fragments, due to
large deletions,like δβ-thalassaemia and HPFH. These syndromes
are rare in Italy and in the Mediterranean area, as reported in
section 4.4 and do not give severe anemia.

2) detection of abnormal restriction fragments produced by
digestion with "specific" restriction enzymes and due to specific
point mutations, which create or abolish restriction enzyme sites.
As shown in fig.7, four β-thalassaemia mutations of this type have
been described and they are recognized respectively by Avr II
(Orkin 1982a), Mst II (Chang 1983), Hph I (Treisman 1982) and Rsa
I (Orkin 1982a). These mutations have the frequency of about 15%
in some Mediterranean areas, as Sicily (Maggio, personal communication) and Campania (Carestia, manuscript in preparation).

3) Specific hybridization to short oligonucleotides
(19mer) whose sequence contains the normal or the mutated base.
Under appropriate conditions the oligonucleotides hybridize only
to the homologous but not heterologous sequence, thus proving the
presence of that sequence in the analyzed DNA. The existence of
a heterozygous state for a specific mutation is demonstrated by
the hybridization to two different oligonucleotides with the
normal and the mutated β-globin gene sequence, respectively.

This technique, whose application is possible in the
diagnosis of all known point mutations, has the principal
limitation to its use in the requirement of extremely specific
conditions of hybridization for each oligonucleotide, due to
to their different base composition.

Oligonucleotides were first used in prenatal diagnosis of
sickle cell anemia (Conner 1983); soon after they have been
successfully used for diagnosis of several β-thalassaemia mutations : β-39 non-sense mutation C→T(Pirastu 1983b); β-IVS I
nt 110 T→C mutation (Orkin 1983b); β-IVS I nt 6 T→C mutation
(Wainscoat 1983b).

In regions where a specific mutation is very frequent or
almost exclusive, such as Sardinia and Cyprus (see section 2.4)
the oligonucleotide technique is commonly used in prental
diagnosis (Rosatelli 1985); in other regions, where multiple
defects coexist the use of this technique is hampered by the need
to preliminary investigate the family with several oligonucleotides.

        The indirect analysis utilizes the polymorphic restriction
enzyme sites of the -globin gene cluster (see Fig.1) as markers
of the chromosomes carrying mutations for a linkage analysis of
the family. The diagnosis is possible when a polymorphic site is
in heterozygous state and when its linkage to a mutation can be
assessed by family analysis, through an affected child or other
informative relatives. With this approach, the knowledge of the
molecular alteration at the DNA level is not required, so this
method can be applied to diagnosis in most patients, and is
essential in regions where a high heterogeneity of mutations
occurs.

        Results by this approach were first achieved in pre-natal
diagnosis of Hb S (Kan 1978b) by the use of the Hpa I site 3'
to the β -globin gene (Kan 1978a), found to be linked with very
high frequency (70-80%) with the β^S-mutation in Blacks of the west
coast of USA.   Later on, in Sardinia β -39 non-sense thalassaemia
mutation was found linked to the absence of a polymorphic site
Bam HI 3' to the β-globin gene (Kan 1980). In both instances the
polymorphisms were also present on chromosomes carrying normal
β -globin genes, so even in these cases they were useful for prenatal diagnosis only when present in heterozygous state in at
least one of the parents.

Type of analysis	Type of approach	Diagnostic aspects
DIRECT analysis	1) Detection of abnormal restriction fragments due to large deletions or riarrangments 2) Detection of abnormal restriction fragments due to mutations that create or abolish restriction sites 3) Hybridization with oligonucleotides reproducing normal and mutated sequences	A)These approaches detect directly the specific mutation B)An extensive familiar study is not necessary C) The knowledge of the molecular basis of the mutations is requested
INDIRECT analysis	It is a familiar linkage analysis based on the use of restriction fragment lenght polymorphism as marker of the chromosome carrying the mutation	A) The knowledge of the the molecular basis of the mutations is not requested B) The familiar study is always requested C) Both parents must be heterozygous for at least one polymorphic site

TABLE IV.   DNA restriction tecniques in β -thalassaemia diagnosis.

Several restriction enzyme polymorphic sites have been used in β-thalassaemia diagnosis (Little 1980, Kazazian 1980, Bohem 1983) and are now currently used in Italy and other countries. Surveys of frequecies of RFLPs carried out in some regions of Italy and Cyprus show that this diagnostic method may be useful to prenatal diagnosis in Cyprus (Waiscoat 1984) and in Campania (Carestia manuscript in preparation) in 75% of cases.

Prenatal diagnosis has decisively changed the strategy of β-thalassaemia prevention. In the past this was essentially carried out by discouraging the marriage between carriers or procreation, a suggestion that was ill acceptable to most populations. Presently, antenatal diagnosis is a valid alternative, accepted in many populations, to the dramatic choice between having no children or taking the risk of generating a homozygote; therefore in regions where educational programs, directed both at the population at risk and at the doctors (general practitioners, obstetricians) have been implemented, considerable decrease of the birth rate of thalassaemic homozygotes has been achieved, for example in Sardinia, Ferrara region, Cyprus and in Britain (Modell 1984).

ACKNOWLEDGEMENTS

This work was supported by CNR Grants 8403465 to Sergio Ottolenghi and 8407543 to C.Carestia (Progetto Finalizzato Ingegneria Génetica e Basi Molecolari delle malattie ereditarie).

REFERENCES

Alter, B.P. (1984). 'Advances in the prenatal diagnosis of
  hematological desease', Blood, 64, 329-340.
Antonarakis, S.W., Boehm, C.D., Giardina, P.J.V., and Kazazian,
  H.H. Jr.(1982a). 'Non-random association of polymorphic
  restriction sites in the β -globin gene cluster', Proc. Natl.
  Acad. Sci. USA, 79, 137-141.
Antonarakis, S.E., Orkin, S.M., Kazazian, H.H. Jr, Goff, S.C.,
  Boehm, C.D., Waber, P.G., Sexton, J.P., Ostrer, H., Fairbanks
  V.F., and Chakravarti, A. (1982b). 'Evidence for multiple
  origins of the $β^E$-globin gene in Southeast Asia', Proc. Natl.
  Acad. Sci. USA, 79, 6608-6611.
Antonarakis, S.E., Orkin, S.M., Cheng, T.C., Scott, A.F., Sexton,
  J.P., Trusko, S., Charache, S., and Kazazian, H. Jr. (1984a).
  ' β -thalassemia in American Blacks: novel mutations in the
  TATA box and an acceptor splice site', Proc. Natl. Acad. Sci.
  USA, 81, 1154-1158.
Antonarakis, S.E., Boehm, C.D., Serjeant, G.R., Theisen, C.E.
  Doyer, G.J., and Kazazian, H.H.Jr. (1984b). 'Origin of the
  $β^S$-globin gene in Blacks: The contribution of recurrent
  mutation or gene conversion or both', Proc. Natl. Acad. Sci.
  USA, 81, 853-856.
Arous, N., Galacteros, F., Fessas, P., Loukopoulos, D., Blouquit,
  Y., Komis, G., Sellaye, M., Boussiou, M., and Rosa, J. (1982).
  'Structural study of hemoglobin Knossos β 27 (B9) Ala→ Ser',
  Febs Lett. 147, 247-250.
Baird, M., Schreiner, H., Driscoll, C., and Bank, A. (1981).
  'Localizatization of the site rccombination in formation of
  the Lepore Boston globin gene', J.Clin.Invest., 68, 560-564.
Bate, C.M., and Humphries, G. (1977).'Alpha-Beta Thalassaemia'
  The Lancet, i, 1031-1033.
Beldjord, C., Lapouméroulie, C., Baird, M.L. Girot, R., Adjrad,
  L., Lenoir, ·G., Benabadji, N., and Labie, D. (1983). 'Four new
  haplotypes observed in Algerian β -thalassemia patients',
  Hum. Genet., 65, 204-206.
Bernards, R., Kooter, J.M., and Flavell, R.A. (1979). 'Physical
  mapping of the globin gene deletion in (δ β)°thalassemia',
  Gene, 6, 265-280.
Boyer, S.H., Belting, K., Margolet, L., and Noyes, A. (1975).
  'Restriction of fetal Hemoglobin to a few erythrocytes
  (F cells) in normal human adult', Science, 188, 361-363.
Bohem, C.D., Antonarakis, S.E., Phillips, J.A.III, Stetten G., and
  Kazazian, H.H. (1983). ' Prenatal diagnosis using DNA polymor-
  phism. Report on 95 pregnancies at risk for sickle cell disease
  or β -thalassaemia', 1054-1058.
Breathnach, R., Benoist, C., O'Hare, K., Gannon, F., and Chambon,
  E. (1978). 'Ovalbuminin gene: Evidence for a leader sequence in
  mRNA and DNA sequences at the exon-intron boundaries',
  Proc. Natl. Acad. Sci. USA, 75, 4853-4857.
Busslinger, M., Moschonas, N., and Flavell, R.A. (1981).
  ' $β^+$-thalassemia: aberrant splicing results from a single point
  mutation in an intron', Cell, 27, 289-298.

Camaschella, C., Ciocca Vasino, M.A., Guerrasio, A., Balegno,
    G.,Barberis, E., Del Ponte, D., and Saglio, G. (1979).
    'Biosynthetic studies and γ -chain composition in the Greek type
    of Hereditary Persistence of Fetal Hemoglobin and in its asso-
    ciation with β -thalassaemia', Acta Haematol., 61, 272-280.
Cao, A., Melis, M.A., Galanello, R., Angius, A., Furbetta, M.,
    Giordano, P., and Bernini, L.F. (1982). ' $\delta\beta$ -Thalassaemia in
    Sardinia', J. Med. Genet., 19, 184-192.
Cappellini, M.D., Fiorelli, G., and Bernini, L.F. (1981).
    'Interaction between homozygous β -thalassemia and the Swiss
    type of hereditary persistence of fetal haemoglobin',
    Br. J. Haematol., 48, 561-572.
Cappellini,M.D., Sampietro, M., Bernini, L., Ranzani, G., and
    Fiorelli, G. (1982). 'F-cells distribution in normals and in
    heterocellular HPFH (Swiss type).', in Advances in Red Blood
    Cell Biology", 271-276. D.J.Weatherall, G.Fiorelli and S.Gorini,
    eds., Raven Press, New York, 1982.
Carè, A., Sposi, N.M., Giampaolo, A., Improta, T., Calandrini,M.,
    Petrini, M., Marinucci, M., Tagarelli, A., and Brancati, C.
    (1984). ' $\delta\beta$ -thalassaemia in Southern Italy: evidence for a
    single mutational event', Journal of Med.Genetics, 21, 117-120.
Chang, J.G., Alberti, A., and Kan, Y.W. (1983). 'A β -thalassaemia
    lesion abolishes the same Mst II site as the sickle mutation',
    Nucl. Acid Res., 11, 7789-7794.
Chang, J.C. and Kan, Y.W. (1981). "Antenatal diagnosis of sickle
    cell anemia by direct analysis of the sickle mutation',
    The Lancet, ii, 1127-1129.
Chakravarti, A., Buetow, K.H., Antonorakis, S.E., Waber, P.G.,
    Boehm, C.D., and Kazazian, H.H. Jr.(1984). 'Non-uniform
    recombination within the human β -globin gene cluster',
    Am. J. Human Genet., 36, 1239-1258.
Cheng, T., Orkin, S.H., Antonarakis, S.E., Potter, M.J., Sexton,
    J.P., Markham, A.F., Giardina, P.J.V., Li, A., and Kazazian,
    H.H., Jr. (1984). ' β -Thalassemia in Chinese: use of in vivo
    RNA analysis and oligonucleotide hybridization in systematic
    characterization of molecular defects', Proc. Natl. Acad. Sci.
    Usa, 81, 2821-2825.
Clegg, J.B., Naughton, M.A., and Weatherall,D.J. (1966). 'Abnormal
    Human hemoglobins. Separation and characterization of the α -
    and β -chains by chromatography.', J. Mol. Biol., 19,91-108.
Clegg, J.B., Metaxatou-Mavromati, A., Kattamis, C., Sofroniadou,
    K., Wood, W.G., and Weatherall, D.J. (1979). 'Occurrence of
    HbF in Greek HPFH: analysis of heterozygotes and compound
    heterozygotes with β -thalassemia', Br. J. Haematol., 43,
    521-536.
Collins, F.S., and Weissman, S.M. (1984a). 'The molecular genetics
    of human hemoglobins', Prog. Nucl. Acid Res., 315-463.
Collins, F.S., Stoeckert, C.J., Serjeant, G.R., Forget, B.G., and
    Weissman, S.M. (1984b). '$^{G}\gamma\ \beta^{+}$ Hereditary persistence of fetal
    hemoglobin: cosmid cloning and identification of a specific
    mutation 5' to the $^{G}\gamma$ -gene', Proc. Natl. Acad. Sci. USA, 81,
    4894-4898.

Collins, F.S., Metherall, J.E., Yamakawa, M., Pan, J., Weissman,
    S.M., and Forget, B.G. (1985). "A point mutation   in the
    $_A$γ-globin gene promoter in Greek hereditary persistence of
    fetal hemoglobin', Nature, 313, 325-326.
Comi, P., Giglioni, B., Ottolenghi, S., Gianni, A.M., Polli,E.,
    Barba, P., Covelli, A., Migliaccio, G., Condorelli, M., and
    Peschle, C. (1980). 'Globin chain synthesis in single erythroid
    bursts from cord blood: studies on γ ⟶ β and ^{G}γ ⟶ ^{A}γ switches',
    Proc. Natl. Acad. Sci., USA, 77, 362-365.
Conner, B.J., Reyes, A.A., Morin, C., Itakura, K., Teplitz, R.L.,
    Wallace R.B. (1983). 'Detection of sickle cell β^S-globin allele
    by hybridization with synthetic oligonucleotides', Proc. Natl.
    Acad. Sci. USA, 80, 278-282.
Dayoff, M.O. (1972). 'Atlas of protein sequence and structure',
    National Biomedical Research Foundation', 5, D51-D85.
David, O., Miniero, R., Sacchetti, L., Cappellini, M.D., Giglioni,
    B., Comi, P., Ottolenghi, S., Saglio, G., Guerrasio, A.,
    Camaschella, C., and Mazza, U. (1982). 'High Hb F levels in a
    Sardinian family: a genetic defect intermediate between HPFH
    Greek type and δβ-thalassaemia?', Haematologica, 67, 499-507.
Del Senno, L., Pirastu, M., Barbieri, R., Bernardi F., Buzzoni,
    D., Marchetti, G., Perrotta, C., Vullo, C., Kan, J.W., and
    Conconi, F. (1984). ' β$^+$-thalassaemia in the Po river delta
    region (northern Italy): genotype and globin synthesis',
    J.Med.Genet., 21, 54-58.
Dierks, P., van Ooyen, A., Cochran, M.D., Dobkin, C., Reiser, J.,
    and Weissman, C. (1983). 'Three regions upstream from the Cap
    site are required for efficient and accurate transcription of
    the rabbit β-globin gene in mouse 3T3 cells', Cell ,32, 695-706.
Dobkin, C., Pergolizzi, R.G., Bahre, P., and Bank, A. (1983).
    'Abnormal splice in a mutant human β-globin gene not at the
    site of a mutation', Proc. Natl. Acad. Sci. USA, 80, 1184-1188.
Dover, G., and Boyer, S.H. (1983). 'Quantitation of hemoglobins
    within individual red cells: asynchronous biosynthesis of fetal
    and adult hemoglobin during erythroid maturation in normal
    subjects', Blood, 56, 1082-1091.
Driscoll, M.C., Baird, M., Bank, A., and Rachmilewitz, E.A.
    (1981). 'A new polymorphism in the human β-globin gene useful
    in antenatal diagnosis', J.Clin.Invest., 68, 915-919.
Efstratiadis, A., Posakony, J.W., Maniatis, T., Lawn, R.M.,
    O'Connell, C., Spritz, R.A., DeRiel, J., Forget, B.J., Weissman,
    S.M., Slightom, J.L., Blechl, A.E., Smithies, O., Baralle,
    F.E., Shoulders, C.C., and Proudfoot, N.J. (1980).'The structure
    and evolution of the human β-globin gene family', Cell, 21,
    653-668.
Ferrari, M., Crema, A., Cantù-Rajnoldi, A., Pietri, S., Travi, M.,
    Brambati, B., and Ottolenghi, S. (1984). 'Antenatal diagnosis
    of haemoglobinopathies by improved method of isoelectric phocu-
    sing of hemoglobins', Brit. J. Haematol., 57, 265-270.
Fessas, P., Loukopoulos, D., Loutradi-Anagnoustou, A., and Komis,
    G. (1982). ' "Silent" β-thalassaemia caused by a "silent"
    β-chain mutant: the pathogenesis of a syndrome of thalassaemia
    intermedia', Brit. J. Haematol., 51, 577-583.

Fritsch, E.F., Lawn, R.M., and Maniatis, T. (1979). 'Characterization of deletions which affect the expression of fetal
globin genes in man', Nature, 279, 598-603.

Fritsch, E.F., Lawn, R.M., and Maniatis, T. (1980). 'Molecular
cloning and characterization of the human β-like globin gene
cluster', Cell, 19, 959-972.

Fukumaki, Y., Ghosh, P.K., Benz, E.J.Jr., Reddy, P.B., Lebowitz,
U., Forget, B.G., and Weissman, S.M. (1982). 'Abnormally
spliced messenger RNA in erythroid cells from patients with
β thalassemia and monkey cells expressing a cloned β -thalas-
semic gene', Cell, 28, 585-593.

Furbetta, N., Galanello, R., Ximenis, A., Angius, A., Melis,
M.A., Serra, P., and Cao, A. (1979). 'Interaction of alpha
and beta thalassaemia genes in two Sardinian families',
Br. J. Haematol., 41, 203-210.

Gale, R.E., Clegg J.B., and Huehns, E.R. (1979). 'Human embryonic
haemoglobins Gower 1 and Gower 2'. Nature, 280, 162-164.

Gallo, E., Massaro, P., Miniero, R., David, D., and Tarella, C.
(1979). 'The importance of the genetic picture and globin synthesis in determining the clinical and haematological features
of thalassaemia intermedia. Br. J. of Haematol. 41, 211-221.

Gelinas, R., Endlich, B., Pfeiffer, C., Yagi, M., and Stamatoyannopoulos, G. (1985). 'G to A substitution in the distal CCAAT
box of the $^A\gamma$-globin gene in Greek hereditary persistence
of fetal haemoglobin', Nature, 313, 323-324.

Giampaolo, A., Mavilio, F., Sposi, N.M., Caré, A., Massa, A.,
Cianetti, L., Petrini, M., Russo, R., Cappellini, M.D., and
Marinucci, M. (1984). 'Heterocellular hereditary persistence
of fetal hemoglobin (HPFH). Molecular mechanisms of abnormal
γ-gene expression in association with β-thalassemia and
linkage relationship with the β-globin gene cluster.',
Hum. Genet., 66, 151-156.

Gianni, A.M., Comi, P., Giglioni, B., Ottolenghi, S., Migliaccio,
A.R., Migliaccio, G., Lettieri, P., Maguire, Y., and Peschle,
C. (1980). 'Biosynthesis of Hb in individual fetal liver bursts:
γ-chain production peaks earlier than β-chain in the erythropoietic pathway', Exp. Cell. Res., 130, 345-362.

Gianni, A.M., Bregni, M., Cappellini, M.D., Fiorelli, G.,
Taramelli, R., Giglioni, B., Comi, P., and Ottolenghi, S.
(1983). 'A gene controlling hemoglobin expression in adults
is not linked to the non-α globin cluster', The EMBO J., 2,
921-925.

Giglioni, B., Casini, C., Mantovani, R., Merli, S., Comi, P.,
Ottolenghi, S., Saglio, G., Camaschella, C., and Mazza, U.
(1984). 'A molecular study of a family with Greek hereditary
persistence of fetal hemoglobin and β-thalassemia',
The EMBO J., 3, 2641-2645.

Gil, N., and Proudfoot, N.S. (1984). ' A sequence downstream of
AAUAAA is required for rabbit β-globin mRNA 3'-end formation',
Nature, 312, 473-475.

Goldsmith, M.E., Humphries, R.K., Ley, T., Cline, A., Kantor, J., and Nienhuis, A.W. (1983). 'Silent nucleotide substitution in a β⁺-thalassemia globin gene activating a cryptic splice site in β-globin RNA coding sequence', Proc. Natl. Acad. Sci. USA, 80, 2318-2322.

Gorski, J., Fiori, M., and Mach, B. (1982). 'A new nonsense mutation as the molecular basis for β°-thalassemia', J. Mol. Biol., 154, 537-540.

Gusella, G., Varsanyi-Breiner, A., Kan, F., Jones, C., Puck, T.T., Keys, C., Orkin, S., and Housman, D. (1979). 'Precise localization of human β-globin gene complex on chromosome 11', Proc. Natl. Acad. Sci. USA, 76, 5239-5243.

Guida, S., Giglioni, B., Comi, P., Ottolenghi, S., Camaschella, C., and Saglio, G. (1984). 'The β-globin gene in Sardinian δβ°-thalassemia carries a G→T nonsense mutation at codon 39', The EMBO J., 3, 785-787.

Hobbins, J., C., and Mahoney, M., J., (1974).'In utero diagnosis of hemoglobinopathies: Technique for obtaining fetal blood', N. Engl. J. Med., 290, 1065-1067.

Humphries, R.K., Ley, T.J., Anagnou, N.P., Baur, A.W., and Nienhuis, A.W. (1984). 'β°-39 thalassemia gene: A premature termination codon causes β-mRNA deficiency without affecting cytoplasmatic β-mRNA stability', Blood, 64, 23-32.

Jagadeeswaran, P., Tuan, D., Forget, B.G., and Weissman, S.M. (1982). 'A gene deletion ending at the midpoint of a repetitive DNA sequence in one form of hereditary persistence of fetal haemoglobin', Nature, 296, 469-470.

Jeffreys, A.J. (1979). 'DNA sequence variants in the $^G\gamma$-, $^A\gamma$-, δ- and β-globin genes of man', Cell, 18, 1-10.

Kan, Y.W., Golbus, M.S., Klein, P., and Dozy, A.M. (1975). 'Successfull application of prental diagnosis in a pregnancy at risk for homozygous β-thalassaemia', N.Engl. J. Med., 1096-1099.

Kan, Y.W., and Dozj, A.M. (1978a). 'Polymorphism of DNA sequence adjacent to human β-globin structural gene: Relationship to sickle mutation', Proc. Natl. Acad. Sci USA, 76, 5631-5635.

Kan, Y.W. and Dozy, A.M. (1978b). 'Antenatal diagnosis of sickle cell anemia by DNA analysis of amniotic-fluid cells', The Lancet, ii, 910-911.

Kan, Y.W., Lee, K.Y., Furbetta, M., Angius, A., and Cao, A., (1980). 'Polymorphism of DNA sequence in the β-globin gene region', The New Engl. J. Med., 302,185-188.

Kanavakis, E., Wainscoat, J.S., Wood, W.G., Wheatherall, D.J., Cao, A., Furbetta, M., Galanello, R., Georgiou,D., and Sophocleous, T. (1982). 'The interaction of α-thalassaemia with heterozygous β-thalassaemia', Br. J. Haematol., 52, 465-473.

Kattamis, C., Metaxotou-Mavromati, A., Wood, W.G., Nash, J.R., and Weatherall, D.J. (1979). 'The heterogeneity of normal Hb A_2 β-thalassaemia in Greece', Br. J. Hematol., 42, 109-123.

Kazazian, H.H. Jr, Phillips III, J.A., Boehm, C.D., Vik, T.A., Mahoney, M.J. and Ritchey, A.K. (1980). 'Prenatal diagnosis of β-thalassaemias by amniocentesis: Linkage analysis using multiple polymorphic restriction endonuclease sites', Blood, 56, 926-930.

Kazazian, H.H. Jr, Orkin, S.H., Markham, A.F., Chapman, C.R.
    Youssoufian, H., and Waber, P.G. (1984a). 'Quantification of
    the close association between DNA haplotypes and specific
    β -thalassaemia mutations in Mediterraneans', Nature, 310,
    152-154.
Kazazian, H.H., Orkin, S.H., Antonarakis, S.E., Sexton, J.P.,
    Boehm, C.D., Goff, S.C., and Waber, P.G. (1984b). 'Molecular
    characterization of seven β -thalassemia mutations in Asian
    Indians', The Embo J., 3, 593-596.
Kohen,G., Philippe, N., and Godet, J. (1982). 'Polymorphism
    of the Hinf I restriction site located 1 kb 5' to the
    human β -globin gene', Hum. Genet., 62, 121-123.
Lebo, R.V., Carrano, A.V., Burkhart-Schultz, K., Dozy, A.M.,
    Yu, L., and Kan,Y.W. (1979). 'Assignment of human γ -, δ -, and
    β -globin genes to the short arm of chromosome 11 by chromosome
    sorting and DNA restriction enzyme analysis', Proc. Natl.
    Acad. Sci. USA, 76, 5804-5808.
Little, P.F.R., Annison, G., Darling, S., Williamson, R., Camba,
    L. and Modell, B. (1980). 'Model for antenatal diagnosis of
    β -thalassaemia and other monogenic disorders by molecular
    analysis of linked DNA polymorphisms', Nature, 285, 144-147.
Longinotti, M., and Masala, B. (1982). ' A tentative systema-
    tization of β -thalassaemia syndromes present in Northern
    Sardinia ' in "Thalassaemia: Recent advances in detection
    and treatment", A.Cao , U.Carcassi and P.T.Rowley eds., by
    Alan R. Liss, Inc.
Mamalaki, A., Gossens, M., Galacteros, F., Ncamoune, S.,
    Monplasir, , Boussou, M., Loukopoulos, D., and Rosa, J. (1983).
    ' Different origins for the β Knossos globin gene in Blacks
    and Mediterraneans', Blood, 62, 68A
Mavilio, F., Giampaolo, A., Carè, A., and Marinucci, M. (1983).
    ' The δβ crossover region in Lepore Boston hemoglobinopathy
    is restricted to a 59 base pairs region around the 5' splice
    junction of the large globin gene intervening sequence',
    Blood, 62, 230-233.
Maeda, N., Bliska, J.B., and Smithies, O. (1983). 'Recombination
    and balanced chromosome polymorphism suggested by DNA sequences
    5' to the human β -globin gene', Proc. Natl. Acad. Sci. USA,
    80, 5012-5016.
Melis, M.A., Pirastu, M., Galanello, R., Furbetta, M., Tuveri,T.,
    and Cao, A. (1983a). 'Phenotypic effect of heterozygous α - and
    β °-thalassaemia interaction', Blood, 62, 226-229.
Melis, M.A., Galanello, R., and Cao, A. (1983b). 'Alpha globin
    gene analysis in a Sardinian family with interacting alpha and
    beta thalassaemia genes', Br. J. Haematol., 53, 667-671.
Modell, B., Petrou, M., Ward, R.H., Fairweather, D.V.I.,
    Rodeck, C., Varnavides, L.A., White, M.J. (1984). ' Effect of
    fetal diagnostic testing on birth-rate of thalassaemia major
    in Britain ',The Lancet, ii, 1383-1386.
Moschonas, N., deBoer, E., Grosveld, F.G., Dahl H.H.M., Wright,
    S., Shewmaker, C.K., and Flavell, R.A. (1981). 'Structure and
    expression of a cloned β -thalassemic globin gene', Nucleic
    Acid Res., 9, 4391-4401.

Moschonas, N., deBoer, E., and Flavell, R.A. (1982). 'The DNA
    sequence of the 5' flanking region of the human β -globin gene:
    evolutionary conservation and polymorphic differences', Nucleic
    Acid Res., 10, 2109-2120.
Mount, S.M. (1982). 'A catalogue of splice junction sequences',
    Nucleic Acid Res., 10, 459-472.
Musumeci, S., Schilirò, G., Romeo, M.A., Di Gregorio, F.,
    Pizzarelli, G., Testa, R., and Russo, G. (1983). 'Heterogeneity
    of thalassemia intermedia', Haematologica, 68, 503-516.
Nienhuis, A.W., Anagnou, N.P., and Ley,T.J. (1984). 'Advances in
    Thalassaemia research', Blood, 63, 738-758.
Old, J.M., Ayyub, M., Wood, W.G., Clegg, J.B., and Weatherall,
    D.J. (1982a). 'Linkage analysis of non deletion hereditary
    persistence of fetal hemoglobin', Science, 215, 981-982.
Old, J.M., Ward, R.H.T., Karagozlu, F., Petrou, M., Modell,
    B., and Weatherall, D.J. (1982b). 'First-trimester fetal
    diagnosis for haemoglobinopathies: three cases', The Lancet,
    1413-1416.
Old, J.M., and Wainscoat, J.S. (1983). 'A new DNA polymorphism in
    the β -globin gene cluster can be used for antenatal diagnosis
    of β -thalassaemia', Br. J. Haematol., 53, 337-341.
Old, J.M., Petrou, M., Modell, B., and Weatherall, D.J. (1984).
    'Feasibility of antenatal diagnosis of β -thalassaemia by DNA
    polymorphism in Asian Indian and Cypriot populations',
    Brit. J. Haematol., 57, 255-263.
Orkin, S.H., and Goff, S.C. (1981). 'Nonsense and frameshift
    mutations in β -thalassaemia detected in cloned   β -globin
    genes', J. Biol. Chem., 256, 9782-9784.
Orkin, S.H., Kazazian, H.H. Jr., Antonarakis, S.E., Goff, S.C.,
    Boehm, C.D., Sexton, J.P., Waber, P.G., and Giardina, P.J.V.
    (1982a). 'Linkage of β -thalassaemia mutations and β -globin
    gene polymorphisms with DNA polymorphisms in human β -globin
    gene cluster', Nature, 296, 627-631.
Orkin, S.H., Kazazian, H.H. Jr., Antonarakis, S.E., Ostrer, H.,
    Goff, S.C., and Sexton, J.P. (1982b). 'Abnormal RNA processing
    due to the exon mutation of β^E -globin gene', Nature, 300,
    768-769.
Orkin, S.H., Sexton, J.P., Cheng, T., Goff, S.C., Giardina,
    P.J.V., Lee J.I., and Kazazian, H.H. Jr. (1983a). 'ATA box
    transcriptional mutation in β -thalassaemia',
    Nucleic Acid. Res., 11, 4727-4734.
Orkin, S.H., Markam, A.F., and Kazazian, H.H. (1983b). 'Direct
    detection of the common Mediterranean β -thalassaemia gene
    with synthetic DNA probes', J. Clin. Invest., 71, 775-779.
Orkin, S.H., Antonarakis, S., and Loukopoulos, D. (1984a).
    'Abnormal processing of β$_{Knossos}$ RNA', Blood, 64, 311-313.
Orkin, S.H. (1984b). 'Prenatal diagnosis of hemoglobin disorders
    by DNA analysis', Blood, 63, 249-253.
Ottolenghi, S., Giglioni, B., Comi, P., Gianni, A.M., Polli, E.,
    Acquaye, C.T.A., Oldham, J.H., and Masera, G. (1979). 'Globin
    gene deletion in HPFH, δ^{0}β^0-thalassemia and Hb Lepore disease',
    Nature, 278, 654-657.

Ottolenghi, S., Giglioni, B., Taramelli, R., Comi, P., Mazza, U., Saglio, G., Camaschella, C., Izzo, P., Cao, A., Galanello, R., Gimferrer, E., Baiget, M., and Gianni, A.M. (1982a). 'Molecular comparison of $\delta\beta$-thalassemia and hereditary persistence of fetal hemoglobin DNAs: Evidence of a regulatory area?', Proc. Natl. Acad. Sci USA, 79, 2347-2351.

Ottolenghi, S., and Giglioni, B. (1982b). 'The deletion in a type of $\delta^\circ\beta^\circ$-thalassemia begins in an inverted Alu I repeat', Nature, 300, 770-771.

Ottolenghi, S., Giglioni, B., Comi, P., Guida, S., Casini, C., Merli, S., Mantovani, R., Terragni, F., Aghib, D., Saglio, G., Camaschella C., and Mazza, U. (1985). 'Heterogeneity of $\delta\beta$-thalassaemia and Hereditary Persistence of Hb F in the Mediterranean area', Annals New York Academy of Sciences, in press.

Pagnier, J., Mears, J.G., Dunda-Belkodja, O., Schaefer-Rego, K.E., Beldjord, C., Nagel, R.L., and Labie, D. (1984). 'Evidence for the multicentric origin of the sickle cell hemoglobin gene in Africa', Proc. Natl. Acad. Sci. USA, 81, 1771-1773.

Papayannopoulou, T., Brice, M., and Stamatoyannopoulos, G. (1977). 'Hemoglobin F synthesis in vitro: evidence for the control at the level of primitive erythroid stem cells', Proc. Nat. Acad. Sci. USA, 74, 2923-2927.

Papayannopoulou, T., Kalamantis, T., and Stamatoyannopoulos, G. (1979). 'Cellular regulation of hemoglobin switching: evidence for inverse relationship between fetal hemoglobin synthesis and degree of maturity of human erythroid cells', Proc. Natl. Acad. Sci. USA, 76, 6420-6424.

Pergolizzi, R., Spritz, R.A., Spence, S., Goossens, M., Kan, Y.W., and Bank, A. (1981). 'Two cloned β-thalassemia genes are associated with amber mutations at codon 39', Nucleic Acid Res., 9, 7065-7072.

Peschle, C., Migliaccio, G., Covelli, A., Lettieri, F., Migliaccio, A.R., Condorelli, M., Comi, P., Porroli, M.L. Giglioni, B., Ottolenghi, S., Cappellini, M.D., Polli, E., and Gianni, A.M. (1980). 'Hemoglobin synthesis in individual bursts from normal adult blood: all bursts and subcolonies synthesize $^G\gamma$ and $^A\gamma$-globin chains', Blood, 56, 218-226.

Peschle, C., Migliaccio, A.R., Migliaccio, A.G., Lettieri, F., Maguire, Y.P., Condorelli, M., Gianni, A.M., Ottolenghi, S., Giglioni, B., Pozzoli, L.M., and Comi, P. (1981). 'Regulation of Hb synthesis in ontogenesis and erythropoietic differentiation: in vitro studies on fetal liver, cord blood, normal adult blood or marrow and blood from HPFH patients', In "Hemoglobins in Development and differentiation"(Eds. G.Stamatoyannopoulos and A.W. Nienhuis), pp. 359-371, Alan R. Liss, New York.

Peschle ,C., Mavilio ,F., Carè, A., Migliaccio, G., Migliaccio, A.R., Salvo, G., Samoggia, P., Petti, S., Guerriero, R., Marinucci, M., Lazzaro, D., Russo, G., and Mastroberardino, G. (1985a). 'Haemoglobin switching in human embryos: asynchrony of $\zeta \to \alpha$ and $\gamma \to \beta$-globin switches in primitive and definitive erythropoietic lineage', Nature, 313, 255-237.

Peschle, C., Migliaccio, A.R., Migliaccio, G., Mavillo, F. Rocca,
    E., Petrini, M., Mastroberardino, G., Comi, P., Giglioni, B.,
    and Ottolenghi, S. (1985b). 'A model for hemoglobin F synthesis
    in adult life: evidence for regulation at the level of erythro-
    blasts', Annals of the New York Academy of Science, (in press).
Pirastu, A., Doherty, M., Galanello, R., Cao, A., and Kan, Y.W.,
    (1983a). ' Frequent crossingover in human DNA generates multi-
    ple chromosomes containing the sickle and β-thalassaemia genes
    and increases Hb F production', Blood, 62 supplement, 75.
Pirastu, , Kan, Y.W., Cao, A., Conner, B.J., Teplitz, R.L., and
    Wallace, R.B. (1983b). 'Prenatal diagnosis of β-thalassaemia.
    Detection of a single nucleotide mutation in DNA', The New
    Engl. J. Med., 284-287.
Pirastu, M., Del Senno, L., Conconi, F., Vullo, C. and Kan Y.W.
    (1984a). 'Ferrara β°Thalassaemia caused by the $\beta-^{39}$ nonsense
    mutation', Nature, 307, 76.
Pirastu, M., Kan, J.W., Galanello, R., and Cao, A. (1984b).
    'Multiple mutations produce $(\delta\beta)$°thalassemia in Sardinia',
    Science, 223, 929-930.
Pirrone, A., Maggio, A., Gambino, R., Hauser, D., Acuto, S.,
    Romano, V., Buttice, G., and Caronia, F. (1982). 'Genetic
    heterogeneity of β-thalassemia in Western Sicily',
    Haematologica, 67, 825-836.
Poncz, M., Ballantine, M., Solowiejczyk, D., Berak, I., Schwartz,
    E., and Surrey, S. (1982). 'β-thalassaemia in a Kurdish Jew',
    J. Biol. Chem., 257, 5994-5996.
Proudfoot, N.J., and Brownlee, G.G. (1976). ' 3' Non-coding region
    sequences in eukaryotic messenger RNA', Nature, 263, 211-214.
Robertson, A., and Hill,W.G. (1983). ' Identity of different
    mutations for deleterious genes', Nature, 301, 176-181.
Rogers, J., and Wall, R. (1980). 'A mechanism for RNA splicing',
    Proc. Natl. Acad. Sci. USA, 77, 1877-1879.
Rosatelli, C., Tuveri, T., Di Tucci, A., Falchi, A.M., Scalas,
    M.T., Monni, G., and Cao, A. (1985). ' Prental diagnosis of
    Beta-thalassaemia with the synthetic-oligomer tecnique',
    The Lancet, i, 241-243.
Rouabhi, F., Chardin, P., Boissel, J.P., Brghoui, F., Labie,D.,
    and Benabadji, M. (1984). ' Silent β-thalassaemia associated
    with Hb Knossos β 27(69) Ala→Ser in Algeria', Hemoglobin, 7,
    555-561.
Saglio, G., Ricco, G., Mazza, U., Camaschella, C., Pich, P.G.,
    Gianni, A.M., Gianazza, E., Righetti, P.G., Giglioni, B.,Comi,
    P., Gusmeroli, M., and Ottolenghi, S. (1979). 'Human $^{A}\gamma$-globin
    chain is a variant of $^{A}\gamma$ chain ($^{A}\gamma$ Sardinia)', Proc. Natl.
    Acad. Sci. USA, 76,3420-3424.
Salditt-Georgieff, M., and Darnell, J.E. Jr.(1983). 'A precise
    termination site in the mouse β^{major} transcription unit',
    Proc. Natl. Acad. Sci. USA, 80, 4694-4698.
Schwartz, S. (1969). 'The silent carrier of beta thalassemia',
    New Engl. Journ. Med., 281, 1327-1333.
Scott, A.F., Phillips III, J.A., and Migeon, B.R. (1979). 'DNA
    restriction endonuclease analysis for localization of
    human β- and δ-globin genes on chromosome 11', Proc. Natl.
    Acad. Sci. USA, 76, 4563-4565.

Semenza, G.L., Delgrosso, K., Poncz, M., Malladi, P., Schwartz, E., and Surrey, S. (1984a). 'The silent carrier allele: β-thalassemia without a mutation in the β-globin gene or its immediate flanking regions', Cell, 39, 123-128.

Semenza, G.L., Malladi, P., Surrey, S., Delgrosso, K., Poncz, M., and Schwartz, E. (1984b). 'Detection of a novel DNA polymorphism in the β-globin gene cluster', The J. of Biol. Chem., 259, 6045-6048.

Shen, S., Slightom, J.L., and Smithies, O. (1981). 'A History of the human fetal globin gene duplication', Cell, 26, 191-203.

Slightom, J.L., Blechl, A.E., and Smithies, O. (1980). 'Human fetal $^G\gamma$ and $^A\gamma$-globin genes: complete nucleotide sequences suggest that DNA can be exchanged between these duplicated genes', Cell, 21, 627-638.

Sofroniadou, K, Wood, W.G., Nute, P.E., and Stamatojannopoulos G. (1975). 'Globin chain synthesis in the Greek type ($^A\gamma$) of Hereditary Persistence of Fetal Haemoglobin', Br. J. Haematol., 29, 137-148.

Spence, S.E., Pergolizzi, R.G., Donovan-Peluso, M., Kosche, K.A., Dobkin, C.S., and Bank, A. (1982). 'Five nucleotide changes in the large intervening sequence of a γ-globin gene in a $β$ thalassemia patient', Nucl. Acid. Res., 10, 1283-1290.

Spritz, R.A., Jagadeeswaran, P., Choudary, P.V., Biro, P.A., Elder, J.T., De Riel, J.K., Manley, J.L., Gefter, M.L., Forget, B.G., and Weissman, S.M. (1981). 'Base substitution in an intervening sequence of $β^+$ thalassemic human globin gene', Proc. Natl. Acad. Sci. USA, 78, 2455-2459.

Spritz, R.A., and Forget, B.G. (1983). 'The thalassaemias: molecular mechanisms of human genetic disease', Am. J. Hum. Genet., 35, 333-361.

Takeshita, K., Forget, B.C., Scarpa, A., and Benz, E.J. Jr. (1984). 'Intranuclear defect in β-globin mRNA accumulation due to a premature translation termination codon', Blood 64, 13-22.

Tamagnini, G.P., Lopes, H.C., Castanheira, M.E., Wainscoat, J.S., and Wood, W.G. (1983). ' $β^+$-thalassemia Portuguese type: clinical, haematological and molecular studies of a newly defined form of thalassaemia', Br. J. Haematol., 54, 189-200.

Trecartin, R.F., Liebhaber, S.A., Cheng, J.C., Lee, K.Y. and Kan, J.W. (1981). ' $β^0$-thalassemia in Sardinia is caused by a nonsense mutation', J. Clin. Invest., 68, 1012-1017.

Treisman, R., Proudfoot, N.J., Shander, M., and Maniatis, T. (1982). 'A single-base change at a splice site in a $β^0$-thalassemic gene causes abnormal RNA splicing', Cell, 29, 903-911.

Treisman, R., Orkin, S.H. and Maniatis, T. (1983). 'Specific transcription and RNA splicing defects in five cloned β-thalassaemia genes', Nature, 302, 591-596.

Troungos, C., Sortino, G., Cacciola, R., Lombardo, T., Cacciola, B., and Labic, D. (1983). 'Haplotype VI associated mild $β^+$-thalassaemia in Sicily: Mediterranean type?', The Lancet, ii, 509-510.

Tuan, D., Murnane, M.J., de Riel, J.K., and Forget, B.J. (1980). 'Heterogeneity in the molecular basis of Hereditary Persistence of Fetal Haemoglobin', Nature, 285, 335-337.

Tuan, D., Feingold, E., Newman, M., Weissman, S.M., and Forget, B.G.(1983). 'Different 3' endpoints of deletions causing δβ -thalassemia and hereditary persistence of fetal hemoglobin: Implications for the control of γ -globin gene expression in man', Proc. Natl. Acad. Sci. USA, 80, 6937–6941.

Wainscoat, J.S., Bell, J.I., Old, J.M., Weatherall, D.J., Furbetta, M., Galanello, R., and Cao, A. (1983a). 'Globin gene mapping studies in Sardinian patients homozygous for β °-thalassaemia', Mol. Biol. Med., 1, 1–10.

Wainscoat, J.S., Old, J.M., Weatherall, D.J., Orkin, S.H. (1983b). 'The molecular basis for the clinical diversity of β -thalassaemia in Cypriots', The Lancet, ii, 1235–1237.

Wainscoat, J.S., Kanavakis, E., Wood, W.G., Letsky, E.A., Huehns, E.R., Marsh, G.W., Higgs, D.R., Clegg, J.B., and Weatherall, D.J. (1983c). 'Thalassemia intermedia in Cyprus: the interaction of β- and α-thalassemia', Br. J. Haematol., 53, 411–416.

Wainscoat, J.S., Thein, S.L., Old, J.M., and Wheatherall, D.J. (1984). 'A new DNA polymorphism for prenatal diagnosis of β -thalassaemia in Mediterranean populations', The Lancet, ii, 1299–1301.

Wainscoat, J.S., Thein, S.L., Higgs, D.R., Bell, S.I., Weatherall, D.J., and Awaruy-Serjeant, G. (1985).' A genetic marker for elevated levels of Hemoglobin F in Homozygous sickle cell disease?', Br. J. Haematol., in press.

Weatherall, D.J., Pressley, L., Wood, W.G., Higgs, D.R., and Clegg, J.B. (1981). 'Molecular basis for mild forms of homozygous beta thalassemia', The Lancet, i, 527–531.

Weatherall, D.J., and Clegg, J.B. (1983a). 'The thalassaemia syndromes', Blackwell Scientific Publications, Oxford.

Wheatherall, D.J. and Old, J.M. (1983b). 'Antental diagnosis of the haemoglobin disorders by analysis of foetal DNA', Mol. Biol. Med., 1, 151–155.

Weatherall, D.J. (1984). 'Prenatal diagnosis of thalassaemia', Brit. Med. J., 288, 1321–1322.

Westaway, D., and Williamson, R. (1981). 'An intron nucleotide sequence variant in a cloned β -thalassaemia globin gene', Nucl. Acid Res., 9, 1777–1788.

Williamson, R., Eskdale, J., Coleman, D.V., Niazi, M., Loeffler, F.E., and Modell, B.M. (1981). ' Direct gene analysis of chorionic villi: a possible tecnique for first trimester antenatal diagnosis of hemoglobinopaties', Lancet, ii, 1125–1127.

Wood, W.G., Stamatoyannopoulos, G., Lim, G., and Nute, P.E., (1975). ' F-cells in the adult: normal values and levels in individuals with hereditary and acquired elevations of Hb F', Blood, 46, 671– 682.

Wood, W.G., Weatherall, D.J., and Clegg, J.B. (1976). 'Interaction of heterocellular hereditary persistence of foetal haemoglobin with β -thalassemia and sickle cell anemia', Nature 264, 247–249.

Human Genes and Diseases
Edited by F. Blasi
© 1986, John Wiley & Sons, Ltd.

THE MOLECULAR GENETICS OF HYPERLIPIDEMIA

Carol C Shoulders & F E Baralle

The principal function of plasma lipoproteins is to transport lipids (primarily cholesteryl esters and triglycerides) from one body organ to another. It has recently become evident, however, that they not only solubilize these hydrophobic lipids, but also dictate the site in the body to which each lipid class is to be delivered. Some defects in lipoprotein structure or metabolism can contribute to the early onset of a number of vascular disorders. This chapter will focus mainly on the characterization of the gene defects underlying the hyperlipidemias, a group of lipid metabolism disorders that are most commonly found associated with high incidence of atheroma formation.

Background

1.1 Structure of Lipoproteins

Plasma lipoproteins are typically spherical in shape, possessing an amphipathic surface coat, consisting of a monolayer of phospholipid embedded with free cholesterol and protein, which encloses a hydrophobic core of triglycerides and cholesteryl esters. They range in diameter from less than 5nm to greater than 500nm. The size and composition of these macromolecular complexes are not fixed; rather, they are in a dynamic flux during circulation in the bloodstream. Variations in the relative amounts of each component in a lipoprotein give rise to a spectrum of particles which vary continuously both in density and size. For conceptual and operational convenience, six main classes of lipoproteins are defined on the basis of their hydrated density in the ultracentrifuge. (Lindgren et al., 1972). Table 1.1 lists the physical properties of each of these classes, which, in ascending

299

TABLE 1.1[*1,2]

Physical Properties of Lipoprotein Particles

	Size of Particle nm.	Molecular Weight Daltons	Density g/ml	Electrophoretic Definition on Paper
CM	75-1200	$>4 \times 10^8$	<0.93	remain at origin
VLDL	30-80	$1-2 \times 10^7$	0.93 -1.006	pre β -LP
IDL	25-35	$5-10 \times 10^6$	1.006-1.019	slow pre β -LP
LDL	18-25	2.3×10^6	1.019-1.063	β -LP
HDL_2	9-12	3.6×10^5	1.063-1.125	α -LP
HDL_3	5-9	1.75×10^5	1.125-1.210	α -LP

[*1] Eisenberg & Levy, 1975
[*2] Schaefer et al, 1978

order of denisty, are: i) Chylomicrons (CM), ii) Very Low Density Lipoproteins (VLDL), iii) Intermediate Density Lipoproteins (IDL or LDL_1),   iv) Low Density Lipoproteins (LDL or LDL_2), v) High Density Lipoprotein-2 (HDL_2),   vi) High Density Lipoprotein-3 (HDL_3).   It should be noted however, that, although it is convenient to visualize six separate classes, numerous studies show that they are all metabolically inter-convertible (see section 1.2).

The proportion of protein in each of these lipoprotein classes is inversely related to particle size (see Table 1.2). Convention-ally, those proteins which are embedded in the phospholipid mono-layer are termed apolipoproteins. At least 10 apolipoproteins have now been identified: apo AI, AII, AIV, B, CI, CII, CIII, D, E and F (for references, see Herbert et al., 1983). The major apolipo-

TABLE 1.2[*1,2,3]

Components of Lipoprotein Particles

	CM	VLDL	IDL	LDL	HDL_2	HDL_3
Protein	1-2	8-10	18-20	21-23	40-45	50-55
Free Cholesterol	2-3	7-12	9-11	8-10	3-6	3-5
Cholesteryl Esters	2-6	10-14	15-18	36-40	14-18	10-12
Phospholipid	5-8	15-20	16-20	21-23	25-30	20-23
Triglyceride	80-90	50-60	30-34	8-10	3-6	3-6
Apo AI	7-10	Minor	-	-	75	65
Apo AII	7-10	Minor	-	-	15	25
Apo B	20	2-10	13	75	Minor	Minor
Apo CI	15	10	26	Trace	1-2	-
Apo CII	15	10-20	9	Trace	1-3	5
Apo CIII	38	50-60	41	22	4-6	1.5-1.8
Apo E	5	10-20	13	3	Minor	-

[*1] % composition by weight
[*2] Schaefer et al, 1978
[*3] Scanu et al, 1982

protein components of human lipoproteins are listed in Table 1.2, while their functional roles are shown in Table 1.3. These include: i) binding and solubilizing of lipid, ii) regulating the activity of enzymes involved in lipid metabolism, and iii) acting as a ligand for specific cell surface receptors.

TABLE 1.3

Functions of Apolipoproteins

	MW Daltons	Origin	Function	Plasma Levels[6] mg/ml
AI	28,000	Intestine	Activate LCAT[1]	90-130
AII	17,000[2]	Intestine Liver	Inhibit LCAT[1] ?Activate HLP[3]	30-50
B-48	264,000	Intestine	Triglyceride trans- port. ?Binding to CM receptor[4]	<5
B-100	549,000	Liver	Triglyceride trans- port. ?Binding to apoB/E receptor[4]	80-100
CI	6,500	Liver Intestine	?Activate LCAT[1,5]	4-7
CII	8,800	Liver Intestine	Activate LPL[5]	3-8
CIII	8,750	Liver Intestine	?Inhibit LPL[5]	8-15
E	35-39,000	Liver	High affinity[4] receptor binding	3-6

[1] Dobiasova, 1983, and references therein; Lecithin Cholesterol
     Acyl Transferase, see section 1.2.1
[2] Fielding et al, 1972a,b; Weight of disulphide linked dimer
[3] Jahn et al, 1983; Hepatic Lipase
[4] Innerarity et al, 1983a and b; Weisgraber et al, 1983
[5] Nestel & Fidge, 1982, and references therein;
     Lipoprotein Lipase, see section 1.2.1 and 1.2.2
[6] Herbert et al, 1983
 ? Controversial function

The importance of apolipoproteins in maintaining the structure and co-ordinating the metabolism of lipoprotein particles has now been well illustrated by studies of the phenotypic abnormalities present in patients possessing altered levels of specific apolipoproteins (Lux et al, 1972; Breckenridge et al , 1978; Carlson & Philipson, 1979; Ghiselli et al, 1981). However, this line of investigation has, to date, failed to resolve the specific function of a number of apolipoproteins (see Table 1.3).

1.2 The Transport of Lipids in Man

For transport, triglycerides and cholesteryl esters are packaged into lipoprotein particles. The former are delivered mainly to adipose and muscle tissue, where the fatty acids are oxidized for energy or re-esterified, while the latter are transported to all body cells, where the unesterified sterol is used as a structural component of plasma membranes. In some specialized cells, the sterol is used for the synthesis of steroid hormones and bile acids. The routes taken by these lipids to reach their destination are outlined in Figure 1. For conceptual simplicity, this complex network of pathways is often subdivided into exogenous and endogenous routes, dealing primarily with the transport of lipids of dietary and hepatic sources, respectively. The following two sections briefly summarise the major events of these pathways. For greater detail, the reader is referred to Brown et al 1981 and references therein, and Myant, 1982.

1.2.1 The Exogenous System

In this pathway, intestinal triglyceride-rich particles, CM, are secreted via the lymphatic system into the bloodstream, whereupon they are rapidly metabolized (4-5 min in man). The triglyceride-rich core is catalytically hydrolysed by Lipoprotein Lipase on the capillaries of muscle and adipose tissue (Scow et al, 1976). As the surface material (mainly phospholipid and free

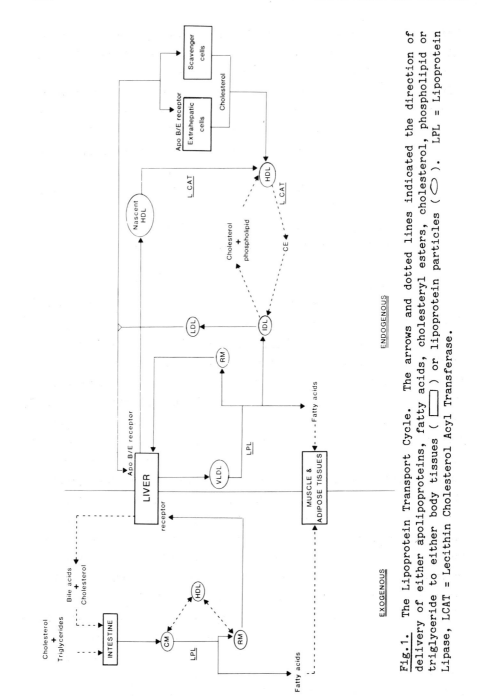

Fig.1. The Lipoprotein Transport Cycle. The arrows and dotted lines indicated the direction of delivery of either apolipoproteins, fatty acids, cholesteryl esters, cholesterol, phospholipid or triglyceride to either body tissues (☐) or lipoprotein particles (◯). LPL = Lipoprotein Lipase, LCAT = Lecithin Cholesterol Acyl Transferase.

cholesterol) becomes in excess of that needed by the CM to solubilize its hydrophobic core, it is either transferred to HDL
particles or taken up by the liver via a receptor mediated pathway
(Kita et al, 1982).   The bulk of the excess free cholesterol
acquired in this procedure is esterified.   HDL particles employ
Lecithin Cholesterol Acyl Transferase (LCAT) (for review, see
Dobiasova, 1983), an enzyme that esterifies cholesterol with free
fatty acids derived from Lecithin for this purpose, while hepatocytes utilize Acyl-CoA-Cholesterol Transferase (ACAT).   In the
former case, the cholesteryl esters are transferred to IDL and LDL
particles (see section 1.2.2) and in the latter they are either
packaged into VLDL particles or stored until required.

1.2.2   The Endogenous Pathway

The liver converts carbohydrates and fatty acids into triglycerides which, together with cholesteryl esters, are
incorporated into VLDL particles.   On entering the circulation the
triglyceride-rich core of these particles is hydrolysed in a
similar fashion to CMs.   The interaction of VLDL with Lipoprotein
Lipase, however, is considerably less efficient, such that the
metabolism of these particles normally takes between 1-5 hours.
This gradual loss of triglyceride, together with redundant surface
materials (see section 1.2.1), gives rise to particles of Intermediate Density (IDL).   These are either removed from the circulation by the liver or metabolised further to form LDL particles.
Thus, through many cycles of triglyceride hydrolysis and cholesteryl ester acquisition, LDL particles are formed from VLDL
particles.   These are removed from the circulation by two major
routes: i) high-affinity receptors (termed either LDL or apo B/E
receptors, which are quite distinct from the receptor responsible
for the uptake of CM remnants [Kita et al, 1982]) present on both
liver and extra-hepatic cells, and ii) by macrophages or scavenger
cells of the reticuloendothelial system (Brown & Goldstein, 1983).
The cellular uptake of LDL cholesterol by the former route, in
contrast to the latter, is under rigorous feedback control in order

to maintain homeostasis of intracellular free cholesterol concentration in the face of fluctuating external supplies. Three of the
major regulatory responses involved are: i) suppression of 3-
hydroxy-3 methylglutaryl CoA reductase synthesis, the rate-
controlling enzyme in endogenous cholesterol synthesis, ii) the
activation of ACAT which re-esterifies free cholesterol, and iii)
suppression of the synthesis of the LDL cell-surface receptors.

1.3   Classification of the Hyperlipidemias

The Hyperlipidemias, with diabetes mellitus and cigarette
smoking, are amongst the major risk factors for the development of
atheroma.   They encompass a heterogeneous group of disorders which
are currently classified by a system proposed by Fredrickson and
colleagues in the 1960s (Fredrickson et al, 1967).   It divides the
hyperlipidemias into six types based on the pattern of elevation of
CM, VLDL and LDL levels in plasma after a twelve-hour fast (see
Table 1.4).   This classification is extremely useful in that it
provides a simple code for communicating various abnormalities in
plasma lipoprotein levels.   However, in the majority of cases, it
does not provide any information about the aetiology of the
condition.

The Genetic Nature of Hyperlipidemia

2.1   The Familial Hyperlipidemias

Six gene defects have now been identified as causing hyperlipidemia.   These diseases, generally referred to as the Familial
Hyperlipidemias, are believed to be inherited by a single gene
mechanism according to Mendelian principles.   They illustrate well
the importance of the components of the lipoprotein system in
maintaining cholesterol and triglyceride plasma levels within
normal limits (see Tables 1.5 and 1.6).

TABLE 1.4*

Conventional Classification of Hyperlipidemia

Type	Lipoprotein Abnormalities	Lipid Changes	
1	CM present and greatly increased; VLDL, LDL, HDL, normal or decreased.	↑C	↑TG
2a	LDL increased; VLDL normal.	↑C	
2b	LDL increased; VLDL increased.	↑C	↑TG
3	Presence of β-VLDL; LDL of abnormal lipid composition.	↑C	↑TG
4	VLDL increased; CM "absent", LDL normal.		↑TG
5	CM present; VLDL increased.	↑C	↑TG

* Adapted from Fredrickson et al, 1978
C Cholesteryl Ester
TG Triglyceride

2.1.1 Type 1 Familial Hyperlipidemia

Two separate molecular forms of this rare, recessively inherited disorder (<1 : 1,000,000) have now been characterized: a deficiency of Lipoprotein Lipase (LPL) itself (Havel & Gordon, 1960), and a deficiency of apo CII (Breckenridge et al, 1978), a co-factor for this enzyme. Both defects are expressed as a result of either inadequate or absent LPL activity, and give rise to similar, but not identical clinical symptoms. In general, apart from more severe attacks of pancreatitis, apo CII-deficient patients are healthier than LPL-deficient patients. At present, there are no adequate experimental data to explain this difference.

TABLE 1.5[*1]

The Familial Hyperlipidemias

Genetic Disorder	Biochemical Defect	Lipoprotein Type
Familial Lipo-protein Lipase Deficiency	a) Deficiency of extra-hepatic triglyceride lipase	1
	b) Deficiency of apo CII activity	1 or 5
Familial Hyper-cholesterolemia	Deficiency of LDL receptor	2a (2b in rare cases)
Familial Hyperlipo-proteinemia Type 3	Unknown (association with an isoform of apoE)	3
Familial Hyper-triglyceridemia	Unknown	4
Familial Hyper-triglyceridemia Type 5	Unknown	5
Familial Combined Hyperlipidemia	Unknown	2a, 2b, or 4 (in rare cases 5)

[*1] Adapted from Fredrickson et al, 1978

One possible reason, however, may be that there is a less severe defect in clearance of CM and VLDL from the circulation in apo CII-deficient patients compared to LPL-deficient patients. For further detailed information on this disorder, the reader is referred to Nikkila, 1983.

11

TABLE 1.6*

Typical Clinical Symptoms of the
Familial Hyperlipidemias

Genetic disorder	Xanthomas	Pancrea-titis	Premature vascular disease		Hyper-glycemia	Obes-ity
			Coron-ary	Peri-pheral		
Familial lipoprotein lipase deficiency	Eruptive	+				
Familial hyperchol-esterolemia	Xanthelasma; tendon		+		+	+
Familial hyperlipo-proteinemia type 3	Xanthelasma; tuberous; palmar creases		+	+	+	+
Familial hypertri-glyceridemia			+		+	+
Familial hyperlipo-proteinemia type 5	Eruptive	+	+(?)		+	+
Familial combined hyperlipidemia			+		+	+

*Adapted from Fredrickson et al, 1978

2.1.2 Familial Hypercholesterolemia

Of the familial hyperlipidemias, this is undoubtedly the one that is best characterized, thanks to the pioneering work of Brown and Goldstein (cf Brown & Goldstein, 1983, Goldstein & Brown, 1984, for reviews). This disease is transmitted as an autosomal dominant trait with a gene dosage effect. Its prevalence, approximately 1 : 500 in most ethnic groups, places it amongst one of the most common errors of metabolism reported to date. Nevertheless, it accounts for only a small proportion (less than 20%) of those persons in the general population with a type 2 hyperlipidemia pattern.

The underlying biochemical defect in this familial condition results from a mutation(s) in the gene specifying the LDL receptor, that regulates LDL degradation and cellular cholesterol synthesis (see section 1.2.2). To date, at least 7 mutant alleles have been identified, exerting their effect in one of 3 possible ways (see Table 1.7). In the first, the mutation appears to disrupt completely LDL-receptor synthesis. In the second, the mutation destroys the processing of the precursor form of this protein, possibly as a consequence of the LDL receptor failing to reach the correct cellular compartment for this to occur. While, in the third, the mutations affect the ability of the mature product to bind to LDL particles.

Individuals with one mutant LDL-receptor gene produce on average one half the normal number of LDL receptors per cell and thus remove both IDL and LDL particles from plasma at a reduced rate. Such individuals normally have moderate hypercholesterolemia (350-550mg/dl, cf normolipidemics 180-220mg/dl) from birth, with tendon xanthomas and coronary atherosclerosis developing soon after the age of thirty. For these individuals, therapeutic measures are now available that stimulate the single normal gene to produce an increased number of hepatic LDL receptors and thus reduce plasma LDL levels. The therapy involves reducing hepatic levels of cholesterol by inhibiting HMG CoA reductase activity (see section 1.2.2) or depleting stores of bile acids (Shepherd et al, 1980), and in some cases, by a combination of the two.

TABLE 1.7

Characterization of 7 mutant LDL-receptor alleles

	Mutant Allele	Molecular Weight of Receptor (KD)		Cell surface LDL binding sites (% of normal) in homozygotes
		Precursor	Mature	
Class I (no immunologically detectable synthesis of precursor)	RO referred to as the NULL allele	ND	ND	< 2 (n = 12)
Class II (failure of precursor to undergo processing)	R100	100	100	< 2 (n = 2)
	R120	120	120	< 2 (n = 3)
	R135	135	135	< 2 (n = 2)
Class III (precursor processed normally)	R100-140	100	140	< 2 (n = 2)
	R120-160	120	160	a) < 2 (n = 2)
				b) 2-30 n = 1)
	R170-210	170	210	

* adapted from Tolleshaug et al, 1983.
Nos. in parentheses state the number of homozygous subjects studied.

Homozygotes for Familial Hypercholesterolemia synthesis have few or no LDL receptors and hence develop severe hypercholesterol-emia (650-12000mg/dl).   Coronary Heart disease and tendon xanthomas are present in childhood and death is not uncommon before the second decade.   In contrast to their heterozygote counterpart, the therapy outlined above is of limited value.   Treatment of the symptoms of this condition can be drastic: recently, for example, an affected six year old girl underwent both a liver and heart transplantation. This combination was carried out in order to provide her with nearly normal numbers of LDL receptors so that her new heart did not become diseased in the same way as her previous one had (Bilheimer et al, 1984).

2.1.3   Familial Type 3 Hyperlipidemia

This disorder is a relatively rare form of Hyperlipidemia, and the probability that a given adult patient with elevated concentration of both cholesterol and triglyceride has Type 3 familial hyperlipidemia is 1 : 100 to 1 : 1000. Clinically, it is associated with unusual lipid deposits in the palmar creases, a high propensity for peripheral vascular disease and a failure to clear efficiently cholesterol-rich remnants, derived from CM and VLDL metabolism from the circulation (see section 1.2 and Brown et al, 1983).

The underlying defect involves a polymorphic genetic locus that specifies the primary structure of apo E, a component of remnant particles (Utermann et al, 1975; Zannis & Breslow, 1981). In the general population, 3 major (in terms of frequency) and at least 6 minor alleles (see Innerarity et al, 1983a, for references) exist at this locus. On the basis of iso-electric focusing behaviour, the protein products of these alleles are divided into 3 major classes, E2, E3 and E4. E2 and E4 differ, by one amino acid, from E3 at position 158 (Arg to Cys) and 112 (Cys 112 to Arg), respectively. In both cases, the amino acid change could be accounted for by a point mutation in the gene encoding this protein. For further detailed information regarding the significance of these changes, the reader is referred to Weisgraber et al, 1983 and Innerarity et al, 1983b.

Most patients symptomatic for type 3 Familial Hyperlipidemia are homozygous for the apo E2 allele (Utermann et al, 1977), a few are heterozygous (Ghiselli et al, 1982) and the remainder completely apo E deficient (Ghiselli et al, 1981). Clearly, this apolipoprotein plays an important role in the regulation of lipid metabolism. However, since only 1% of the persons in the population with the apo E2 homozygous phenotype, and probably less with the heterozygous phenotype, exhibit type 3 Familial Hyperlipidemia symptoms, the presence of an apo E2 genotype appears a necessary but not in itself a sufficient condition for the development of the clinical symptoms of this disease. Another factor causing hyperlipidemia, eg thyroid deficiency or defective receptor binding of LDL (Hazzard et al, 1981a), needs to be present. This has now been demonstrated by several pedigree studies where the mode of inheritance of an additional "hyperlipidemic gene" proceeds in an autosomal dominant fashion. For example, in a large family studied by Hazzard and colleagues (Hazzard et al, 1981b), it was found that those hyperlipidemic members who were homozygous for the apo E2 allele exhibited type 3 familial hyperlipidemia, while those who were not had the clinical picture of type 2a, 2b or IV hyperlipidemia.

It should be noted, however, that the apo E2 allele may prove to contain a more heterogeneous population of apo E isoforms than presently reported and that this provides an alternative explanation why most individuals homozygous for this allele do not express type 3 familial hyperlipidemic symptoms. On this point, the reader is referred to Rall et al, 1982.

2.1.4 Familial Hypertriglyceridemia

Among the simply inherited disorders of lipoprotein metabolism, familial hypertriglyceriemia is one of the least well understood (for reviews see Fredrickson et al, 1978). The syndrome is generally believed to be transmitted as an autosomal dominant trait at a fairly high frequency, but to date no homozygotes for this condition have been identified. The reason for this is unknown; however, explanations for failing to recognize such individuals could include any of the following: i) the homozygous phenotype may not differ from the heterozygous phenotype and thus be, unlike Familial Hypercholesterolemia where there is gene dosage effect, a truly dominant inherited disorder; ii) plasma triglyceride levels have not been measured frequently enough in offspring of affected individuals to identify the homozygotes that exist, and iii) the disorder is genetically heterogeneous and thus the apparent high frequency may be an overestimate. In this case, the mating between two presumed heterozygotes would produce genetic compounds, rather than homozygotes.

In contrast to the familial hyperlipidemic conditions discussed previously, Familial Hypertriglyceridemia individuals possess no unique clinical or biochemical features to distinguish them from those hypertriglyceridemic individuals whose symptoms derived from other factors. Thus, at present the diagnosis of Familial Hypertriglyceridemia can only be made with certainty if the affected individual has a sufficiently large enough family to show the trait segregates in an autosomal dominant fashion. However, even with this criteria, it is still probable that this type of hyperlipidemia will encompass a wide variety of as yet uncharacterized biochemical defects that produce a similar phenotype.

Patients with Familial Hypertriglyceridemia have an elevation in the plasma concentration of VLDL carrying endogenous synthesized triglyceride and hence exhibit a type 4 lipoprotein pattern. Triglyceride levels typically range between 200-500mg/dl compared to 78 ± 6mg/dl in normal individuals. In those cases where the plasma triglyceride level rises above 1000mg/dl, particles characteristic of CM may appear, and the patient could then exhibit a type 5 lipoprotein pattern. In fact, any patient with a type 4 lipoprotein pattern for whatever reason may develop hyperchylomicronemia and a type 5 pattern in any of the following ways: i) the consumption of a diet extremely high in carbohydrate, ii) excessive alcohol intake, iii) development of diabetes mellitus or hypothyroidism, iv) the ingestion of oral contraceptives.

The pathology of Familial Hypertriglyceridemia is not understood. An overproduction of VLDL and a reduced efficiency of VLDL removal from the plasma by extra hepatic tissues are just two of the proposals raised. However, the monogenic nature of the disorder does imply that eventually its underlying defect will be traced back to a single protein molecule. Furthermore, because the defect is expressed as a dominant trait, this suggests that the mutant protein is likely to be non-enzymatic and play a regulatory or structural role.

Clinically many hypertriglyceridemic patients, whether of the familial type or not, possess a wide range of additional metabolic disorders (moderate obesity, insulin resistance, fasting insulinemia, glucose intolerance and hyperuricemia). However, insufficient data are available to assess if any of these conditions play an important role in the aetiology of Familial Hypertriglyceridemia. There is some evidence though to suggest that hypertriglyceridemia in general predisposes the affected individuals to premature coronary heart and peripheral vascular disease. Hence, whenever possible, regimes that reduce the hyper-lipidemia are put into action. In some cases, weight reduction and carbohydrate restriction achieve this end, but in others drug therapy (clofibrate or nicotinic acid) is required.

2.1.5   Type 5 Hyperlipoproteinemia

The familial nature of this disorder is well documented but nevertheless its mode of transmission is still uncertain (see Nikkila, 1983, for review). While primary type 5 hyperlipo-proteinemia appears to be a genetic disorder completely distinct from type 4 hypertriglyceridemia, affected families with the former syndrome have members exhibiting a lipoprotein pattern characteristic of the latter. Thus, the presence of individuals with a type 5 phenotype is used as the marker for type 5 disorder, distinguishing it from Familial Hypertriglyceridemia in which, under standard physiological conditions (see previous section), all affected family members show a type 4 pattern.

Generally, confirmation of the diagnosis of type 5 Familial Hyperlipidemia in an affected individual is based upon the lipid and lipoprotein analysis of their first degree relatives. The transmission of an abnormal trait, in an autosomal dominant fashion, that gives rise to a type 5, and in some members a type 4

lipoprotein pattern, is the hallmark of this syndrome. Clinically
the symptoms of type 5 patients are related to their chylo-
micronemia and are hence similar, but not identical, to those
experienced by patients with Familial LPL deficiency (see Table
1.6). Treatment of type 5 patients by diet alone or in combination
with drugs alone is not successful in reducing plasma lipoprotein
levels to within normal limits. However, it is generally adequate
to reduce the hypertriglyceridemia so that attacks of pancreatitis
are prevented.

TABLE 1.8*

Distribution of apoE isoforms in normolipidemics
and hyperlipidemic individuals

Genotype	Normolipidemics	Familial Type 3 hyperlipidemia	Familial Type 5 hyperlipidemia
E4/E4	4.1% (3)	0.0% (0)	31.4% (11)
E3/E4	21.6% (16)	0.0% (0)	37.2% (13)
E2/E4	0.0% (0)	0.0% (0)	5.7% (2)
E3/E3	55.4% (41)	0.0% (0)	5.7% (2)
E2/E3	18.9% (14)	25.0% (3)	17.1% (6)
E2/E2	0.0% (0)	75.0% (9)	2.9% (1)
Total	100% (74)	100% (12)	100% (35)

* adapted from Ghiselli et al, 1981

Nos. in parentheses state the number of subjects studied.

The biochemical defect(s) underlying this disorder have not yet been identified, and many speculations still remain to be tested. Given that, in common with type 3 familial hyperlipidemia, type 5 familial hyperlipidemia is characterised by altered VLDL and CM metabolism, it is perhaps of interest to note the unusual distribution of the major apo E alleles in this latter group in contrast to both the former and a normolipidemic population (see also section 2.1.3).

2.1.6   Familial Combined Hyperlipidemia

This disease appears to be transmitted via a defective, dominant autosomal gene, but because of its relatively recent characterization, this has yet to be confirmed (Goldstein et al, 1972; Fredrickson et al, 1977). In Familial Combined Hyperlipidemia, affected members in the same family may exhibit elevation of either: i) VLDL (type 4 lipoprotein pattern); ii) LDL (type 2a lipoprotein pattern), or both. In spite of this, there are several lines of evidence that suggest that this disorder is genetically distinct from both Familial Hypercholesterolemia and Familial Hypertriglyceridemia. These include: i) the demonstration that plasma lipid levels are not elevated to the same magnitude seen in Familial Hypercholesteridemia and Familial Hypertriglyceridemia, ii) the distribution pattern of lipid abnormalities in affected relatives is distinctive: about one-third manifest a type 2a pattern, another third a type 2b and the remainder type 4, iii) in contrast to Familial Hypercholesterolemic patients, Familial Combined Hyperlipidemic patients do not exhibit hypercholesterolemia in childhood, iv) in families with this syndrome, a hypercholesterolemic mating with a normolipidemic partner generally gives rise to a hypertriglyceridemic offspring, and vice versa.

The clinical features of Familial Combined Hyperlipidemia are not yet well characterized. In affected families it seems that the full-blown expression of the hyperlipidemia is delayed until the age of 25-30, and that tendon xanthomas do not develop even amongst the hypercholesterolemic members (see section 2.1.2). However, affected individuals show an apparent increased frequency of coronary heart disease.

The underlying biochemical defect(s) in Familial Combined Hyperlipidemia still remains unidentified, although it seems that overproduction of VLDL may be involved. The interaction of such factors as an excessive fat load, stress or a polymorphic protein with the major gene defect have been postulated to account for the observed variability among affected members of the same family.

2.2 The Prevalent Forms of Hyperlipidemia

In the general population 5% of individuals are hyperlipidemic. The gene defects underlying the familial hyperlipidemias discussed in the previous section are however responsible for causing less than 20% of these disorders. The remaining 80% are believed to occur as a result of complex interactions between polygenic and environmental factors, but the nature of these defects has yet to be identified. This is perhaps not surprising, given the complex nature of lipid transport and metabolism which employs many gene products.

In such a situation, a pedigree analysis of hyperlipidemic individuals can prove to be of limited value in identifying the genetic components underlying the pathological condition. For example, if the hyperlipidemia results from the combined action of two genes alone, one recessive and one dominant, it is possible that only one member in three generations of a family will inherit the appropriate set of alleles for exhibiting the pathological condition and it will thus appear to be non-familial in character.

In an attempt to overcome such difficulties, serum from large samples of the population are continually being screened for variant forms of apolipoproteins that may associate with hyperlipidemic conditions (Utermann et al, 1982; Schumaker et al, 1984). Although it is still too early to assess the success of this approach, one can envisage that it could also prove limited, in that there may exist in the population polymorphic alleles that produce normally functional protein but respond inadequately to such factors as a high fat diet and, as a consequence, alone or together with other factors, cause certain forms of hyperlipidemia. We and others have therefore chosen to use recombinant DNA technology as a method both to study the regulation of genes involved in lipid transport and metabolism, and to identify polymorphic sequences in the population that associate with hyperlipidemia. The results of these recent investigations will be described in the remainder of this chapter.

2.2.1   The Isolation of Genes by Recombinant DNA Technology

In order to study a gene in any detail, it needs to be purified, free from the many other genes and transcripts present in the organism, in a form where it can be readily prepared in large quantities. Molecular cloning now routinely achieves such ends by: i) individually recombining large collections of cDNA (derived from mRNA) or DNA fragments with a vector,   ii) placing the resulting recombinant molecules singly into host cells for autonomous replication,   iii) using a specific probe to identify which of these host cells contains the recombinant DNA of interest, and finally, iv) propagating this host cell to produce as many copies of this recombinant as required (see Figure 2 and Maniatis et al, 1978, 1982).

Such collections of clones prepared from heterogeneous populations of cDNA and DNA are called genomic and cDNA libraries, respectively.   Usually genomic libraries contain a complete collection of cloned fragments which comprise the entire genome of the organism used, while cDNA libraries contain cDNA molecules produced from a heterogeneous mRNA preparation. The total number of different cDNA clones present in such a library will, to a large extent, reflect the concentration of the different mRNA species present in the original tissue and influence the ease with which a particular clone may be isolated.   Thus, clones containing sequences corresponding to a relatively abundant mRNA species will,

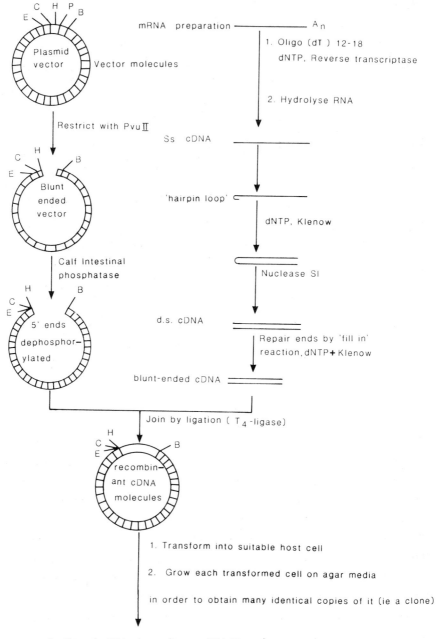

Fig. 2

<u>Fig.2.</u> Schematic illustration of the procedure for construction of a cDNA library. mRNA from a donor tissue (e.g. liver, brain) is reverse transcribed into a single-stranded DNA copy (complementary DNA, cDNA) utilizing oligo(dT)$_{12-18}$ as a primer for the reaction. The template RNA is then hydrolysed and the single-stranded copy converted to a double-stranded molecule using enzyme directed replication (large fragment of DNA polymerase I, referred to as "Klenow"). The primer for this reaction is a loop formed within the single-stranded cDNA by intra-strand hybridization. After 2nd strand formation, the loop structure is selectively digested away using a single-stranded specific nuclease, and any double-stranded cDNA molecules left with 5' overhanging ends repaired by a 'fill in' reaction using deoxyribonucleotides and Klenow for this purpose. Finally, the double-stranded blunt-ended cDNA molecules are joined to previously prepared vector molecules and then transformed into suitable bacterial hosts. Generally conditions for these reactions are optimized so that in the majority of cases only one DNA fragment joins with one vector molecule and each transformed cell carries one recombinant molecule.

   The figure shows a typical plasmid vector; it has a number of unique restriction enzyme sites in close proximity to each other. (C = <u>Cla</u>I, E = <u>Eco</u>RI, H = <u>Hind</u>III, B = <u>Bam</u>HI, P = <u>Pvu</u>II). The vector has been cleaved with <u>Pvu</u>II and its 5' ends dephosphory-lated. The former reaction creates a blunt-ended vector, while the latter ensures that it is unable to recircularize in the absence of phosphorylated DNA (or cDNA) molecules upon addition of T$_4$-ligase.

on average, be isolated by screening far fewer recombinant clones than would be the case for a sequence corresponding to a rare mRNA species. However, current methods of screening are sufficiently powerful that such rare sequences will stand a good chance of being detected. These methods include utilizing specific antibodies (Stanley and Luzio, 1984) or synthetic oligonucleotides as probes. The latter technique is particularly favoured when the clone of interest codes for a protein where some amino acid sequence data, 5 contiguous amino acids or more, are known, and indeed was employed to isolate apolipoprotein and LDL-receptor cDNA clones (Sharpe et al, 1984; Russell et al, 1983).

2.2.2   Identifying Polymorphic Alleles in the Population by Recombinant DNA Technology

This technology makes it possible to readily analyse large population samples for alterations in nucleotide sequence in the genome that may associate with pathological conditions. Generally, these alterations will be detected as changes in the length of characteristic fragments, produced from the digestion of genomic DNA with specific restriction enzyme (Roberts, 1983) (Restriction Fragment Length Polymorphism, RFLP), that hybridizes to a specific DNA probe. A RFLP will arise when the region under study has either a rearrangement in its sequences (addition, deletion, translocation or inversion), or a substitution of a nucleotide such that either a new restriction enzyme site is created or a previously existing one is lost.

The procedure for detecting a RFLP is as follows. Genomic DNA is usually derived from peripheral blood leukocytes, purified, and then specifically digested with a restriction endonuclease appropriate for the gene(s) under study. The cleaved DNA fragments are then fractionated by size in an agarose gel, denatured, and transferred onto a filter using the procedure of 'genomic blotting' (Southern, 1975). The filter-bound DNA is incubated with a radio-active probe (typically prepared from cloned cDNA or genomic DNA) and then extensively washed so that the probe will remain bound only to those restriction fragment(s) containing complementary sequences; the latter are visualized by autoradiography (Laskey & Mills, 1977).

Prior to searching for RFLPs in population samples of DNA, one normally constructs a restriction enzyme map of the DNA region under study so that one is provided with some clues to which might be the most appropriate endonuclease to use for this purpose. A small amount of DNA from a normal individual is digested with several different restriction endonucleases in combination and singly, then size fractionated, blotted and incubated with a suitable probe to determine the length of the fragment(s) that contain the complementary sequences. In this type of analysis, it is also of interest to determine the orientation of the probe sequences with respect to the restriction enzyme map. This can be readily deduced when the 5' and 3' ends of a DNA probe hybridize to restriction fragments of different sizes that are derived from a single endonuclease digest of genomic DNA (for example, see Figure 3). On the basis of all these data, one is then able to choose which are the best restriction enzymes to use for searching for possible DNA polymorphisms that associate with a particular pathological state. Ideally the enzyme of choice would allow one to

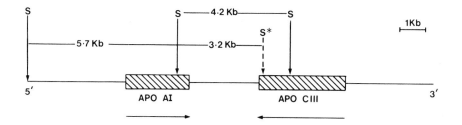

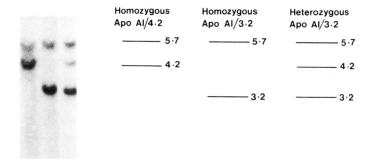

Fig.3. Map of the apoAI and CIII genes with the SstI restriction enzyme sites shown. S* denotes the location of the SstI site present in individuals possessing the apoAI/3.2 allele. (For more details of the structure of the two genes, see section 2.2.3 and Fig.4.) The lower part of the figure shows an autoradiograph of a genomic blot of SstI restricted human DNA obtained from 3 unrelated Caucasian hypertriglyceridemic patients. The adjacent line diagrams show the genotype of the 3 individuals.

examine a large region of the genome with a single blot and yet maintain high resolution for detection of variation in molecular weight.

In the genome two types of polymorphic sequences are present: functional and neutral. The former may give rise to RFLPs that are directly responsible for the disorder and in the majority of cases will be intragenic, occurring in regions that affect either the transcriptional or translational regulation of the gene, or the quality of the final product (Geever et al, 1981). The latter, in contrast, are scattered throughout the genome and, while they do not directly cause the disorder, can be used as genetic markers to follow the inheritance of defective alleles in linkage analysis studies (Kan & Dozy, 1978; Murray et al, 1982; Gusella et al, 1983). Briefly, to be "informative", these markers need to be physically closely linked to the defective loci so that they are inherited together as a unit. Clearly, the further apart the RFLP site and the defective loci under study on the chromosome, the more likely it is that recombination will occur in any meiosis, and hence the less useful they will be in diagnosis of any given disorder (see Weatherall, 1982, for further discussion).

Thus, two different strategies can be used to identify DNA polymorphic sequences in the genome that associate with hyper-lipidemia. In the first, "candidate" genes, such as those coding for apolipoproteins, LDL receptor and LPL, are used as probes to search for functional mutations (Kan & Dozy, 1978; Orkin et al, 1982). In some cases these gene probes will find neutral poly-morphisms that are closely linked to functional mutations. Whilst these generally tell one nothing about the molecular basis of the disorder, they can be used to follow the inheritance of the defective allele in family studies. Furthermore, they can indicate where to look along the chromosome for the functional mutation, which may in some cases prove to involve some unexpected or uncharacterized protein. The second strategy, which is described in greater detail by Botstein et al 1980, involves using a number of random gene probes to identify RFLPs that define genetic

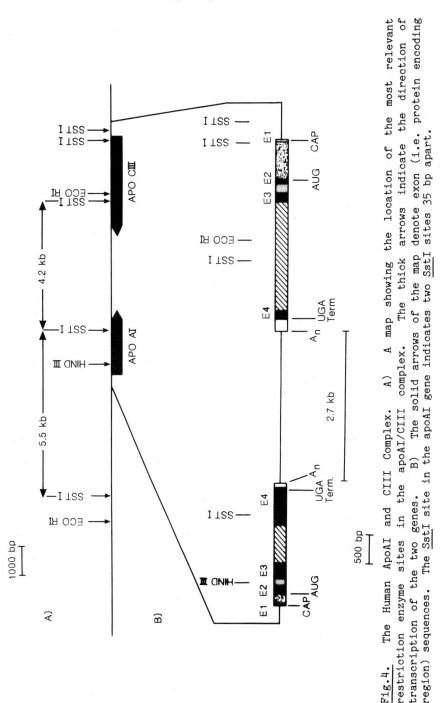

Fig. 4. The Human ApoAI and CIII Complex. A) A map showing the location of the most relevant restriction enzyme sites in the apoAI/CIII complex. The thick arrows indicate the direction of transcription of the two genes. B) The solid arrows of the map denote exon (i.e. protein encoding region) sequences. The SstI site in the apoAI gene indicates two SstI sites 35 bp apart.

markers.   Such methodology is based on calculations which suggest
that, if 500-1000 random DNA markers were available, it would be
nearly certain that in a family study, at least one of them would
show linkage to any given monogenic disorder.

2.2.3   Characterization of Apolipoprotein AI and CIII Genes.

It has now been demonstrated that apo AI and CIII genes are
closely linked in the genome by characterizing the insert of a
genomic clone, apo AI 6, previously shown to contain apo AI
sequences (Karathanasis et al, 1983a).   DNA from this clone was
specifically digested with several different restriction endonucleases, in combination and singly, size fractionated, blotted
onto a nitrocellulose filter and then incubated with an apo AI cDNA
probe to determine the length of the fragments that contained
complementary sequences.   The filter was then extensively washed to
remove all traces of the hybridizing apo AI probe and then rehybridized with an apo CIII cDNA probe.   In this way, the fragments
resulting from restriction enzyme digestion that contained both apo
AI and CIII sequences could be detected and a partial restriction
map of the two genes constructed.   From such data it was possible
to conclude that apo AI and CIII genes are separated by 2.7 $\pm$ 0.1
kilobase pairs (Kb), while further blotting experiments utilizing
5' and 3' ends of the two cloned cDNA inserts as hybridizing probes
showed the two genes are convergently transcribed (see Figure 4)
and thus in the chromosome lie on opposite strands of the DNA
molecule.

That the two genes might be linked was suspected from: i)
protein sequence data which suggested they arose from a common
evolutionary precursor by gene duplication (Barker & Dayhoff,
1977), and ii) by the recent observation that a mutant apo AI
allele in two sisters correlated with a plasma deficiency in both
AI and CIII apolipoproteins (Norum et al, 1982).   This mutant apo
AI allele is inherited as a Mendelian trait and manifests as
premature atherosclerosis.   In preliminary genomic blotting experiments, using a number of restriction enzymes, it was shown that
this allele is contained in different size restriction enzyme

fragments from those of the normal allele (Karathanasis et al, 1983b). Such results clearly demonstrated that the mutant allele has not arisen by acquisition of a point mutation but rather by the acquisition or loss of sequences in the apo AI gene locus. By more extensive genomic blotting experiments using a series of different hybridization probes spanning the normal apo AI gene locus, it was possible to deduce that an insertion of approximately 6,500 nucleotides in the 3' end of the apo AI gene is responsible for the creation of this mutant allele (Karathanasis et al, 1983b).

To characterize the mutant allele further, a library of clones containing DNA fragments from one of the affected individuals has been created and the clone containing apoAI/CIII sequences isolated (Karathanasis et al, 1984). Analysis of the latter, by restriction enzyme mapping and some DNA sequence determination, has made it possible to show that at least part of the 6.5 Kb insert in this mutant apoAI gene resides in normal individuals approximately 5 Kb downstream of the apoAI gene, and furthermore that this insert is deleted from its normal position in the genome of patients possessing the mutant allele. Thus, at present, it seems probable that this deletion may have removed the promoter region of the apo-CIII from its normal location and may well explain the molecular basis of plasma apoCIII deficiency in patients homozygous for this mutant apoAI/CIII complex.

In the affected family, the homozygotes have skin and tendon xanthomas, corneal clouding and severe premature atherosclerosis, with very low plasma HDL levels and deficiencies of apo AI and CIII. The heterozygote members of the family, in contrast, are asymptomatic, but have lower than average levels of plasma HDL, apo AI and CIII. These levels fall into the range normally associated with an increased risk of cardiovascular disease (Miller & Miller, 1975; Willet et al, 1980). However, whether the above described mutant allele is a common cause of low plasma HDL in the general population is at present unknown. The parents of the homozygous probands are heterozygous for this allele and, whilst they are both descended from ancestors of Scottish origin, they are believed to be unrelated. Nevertheless, as yet, this mutant allele has not been found in any other family or population group analysed to date (Rees et al, 1983, 1985). Whatever the case, as seen from the preceding discussion, this mutant allele can be readily detected in carriers by the genomic blotting procedure and therefore such methodology should prove useful in the genetic counselling of this affected family and its future progeny.

2.2.4   A RFLP linked to the apo AI/CIII complex associates with hypertriglyceridemia

A RFLP has now been found that distinguishes between two alleles (apoAI/4.2, apoAI/3.2) of the apoAI/CIII gene complex in the population (Rees et al, 1983, 1985). On restriction with endonuclease SstI, allele apoAI/4.2 yields two fragments, of 5.7 and 4.2 Kb, which hybridize to an apoAI cDNA probe, whereas apoAI/3.2 yields fragments of 5.7 and 3.2 Kb (Figure 3). The finding of individuals homozygous and heterozygous for both of the alleles in the population rules out the possibility that the two alleles represent experimental artefacts, two common causes of which arise from partial digestion of DNA and the presence of contaminating plasmid DNA in the genomic DNA sample. Furthermore, it suggests that the sequence variation giving rise to the RFLP is indeed inherited and not a consequence of spontaneous de novo mutation.

The frequency of the two apoAI alleles has been estimated for both normolipidemic and type IV/V hypertriglyceridemic Caucasian patients. Interestingly, none of the normolipidemics (n = 52) analysed has the apoAI/3.2 allele, whereas 26 out of 74 of the hypertriglyceridemics did (see Table 1.9). Thus, in this population sample, the apoAI/3.2 allele associates with certain forms of hypertriglyceridemia. This result led us to postulate, given the known central role that apoAI and CIII play in triglyceride metabolism, that these patients with the apoAI/3.2 allele would have lower plasma levels of apoAI or CIII than those without it. Recent results however have shown that this is not the case; there is no correlation between plasma levels of either apoAI or

TABLE 1.9

Frequency of the apoAI/3.2 allele in various racial groups and

in normolipidemic and hypertriglyceridemic Caucasians

Group	Number of Individuals	Number with 3.2 Kb allele		Frequency of 3.2 Kb allele
		Homo-zygotes	Hetero-zygotes	
Normolipidemic:				
Chinese	20	6	7	0.475
Japanese	21	0	8	0.19
Africans	20	0	6	0.15
Indian Asians	28	0	10	0.18
Caucasians	52	0	0	0
Hypertriglyceri-demic:				
Caucasians	74	3	23	0.19

CIII in the fasting state and the genotype of the hypertrigly-ceridemic patients (Rees et al, 1985). Nevertheless, it is still not possible to conclude that there is no difference between these two groups of hypertriglyceridemic patients with regard to, for example, their rate of clearance of fat from the circulation after fat adsorption.

Our understanding of the defect associated with the apoAI/3.2 allele is further complicated by the extreme variation seen in its distribution in different normolipidemic racial groups (Rees et al, 1985). For example, 65% (13/20) of normolipidemic Chinese and 35% (8/21) of Japanese possess this allele. Possible explanations for this apparent discrepancy could include any of the following:

i)   In the Caucasian population, in common with other racial
      groups, the polymorphism giving rise to the apoAI/3.2 allele
      is in itself neutral; but in the former population this allele
      is linked to a functional mutation, while in the latter it is
      not.

ii)   the expression of the phenotype symptoms only occurs under
      environmental stress.

iii) the sequence variation giving rise to the apoAI/3.2 allele in
      the Caucasian hypertriglyceridemic patients and the racial
      groups are different.

Whatever the case, clearly, the apoAI/3.2 allele cannot be used as
a genetic marker for hypertriglyceridemia in non-Caucasians.

In the Caucasian populations, in order to use the RFLP directly as
a genetic marker for hypertriglyceridemia, it is important to
establish that no Caucasian normolipidemic individual possesses the
apoAI/3.2 allele.   This requirement will clearly be met if the
sequence variation giving rise to the RFLP is directly responsible
for the aetiology of the disorder.   However, if this allele should
prove to be a marker for some abnormality present in a neighbouring
region of the chromosome which directly causes the hypertriglyceridemic symptoms, the frequency of meiotic crossing over between
the RFLP and the linked mutant locus will need to be considered.
Meiotic crossing over is estimated to occur in about 1 per 10^8 bp
per meiosis.   Therefore, if the mutant gene and the genetic marker
are about 20 Kb apart, the possibility of crossing over occurring
between them and hence the diagnostic error rate, will be in the
order of 1-5000 (McKusick & Ruddle, 1977).

   At present, this genetic marker tells us nothing about the
cause of the abnormal lipoprotein levels seen in those Caucasian
patients who bear it.   In order to improve upon this situation, we
have clarified the role of the sequence variation giving rise to
it.   Genomic blotting experiments using a number of different
restriction enzymes ruled out the possibility that the apo AI/3.2
allele arises from deletion of sequences in the complex and
provided good evidence that a polymorphic nucleotide 2.7 ± 0.1 Kb

downstream of the apo AI gene and therefore within or flanking the
3' non-coding region of the apoCIII gene, is responsible for its
creation in both Caucasian and other racial populations. With
these data, it has been possible, by DNA sequence analysis of a
genomic clone constructed from Caucasian DNA homozygous for the apo
AI/3.2 allele, to: i) locate precisely the position of the polymorphic nucleotide, in the apo AI/CIII complex (the substitution of
a cytosine residue for a guanosine at nucleotide 40 of the 3' non-
coding region of the apoCIII gene, and ii) determine that the
encoded protein product of the CIII gene of this apoAI/3.2 allele
is identical to the encoded product of the apoAI/4.2 allele.

These data suggest that the polymorphic nucleotide giving rise
to the apo AI/3.2 allele is in itself neutral and constitutes a
genetic marker for a functional mutation (see section 2.2.2). This
functional mutation could affect the level of expression of either
apo AI or CIII, or cause some structural modification in the
former. These possibilities can be checked by the sequence
analysis of further relevant sections of genomic classes obtained
from hypertriglyceridemic individuals who bear this allele.
However, in the event that no differences can be discerned in any
known functionally important regions of the two apoAI alleles, it
would then be necessary to examine their regulation of expression
under physiological conditions, to investigate whether any other
sequences in the apoAI/CIII complex are important for optimal
functioning. This investigation would be most readily carried out
by inserting the gene complex from normal and hypertriglyceridemic
individuals, of both genotypes, via a suitable vector into
eukaryotic cells and evaluating the relative levels of expression
of both apoAI and CIII (Hames & Higgins, 1984). This type of
experiment may allow us to determine whether, in liver and
intestine cells, different but specific sequences in the apo
AI/CIII complex regulate the transcription of these 2 genes and if
a difference in these specific sequences between the two allelic
complexes correlates with the differences observed in phenotypes.

As described by Weatherall (1982), however, a RFLP can act as a genetic marker for a defective loci up to 10 to 20 x 10^6 bp away and therefore, clearly the above described methods may not find any functional differences between the two alleles of the apoAI/CIII complex. In this case, to isolate the particular area of the genome associated with the defective phenotype, such techniques that allow one to 'walk' or 'jog' the chromosome, may be of value (Lawn et al, 1978). In principle, this procedure involves isolating a series of overlapping clones from a genomic library upstream and downstream from a given point, mapping them with respect to one another and then isolating from the 5' and 3' extremities of the resulting cluster further fragments that can be used to isolate further overlapping clones. This procedure is repeated as many times as required to reach the desired destination.

Finally, in a broader context, the association of a RFLP with hyperlipidemia can open up many avenues for future clinical research. It should now be possible, for example, to examine whether there is any significant difference in either the clearance rate of dietary triglyceride from the circulation, or in the severity and incidence of cardiovascular disease between hyper-triglyceridemic patients of different genotypes. Previously, the results from such investigations may well have been masked by the heterogeneous nature of such a disorder. Moreover, by using two or more genetic markers, the polygenic origin of disorders such as atherosclerosis can be evaluated. Already in this respect, preliminary results from Jowett et al (1984) indicate that the severity of hypertriglyceridemia may depend upon the additive effect of two genetic defects. In their studies, they divided 33 hypertriglyceridemic patients into 6 categories on the basis of their genotype (see Table 1.10). Two genetic markers were used for this purpose; the RFLP associated with the apoAI/CIII complex, and the RFLP associated with the insulin gene (designated class 1 and 3) (Bell et al, 1981). Significantly perhaps, in diabetic patients homozygous for the insulin class 3 allele, they noted a correlation

TABLE 1.10

Genotype of hypertriglyceridemic patients

Patient Group	Insulin Genotype			ApoAI Genotype	
	3-3	3-1	1-1	Homozygous ApoAI/4.2	Heterozygous ApoAI/3.2
Diabetic Hypertriglyceridemic (n = 12)	6 (50%)	5 (42%)	1 (8%)	6 (50%)	6 (50%)
Non-diabetic Hypertriglyceridemic (n = 21)	0 (0%)	16 (76%)	5 (24%)	7 (33%)	14 (67%)
Total (n = 33)	6 (18%)	21 (64%)	6 (18%)	13 (39%)	20 (61%)

between the severity of hypertriglyceridemia and the genotype of the apoAI/CIII complex (Figure 5).

Acknowledgements

We gratefully acknowledge the support of the British Heart Foundation.

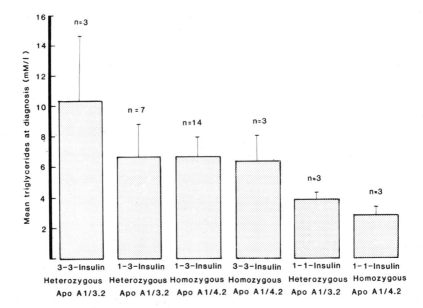

<u>Fig.5.</u>   Distribution of the polymorphisms of the insulin and apo-
lipoprotein AI genes in hypertriglyceridemic patients.   Insulin 1
and 3 indicate the genotype as defined by Bell et al (1981).
ApoAI/3.2 and 4.2 denote the apoAI/CIII complex genotype as defined
in Fig. 3.   n denotes the number of patients analysed.   The y
ordinate denotes the triglycerides levels observed at diagnosis.

References

Barker W C and Dayhoff M O (1977) 'Evolution of lipoproteins
    deduced from protein sequence data', <u>Comp.Biochem.Physiol.</u>
    <u>57</u>B, 309-315.

Bell G I, Karan J H and Rutter W J (1981) 'Polymorphic DNA region
    adjacent to the 5' end of the human insulin gene', <u>Proc.Natl.</u>
    <u>Acad.Sci.USA</u>, <u>78</u>, 5759-5763.

Bilheimer D W, Goldstein J L, Grundy S M, Starzi T E and Brown M S
    (1984) Liver transplantation to provide low-density-lipo-
    protein receptors and lower plasma cholesterol in a child with
    homozygous familial hypercholesterolemia, <u>N.Eng.Jnl.Med.</u>   <u>311</u>,
    1658-1663.

Botstein D, White R L, Skolnick M and Davis R W (1980) 'Construction of a genetic linkage map in Man using RFLP', Am.J.Hum.Genet. 32, 314-331.

Breckenridge W C, Little J A, Steiner G, Chow A and Poapst M (1978) 'Hypertriglyceridemia associated with a deficiency of apolipoprotein CII', New Engl.Jnl.Med. 298, 1265-1272.

Brown M S and Goldstein J L (1983) 'Lipoprotein metabolism in the macrophage: implications for cholesterol deposition in atherosclerosis' in Ann.Rev.Biochem. (eds. E E Snell, P D Boyer, A Meister and C C Richardson), pp 223-261.

Brown M S, Goldstein J L and Fredrickson D S (1983) 'Familial type 3 hyperlipoproteinemia' in Metabolic Basis of Inherited Disease, (eds. J B Stanbury, J B Wyngaarden and D S Fredrickson), pp 655-671.

Brown M S, Kovanen P T and Goldstein J L (1981) 'Regulation of plasma cholesterol by lipoprotein receptors', Science 212, 628-635.

Carlson L A and Philipson B (1972) 'A new familial condition with massive corneal opacities and dyslipoproteinemia', Lancet 2, 921-923.

Dobiasova M (1983) 'Lecithin: cholesterol acyltransferase and the regulation of endogenous cholesterol transport', in Advances in Lipid Research (eds. R Paoletti & D Kritchevsky), pp 107-194, Academic Press.

Eisenberg S and Levy R I (1975) 'Lipoprotein metabolism', in Adv. Lipid Research, (eds. R Paoletti & D Kritchevsky), vol. 13, pp 2-89.

Fredrickson D S, Levy R I and Lees R S (1967) 'Fat transport in lipoproteins - an integrated approach to mechanisms and disorders', N.Eng.J.Med. 276, 6 articles.

Fredrickson D S, Goldstein J L and Brown M S (1978) 'The familial hyperlipoproteinemias' in The Metabolic Basis of Inherited Disease (eds. J B Stanbury, J B Wyngaarden & D S Fredrickson) 4th edn, pp 604-655, McGraw-Hill,NY.

Geever R F, Wilson L B, Nallaseth F S, Milner P F, Bittner M and Wilson J T (1981) 'Direct identification of sickle cell anemia by blot hybridization', Proc.Natl.Acad.Sci.USA 78, 5081-5085.

Ghiselli G, Schaefer E J, Gascon P and Brewer H B Jr (1981) 'Type III hyperlipoproteinemia associated with apolipoprotein E deficiency', Science 214, 1239-1241.

Ghiselli G, Gregg R E, Zech L A, Schaefer E J and Brewer H B Jr
    (1982) 'Phenotype study of apolipoprotein E isoforms in hyper-
    lipoproteinaemic patients', Lancet ii, 405-407.

Goldstein J L and Brown M S (1983) 'Familial Hypercholesterolemia'
    in The Metabolic Basis of Inherited Disease (eds. J B
    Stanbury, J B Wyngaarden & D S Fredrickson) 4th edn, pp 672-
    712, McGraw-Hill,NY.

Goldstein J L, Schrott H G, Hazzard W R, Bierman E L and Motulsky A
    G (1973) 'Hyperlipidemia in coronary heart disease. II.
    Genetic analysis of lipid levels in 176 families and
    delineation of a new inherited disorder, combined hyper-
    lipidemia', J.Clin.Invest. 52, 1544-1568.

Gubler U and Hoffman B J (1973) Gene 25, 263-269.

Gusella J F, Wexler N S, Connealy P M, Naylor S L, Anderson M A,
    Tanzi R E, Watkins P C, Ottina K, Wallace M R, Sakaguchi A Y,
    Young A B, Shoulson I, Bonilla E and Martin J B (1983) Nature
    306, 234-238.

Hames B D & Higgins S J (eds.) (1984) 'Transcription and trans-
    lation: a practical approach', in Practical Approach Series,
    several articles, IRL Press.

Havel R J and Gordon R S Jr (1960) 'Idiopathic hyperlipidemia:
    metabolic studies in an affected family', J.Clin.Invest., 39,
    1777-1790.

Hazzard W R, Miller N, Albers J J, Warnick G R, Baron P and Lewis B
    (1981a) 'Association of isoapolipoprotein E3 deficiency with
    heterozygous familial hypercholesterolaemia: implications for
    lipoprotein physiology', Lancet i, 298-301.

Hazzard W R, Warnick G R, Utermann G, Albers J J (1981b) 'Genetic
    transmission of isoapolipoprotein E phenotypes in a large
    kindred: relationship to dysbetalipoproteinemia and hyper-
    lipidemia', Metabolism 30, 79-88.

Herbert P H, Assmann G, Gotto A M and Fredrickson D S (1983) In The
    Metabolic Basis of Inherited Disease, 5th edn. (ed. J
    Stanbury, J B Wyngaarden, D S Fredrickson, J L Goldstein & M S
    Brown), pp 589-621.

Holdsworth G, Stocks J, Dodson P and Galton D J (1982) 'An abnormal
    triglyceride-rich lipoprotein containing excess siacylated
    apolipoprotein CIII', J.Clin.Invest. 69, 932-939.

Innerarity T L, Weisgraber K H, Arnold K S, Rall S C Jr and Mahley
    R W (1983a) 'Normalization of receptor binding of apolipo-
    protein E2', Jnl.Biol.Chem. 259, 7261-7267.

Innerarity T L, Friedlander E J, Rall S C Jr, Weisgraber K H and Mahley R W (1983b) 'The receptor-binding domain of human apolipoprotein E', Jnl.Biol.Chem. 258, 12341-12347.

Jackson R L, Morrisett J D and Gotto A M, Jr (1976) 'Lipoprotein structure and metabolism', Physiol.Revs. 56, 259-316.

Jahn C E, Osborne J C, Schaefer E J, Jr and Brewer H B, Jr (1983) 'Activation of the enzymic activity of hepatic lipase by apolipoprotein AII', Eur.J.Biochem. 131, 25-29.

Jowett N I, Rees A, Williams L G, Stocks J, Vella M A, Hitman G A, Katz J and Galton D J (1984) 'Insulin and apolipoprotein AI-CIII gene polymorphism relating to hypertriglyceridaemia and diabetes mellitus', Diabetologia, in press

Kan Y W and Dozy A M (1978) 'Antenatal diagnosis of sickle-cell anaemia by DNA. Analysis of amniotic fluid cells', Lancet ii, 910-912.

Karathanasis S K, McPherson J, Zannis V I and Breslow J L (1983) 'Linkage of human apolipoproteins AI and CIII genes', Nature 304, 371-373.

Karathanasis S K, Norum R A, Zannis V I and Breslow J L (1983) 'An inherited polymorphism in the human apolipoprotein AI gene locus related to the development of atherosclerosis', Nature 301, 718-720.

Karathanasis S K, Zannis V I and Breslow J L (1983) 'A DNA insertion in the apolipoprotein AI gene of patients with premature atherosclerosis', Nature 305, 823-825.

Karathanasis S K, Zannis V I and Breslow J L (1984) Fed.Proc. 43, part 2, p 1815 (abstract).

Kashyap M L, Scrivastava L S, Hynd B A, Gartside P S and Perisutti G (1981) J.Lipid Res. 22, 800-810.

Kita T, Goldstein J L, Brown, M S, Watanabe Y, Hornick C A and Havel R J (1982) 'Hepatic uptake of chylomicron remnants in WHHL rabbits: a mechanism genetically distinct from the low density lipoprotein receptor', Proc.Natl.Acad.Sci.USA 79, 3623-3627.

Lawn R M, Fritsch E F, Parker R C, Blake G and Maniatis T (1978) 'The isolation and characterization of linked δ - and β - globin genes from a cloned library of human DNA', Cell 15, 1157-1174.

Laskey R A and Mills A D (1977) 'Enhanced autoradiographic detection of ^{32}P and ^{125}I using intensifying screens and hypersensitized film', FEBS Lett. 82, 314-316.

Lindgren F T, Jensen L C and Hatch F T (1972) In Blood Lipids and Lipoproteins, Quantitation, Composition and Metabolism (ed. G J Nelson) pp 181-274, Wiley-Interscience, New York.

Lux, S E, Levy R I, Gotto A M and Fredrickson D S (1972) 'Studies on the protein defect in Tangier Disease. Isolation and characterization of an abnormal high density lipoprotein', J.Clin.Invest. 51, 2505-2519.

Maniatis T, Fritsch E E and Sambrook J (1982) Molecular Cloning: a laboraotry manual, Cold Spring Harbor, New York.

Maniatis T, Hardison R C, Lacy E, Lauer J, O'Connell C, Quon D, Sim G K and Efstratiadis A (1978) 'The isolation of structural genes from libraries of eucaryotic DNA', Cell 15, 687-701.

Miller G J and Miller N E (1975) 'Plasma high density lipoprotein concentration and development of isochaemic heart disease', Lancet i, 16-19.

McKusick V A and Ruddle F H (1977) 'The status of the gene map of the human chromosomes', Science 196, 390-405.

Myant N B (1982) 'Cholesterol transport through the plasma', Clin. Sci. 62, 261-271.

Murray J M, Davies K E, Harper P S, Meredith L, Mueller C R and Wiliamson R (1982) 'Linkage relationship of a cloned DNA sequence on the short arm of the X chromosome to Duchenne muscular dystrophy', Nature 300, 69-71.

Nestel P J and Fidge N H (1982) 'Apoprotein C metabolism in man', Adv.in Lipid Res. (eds. R Paoletti & D Kritchevsky), pp. 55-83, Acad.Press.

Nikkila E A (1983) 'Familial lipoprotein lipase deficiency and related disorders of chylomicron metabolism', In The Metabolic Basis of Inherited Disease, 5th Edn. (eds. J B Stanbury, J B Wyngaarden, D S Fredrickson, J L Goldstein & M S Brown), pp 622-642.

Norum R A, Lakier J B, Goldstein S, Angel A, Goldberg R B, Block W D, Woffze D K, Dolphin P J, Edelglass J, Bogorand D D and Alanpovic P (1982) 'Familial deficiency of apoAI and CIII and precocious coronary artery disease', New Eng.J.Med. 306, 1513-1519.

Orkin S H, Kazazian H H Jr, Antonarakis S E, Goff S C, Boehm C D, Sexton J P, Waber P G and Giardina P J V (1982) 'Linkage of β-thalassaemia mutations and β-globin gene polymorphisms with DNA polymorphisms in human β-globin gene cluster', Nature 296, 627-631.

Rall S C, Jr, Weisgraber K W, Innerarity T L and Mahley R W (1982) 'Structural basis for receptor binding heterogeneity of apo-lipoprotein E from type III hyperlipoproteinemia subjects', Proc.Natl.Acad.Sci.USA 79, 4696-4700.

Rees A, Shoulders C C, Stocks J, Galton D J and Baralle F E (1983) 'DNA polymorphism adjacent to the human apoprotein AI gene: relationship to hypertriglyceridaemia', Lancet i, 444-446.

Rees A, Stocks J, Sharpe C R, Vella M A, Shoulders C C, Katz J, Baralle F E and Galton D (1985) 'DNA polymorphism in the apoAI-CIII gene cluster, association with hypertrigly-ceridaemia', J.Clin.Invest. in press.

Roberts R J (1983) 'Restriction and modification enzymes and their recognition sequences', Nucl.Acids Res. 11, r167-r204.

Russell D, Yamamoto T, Schneider W J, Slaughter C J, Brown M S and Goldstein J L (1983) 'cDNA cloning of the bovine low density lipoprotein receptor: feedback regulation of a receptor mRNA', Proc.Natl.Acad.Sci.USA, 7501-7505.

Scanu A M, Byrne R E, Mihovilovic M (1982) 'Functional roles of plasma high density lipoproteins', Crit.Rev.Biochem. p 109-140.

Schaefer E J, Eisenberg S and Levy R I (1978) 'Lipoprotein apo-protein metabolism', Jnl.Lipid Res. 19, 667-687.

Scow R O and Egelrud T (1976) 'Hydrolysis of chylomicron phosphate-dylcholine in vitro by lipoprotein lipase, phospholipase A2 and phospholipase C', Biochem.Biophys.Acta 431, 538-549.

Sharpe C R, Sidoli A, Lucero M A, Shelley C S, Shoulders C C and Baralle F E (1984) 'Human apolipoproteins AI, AII, CII, and CIII DNA sequences and mRNA abundance', Nucl.Acids Res. 12, 3917-3932.

Schumaker V N, Robinson M T, Curtiss L K, Butler R and Sparkes R S (1984) 'Anti-apoprotein B monoclonal antibodies detect human low density lipoprotein polymorphism', Jnl.Biol.Chem. 259, 6423-6430.

Shepherd J, Packard C J, Bicker S, Lawrie T D V and Morgan H G (1980) 'Cholestyramine promotes receptor-mediated low density lipoprotein catabolism', N.Eng.J.Med. 302, 1219-1222.

Southern E M (1975) 'Detection of specific sequences among DNA fragments separated by gel electrophoresis', J.Mol.Biol. 98, 503-517.

Stanley K K and Luzio J P (1984) 'Construction of a new family of high efficiency bacterial expression vectors: identification of cDNA clones coding for human liver proteins', EMBO J., 3, 1429-1434.

Tolleshaug H, Hobgood K K, Brown M S and Goldstein J L (1983) 'The LDL receptor locus in familial hypercholesterolemia: multiple mutations disrupt transport and processing of a membrane receptor', Cell 32, 941-951.

Utermann G, Jaeschke M and Menzel J (1975) 'Familial hyperlipo-proteinemia type III: deficiency of a specific apolipoprotein (apoE-III) in very low density lipoproteins', FEBS Lett. 56, 352-355.

Utermann G, Hees M and Steinmetz A (1977) 'Polymorphism of apolipo-protein E and occurrence of dysbetalipoproteinemia in man', Nature 269, 604-607.

Utermann G, Feussner G, Franceschini G, Haas J and Steinmetz A (1982) 'Genetic variants of group A apolipoproteins', Jnl. Biol. Chem. 257, 501-507.

Weisgraber K H, Innerarity T L, Harder K J, Mahley R W, Milne R W, Marcel Y L and Sparrow J T (1983) 'The receptor-binding domain of human apolipoprotein E', Jnl.Biol.Chem. 258, 12348-12354.

Weatherall D J (1982) 'The new genetics and clinical practice', The Nuffield Provincial Hospitals Trust, London.

Willet W, Hennekens C H, Siegel A J, Adner M M and Castelli W P (1980) New Engl.J.Med. 303, 1159-1161.

Zannis V I and Breslow J L (1981) 'Human very low density apolipo-protein E isoprotein polymorphism is explained by genetic variation and post-translational modification', Biochemistry 20, 1033-1041.

HUMAN COLLAGENS: BIOCHEMICAL, MOLECULAR AND GENETIC FEATURES

IN NORMAL AND DISEASED STATES

Francesco Ramirez, Frank O. Sangiorgi and Petros Tsipouras

The term "collagen disease" first appeared in the literature

when Klemperer et al. (1) described the histopathological abnor-

malities of the connective tissues in patients with lupus erythe-

matosus and scleroderma. Although no biochemical evidence was

presented to correlate structural alterations of collagen with

the "fibrinoid degeneration" observed in the affected tissues,

the term "collagen disease" not only was widely accepted but it

was also applied to several other systemic disorders of the con-

nective tissues. It was only later that investigators began to

recognize that this terminology did not apply for many of these

disorders, and that only a smaller group of pathological entities

were indeed caused by structural or metabolic defects in colla-

gen. Since then our understanding of the pathobiology of the

true "collagen diseases" has been achieved slowly and it is still

far from being completed. In part this was because of the biolo-

gical complexity of the connective tissues and in part because of

our lack of knowledge of collagen metabolism. To a certain

extent the latter problem was eased, in the past fifteen years,

by several fundamental discoveries such as the realization of the

tissue heterogeneity of the collagens and the elucidation of the

numerous biosynthetic steps leading to the ultimate process of
fibrillogenesis (2).

Besides collagen fibrils, the biomechanical properties of any
given connective tissue depend on the relative abundance, physical organization and coordinated interrelationship of many other
components of the extracellular matrices including elastic
fibres, glycoproteins and protein-polysaccharides. Thus structural and/or biosynthetical alterations of one or more of these
macromolecules may lead to similar pathological phenotypes or to
a wide spectrum of abnormalities. It is, therefore, conceivable
to predict that inherited connective tissue disorders may encompass a number of systemic syndromes with variable degrees of
expressivity and etiological complexity. It is the purpose of
this review to outline some of the investigations which have
related collagen defects to genetically inherited pathological
entities. Future studies however may prove the independent or
synergistic involvement of other mutations in the genesis of connective tissue disorders.

COLLAGEN STRUCTURE

Most of the collagen in the body is accounted for by one species of collagen, called Type I (2). There are, however, more
than ten different types of collagens in different tissues and at
different developmental stages in the same tissue (2). In
general, a collagen molecule is comprised of three α chains each
containing about 1,000 amino acids. Except for short regions of

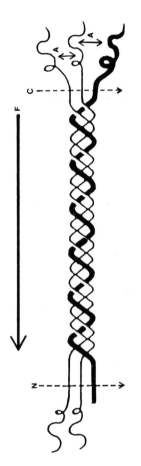

Figure 1: Schematic representation of the Type I procollagen
molecule. In black is the proα2(I) chain, whereas the two
proα1(I) chains are in white. The approximate location of the N-
and C-propeptides cleavage sites are indicated by the letters N
and C respectively. In the C-propeptide domain the letters A
signify the assembly of the procollagen molecules through the
establishment of disulfide bonds. The letter F indicates the
direction of the triple-helical formation.

about 20 amino acids at the ends (telopeptides), each of the α chains contains glycine as every third amino acid. The amino acid immediately following glycine is usually proline and the amino acid preceding glycine is frequently hydroxyproline. Therefore, the structure of each polypeptide chain can be approximated with the formula $(Gly-X-Y)_{333}$ (4). Additional domains present in the native collagens (procollagens) are the signal peptide and the two terminal propeptides. Three proα chains form a triple-helical molecule where glycine occupies the restricted space in which the chains come together (Fig. 1). Proline and hydroxyproline residues add rigidity to the triple-helical conformation, whereas clusters of charged amino acids located on the surface of the triple-helix provide the properties necessary for self-assembly of the collagen molecules in the quarter-staggered array found in many collagen fibers (5).

According to the type of supramolecular aggregates formed, the different collagen types can be divided into two major groups: fibrillars and nonfibrillars (4). The former, also called interstitial collagens, are found in the extracellular matrix of the major connective tissues such as: skin, bone, tendon, ligaments and cartilage. The latter, exhibiting a finer texture are part of the basement lamina and the pericellular lumen and constitute the cellular exoskeleton (6). In the fibrillar collagens each molecule is either an homopolymer or an heteropolymer. This results in four distinct loci coding for the constituents of the mature fibrillar collagens. The molecular

structure of these proteins is formulated as follows: Type I
$[(\alpha1(I))_2\alpha2(I)]$, Type II $[\alpha1(II)]_3$ and Type III$[\alpha1(III)]_3$.

The dissimilarities in the amino acid composition of the different α chains and consequently the variable degree of post-
translational modifications probably explains why the fibers of
the interstitial collagens are of different diameters and morphology (5). Interspecies variations for each chain are less pronounced than intraspecies differences among the various chains
(2,5). This notion has led to the proposal that the genes coding
for the $\alpha1(III)$, $\alpha2(I)$, $\alpha1(II)$ and $\alpha1(I)$ arose in that order by
successive duplications (2).

The expression of the fibrillar collagen genes is highly specific; Type II collagen is found only in hyaline cartilage,
whereas osteoblasts express exclusively Type I collagen; Type III
collagen is usually co-expressed with Type I, but at variable
ratios in different organs and tissues (3).

Type IV collagen represents the most important member of the
class of "nonfibrillar" collagens in that it is a major component
of most basement membranes (4). It is comprised of two different
kinds of α chains and it resembles other types of collagens in
having a long central triple-helical domain, and a large globular
domain (7). The binding of globular domains in adjacent molecules allows the Type IV collagen to form a mesh-like network that
serves as the basic framework of basement membranes (6). Type V
collagen appears to serve as an exoskeleton between cells such as
smooth muscle cells and other mesenchymal structures (8).

Several additional kinds of collagens are present in small amounts in many tissues. For example, cartilage contains a variety of minor collagens which include types that are similar to, but distinct from, Type V collagen (2,4). Thus far, no conclusive evidence has been presented demonstrating the existence of inherited disorders affecting the expression of the non-fibrillar collagen genes. Therefore, for the remainder of this review, we will limit our discussion only to the better known fibrillar collagens.

COLLAGEN METABOLISM

The subunits of Type I, II and III collagens are first synthesized on the membrane-bound polysomes of the rough endoplasmic reticulum (RER) as larger precursor proteins (procollagens) containing N- and C-terminal propeptides and a short signal peptide common to most secretory proteins. As the nascent polypeptides appear in the lumen of the RER, the post-translational modifications of the molecules begin to take place: hydroxylation of the proline and lysine residues, glycosylation of the hydroxylysine residues and glycosylation of the C-propeptides.

The presence of 4-hydroxyproline residues has been proven to stabilize the triple-helical structure at body temperature (9,10), whereas the hydroxylysine residues are important for the formation of intra and intermolecular cross-links. The glycosylation of the hydroxylysine residues is mediated by lysyl

hydroxylase and it appears to influence the interaction of the
collagen with other macromolecules of the extracellular matrix.
Other experimental evidence has suggested that this post-
translational modification is also needed for the proper assembly
of collagen molecules into higher order structures (fibrils)(13).
Unlike the previous process, the glycosylation of the C-propeptides occurs by the transfer of pre-formed oligosaccharides to
specific asparagine residues occurring in the sequences
Asn-X-Ser/Thr (14,15). These intracellular modifications take
place while the proα chains assemble through the establishment of
disulfide bonds in the C-propeptides, thereby defining this protein domain as the "registration peptide" (16).

Environmental and/or structural factors may change the rate
of triple-helix formation, greatly influencing the degree of the
post-translational modifications, which in turn result in abnormalities in the secretion, fibrillogenesis, cross-linking or
degradation of the collagens.

Three important post-translational modifications seem to occur
extracellularly: removal of the N- and C-propeptides (17,18) and
the oxidation of lysine residues by the enzyme lysyl-oxidase
(19). These processes are necessary for the formation and stabi-
lization of the collagen fibers (2). It has been shown that
variations of the biomechanical properties of the various connective tissues at different developmental stages reflect the differential nature and the extent of the post-translational modifications within each of the collagen types (5).

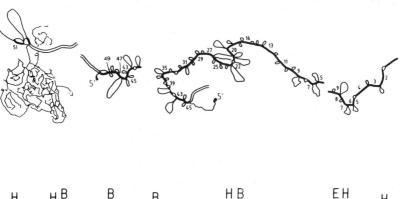

Figure 2: Structure of the human proα1(I) collagen gene. On the top is the tracing of the R-looping analysis of three overlapping genomic clones containing the entire gene, which is 18 kilobases long. On the bottom is a schematic representation of the exon (black boxes) intron arrangement of the gene. The hatched areas of the first and last exons indicates the non-coding regions. Some of the restriction sites are indicated by the letters B (BamHI), E (EcoRI), and H (HindIII). The letter r below intron XXVI signifies the presence of a short AluI repeat. The exons are numbered beginning from the 3' end of the gene. (Modified from Chu, M.-L., et al. (27) and Chu, M.-L. et al (44)).

COLLAGEN GENE STRUCTURE

A considerable amount of information has recently been
generated from the isolation and molecular characterization of
the fibrillar collagen genes from various species (20). Like
most eucaryotic genes, the coding elements of the collagens are
interrupted by intervening sequences (introns). However, the
number of the introns found in the collagen genes by far exceeds
any thus far reported. A dramatic example of this phenomenon is
shown by the visual representation of the human proα1(I) collagen
gene structure obtained by R-looping analysis of three overlapping genomic clones (Fig. 2). The 4,392 bp coding for the
1,464 amino acid residues of the pre-proα1(I) chain are in fact
divided into 51 exons embedded within more than 18 kb of chromosomal DNA. Similarly, 51 introns interdisperse the 35 kb of the
proα2(I), proα1(II) and proα1(III) genes (20-22). The disparity
observed between the lengths of the proα1(I) and the other
fibrillar collagen genes is primarily due to the average larger
size of the introns, which otherwise are located in homologous
position. This common structure appears to have been concertedly
maintained by the four fibrillar collagen genes during the
divergence of mammals and birds, which occurred more than 300
million years ago (20,22). An important, practical consideration
which stems from these studies is that the primary transcripts of
the collagens must undergo a series of accurate and efficient
splicing mechanisms in order to produce a correct messenger RNA
capable in turn of translating a functional collagen chain.
Thus, it is conceivable to predict that such an elaborate gene

structure may greatly enhance the possibilities for different
splicing mutations, leading to a wide spectrum of quantitative
abnormalities of collagen biosynthesis. This notion in turn
generates a new kind of problems associated with the characterization of pathological conditions both at the molecular and
clinical level. It is in fact obvious the formidable task that
the collagens offer for the molecular assessment of splicing
mutations. For the clinician, on the other hand, the heterogenous expression of the various splicing mutations makes it even
harder to judge the mode of inheritance of the abnormal
phenotype(s). Following is a more detailed description of the
collagen gene structure in relationship to the various domains of
the protein molecule: N- and C-propeptides and triple-helical
region.

(a) C-propeptide and C-terminal telopeptide domains

The finding that four exons encode the C-propeptide and C-
terminal telopeptide domains of the fibrillar collagen genes
strongly supported the idea that they define functionally
distinct subdomains (21). It was, therefore, of interest for
several investigators to subject the DNA sequences of the four C-
propeptide domains and their amino acid translation to interspecies and pairwise comparisons. The rationale for these analyses
was to identify portions of the C-propeptide domains which were
evolutionarily conserved and consequentially suggestive of being
essential for some of the many biological functions proposed for
the "registration peptides."   Potentially this information could
be of great importance in predicting areas of the genes where

even small mutations may dramatically affect the functional behavior of the collagen protein.

Pairwise comparison of the four C-propeptides revealed that,
although the rate of total nucleotide replacements in the different chains is almost equal, the amino acid divergence of the
proα1(I) and proα1(III) are clearly at the opposite end of a wide
spectrum of values (22). This observation is in agreement with
the evolutionary tree proposed for the four genes (2). Interspecies comparison within the four C-propeptide exons identified
highly conserved sequences suggesting that they may code for
regions of biological importance. For example, conserved are the
regions of exons 3 and 2 where the cysteinyl residues involved in
intrachain and interchain bonding are located (20). Similarly
conserved is a short segment, 54 bp long, around the carbohydrate
attachment site in exon 2 (23). On the other hand the com-
paritive analysis of exon 4 refuted the hypothesis that this
region of the C-propeptide could have a specific structural role
in the assembly of the different fibrillar collagen molecules
(24). In fact, one of the implications of this notion is that
such specificity should be conserved in the same chain between
different species. However, unlike the proα1(I) and, to a certain extent, the proα2(I) and proα1(II) telopeptides, the
proα1(III) C-terminal telopeptide is highly divergent between
birds and mammals (22).

(b) Triple-helical domain

Most of the collagen exons code for the large triple-helical

domain of the polypeptide chain and they range in size from 45 bp
to 162 bp, thus appearing to be related to a basic 54 bp unit
which codes for an in-phase six-fold Gly-X-Y repeat. This observation has led to the hypothesis that the triple-helical domain
may have evolved by tandem duplication of an ancestral 54 bp element (25).

The detailed analysis of the fibrillar collagen genes in different species has revealed only two major differences in the
arrangement of the exons coding for the triple-helical domains.
The first dissimilarity was observed in the region of exon 4
bearing triple-helical domain sequences (22). In fact, in the C-
terminal junction exon, unlike the N-terminal junction exon, the
number of base pairs coding for the triple-helical domain segment
are either 45 or 54. These values are consistent with the
observed length of the other triple-helical domain exons. In
exon 4 of human Type III gene, however, an extra Gly-X-Y triplet
is present, resulting in an unusual 63 bp triple-helical element
(22). A second, important difference was observed for amino acid
residues 568-603, which are encoded by a single 108 bp exon
(number 19) in the $proαl(I)$ gene and by two exons (numbers 19 and
20) in all the other collagens (26,27). Because of the considerations previously discussed about the temporal divergence of
the collagen genes, this result indicated that the fusion of the
two 54 bp exons occurred in the $proαl(I)$ gene sometime after the
duplications and divergency of the other fibrillar loci. Thus,
these studies have conclusively shown that the vertebrate
fibrillar collagen genes evolved by duplication of a common

multiexon structure, which in turn originated from a simpler 54 bp unit. During evolution this elaborate structure was concertedly maintained by the four genes under rigorous structural and functional constraints commonly exerted upon their products. Hence, it is conceivable that the numerous introns in the triple-helical region may have been conserved in order to minimize the frequency of recombinations within the repetitive exons, as well as demarcate multiexon subdomains (27).

(c) N-propeptide, N-terminal telopeptide and signal propeptide domains

The information available for the three domains of the N-terminal portion of the collagens indicate that this region of the protein molecule exhibits a high degree of structural variations among the different chains as well as in the same molecule between different species (2).

Regardless of these differences, the N-terminal domains of all the collagen genes analyzed thus far are encoded by six exons, which are variable in sizes and composition (27,28). This observation was dramatically confirmed by the finding in the proα2(I) gene of two very small exons, 11 bp and 18 bp, (28). It has been proposed that these two small exons represent vertiges of larger coding segments absent in the proα2(I) gene (27). Presently there are no conclusive explanations for this unusual fixation of the exon/intron arrangement of the N-terminal domains of the fibrillar collagen genes.

Like the C-propeptide, a junction exon codes for the transition from the central triple-helical domain to the N-propeptide. This exon contains the sequences for the amino terminal peptidase cleavage site (20,22). Likewise, the highly hydrophobic signal peptide and a short stretch of acidic amino acid residues are found together within the last exon of the collagen genes (27). In the same exon of the proα1(I), proα2(I) and proα1(III) genes a uniquely conserved structure has been identified around the initiation site of translation (29). Because of its quasipalindromic nature this short sequence can lead to the formation of a hairpin structure, whose thermodynamic stability is not influenced in the different genes by the few substitution sites, mostly clustered in the loops (30,31). It has been proposed that these secondary structures may represent a functional domain important for collagen-specific ribosomal binding and which may, therefore, regulate post-transcriptionally the 2:1 stoichiometry of the Type I chains (31).

Chromosomal location

The studies we have outlined suggest that the elaborate structure of the fibrillar collagen genes is ancient and has preceded the divergence between mammals and birds. Furthermore, these genes diverged from each other both at the level of the coding sequences and the size of the introns which are otherwise located in identical positions. Although gene families are considered to evolve in concert through gene conversion, genes on non-homologous chromosomes may differentiate more than those on homologous chromosomes (32). This notion is supported by

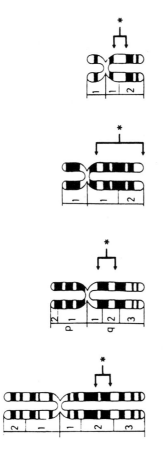

Figure 3: Chromosomal localization and regional mapping of the four fibrillar collagen genes. From left to right are: chromosome 2, proα1(III) gene; chromosome 7, proα2(I) gene; chromosome 12, proα1(II) gene; chromosome 17, proα1(I) gene.

experiments of gene mapping which demonstrated that the four
fibrillar collagen loci are dispersed on different chromosomes.
More precisely, by a combination of DNA analysis of human/rodent
hybrid cell lines and in situ hybridization of human metaphasic
chromosomes the following chromosomal assignments have been
obtained: proα1(I):17q21-17q22;proα2(I):7q21-7q22;proα1(II):
12q131→12q132;proα1(III): 2q31→2q323 (33-38) (Fig. 3).

Genetic disorders of collagen

Connective tissue disorders include a large variety of
syndromes with heterogeneous and sometimes overlapping modes of
expression. Abnormalities in the structure and/or biosynthesis
of the fibrillar collagens have been associated with some of the
connective tissue disorders such as: Osteogenesis imperfecta
(OI), Marfan syndrome (MS), Ehlers-Danlos syndrome (EDS) and
possibly chondrodystrophies.

Osteogenesis imperfecta (OI)

The term osteogenesis imperfecta describes a group of disorders which are highly heterogeneous in their phenotypic manifestations. Bone fragility is the cardinal clinical manifestation of the OI syndrome. In addition, short stature, joint
laxity, easy bruising, blue sclerae, presenile hearing loss and
dentinogenesis imperfecta may also be present. This phenotypic
diversity has prompted numerous classification attempts, the most
recent and widely accepted one having been proposed by Sillence
et al., (38) (Table 1). According to this classification, OI is
divided into four phenotypic groups. However, it has been

Table I

Classification of osteogenesis imperfecta syndromes*

Clinical findings	Inheritance	OI Type
Bone fragility, blue sclerae ± dentinogenesis imperfecta ± presenile hearing loss	Autosomal dominant	I
Extreme bone fragility expressed prenatally, perinatal death	Autosomal dominant or recessive	II
Severe bone fragility, severe growth retardation and skeletal deformity ± blue sclerae ± dentinogenesis imperfecta	Autosomal dominant recessive	III
Bone fragility, normal sclerae ± dentinogenesis imperfecta ± presenile hearing loss	Autosomal dominant	IV

*Modified from Sillence et al., J. Med. Genet. 16:101-106 (1974)

pointed out by several investigators that many affected individuals do not fit any of the four types.

A number of OI variants have been extensively studied at the
biochemical and molecular levels and in the following paragraphs
we will try to correlate the clinical phenotype and the laboratory findings in some of these variants.

OI Type I is inherited as an autosomal dominant trait and it
is characterized by postnatal onset of fractures, blue sclerae,
joint laxity and in some cases by hearing loss and dentinogenesis
imperfecta. Barsh et al., (40) have reported decreased synthesis
of proα1(I) chains intracellularly, which results in decreased
synthesis of mature triple-helical Type I collagen molecules.
The synthesized proα1(I) chains are approximately 50% of the normal amount and this observation correlates with the finding that
the proα1(I) to proα2(I) mRNA ratio in the three cell lines
studied is 1:1, rather than the expected 2:1 value (Byers, P.H.
and Rowe, D.W., personal communication).

OI Type II is the lethal-perinatal type, which is characterized by extreme bone fragility resulting in death in utero or
shortly after birth. It had been presumed that this type of OI
is inherited as an autosomal recessive trait. Recent clinical,
genetic, biochemical and molecular studies have shown the heterogeneity of this particular OI phenotype as having both recessive
and dominant forms, the latter representing fresh mutations.
Since a number of individuals with OI Type II have been extensively studied, we will refer to some of them in detail. Defects
in both the proα1(I) and proα2(I) chains have been shown in

variants with OI type II. Furthermore, both amino acid deletions and amino acid substitutions have been described.

The most extensively studied variant with OI Type II (CRL-1262) has been found to be heterozygous for an internal deletion of approximately 0.5 kb in the proα1(I) procollagen gene (41). This observation is consistent with previous findings which indicated that the proband's cultured fibroblasts synthesized two different species of proα1(I) chains, a normal and a shortened one (42,43). These investigations also showed that the secreted amounts of Type I procollagen were substantially reduced. Using S1 nuclease mapping experiments, Chu et al. (41) located this proα1(I) deletion between amino acid residues 325 to 410. More recently the same authors by cloning and sequencing the mutant proα1(I) allele, showed that the deletion (643 bp long) is contained between two introns of the triple-helical domain and it eliminated exons 27, 28 and 29 (aa 411-328). Interestingly, the breakpoints of the rearrangement were found to be located in two almost perfect inverted repeats (AGAGCC ACA---- -TG↓TGGCCACT). This observation strongly suggests that the self-complementary nature of the quasipalindromic sequence may have favored the formation of a secondary structure (hairpin) in turn used as substrate for the deletion (44) (Fig. 4).

Another OI Type II variant has been reported recently where disulfide bonded α1(I) dimers have been found in the pepsin extracted dermis collagen. A Gly to Cys substitution in the cyanogen bromide peptide α1(I)-CB6 (aa 816-1014) resulted in an interruption of the (Gly-X-Y) triplet periodicity and subsequent

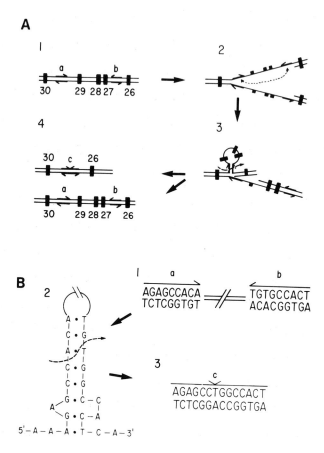

Figure 4: Possible mechanism leading to the genesis of the in-phase exons deletion in the Type II OI patient CRL-1262 (Panel A). The arrows indicate the locations of the complementary sequences (a,b) of the quasipalindrome. The numbers identify the proα1(I) exons. In Panel B is a more detailed representation of the same deletion showing the sequences of the two complementary elements (a,b) of the quasipalindrome (1), the possible mechanics of the rearrangement, (2), and the resulting palindrome in the affected allele (3). (Modified from Chu, M.-L., et al. (44) and Ramirez, F., et al., N.Y. Acad. Sci. (1985) (in press)).

impairment of triple-helix formation. Biochemical studies on the
parents' cultured cell fibroblasts did not reveal any abnormalities, although it should be noted that one parent is affected
with Marfan syndrome (45). It can, therefore, be assumed that
the phenotype OI is due to a spontaneous mutation in the proα1(I)
collagen gene. The two OI variants we discussed share a common
pathogenetic mechanism whereby 75% of the triple-helical molecules are virtually non-functional since they cannot form a stable
triple helix and subsequently cannot be secreted.

Biochemical and molecular studies from cultured fibroblasts
from a third OI Type II patient have shown this individual to be
a compound heterozygote for two different mutations in both
proα2(I) alleles (46). Quantitation of the relative
proα1(I):proα2(I) mRNA ratio showed that the fibroblasts of the
proband and the father transcribed reduced amounts of proα2(I)
mRNA. Moreover, unlike the asymptomatic parents, the fibroblasts
of the proband also synthesized a shortened proα2(I) chain, the
result of a spontaneous mutation. R-looping analysis of the proband's fibroblasts RNA showed only one kind of RNA:DNA hybrid
molecule, lacking the sequences corresponding to exon 25
(aa 448-465) (de Wet W. and Ramirez, F., unpublished data). Based
on the previous considerations it is clear that the lethal phenotype of this variant is due to the sole presence of shortened
proα2(I) chains which interfere with the triple helix formation
by affecting its rate, its stability, the extent and type of
post-translational modifications, and its secretion.

OI Type III is also heterogeneous and it is clinically

characterized by significant bone fragility with fractures frequently occurring in utero, short stature and usually significant
skeletal deformity, dentinogenesis imperfecta, hearing loss and
blue sclerae. Reports have suggested both autosomal dominant and
autosomal recessive modes of inheritance. Biochemical studies on
the fibroblasts derived from a patient with OI Type III showed
secretion of Type I procollagen molecules consisting of proα1(I)
trimers, despite the presence of proα2(I) chains intracellularly
(47). Northern blot hybridizations revealed that the patient's
fibroblasts contained normal levels of proα2(I) mRNA, efficiently
translatable in a heterologous cell-free system (48). Sl nuclease
mapping using a specific cloned proα2(I) probe indicated the presence of an homozygous mutation in the C-propeptide domain of the
proα2(I) chain (49). Cloning and sequencing of the affected
allele showed that a 4 bp frame-shift deletion eliminated the
last cysteine in the C-propeptide (50). This observation dramatically stresses the importance of the intrachain disulfide bonds
for the chains association in the RER. The mutation was
inherited as a recessive trait from the parents, who were third
cousins and phenotypically normal.

OI Type IV is inherited as an autosomal dominant trait and it
is characterized by mild to moderate bone fragility, usual postnatal onset of fractures, short stature, white sclerae, hearing
loss and dentinogenesis imperfecta. Tsipouras et al. (51), using
a restriction fragment length polymorphism (RFLP) associated
with the proα2(I) collagen gene, demonstrated co-segregation of
the molecular marker with the OI phenotype. This suggested that

the abnormal phenotype was due to a mutation in the proα2(I) gene. Further studies from cultured skin fibroblasts of an affected individual of that family showed a small (approximately 10 aa) deletion in the amino terminal end of α2(I)-CB3,5 peptide between amino acid residues 402 and 776 (R. Wenstrup, P. Tsipouras and P.H. Byers, unpublished data).

Analysis of an atypical OI variant revealed a deletion of eighteen amino acids in the proα2(I) chain (52). The phenotype of this individual was characterized by significant joint laxity, scoliosis, pectus excavatum, blue sclerae but no fractures. The abnormal phenotype was inherited as an autosomal dominant trait. Biochemical analysis of the collagen synthesized by the fibroblasts of the patient and his mother revealed the presence of a normal and a shortened proα2(I) chain. Two-dimensional electrophoresis of the proα2(I) CNBr peptide pattern demonstrated the change in the peptide α2(I)-CB4 (aa 6-327). This observation was indirectly confirmed by R-looping experiments which showed that half of the RNA molecules failed to hybridize with the DNA sequences corresponding to exon 42 (aa 73-90) (de Wet, W. and Ramirez, F., unpublished data). This patient has also some of the features of an individual affected by EDS VII. Thus the analysis of this particular OI mutation raises questions as to whether a correlation exists between the topology of the defect within the triple helix and a specific phenotype.

Marfan Syndrome

The Marfan Syndrome is now recognized to be a group of disor-

ders which can be distinguished on a clinical basis, like the
aesthenic and nonaesthenic varieties, the contractural arach-
nodactyly and the Marfanoid hypermobility syndrome (54). The
overlapping clinical phenotypes are characterized by arachyodac-
tyly, dolichostenomelia, pectus deformities, scoliosis, aortic
root dilatation, mitral valve prolapse, myopia and lens dislocation or sublaxation. All four clinical disorders are transmitted
as autosomal dominant traits.

In at least one case the phenotype has been linked to a
structural defect of Type I collagen; more precisely, a 20 amino
acid in-frame insertion in 50% of the proα2(I) chains, located
within the α2(I)-CB4 peptide. This defect seems to interfere
with collagen crosslinking by shifting the normal registering of
the procollagen chains (55). However it is still unclear why
such a mutation did not result in an OI rather than Marfan phenotype.

Some of the earliest studies of tissues and cells in culture
from patients with Marfan syndrome demonstrated an increase in
collagen solubility, suggestive of altered cross-linking.
However, no enzymatic basis for this solubility could be determined (56). Recently, Boucek et al. (57) have shown that skin
collagen from patients with Marfan syndrome have a decrease in
non-reducible cross-links and increased extractability. This may
be secondary to a defect(s) in the primary structure of either
the proα1(I) or the proα2(I) collagen chains. Elastin has also
been implicated in the etiology of Marfan Syndrome. Increased
urinary excretion of desmosine, an amino acid cross-link unique

to elastin, and decrease of desmo- and iso-desmosine in elastin extracted from aortic tissue has been shown by Abraham et al. (58). Elastin and collagen are very closely related in the aorta, and it is possible that a mutation which alters the structure of either of these macromolecules could change the morphology or function of the other.

There have, in fact, been a few observations suggesting that still another component of the extracellular matrix, hyaluronic acid, may be involved in the Marfan syndrome (59). Fibroblasts and cell extracts from some patients were shown to synthesize increased amounts of hyaluronic acid. This particular observation, however, is difficult to interpret, since similar observations of increased rates of hyaluronic acid have been seen in fibroblasts from some patients with OI.

Ehlers-Danlos Syndromes

EDS appears to be an even more heterogeneous group of disorders than OI. The clinical manifestations of these disorders include laxity of joints, soft and extensible skin which may also be fragile, and excessive bruising (54). The syndrome has been classified into at least eleven clinical types.

The most severe form of EDS is clearly the Type IV. Patients with this disease have minimal joint laxity and very thin transparent skin. The most dramatic manifestation of this disorder is the spontaneous rupture of large vessels and hollow organs such as the intestines. Biochemical studies in a number of variants to date show a decreased amount of Type III procolla-

gen secreted into the medium. This may be due to a structural
alteration in the proα1(III) chains interfering with the triple
helix formation (60).

Evidence for a structural alteration in a proα1(III) chain
was recently obtained from one variant of EDS Type IV. Here the
rate of secretion of Type III procollagen into the medium of
cultured fibroblasts was about normal. However, the proα1(III)
chains were found to consist of both the normal species and a
species which migrated more slowly on polyacrylamide gel (61).

Type VII EDS is characterized by marked laxity of joints.
Most patients are born with bilateral dislocation of the hips.
In three patients with Type VII EDS, a structural defect of the
collagen molecules has been identified. In one case the proband
was born with bilateral hip dislocations, generalized laxity of
all joints and muscle hypothonia. Originally it was thought that
this individual had a deficiency of procollagen N-proteinase.
Incidentally, both parents were normal (62). Subsequently,
however, this patient was shown to have a structural alteration
of the proα2(I) chains which prevented cleavage of the N-
propeptide by the procollagen N-proteinase. The structural
alteration has not been defined in detail, but clearly it leads
to a persistance of partially processed collagen molecules containing the N-propeptide. Hence, the clinical syndrome is very
similar to that produced by a presumed deficiency of procollagen
N-proteinase (EDS VII A Type).

Similarly biochemical analysis of the tissue collagens of a
second EDS VII patient have indicated that one of the two

proα2(I) chains carries a small deletion of approximately 20
amino acids in the telopeptide sequence between the N-propeptide
and the triple helical domain of the collagen molecule (53).
Finally, biochemical studies have also revealed a small heterozygous deletion in the homologous region of the proα1(I) chain in a
third individual affected with EDS VII (Cole, W., personal
communication).

Chondrodystrophies

This term defines an extremely heterogeneous group of disorders affecting cartilage or bone or both (54). Very little is
known about the etiology and pathogenesis of the chondrodystrophies because of the complex and sequential nature of the
ossification processes. Indeed, defects at different levels
have been suggested, but thus far only in one case of diastrophic
dysplasia has a structural abnormality in Type II collagen been
reported. The patient was a nine year old child affected with
diastrophic dysplasia. Structural and biochemical analyses of
the cartilage collagen suggested an abnormal cross striation pattern of SLS of Type II collagen in a position between bands
42-145 corresponding to the position of the α1(II)-CB10,5 peptide
(aa 557-875) (63). Qualitative defects in Type II collagen
expression have been only suggested in thanatophoric dysplasia
(64).

RFLPs associated with collagen genes

Traditionally, genetic variations in man have been associated

with structural changes leading to allelic forms of the same protein product. The incidence of these structural polymorphisms is
obviously very low and greatly limited by the availability of
proper biochemical assays and by the adverse selection for lethal
mutations. More recently, however, our knowledge of the heterogeneity of the human genome has been greatly enhanced by the fine
analysis of cloned genes by restriction endonuclease mapping.
Several investigations have in fact revealed the existance of
sequence variations mostly located within the non-coding elements
of many structural genes. Essentially, a great number of these
variations are the result of single nucleotide changes, which in
turn lead to changes in the length of the fragments generated by
a given restriction enzyme. Unlike the structural polymorphisms,
these restriction fragment length polymoprphisms (RFLP), are evolutionarily maintained because they do not destabilize the expression of the gene to which they are associated. Moreover, when
they are present at high frequency in the population, they become
sensitive markers for following the segregation of a given phenotype in large pedigrees. Tsipouras et al., have first shown the
presence of a highly polymorphic Eco RI site associated with the
proα2(I) collagen gene. When the Eco RI site is absent a specific probe identifies a single band 14 kb in size; on the other
hand, its presence results in the generation of two bands, 10.5
kb and 3.5 kb in size (Fig. 5). Because the allelic frequency of
this polymorphism is 38%, it is possible to use this marker to
study the co-segregation of the proα2(I) gene in large families
affected by connective tissue disorders, such as the type III OI

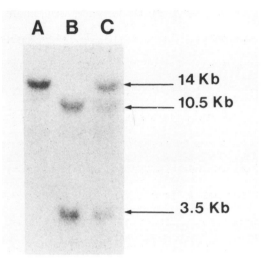

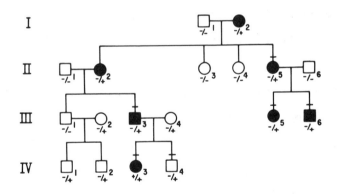

<u>Figure 5</u>: Eco RI restriction fragment length polymorphism asso-
ciated with the proα2(I) collagen. Top : Southern blotting
hybridization of the DNA from three individuals who are (A) homozygous for the absence of the Eco RI site (-/-); (B) homozygous
for the presence of the Eco RI site (+/+) and (C) heterozygous
for the site (-/+). Bottom: Segregation of the RFLP in a family
with an autosomal dominant form of OI (OI Type II). The
blackening of the symbols in the pedigree indicates the affected
individuals. Underneath each member is the molecular analysis of
the RFLP (modified from Tsipouras et al. (51).

phenotype of the family depicted in Figure 5. With the exception
of the proα1(I) gene, specific DNA markers have been identified
in the proα2(I), proα1(II) and prol(III) collagen genes. The
majority of these markers are single base substitutions leading
to a change in the restriction pattern. In addition, in the 3'
flanking sequence of the proα1(II) gene a DNA insertion of 200 bp
has been discovered. This polymorphic change seems to have a
higher frequency in certain ethnic groups (Tsipouras, P. and
Ramirez, F., unpublished data).

The presence of polymorphic markers associated with the large
collagen genes has in turn led to the identification of molecular
haplotypes in linkage disequilibrium. Tsipouras et al. (51) have
used this approach to demonstrate the molecular heterogeneity of
the mild autosomal dominant forms of OI.

It is tempting to speculate that, like for the globin genes,
(65) future linkage studies of these molecular haplotypes may
prove the existance of fixed combinations associated with different abnormal phenotypes. This will result in a new and powerful tool for the a-priori categorization of the connective
tissues disorders at the molecular level.

Concluding Remarks

In conclusion, we have outlined some of the progress recently
made in our understanding of the genetics, evolution and pathobiology of the collagens. It is clear that many questions remain
unanswered and that new problems have arisen. The clinical cases

we have described should be considered more as prototypical

examples of the heterogeneity of the connective tissue disorders

than as specific paradigms of well-defined pathological con-

ditions. It is however conceivable to predict that the analysis

of the collagen gene structure and expression in normal and

diseased states will eventually be beneficial not only for the

molecular categorization of these disorders but also for our

general knowledge of the factors leading to gene abnormalities in

higher eucaryotes.

Acknowledgements

The authors are very much in debt to Ms. L. Bertha and Ms. K.

Valentine for the excellent typing of this manuscript.

REFERENCES

1. Klemperer, P., Pollack, D., Baher, G., J.A.M.A. 119:331-332
   (1972).

2. Bornstein, P., and Sage, H., Ann. Rev. Bioch. 44:957-1003
   (1980).

3. Ninomiya, Y., and Olsen, B.R., Proc. Natl. Acad. Sci. USA
   81:3014-3018 (1984).

4. Eyre, D., Science 207:1315-1322 (1980).

5. Minor, R., Am. J. Path. 48:228-277 (1980).

6. Timpl, R., Martin, G.R., In: Immunochemistry of the extracellular matrix VII pp 119-150, C.R.C. Press (1982).

7. Treub, B., Grobli, B., Spiess, M., Odermatt, B.F., and Win-
   terhalter, K.H. J. Biol. Chem. 5239-5245 (1982).

8. Rhodes, R.K. and Miller, E.J., Biochemistry 17:3342-3348
   (1978).

9. Berg, P., and Prockop, D.J., *Biochem. Biophys. Res. Commun.* 52:115-129 (1973).

10. Rosenbloom, J., Hirsch, M., Jimenez, S.A., *Arch. Biochem. Biophys.* 158:478-484 (1973).

11. Miller, E.J. and Robertson, P.B., *Biochem. Biophys. Res. Commun.* 54:432-438 (1973).

12. Fujii, K., Tanzer, M.L., Cooke, P.H., *J. Mol. Biol.* 106:223-227 (1976).

13. Bailey, A.J., Robbins, S.P., Balian, G., *Nature* 251:105-109 (1974).

14. Neuberger, A., Gottschalk, A., Marshall, R.O. and Spiro, R.G., In: "The Glycoproteins their composition structure and function" A. Gottschalk, ed. pp. 450-490, Elsevier, Amsterdam.

15. Olsen, B.R., Guzman, N.A., Engel, J., Condit, C. Aise, S., *Biochemistry* 16:3030-3036 (1977).

16. Uitto, J., and Prockop, D.J., *Biochem. Biophys. Res. Commun.* 55:404-411 (1978).

17. Kohn, L.D., Iserky, C., Zupnek, Lenaers, A., Lee, G., Lapiere, C.M., *Proc. Natl. Acad. Sci. USA* 71:40-44 (1974).

18. Davidson, J.M., McEneeny, L.S.G., and Bornstein, P., *Biochemistry* 14:5188-5194 (1974).

19. Tanzer, M.D., *Science* 180:561-566 (1973).

20. Tate, V., Finer, M., Boedtker, H., Doty, P., Cold Spring Harbor Symp. Quant. Biol. 47:1039-1049 (1982).

21. Wozney, J., Hanahan, D., Tate, V., Boedtker, H., Doty, P., *Nature* 294:124-135 (1981).

22. Chu, M.-L., Weil, D., de Wet, W., Bernard, M., and Ramirez, F., *J. Biol. Chem.* (in press).

23. Yamada, Y., Kuhn, K., and de Crombrugghe, B., *Nucl. Acids Res.* 11:1733-1744 (1983).

24. Ninomiya, Y., Showalter, A.M., van der Rest, M., Seidah, N.G., Chretran, M., and Olsen, B.R., *Biochemistry* 23:617-624 (1984).

25. Yamada, Y., Avvedimento, V.E., Mudryi, M., Ohkubo, H., Vogeli, G., Irani, M., Pasten, I., and de Crombrugghe, B. *Cell* 22:887-892 (1980).

26. Monson, J.M., and McCarthy, B.T., DNA 1:59-69 (1981).

27. Chu, M.-L., de Wet, W., Bernard, M., Ding, J.F., Morabito, M., Myers, J.C., Williams, C.J., and Ramirez, F., Nature 310:337-340 (1984).

28. Tate, V., Finer, M., Boedtker, H., and Doty, P., Nucl. Acids Res. 11:91-104 (1983).

29. Vogeli, G., Ohkubo, H., Sobel, M.E., Yamada, Y., Pastan, I., and de Crombrugghe, B., Proc. Natl. Acad. Sci. USA 78:5334-5338 (1981).

30. Yamada, Y., Mudryi, M., and de Crombrugghe, B., J. Biol. Chem. 258:14914-14919 (1983).

31. Chu, M.-L., de Wet, W., Bernard, M., and Ramirez, F., J. Biol. Chem. (in press) (1985).

32. Ohta, T., and Dover, G.A., Proc. Natl. Acad. Sci. USA 80:4079-4083 (1983).

33. Junien, C., Weil, D., Myers, J.C., Van Cong, N., Chu, M.-L., Foubert, C., Gross, M.S., Prockop, D.J., Kaplan, J.C., and Ramirez, F., Am. J. Hum. Genet 34:381-387 (1982).

34. Huerre, C., Junien, C., Weil, D., Chu, M.-L., Morabito, M., Foubert, C., Myers, J.C., Van Cong, N., Gross, M.S., Prockop, D.J., Bow, A., Kaplan, J.C., de la Chapelle, A., and Ramirez, F., Proc. Natl. Acad. Sci. USA 78:6627-6630 (1982).

35. Solomon, E., Hiorns, L., Delgleish, R., Tolstoshev, P., Crystal, R., and Sykes, B., Cytogen. Cell Genet. 35:64-66 (1983).

36. Henderson, A.D., Myers, J.C., and Ramirez, F., Cytogen. Cell Genet. 36:586-587 (1983).

37. Strom, C.M., Eddy, R.L. and Shans, T.B., Somatic Cell Mol. Gen. 10:651-655 (1984).

38. Huerre-Jeanpierre, C., Mattei, M.G., Weil, D., Grzeschik, K.H., Chu, M.-L., Sangiorgi, F.O., Sobel, M.E., Ramirez, F., and Junien, C., (submitted).

39. Sillence, D.O., Senn, A., and Danks, D.M., J. Med. Genet. 16:101-116 (1979).

40. Barsh, G.S., David, K.E., and Byers, P.H., Proc. Natl. Acad. Sci. USA 79:3838-3842 (1982).

41. Chu, M.-L., Williams, C.J., Pepe, G., Hirsch, J., Prockop, D.J. and Ramirez, F., _Nature_ 304:48-50 (1983).

42. Barsh, G.S., and Byers, P.H., _Proc. Natl. Acad. Sci. USA_ 78:5142-5146 (1981).

43. Williams, C.J., and Prockop, D.J., _J. Biol. Chem._ 258:5915-5921 (1983).

44. Chu, M.-L., Gargiulo, V., Williams, C.J., and Ramirez, F., _J. Biol. Chem._ 260:691-694 (1985).

45. Steinman, B., Rao, V.H., Vogel, A., Bruckner, P., Gitzelman, R., and Byers, P.A., _J. Biol. Chem._ 259:11129-11138 (1984).

46. de Wet, W.J., Pihlajaniemi, T., Myers, J.C., Kelly, T.E., and Prockop, D.J., _J. Biol. Chem._ 258:7721-7728 (1983).

47. Deak, S.B., Nicholls, A.C., Pope, F.M., and Prockop, D.J., _J. Biol. Chem._ 258:15142-15147 (1983).

48. Chu, M.-L., Nicholls, A.C., Pope, F.M., Rowe, D., and Prockop, D.J., _Collagen Rel. Res._ 4:388-394 (1984).

49. Dickson, L.A., Pihlajaniemi, T., Deak, S., Pope, F.M., Nicholls, A., Prockop, D.J. and Myers, J.C., _Proc. Natl. Acad. Sci. USA_ 81:4524-4528 (1984).

50. Pihlajaniemi, T., Dickson, L.A., Pope, F.M., Korhonen, V.R., Nicholls, A., Prockop, D.J., and Myers, J.C., _J. Biol. Chem._ 259:12941-12944 (1984).

51. Tsipouras, P., Myers, J.C., Ramirez, F., and Prockop, D.J., _J. Clin. Invest._ 72:1262-1267 (1983).

52. Sippola, M., Kaffe, S., and Prockop, D.J., _J. Biol. Chem._ 259:14094-14100 (1984).

53. Eyre, D.R., Aldridge, J., Shapiro, F.D., Millis, M.B., Lewi, L., and Ellezian, L.Y., _Transactions of Orth. Res. Soc._ Vol. 9, (1984).

54. Hollister, D.W., Byers, P.H., and Holbrook, K.A., _Adv. in Human Genetics_ 121-87 (1982).

55. Byers, P.H., Siegel, R.C., Peterson, K.E., Rowe, D.W., Holbrook, K.A., Smith, L.T., Chang, Y.H., and Fu, F.C.C., _Proc. Natl. Acad. Sci. USA_ 78:7745-7749 (1981).

56. Macek, M., Hurych, J., Chvapil, and Kadlecova, V., _Humangenetik_ 3:87-97 (1966).

57. Boucek, R.J., Noble, N.L., Gunja-Smith, Z., and Butler, W.T., N. Engl. J. Med. 305:988-991 (1981).

58. Abraham, P.A., Perejda, A.J., Carnes, W.H., and Uitto, J., J. Clin. Invest. 70:1245-1252 (1982).

59. Appel, A., Horwitz, A.L., and Dorfman, A., J. Biol. Chem. 254:12199-12203 (1979).

60. Prockop, D.J., and Kivirikko, K., N. Engl.J. Med. 311:376-386 (1984).

61. Stolle, C.A., Pyeritz, R.E., Myers, J.C., and Prockop, D.J., J. Cell Biol. 97:2280 (1983).

62. Steinman, B., Tuderman, L., Pentonen, L., Martin, G.R., McKusick, V.A., and Prockop, D.J., J. Biol. Chem. 255:8887-8893 (1980).

63. Stanescu, R., Stanescu, V., and Maroteaux, P., Coll. Rel. Res. 2:111-116 (1982).

64. Hollister, D.W., Burgeson, R.W., and Rimoin, D.L., Am. J. Hum. Gen. 27:783A (1975).

65. Orkin, S., and Kazazian, H.H., Ann. Rev. in Genet. 18:131-176 (1984).

HUMAN PLASMINOGEN ACTIVATORS. GENES AND
PROTEINS STRUCTURE.

Francesco Blasi, Andrea Riccio and Gianfranco Sebastio

1. INTRODUCTION: EXTRACELLULAR PROTEOLYSIS.

Activation of plasminogen to plasmin represents a key step in
controlling extracellular proteolysis. Plasminogen is a 92,000
daltons zymogen which is proteolytically activated to the
trypsin-like protease, plasmin, which is capable to act on several
substrates. Among the reactions catalyzed by plasmin, intravascular
fibrinolysis is relatively well understood. The proteolytic action
of circulating plasmin is restricted to the fibrin-deposition sites
for the occurrence of two conditions: 1) activation of plasminogen
takes place on polymerized fibrin, where a ternary complex forms
between fibrin, plasminogen and plasminogen activator (Hoylaerts et
al., 1982). 2) Once plasmin is formed, it remains bound to fibrin
until dissolution of the polymer; any plasmin released from the
ternary complex is then rapidly inactivated by circulating alpha-2
antiplasmin (Wiman and Collen, 1978; Collen & Verstraete, 1983).
This prevents degradation of other circulating proteins. Similar
mechanisms can be envisaged for other plasmin-mediated extracellular
proteolysis reactions, occurring outside of the circulation in the
interstitial spaces.

The rate-limiting step of the reactions outlined above appears
to be the formation of plasmin by plasminogen activator (PA). The
highly specific PA (so far plasminogen is the only known substrate)
is able to activate plasminogen by hydrolyzing specific peptide
bond(s) of the zymogen. The product of this reaction, plasmin, has
a wide specificity; tight regulation of its activity is therefore

required. This can be accomplished on one side by regulating the
activation of plasminogen, or inhibiting the activity of plasmin on
the other. Regulation of the synthesis and/or activity of both PA
and plasmin-inhibitors permits a rapid expansion or restriction of
the extracellular proteolytic potential of a whole tissue as well as
of individual cells. Since also PAs are synthesized as inactive
zymogens like plasminogen (see below), activation of PA constitutes
another potential regulatory step in the extracellular proteolysis
chain. Hence two reactions in cascade are required to produce
active plasmin (Fig. 1), with an impressive potential for rapid and
large signal amplification or restriction. The key position and
function of PA in this cascade gives this molecule an unique role in
controlling vital cellular functions which depend on extracellular
proteolysis.

Extracellular proteolysis may be directed to a variety of
proteins. Aside from fibrin polymers which are the best known
targets in the circulatory bed, proteolysis of basement membrane
proteins must be required for cellular migration and invasiveness.
Basement membrane collagen and other basement membrane and
extracellular matrix proteins (fibronectin, laminin, etc.) can be
degraded by plasmin (Liotta et al., 1981), a process mediated by PA
(Sheela & Barrett, 1982). Moreover, plasmin inhibitors are known to
inhibit cell migration (Ossowski et al., 1973).

Extracellular proteolysis may also contribute to changes in
cell-surface proteins like antigens, receptors, etc. Thus, PA
molecules might be in more or less direct control of a wide variety
of cellular functions like responses to growth factors and hormones,
immunologic and inflammatory stimuli, cellular migration, tissue
degradation and involution, expansion of individual cell clones,
invasion of normal tissues by monocyte-macrophage cells or by cancer
cells, etc. These possibilities have been suggested by several
authors, and are summarized elsewhere (Reich, 1978; Danø et al.,
1985). In this paper we only discuss those data which seem
significant in the context of the present discussion.

The level of PA activity does in fact correlate with potential functions of PA in the different cells. For example, it has been shown that in the developing chicken embryo, the PA synthesis in the bursa of Fabricius starts at the stage where B-lymphocytes appear and begin to migrate out into the circulation (Valinsky et al., 1983). Another example is angiogenesis: tumor angiogenic factors appear to induce the synthesis of PA in bovine endothelial cells (Gross, et al., 1983). In view of the relevance of angiogenesis in tumor growth and invasiveness, this results assumes a particular significance. However, it remains to be shown that the PA inducing- and the angiogenic-factor is indeed the same. Another correlation appears to us as particularly striking. Immunocytochemical staining of uPA (one of the two species of PA, see below) in rat Lewis lung carcinomas shows that the enzyme is synthesized only by those tumor cells that are in close vicinity to or in contact with normal cells of the surrounding tissue being invaded by the tumor (Skriver et al., 1984). The location of uPA molecule at the boundary between tumor and normal tissue supports the biochemical data indicating PA as a major regulatory function in cellular invasiveness.

The former examples represent just a minority of the observations that indicate PA in control of such important biological processes (See Danø et al., 1985 for other references). It must be emphasized, however, that the involvement of PA in regulating extracellular proteolysis and thus cell migration, invasiveness, etc., is based largely on correlation data. Experiments designed to prove PA involvement directly are now possible since antibodies and cloned genes are available. One such experiment designed to study the role of uPA in metastatis has been carried out. The human tumor cells Hep-3 can be inoculated in the chorion-allantoid chamber of the developing chicken embryo, and there they proliferate, invade the circulation and metastasize different organs, including lung. Inoculation of antibodies to human uPA (which is synthesized by the tumor cells), greatly and specifically decrease the number of metastasis by inhibiting the activity of uPA (Ossowski and Reich, 1983). This experimental

system, while allowing tight control of the experimental conditions
and thus offering high confidence in the results, suffers of the
limitation that Hep-3 tumor is the only human tumor with which this
experiment is feasible. In addition, Hep-3 cell line has been
maintained in culture for several decades. Despite these
limitations Ossowski and Reich's experiment clearly indicates that
uPA is involved in tumor metastasis and show the direction to be
followed in trying to prove or disprove the role of PA not only in
the human cancer phenotype, but in general in any normal or
pathological process. In any case, deregulation of the reactions
cascade (Fig. 1) may have a profound effect on the structure of
entire tissues and on the properties of many cell types.

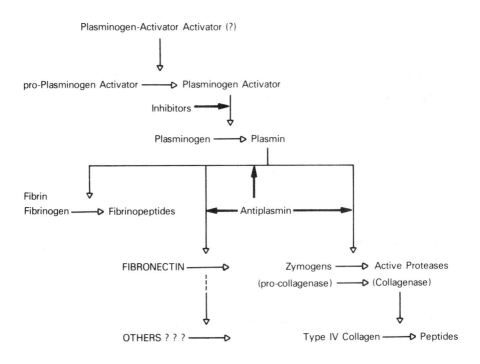

Figure 1. Activation cascade of the extracellular proteolysis
pathway. While the reactions have been shown to occur in vitro,
their in vivo significance is not yet proven.

All these reports are obscured by one major discrepancy. The action of PA in cell migration is believed to be dependent on its ability to activate plasminogen to plasmin. Recent data, show that urokinase PA is present in tissues and cultured cells media in an inactive form (Kielberg et al., 1985; Skriver et. al., 1984) (see Fig. 2). The urokinase PA (uPA, see below) zymogen can be activated by conventional proteolytic enzymes, like trypsin, plasmin, etc., (Nielsen et al., 1982; Skriver et al., 1982; Wun et al., 1982). So far no evidence is available as to how and where in vivo plasminogen activator activity is gained. In order to bring previous correlations between PA synthesis and cellular migration into a physiologically credible and meaningful picture, the discrepancy of the synthesis of the inactive PA must be solved.

BIOSYNTHESIS AND ACTIVATION OF HUMAN uPA

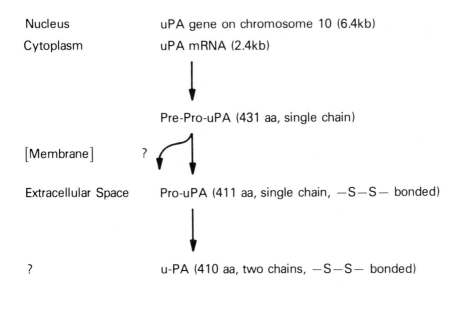

| Nucleus | uPA gene on chromosome 10 (6.4kb) |
| Cytoplasm | uPA mRNA (2.4kb) |

Pre-Pro-uPA (431 aa, single chain)

[Membrane]   ?

Extracellular Space   Pro-uPA (411 aa, single chain, −S−S− bonded)

?   u-PA (410 aa, two chains, −S−S− bonded)

Figure 2. Biosynthesis and activation of human uPA.

In our mind, two possible ways must be explored: 1) plasminogen is not the physiological substrate of PA, at least not of uPA; it is therefore possible that extracellular fluids do contain another specific substrate for PA which recognizes also the so-called inactive uPA precursor. 2) Activation of zymogen uPA takes place on the surface of the cell. Vassalli et. al., (1985) and Stoppelli et. al. (1985) have shown that monocytes have specific urokinase (uPA) membrane receptors. These receptors bind uPA without interfering with its catalytic property. uPA zymogen also binds to the same receptor. It is not yet known whether secreted uPA precursor acquires a PA activity when bound to the uPA receptor. This is not inconceivable, since uPA zymogen has all parameters of a plasminogen activator, like fibrin-affinity (Kohno et al., 1984), receptor binding affinity (Corti, Stoppelli and Blasi, unpublished), but no PA activity. It is possible that binding to the cell surface receptor induces a conformational change similar to that caused by the single proteolytic cleavage (or the cleavage itself) that activates pro-uPA to uPA. Experiments are in progress in different laboratories to address both these questions.

In the following sections we shall analyze the structure of PAs and their genes, as well as a short outline of the diseases in which an active role of PA has been proposed or in which abnormal levels of enzyme have been demonstrated.

2. PLASMINOGEN ACTIVATORS: PROTEINS

2.1 <u>Molecular types of PA and tissue distribution</u>. Most of the published work on PA was carried out before the occurrence of two different types of PA was recognized. We distinguish in fact the urinary PA (uPA), also called urokinase, from the tissue PA (tPA). Despite their names, both uPA and tPA are found in urines and in some tissues (see below). The two proteins have different aminoacid sequence, molecular weight, subunit structure and catalytic parameters (Table 1).

2.2 <u>The uPA form</u>.

a) <u>Biosynthesis</u>. The product of human, pig and mouse uPA mRNA has been identified by RNA-directed cell-free protein synthesis <u>in</u> <u>vitro</u> (Salerno et al., 1984), microinjection of mRNA into Xenopus

oocytes (Miskin and Soreq, 1981; Belin et al, 1984) and by cDNA
cloning and sequencing (Verde et al., 1984; Nagamine et al., 1984)
(Figure 2). We have shown that monoclonal antibodies directed
towards different portions of active human 2-chains urokinase
immunoprecipitate the same single-chain product of human kidney uPA
mRNA (Salerno et al., 1984). This biosynthetic precursor has a
molecular weight of about 50,000 daltons based on electrophoretic
migration rate and is the product of a 2,400 nucleotides mRNA

Table 1

A comparison of uPA structure and properties[#]

	uPA	tPA
M.W. of biosynthetic precursor	52,000	70,000
N. aminoacids in the pro-form	431	562
M.W. of active enzyme	48,000[*]	65,000
N. amino acids residues in the active form	411	527
N. chains of active enzyme	2	2
Fibrin Affinity	no	yes
Activation by Fibrin	?	yes

[#]Appropriate references are given in the text.
[*]The M.W. of active uPA is 48,000 based on the amino acids sequence.
 The enzyme runs at about 54,000 on SDS polyacrylamide gel electro-
 phoresis under non-reducing conditions.

(Ferraiuolo et al., 1984). The nucleotide sequence (Verde et al.,
1984; Heyneker, et al., 1983; Riccio et al., 1985) agrees with the
aminoacids sequence determined on the active urinary urokinase
(Guenzler et al., 1982 a, b; Steffens et al., 1982), with the
exception of a 20 aminoacids signal peptide. The 431 residues
sequence is shown in Fig. 3. The biosynthetic precursor has never
actually been observed, in agreement with the loss of the signal
peptide during the synthesis of secreted proteins. However, a
single-chain extracellular human pro-uPA of 48,000 daltons has been
observed (Nielsen et al., 1982; Wun et al., 1982; Ferraiuolo et al.,
1984) and sequenced (Kohno et al., 1984). The single-chain human
pro-uPA is an inactive zymogen unable to attack either plasminogen

HUMAN PRO-UROKINASE

411 AMINOACIDS
12 DISULFIDE BONDS
20 AMINO-ACIDS SIGNAL PEPTIDE
ACTIVE SITE AMINO-ACIDS: HIS 204,
ASP 255; SER 356 AND ASP 344

Figure 3. Scheme of the primary and secondary structure of human uPA precursor, including the signal peptide. Arrows and roman numerals indicate the position of intervening sequences. Upon activation, residue lys 158 is lost. The disulfide bonds in the region of residues 50-411 have been assigned according to Guenzler et al. (1982b); in the Growth Factor domain (residues 1-45), disulfide bonds have not been assigned, since the position of the cysteines is not really homologous to that of EGF. Amino acids are indicated using the one-letter code.

or synthetic substrates. It can be activated by a serine protease, like trypsin and plasmin. As already stated above, the enzyme which physiologically converts pro-uPA into active uPA is not yet known. Activation of pro-uPA consists in the introduction of a proteolytic cleavage at position 157-159 (Figure 4A) which hence converts the single-chain pro-uPA into a two-chains active enzyme.

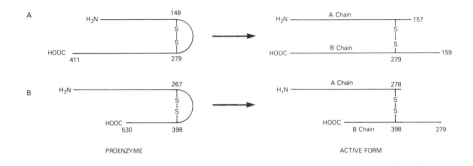

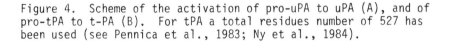

Figure 4.   Scheme of the activation of pro-uPA to uPA (A), and of
pro-tPA to t-PA (B).   For tPA a total residues number of 527 has
been used (see Pennica et al., 1983; Ny et al., 1984).

Another sequence difference between the urinary protein and the
cDNA predicted sequence is the presence of a lysine residue (lys
158) exactly at the junction of the two chains.   The presence of lys
158 poses the interesting question of the nature of pro-uPA
activating enzyme.   In fact, the cleavage involves the sequence
phe-lys-ile-ile and the production of a carboxyl terminus lacking
lys 158 and sometimes also phe 157 (Guenzler et al., 1982a).

    b)   <u>Subcellular localization</u>.   In addition to the secreted
form of uPA, the enzyme is also found to be cell-associated, in both
soluble and particulate forms.   Some of the studies suggest that the
particulate form may be an integral membrane protein since it
co-purifies with plasma membrane-containing cellular fractions
(Quigley, 1976).   Immunofluorescence studies with monoclonal
antibodies have been carried out with human glioblastoma (Danø et
al., 1982) and kidney carcinoma (Ferraiuolo et al., 1984) cells.   In
both cases no evidence of membrane form of uPA has been obtained,
fluorescence being localized mostly around the nucleus, in agreement
with uPA being a secreted protein.   Outer membrane reactivity, on
the other hand, has been observed on non-permeabilized preparations
of A431 human carcinoma (C. Tacchetti and F. Blasi, unpublished) and
on retinoic acid differentiated HL60 promyelocytic leukemia (F.
Pegoraro, F. Ghezzo and F. Blasi, unpublished observations) cells.

By using an ELISA method or by iodinating whole cells of mouse melanoma B16 F10 we have obtained additional evidence of the presence of an external form of uPA, membrane bound (Hearing, Appella and Blasi, unpublished). Whether this membrane reactivity is due to an integral membrane uPA form or to an interaction of secreted uPA with the cell surface receptor, remains to be established.

The presence of urokinase onto the cell membrane may be due to interaction of this protein with plasma-membrane receptors. Vassalli et. al. (1985), in fact, have shown that monocytes and monocyte-like cells possess a specific receptor for urokinase. We have been able to reproduce their finding and to extend it in showing that the binding function is retained by the purified amino terminal fragment of urokinase (residues 1-135) which is not required for activity; the catalytic carboxy-terminal portion of the enzyme being not required for binding. Moreover, we have shown that the bound urokinase or the aminoterminal fragment are exposed on the outer membrane surface for at least three hours after binding (Stoppelli et al., 1985). These data suggest that the monocyte-like cells use the membrane-bound urokinase to localize its proteolytic activity for cell migration. The fact that the urokinase receptor has so far been found on monocytes and on monocytes-like cells may indicate a role of this receptor in the function of cell migration.

The demonstration of a specific membrane receptor for uPA raises the possibility that the membrane-associated uPA may not be an integral membrane protein, but rather represent enzyme bound to the receptor. Experiments are being carried out to discriminate between these two possibilities.

2.3 The tPA form

A second plasminogen activator form, different from urokinase, is tissue-type PA (tPA). tPA has been purified from several tissues and tumor cell lines. Bowes melanoma cell line has been a particularly valuable source of tPA.

Two different forms of tPA, having a MW of 65,000 and 63,000 daltons respectively, immunoprecipitate from Bowes melanoma cell

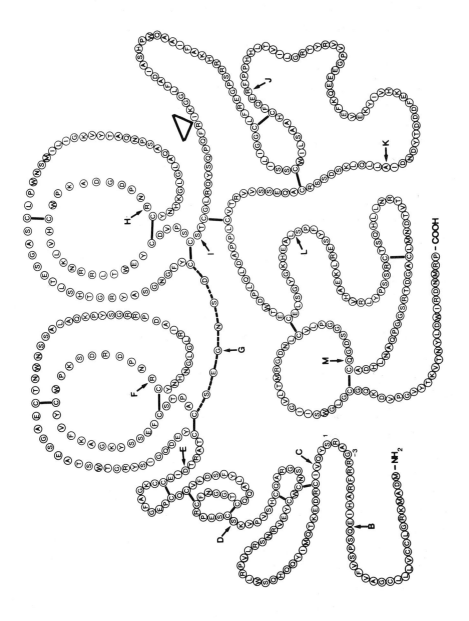

Figure 5. Secondary structure of human tPA precursor (courtesy of Dr. Thor Ny) (Ny et al., 1984), including signal peptide and pro-sequence. Black bars indicate potential disulfide bridges. The arrows at positions B-M indicate the intervening sequences in the tPA gene. The triangle indicates the cleavage site between the heavy and light chain.

medium. The molecular weight difference seems to be due to a
variation in carbohydrate composition (Pennica et al., 1983). tPA
is secreted as a single-chain proenzyme, which is converted to a
double chain form through limited proteolysis of the Arg 275-Ile276
bond (Fig. 4B) (Andreasen et al., 1984). The active form of tPA
consists of an A-chain of 278 aminoacids and of a B-chain, 252
residues long, connected by a disulphide bond (Fig. 4B). The
complete aminoacid sequence of tPA has been derived from the
nucleotide sequence of a cDNA clone isolated from a library prepared
from Bowes melanoma cell mRNA (Pennica et al., 1983). Fig. 5 shows
a bidimensional model of tPA structure. Although no secondary
structure studies have been carried out on tPA, its disulfide
bonding has been derived on the basis of the aminoacids sequence
homology (see below) with proteins of known secondary structure
(i.e., plasminogen, prothrombin, etc.) (Pennica et al., 1983).
Starting from the aminoterminus, residue n°1, tPA contains a first
domain homologous to the fibronectin finger-like domain, (residues
1-32); the second domain, growth factor domain (residues 32 to 90)
is homologous to urokinase, EGF and TGF alpha (Guenzler et al.,
1982b; Derynck et al., 1984). Two Kringle structures (triple
disulfide-linked regions, in which -S-S- bonds hold together an
onion-like structure of two aminoacids sheets invaginating into one
another) follow (resides 91-138 and 149-232). These sequence have
high sequence homology with similar structures in plasminogen,
prothrombin and uPA (Figure 6), in particular around the cysteines.
In analogy to the functions of Kringle in plasminogen, tPA Kringles
have been tentatively assigned the function to bind fibrin. The rest
of the molecule contains the active site. Similarly to uPA (Sumi
and Robbins, 1983), tPA contains one single disulfide bond which
holds together the two chains (cys 278 -S-S-cys 398) after
proteolytic activation at the 265-266 peptide bond. The structural
consequences of this activation are not known. The catalytic moiety
of uPA and tPA also share considerable sequence homology to other
serine proteases like chymotrypsin, trypsin and elastase (Fig. 7).

```
                                          45
Urokinase .....................DKSKTCYEGNGHFYRGKASTDTMGRPCLPW
                                        87
t-PA ( Kringle 1 )..............DTRATCYEDQGISYRGTWSTAESGAECTNW
                                       173
t-PA ( Kringle 2 )..............ECNSDCYFGNGSAYRGTHSLTESGASCLPW
                                       60
Prothrombin ( Kringle 1 ).........CLEGNCAEGLGTNYRGNVSITRSGIECQLW
                                      165
Prothrombin ( Kringle 2 ).........PPLEQCVPDRGQQYQGRLAVTTHGLPCLAW
                                      352
Plasminogen ( Kringle 4 ).........PVVQDCYHGDGQSYRGTSSTTTTGKKCQSW
```

```
                                                               134
NSATVLQQTYHAHRSDALQLGLGKHNYCRNPDNR  RRPWCYVQVGLKPLVQECMVHDCADG
                                                               176
NSSALAQKPYSGRRPDAIRLGLGNHNYCRNPDRD  SKPWCYVFKAGKYSSEFCSTPACSEG
                                                               262
NSMILIGKVYTAQNPSAQALGLGKHNYCRNPDGD  AKPWCHVLKNRRLTWEYCDVPSCSTC
                                                               146
RSRYPHKPHINSTTHPGADLQENF  CRNPDSSITGPWCYTTDPTARR QECSTPVCGQD
                                                               251
ASAQAKALSKHQDFNSAVQLVENF  CRNPDGDEEGVWCYVAGKPGDF GYCDLNYCEEA
                                                               437
SSMTPHRHQKTPENYPNAGLTMNY  CRNPDADKGCPWCFTTDPSVRW EYCNLKKCSGT
```

Figure 6. Comparison of the amino acids sequence and genomic structures in the kringle region: uPA data are from Riccio et al. (1985); tPA data are from Ny et al. (1984), prothrombin and plasminogen data are from Magnusson et al., 1976. Arrows and vertical bars indicate the position of intron in each gene. The residue number in each sequence is indicated.

3.   COMPARATIVE STRUCTURE-FUNCTION RELATIONSHIP IN tPA AND uPA

Comparison of the aminoacids sequence of tPA and uPA (Fig. 8) immediately shows that the two proteins are highly homologous (50% on the average) (Strassburger et al., 1983). uPA is shorter than tPA by about 120 amino acids largely due to the absence of two entire domains. The first missing domain is the extreme N-terminal portion (corresponding to aminoacids 1-40 of tPA). This is the finger-like structure of tPA which may be responsible for the high affinity of tPA to fibrin (Banyai et al., 1983). The second domain of tPA which is missing in uPA is the second kringle (aminoacids 177-262 of tPA). uPA has one single kringle structure (residues 50-131) which is highly homologous to the first one of tPA (102-160). Kringles are present also in other zymogens like prothrombin (Sottrup-Jensen et al., 1975; Magnusson et al., 1976) and plasminogen (Sottrup-Jensen et al., 1978).

```
                                     158              ★     180                        200       ☆ ★
TISSUE-TYPE PLASMINOGEN ACTIVATOR.......... IKGGLFADIASHPWGAAIFAKHRRSPGERFLCGGILISSCWILSAAHCFQEREFPPHIL
UROKINASE................................. IIGGEFTTIENQPWFAAIYRRHR GGSVTYVCGGSLISPCWVISATHCFIDYPKKEDY
CHYMOTRYPSIN.............................. IVNGEEAVPGSWPWQVSLQDKT  GFHFCGGSLINENWVVTAAHCKVT    TSD
TRYPSIN................................... IVGGYTCGANTVPYQVSLNS    CYHFCGGSLINSQWVVSAAHCYKS    GI
ELASTASE.................................. VVGGTEAQRNSWPSQISLQYR S GSSWAHTCGGTLIRQNWVMTAAHCVDRE  LTF

    220                 ★240            ☆ ★260              280                      300        ★
TVILGRTYRVVPGEEEQKFEVEKYIVHKEFDDTY DNDIALLQLKSDSSRCAQESSVVRTVCLPPADLOLPDWTECELSGYGKHFALSPFYSF
IVYLGRSRLNSNTQGEMKFEVENLILHKDYSADTLAHHNDIALLKIRSKEGRCAOPSRTIQTICLPSMYNDPOFGTSCFITGFCKENSTDYLYPE
VVVAGEFDQGSSSEKIQKIKIAKVFKNSKYNSLTI NNDITLLKLSTA ASFSQTVSAVCLPSASDDFAAGTTCVTTGWGLTRYTSANTPD
QVRLGQDNINVVEGNQQFISASKSIVHPSYNSNTL NNDIMLIKLKSA ASLNSRVASISLPT SCASAGTQCLISCWGNTKSSGTSYPD
RVVVGEHNLNQNNGTEQYVGVQKIVVHPYWNTDDVAACYDIALLRLAQS VTLNSYVQLGVLPRAGTIIANNSPCYITGWGLTRTNQGIAQ

    320                 340                    ★ ☆     360                    380
RLKEAHVRLYPSSRCTSQHLLNRTVTDNMLCAGDTRSGGPQANLHDACQGDSGGPLVCLNDGRMTLVGIISWGLG   CGOKDVPGVYTKVTNYLD
QLKMTVVKLISHRECQQPHYYGSEVTTKMLCAAD    PQ WKTDSCQGDSGGPLVCSLQGRMTLTEIVSWGRG   CALKDKPGVYTRVSHELP
RLQQASLPLLSNTNCKK  YWGTKIKDAMICAGA     SGVSSCMGDSGGPLVCKKNGAWTLVGIVSWGSS  TCSTS TPCVYARVTALVN
VIKCLKAPIISNSSCKS  AYPGQITSNMFCAGY     LQ GCKDSCQGDSGGPVVCS    GKLQGIVSWGSG  CAQKNKPGVYTKVCNYVS
TLQQAYLPTVDYAICSSSSYWGSTVKNSMVCAGG     DG  VRSGCQGDSGGPLHCLYNCQYAVHGVTSFVSRLGCNVTRKPTVFTRVSAYIS

    400
WIRDNMRP
WIRSHTKEENGLAL
WVQQTLAAN
WIKQTIASN
WINNVIASN
```

Figure 7. Comparison of amino acids sequence of tPA, uPA, and three serine proteases, in the region of the heavy chain of human uPA. Numbers indicate aminoacids position for human uPA. Full asterisks point to the position of introns; empty asterisks indicate the position of the residues involved in the catalytic site (data from Riccio et al., 1984; Ny et al, 1985; Craik et al., 1984; Bell et al, 1984; Swift et al., 1984).

The N-terminal portion of uPA (residues 1-49) is homologous not only to tPA (residues 44-91), but also to epidermal growth factor, transforming growth factor alpha (Gunzler et. al., 1982b; Derynck et. al., 1984) and to coagulation factors IX and X, prothrombin and protein-C (Dayhoff, 1978; Anson et. al., 1984; Fernlund et. al., 1982). Since this region is known to be completely dispensable for uPA catalytic activity (Sumi and Robbins, 1983), the possibility that it serves another hitherto unknown function must be considered. The homology to growth factors might suggest a growth promoting or growth-modulating activity. So far, however, no such function has been demonstrated. As discussed above, the amino terminal 137 residues of uPA have all the information for binding membrane receptors on human monocytes (Stoppelli et al., 1985). Whether the

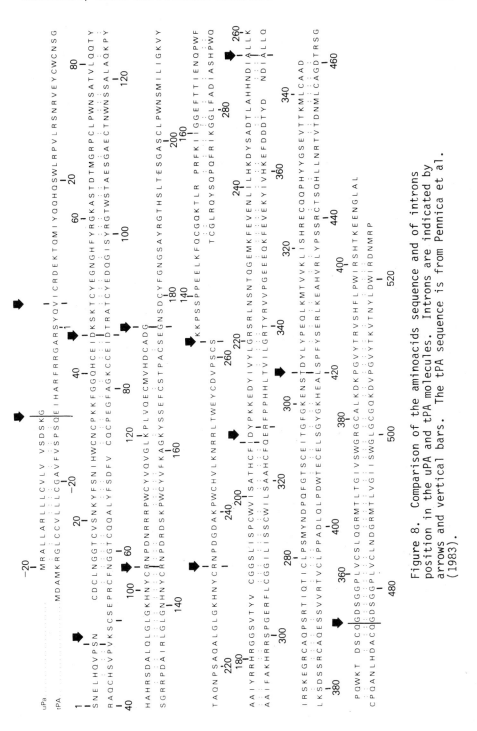

Figure 8. Comparison of the aminoacids sequence and of introns position in the uPA and tPA molecules. Introns are indicated by arrows and vertical bars. The tPA sequence is from Pennica et al. (1983).

"growth factor" domain is responsible for this property or not, is
not yet known. The homology to EGF and TGF-alpha (Guenzler et al.,
1982b; Derynck et al., 1984) includes 5 of the first 6 cysteine
residues of urokinase (Fig. 9). While the exact location of the

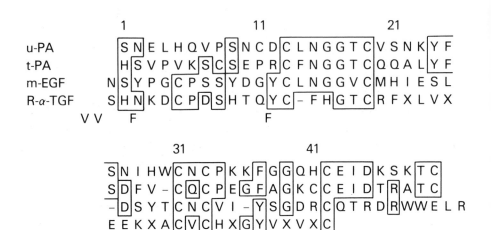

Figure 9. Comparison of amino acids sequence of uPA and tPA in the
Growth Factor domains: m-EGF is mouse epidermal growth factor;
R-α-TGF is the rat transforming growth factor alpha (Derynck et al.,
1984). Residues are numbered according to the uPA sequence
(Guenzler et al. 1982b).

disulfide bonding has not yet been determined, a disulfide bond
pattern has been proposed (Guenzler et al., 1982b) homologous to
that of the epidermal growth factor. This bonding pattern, if
applicable, would however result in a secondary structure quite
different from that of EGF and TGF alpha. It is thus not surprising
that no growth promoting activity can be observed in the amino
terminal fragment (A. Corti and F. Blasi unpublished results), or in
the entire uPA molecule except at very high concentration (Cohen et
al., 1981). Neither uPA nor its amino terminal fragments is able to
compete with EGF for the EGF receptor (Stoppelli, Blasi and Assoian,
unpublished results). No experiment has so far been reported on the
growth factor domain of tPA.

The B-chains of tPA (residues 276-530) and uPA (residues 159-411) share about 45% homology. This portion carries the typical active site (residues his 204, asp 344 and ser 376 in uPA; his 322, asp 460 and ser 478 in tPA) of the serine protease, as well as the residue responsible for the hydrolysis of basic aminoacids (asp 472 and asp 344 for tPA and uPA, respectively). By analogy to the structure of other serine proteases, it has been possible to suggest a common secondary structure for both uPA and tPA (see Figs. 3 and 5). While this structure is probably correct for the Kringle and the B-chain regions, that of the A chain is questionable in the region preceding the Kringles. For this reason no disulfide-bonding is shown for the first portion of uPA (Figure 3). Fig. 9 shows that 5 of 6 cysteines fall on identical position in the EGF and in the growth factor domain of uPA and tPA. However, this single difference may lead to profound differences in the resulting structures or may cause a quite different secondary structure. Experiments are required to solve this point.

3.1 Different physiological functions for tPA and uPA?

Human urinary urokinase has been used for many years in therapy of acute myocardial infarction and of deep vein thrombosis. Cloning of tPA cDNA and its expressions (Pennica et. al., 1983) has provided tPA for use in clinical trials. The results of such attempts (Van de Werf et al, 1984) and previous biochemical data (Rijken & Collen, 1981) indicate that tPA is a more physiological and rather efficient agent in dissolving blood clots. This is probably due to the fact that tPA has a much higher affinity for fibrin than uPA (see Table 1). As a result, tPA has a localized thrombolytic activity (Van de Werf et al., 1984) while uPA administration is accompanied by systemic activation of the fibrinolytic system, with generalized decrease of blood levels of plasminogen, fibrinogen, etc. (Collen and Verstraete, 1983). This has led to the proposal that tPA is the enzyme that is physiologically important in homeostasis of the fibrinolytic state (Matsuo et al., 1981), uPA being rather involved in other functions like cell migration etc. While this distinction

may actually be true, its biochemical basis might be questioned.
The low affinity of uPA for fibrin is in fact true only for the
urinary active uPA. uPA is synthesized and secreted as a zymogen
pro-urokinase (single-chain, 411 residues) having in fact affinity
for fibrin (Kohno et. al., 1984; Corti et al., 1985). Cleavage of
the single chain pro-uPA at residue 157 results in the loss of
fibrin affinity and in the acquisition of plasminogen-dependent
fibrinolytic activity. The zymogen pro-urokinase, therefore, might
turn out to be as good and safe a thrombolytic agent as tPA, because
of its high fibrin affinity, provided a physiological or
pharmacological method is found to activate it (see below).
Pro-urokinase has affinity for the specific monocyte receptors
(Stoppelli, Cubellis and Blasi, to be published).

4.   PA: THE GENES

     Both human uPA and tPA genes have been isolated and
characterized. The human uPA gene is 6.4kb long and is split into
eleven exons (Riccio et al., 1985), the human tPA gene spans more
than 20kb and consists of at least fourteen exons (Ny et al., 1984)
(Fig. 10). Not only the proteins, but also the genes for the two

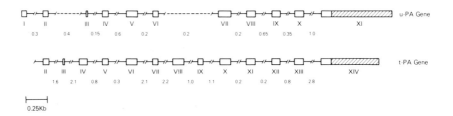

Figure 10. Comparison of the gene structure for human uPA (Riccio
et al., 1985) and tPA (Ny et al., 1984) genes. Open boxes indicate
exons and are indicated by roman numerals. The striped boxes
indicate the 3' untranslated region. The numbers below the symbols
indicate intron size in kilobases. Exons size is on scale.

forms of PAs are highly homologous in their general structure. The
intron-exon organization of the uPA gene, in fact, closely resembles
the structure of the tPA gene (Figs. 8 and 10). Exons VIII and IX

of tPA have no correspondence in uPA, they code for the second
Kringle which is missing in uPA. Other general differences are the
absence in uPA of exons III and most of exon IV, which correspond to
the domain homologous to the finger-like structure of fibronectin.
Finally, the 3' untranslated region of uPA is longer than the
corresponding one of tPA. A correlation between exons positions and
structural and functional uPA domains is shown in figure 11. Most
of the 5' untranslated sequence, the signal peptide and the "growth
factor" domain are encoded by different exons in both PA genes.
Each "kringle" domain is encoded by two exons and the interposed
intron splits exactly the Arg codon in all three coding sequences
(Fig. 6 and 10). Other genes coding for these structures, such as

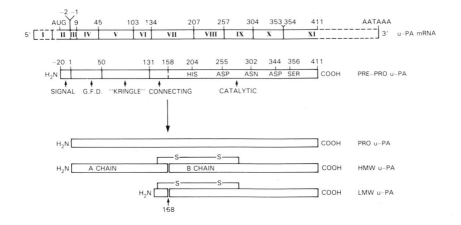

Figure 11. Scheme of the correlation between uPA genomic structure,
functional domains and activation of the gene product.

prothrombin and plasminogen, present a somewhat different
intron-exon organization (Fig. 6).

The protease-part (mostly B-chain) of both PAs is encoded by
five exons. Introns separate each of the important catalytic
residues of the PAs: the three residues (His, Asp and Ser) common
to all the serine protease active sites and the Asp residue
responsible for the trypsin-like specificity (Fig. 11). The
connecting peptide domain of uPA (res. 131-157) (Fig. 11) has no

counterpart in tPA and is encoded by the extreme 5' portion of exon
VII, which is on this side 39 bp longer than the homologous exon (X)
of the tPA gene. Other internal-polypeptide length differences
among serine proteases have been correlated with the position of
intervening sequences and an evolutionary mechanism based on the
shift of splice junctions causing the expansion of exons in the
adjacent introns (junctional sliding) (Craik et al., 1983) has been
proposed to explain these differences (Ny et al., 1984). The
homology in primary and secondary structure described between the
protease part of PAs and several simpler serine proteases, such as
trypsin, chymotrypsin, elastase (Fig. 8) and kallikrein
(Strassburger et al., 1983) can be observed also in the
corresponding gene structure (Riccio et al., 1985; Ny et al., 1984;
Craik et al., 1984; Bell et al., 1984; Swift et al., 1984; Mason et
al., 1983). Intervening sequences occur in all these genes at
positions corresponding to urokinase introns G, I and J. Intron H
of uPA has an equivalent only in the tPA and elastase genes (Fig.
8). The finding of such conserved gene structure reinforces the
hypothesis that the serine protease genes belong to a gene family
and have evolved from a common ancestor. It has been proposed that
exons correspond to protein functional and structural domains which
can be reshuffled by recombination within introns to produce new
kinds of proteins from parts of existing ones (Gilbert, 1978). A
more recent view suggests that also multi-exon units could be
reshuffled to form more complex transcription units for multi-domain
proteins (Blake, 1983). This is in accord with the finding that
functional domains shared by different proteins are usually encoded
by more than one exon. The serine protease family and the PA genes
in particular seem to confirm this rule.

       Comparison of human uPA gene and its porcine homologue
(Nagamine et al., 1984) shows an interestingly high degree of
homology. Although the porcine gene is slightly smaller than the
human gene (6.0 and 6.4 kb respectively), the number of exons and
the position of introns, with the exception of one, is identical.
The major differences between the two proteins (79% nucleotide-
sequence homology) consist in insertions of two and nine aminoacid

residues at positions 9 and 135 in the porcine urokinase (Nagamine et al., 1984). Both insertions map close to an intron position and, as other length variations among similar genes, could have been generated by junctional sliding. The high homology found between the coding sequences of the extra nine aminoacids of porcine uPA and the 5' end of intron F of human uPA (Fig. 12) supports this

```
                          Exon VI                        Intron F

porcine u-PA   TCTGGTGGTGAGAGTCACCGGCCTGCTTATGATGGTGGGTAGAAAGGGACA
               : :::::::  :::: : :::::: :::: : :: : :::::
human u-PA     GCAGATGGTGAGCATCACTGACCTGCTGATGACAGGTGGGTGGAAGGGGAC

               Exon VI                    Intron F
```

Figure 12. Comparison between the nucleotide structure of porcine and human uPA at the exon VI-Intron F junction. Underlined sequences are present in the mature message and are translated into protein. Note that 12 bases after the translated (underlined) portion in the human sequence a termination triplet TGA is encountered, in phase (GGT.GAG.CAT.CAC.TGA).

hypothesis. Since the consensus sequences for both human and porcine splice junctions are present in both genes, a mechanism of alternative splicing cannot be excluded. Comparison of human and porcine uPA genes by the dot matrix method shows high homology in the coding sequence and the junctional regions. Considerable but lower homology is found in the noncoding sequences and in the 5'-flanking region up to at least 800 bp from the transcription initiation site. The size of the homologous introns in the two species is very close. Human uPA introns D and J contain Alu sequences, which are not present in the porcine uPA gene. The length of the non-repetitive sequences for these two introns is identical in the two species. Comparison of intron sequences show a considerable number of homologous 10-20 bp long stretches. The overall degree of divergence in intron size and sequence between human and porcine uPA genes appears to be lower than what previously observed among globin genes in man, mouse, chicken and rabbit (Efstratiadis et al., 1980) and for DHFR genes in man, mouse and chinese hamster (Yang, Masters and Attardi, 1984).

The chromosomal location of both uPA and tPA genes have been determined by analyzing DNA of human/mouse and human/hamster somatic cell hybrids and by in situ-hybridization. The uPA gene is located on the long arm of chromosome 10 (Tripputi et al., 1985) whereas tPA is located onto chromosome 8 (Tripputi, Blasi, Ny and Croce, to be published).

Restriction fragments length polymorphisms (RFLP) have become an extremely useful tool for mapping the human genome. Many RFLPs have been discovered, mostly by random screening, and mapped onto the human chromosomes. A BamH1 RFLP is linked to the human uPA gene, and is located a few hundred base pairs downstream of the polyadenylation site, within an Alu repeated sequence. The minor allele is observed in 20-30% of the population (G. Sebastio and F. Blasi, unpublished).

5. GENETIC REGULATION

A very large literature has dealt with regulation of PA synthesis and in particular of uPA. The reader is referred to the review of Danø et. al., (1985) for an extensive reference list. In this chapter, we shall simply confine ourselves to the most recent work dealing with the mechanisms underlying the stimulation of uPA synthesis by hormones, tumor-promoting phorbol esters and growth factors.

The increase in uPA synthesis by tumor promoting phorbol esters, originally described by Wigler and Weinstein (1976) has been largely confirmed by many groups in a variety of cell lines (see Danø et. al., 1985). This finding has been often related to a possible role of PAs in tumor invasiveness and metastases. In view of the fast growing knowledge on the mechanisms by which tumor promoting phorbol esters act in modifying cell growth, understanding of the molecular details of the regulation of the uPA gene may shed light on the role of uPA in cellular physiology and of how tumor promoters can interfere with it.

Phorbol esters while being not tumorigenic per se, are able to promote tumors by conventional carcinogens (Hecker, 1968). However,

quite opposed effects have been obtained on model cell-culture
systems. For instance while increasing the tumorigenic capacity of
carcinogens in the mouse skin, tumor promoters fail to act on
differentiated mouse epidermal cells (Hennings et al., 1982) or
even are able to differentiate tumor cells. For instance, phorbol
esters differentiate human leukemic cells into macrophages (Rovera
et al., 1979; Ralph et al., 1982). The most used phorbol ester, PMA
(phorbol myristate acetate), stimulates DNA replication and growth
in some cells (Lichti et al., 1978; Dicker and Rozengurt, 1978)
while inhibiting it in others (Cohen et al., 1977; Rovera et al.,
1979). Addition of PMA to mouse NIH-3T3 cells induces a strong
increase of the synthesis of myc mRNA (Kelly et al., 1983) a nuclear
oncogene protein. PMA enhances the synthesis of uPA mRNA in human
carcinoma cells (Ferraiuolo et al., 1984; Stoppelli et al., 1984),
and in mouse and human fibroblasts transformed by oncogenic viruses
(Belin et al., 1984; Grimaldi, Locatelli, and Blasi, unpublished).
In A1251 kidney carcinoma cells increase in uPA mRNA becomes
apparent as early as 30 min. after addition of PMA, reaches a
maximum 1-2 hrs thereafter and does not require protein synthesis.
Cycloheximide, in fact, superinduces uPA mRNA synthesis (Stoppelli
et al, unpublished). The level of uPA mRNA in kidney carcinoma
cells is increased about 20 fold by PMA and about 5 fold by
cycloheximide (30 min exposure).

In human carcinoma cells, epidermal growth factors has given
conflicting results on uPA mRNA synthesis. No effect has been
observed on A1251 kidney carcinoma cells, although these cells are
able to bind EGF (Stoppelli & Blasi, unpublished). On the other
hand, A431 vulva epidermoid carcinoma cells which have a very large
number of EGF receptors (Fabricant, et al., 1977) respond to 20nM
EGF by increasing both uPA mRNA (2-3 fold) and secreted uPA protein
(Stoppelli, Verde, Locatelli & Blasi, unpublished). These cells
respond much better to PMA (see below) than to EGF. However, the
basal level of uPA mRNA is much higher than in other transformed
cells, like SV40-transformed human fibroblasts. The basal level

appears to be correlated to the number of EGF receptors present. In
fact the A431 variant clone 18, which has about 10 fold less EGF
receptors than A431 (Buss et. al., 1982) does not respond to EGF and
its basal uPA mRNA and uPA protein level appears to be about 10 fold
less than in the parental A431. However, clone 18 cells still
retain the full ability to respond to PMA by increasing uPA mRNA
10-20 fold (Stoppelli et. al., unpublished). Since many tumor cells
produce EGF or EGF-like transforming growth factors (Sporn and
Todaro, 1980) the failure of EGF to induce uPA mRNA in A1251 cells
may be due to the presence of such growth factors, leading to an
already maximal stimulation of the uPA gene. These preliminary
experiments may indicate that the DNA sequences of the uPA gene
responsible for PMA and EGF induction are different.

While EGF appears to have little or no effect on uPA synthesis
in human carcinoma cells, it appears to greatly stimulate tPA and
tPA mRNA synthesis in human Hela cells (Tosic and Schleuning, 1984).
PMA also stimulates tPA mRNA synthesis in human melanoma cells
(Opdenakker et al., 1983).

In all the experiments described, it is not yet absolutely
clear whether the effect of PMA and EGF is actually on gene
transcription or rather on mRNA stability. This information is of
the utmost importance for the future investigations on the mechanism
of EGF and PMA enhancement of uPA and tPA synthesis.

6. PA AND DISEASE

A role of PA in several human diseases has been proposed,
largely on the basis of correlative evidence. In fact, the very
nature of PA enzymatic activity puts this protein in the middle of
key phonomena involving cell-cell and cell-substrates interactions.
It was first recognized that a variety of inflammatory agents could
stimulate production of PA by macrophages (Unkeless, Gordon and
Reich, 1974) and that this effect was mediated by lymphokines
(Gordon, Unkeless and Cohn, 1974; Vassalli and Reich, 1977). It was
also recognized that transformation with oncogenic viruses induced
the synthesis of PA and that this effect was transformation-

dependent (Unkeless et al., 1973; Ossowski et al., 1973). Certain features of the inflammatory reactions are similar to the invasion of normal tissue by malignant cells, like neoangiogenesis, tissue disintegration, basement membrane dissolution and naturally penetration by "foreign" cells. This similarity appears particularly marked in the chronic inflammatory lesion of rheumatoid arthritis. The way rheumatoid pannus erodes and replaces cartilage has been likened to the way malignant cells invade and destroy normal surrounding tissue (Hamilton 1983). The role of PA in tumors has been already reviewed (Reich, 1978; Mullins and Rohrlich, 1983; Danø et al., 1985). We shall not therefore cover this aspect.

Although the amount of data involving PA in diseases other than cancer is quite limited, and despite the fact that these data will have to be re-evaluated in light of the discovery of receptors (Vassalli et al., 1985; Stoppelli et al., 1985) and inhibitors (Seifert and Gelehrter, 1978), we feel that some of them are worth being summarized since they uncover some unique aspects of PA physiology and pathology.

6.1. Pemphigus

Pemphigus is a severe blistering disease of the skin and mucous membranes characterized by a rounding up of epidermal cells with loss of epidermal cell adhesion (acantholysis). It is an auto-immune disease with circulating auto-antibodies reacting with the surface of epidermal cells. An experimental system is available in which addition of pemphigus antibodies to an organ culture of human skin from a normal subject results in histologic changes identical to those seen in biopsies of the skin of affected patients (Michel and Ko, 1974). The use of this system has allowed investigators (Hashimoto et al., 1983) to demonstrate that an excess production of uPA might be the pathogenetic cause of the skin lesion. In fact, 1) uPA addition to a normal skin explant results in acantholytic lesion; 2) the establishment of acantholysis is plasminogen-dependent and is inhibited by plasmin inhibitors; 3) pemphigus epidermis produces much more uPA than normal; 4) addition

of purified pemphigus antibodies to normal skin explant results in
the induction of uPA synthesis; 5) glucocorticoids, which are known
to depress uPA synthesis (Nagamine et al., 1983) and induce PA
inhibitors synthesis (Seifert and Gelehrter, 1978), reverse the
effect of the auto-antibodies. Interestingly, glucocorticoids are
used successfully in pemphigus therapy.

The pemphigus lesion can be ascribed to an excess of plasmin,
brought about by increased synthesis of uPA due to a chronic
stimulation of the uPA gene activity by the auto-antibody. The
molecular nature of the stimulus is not known. However, many
membrane signals like peptide hormones (Lee and Weinstein, 1978;
Stoppelli et al., in preparation), antibodies (Becker, Ossowski and
Reich, 1981) have been shown to induce the synthesis of uPA. Many
such signals are transduced to the gene level using cAMP as a second
messenger (Sutherland et al., 1965) in as yet undetermined way. It
is very interesting also in this respect that the expression of uPA
genes is controlled positively by cAMP and negatively by
glucocorticoids (Nagamine et al., 1983). Pemphigus lesion is in
fact induced by a membrane signal (possibly via cAMP) and
counteracted by gluco-corticoids. However, the fact that
glucocorticoids depress the immune system, and hence also the
synthesis of auto-antibodies, must not be underestimated.

The detailed pathogenetic description of uPA role in
acantholysis is far from being understood. It is possible that in
addition to the increase in uPA synthesis, pemphigus epidermis also
experiences an unbalance among uPA and uPA inhibitor and/or between
receptor bound and free uPA. In any case pemphigus is the first
human disease where little doubt exists on a direct pathogenetic
involvement of uPA.

6.2 Xeroderma Pigmentosum

Xeroderma pigmentosum (XP) is an autosomal recessive skin
disease caused by a defect in the DNA repair mechanism. In addition
to quite characteristics skin lesions, homozygous individuals show a
marked tendency to develop skin cancers from sunlight exposure (see

detailed study of uPA in other diseases in which DNA repair is
defective, would also be very useful.

6.3 Other diseases

The involvement of PA in dermatological diseases is not limited
to pemphigus and XP. High levels of PA have been observed in
psoriatic scales (Piper et al., 1967). Psoriasis Vulgaris is a
common chronic skin disease showing hyperproliferative epidermis and
shortening of the transit time of epidermal cells with resulting
desquamation. The typical lesion shows erythematous papules which
cohalesce to form plaques with sharply demarcated irregular borders.
In most patients the disease starts within the first two decades of
life, and in about half of the cases a positive family history for
the disease is evidenced. Nothing is known of the mode of
transmission, possibly multi-factorial, nor of its pathogenesis.
The possible involvement of PA in psoriasis vulgaris has been made
more realistic by a recent experiment of Fraki, Briggaman and
Lazarus (1982). These investigators observed that the uninvolved
skin of psoriatic patients develops signs of psoriasis after being
grafted onto nude mice. This activation is characterized by a very
early increase of PA activity, even before any detectable sign of
inflammatory infiltrate in the transplanted tissue.

As mentioned before, another disease for which a direct
pathogenetic role of PA has been proposed, is rheumatoid arthritis.
In this disease a synovial pannus is in part responsible for the
destruction of joints which is often seen in the chronic cases. The
lesion is inflammatory, hyperplastic and invasive and is made up of
connective tissue, small blood vessels, hyperplastic mesenchymal
cells of the synovium, lymphocytes, plasma cells and macrophages.
The final result can be an extensive damage to the cartilage,
subchondrial bone and capsular soft tissue. The way in which the
rheumatoid synovial pannus erodes and replaces the cartilage has
been linked to the invasiveness of tumor cells that invade and
destroy normal surrounding tissues (Hamilton, 1983). Local release
of proteolytic enzymes such as plasmin and collagenase could be

Cleaver, 1983, for a review on XP). Fibroblasts from XP patients
can be induced to synthesize PA, notably uPA, by irradiation with
uv-light (Miskin and Ben-Ishai, 1981). The largest effect is
observed with XP fibroblasts of the A complementation group, i.e.
the one with the highest deficiency in DNA repair. Ultraviolet
irradiation induces uPA synthesis also in XP-heterozygotes and
normal human embryonic fibroblasts but at 10 fold higher uv
fluences. By contrast, uv irradiation does not turn on the uPA gene
in normal adult fibroblasts (Miskin and Ben-Ishai, 1981).

Induction of uPA synthesis by DNA damage can be also
accomplished in both human and murine cell lines by treatment with
reagents known to chemically modify and damage the DNA, like
hydroxyurea, intercalating or alkylating agents (Miskin and Reich,
1980).

The events following DNA damage (and more generally any
cellular damage) have been deeply studied in the microorganism
Escherichia coli (see Walker, 1984, for a review). Also in this
organism a highly specific protease, the rec A protein, is induced
upon DNA damage. This protease is able to specifically attack and
split the lex A protein, the repressor of its own synthesis and of
that of other protein required for DNA repair. The similarity
between uPA and rec A, both specific proteases, both induced by DNA
damage, however obvious, ends here. In fact no experiments have
been reported supporting or suppressing the hypothesis of uPA being
an eucariotic rec A protein (Miskin and Ben-Ishai, 1981). That uPA
is not part of a general SOS response is also shown by the lack of
induction of this gene by heat shock in normal or SV40 transformed
human fibroblasts (Grimaldi, Locatelli, Blasi, unpublished). While
a specific role for uPA in generating the pemphigus lesion is well
documented (see previous paragraph), nothing is known whether uPA
has any role in determining the XP phenotype (particularly the
sunlight dependent symptoms). A study of uPA production and
distribution in the skin of XP patients is missing and is required
before any assessment can be made of uPA role in XP. Moreover, a

responsible for some of the tissue damage. Cultured cells from the
rheumatoid tissue produce PA (Werb, Mainardi, and Harris, 1977).
Plasmin is known to be able to degrade basement membrane and
activate latent collagenase (Liotta et al., 1977; 1980), or the
enzyme cascades of complement, clotting and kynin system. It has
been shown that the conditioned medium of concanavalin A-stimulated
human peripheral blood mononuclear cells stimulates PA production as
well as reversible morphological changes when added to normal
synovial fibroblasts (Hamilton and Slywka, 1981). Similar enzymatic
and morphological effects are obtained with synovial fibroblasts
from rheumatoid patients. Modulation of PA activity of synovial
fibroblasts by mononuclear cells has been proposed to be part of the
aggressive and invasive behavior of the rheumatoid pannus.
Sonicates of the cell wall of group A streptococci, possible
etiologic agent for rheumatoid arthritis, and muramyl dipeptide
stimulate mononuclear cells to produce the substance activating
synovial fibroblasts (Hamilton et al., 1982). These findings
although provocative, do not demonstrate, however, a direct
pathogenetic involvement of PA in the disease. It will be important
to complete this initial study with a) identification of the type of
PA involved and b) a correlation with the levels of inhibitors and
specific surface receptors.

7. CONCLUSIONS

The new information on the structure of plaminogen activators
are starting to shed some light on the role of the enzyme in
cellular physiology. The reagents are now available to ask
questions in a rigorous way and to pass from correlations to proof.
The genetics approach should be very fruitful, especially if PA
genes will be isolated in other organisms (like Drosophila or yeast)
where mutants can be selected (if not already available) and their
phenotype studied. In addition, transfection of cells with normal
or modified PA genes should also answer the question of which
function are under PA control in vivo. We expect therefore, in the
near future, a large body of knowledge to be gained by such

experiments, and hence a relatively fast collection of data on the
role of PA in human diseases. Finally, the novel discoveries of
receptors and inhibitors will be certainly exploited in terms of
modulating PA function in vivo, an approach that will prove useful
not only to gain knowledge, but also to influence pathological and
physiological processes.

REFERENCES

Andreasen, P.A., Nielsen, L.S., Grøndahl-Hansen, J., Skriver, L.,
Zeuthen, J., Stephens, R.W., Danø, K. 1984. Inactive proenzyme to
tissue plasminogen activator from human melanoma cells, identified
after affinity purification with a monoclonal antibody. EMBO J. 3,
51-56.

Anson, D.S., Choo, K.H., Rees, D.J.G., Giannelli, F., Gould, K.,
Huddleston, J.A. and Brownlee, G.G. 1984. The gene structure of
human antihaemophilic factor IX. EMBO J. 3, 1053-1060.

Banyai, L., Varadi, A. and Patthy, L. 1983. Common evolutionary
origin of the fibrin-binding structures of fibronectin and
tissue-type plasminogen activator. FEBS Letters 163, 37-41.

Becker, D., Ossowski, L. and Reich, F. 1981. Induction of
plasminogen activator synthesis by antibodies. J. Exp. Med. 154,
385-396.

Belin, D., Godeau, F. and Vassalli, J.D. 1984. Tumor promoter PMA
stimulates the synthesis and secretion of mouse pro- urokinase in
MSV-transformed 3T3 cells: this is mediated by an increase in
urokinase mRNA content. EMBO J. 3, 1901-1906.

Bell, G.I., Quinto, C. Quiroga, M., Valenzuela, P., Craik, C.S.
and Rutter, W.J. 1984. Isolation and sequence of a rat
chymotrypsin B gene. J. Biol. Chem. 259, 14265-14270.

Blake, C. 1983. Exons. Present from the beginning? Nature
306, 535-537.

Buss, J.E., Kudlow, J.E., Lazar, C.S. and Gill, G.N. 1982.
Altered epidermal growth factor (EGF)-stimulated protein kinase
activity in variant A431 cells with altered growth responses to EGF.
Proc. Natl. Acad. Sci. USA, 79, 2574-2578.

Dicker, P. and Rozengurt, E. 1978. Stimulation of DNA synthesis by tumor promoter and pure mitogenic factors. Nature, 276, 723-726.

Efstratiadis, A., Posakony, J.W., Maniatis, T., Lawn, R.M., O'Connell, C., Spritz, R.A., DeRiel, J.K., Forget, B.G., Weissmann, S.M., Slightom, J.L., Blechl, A.E., Smithies, O., Baralle, F.E., Shoulders, C.C. and Proudfoot, N.J. 1980. The structure and evolution of the human beta-globin gene family. Cell 21, 653-668.

Fabricant, R.N., DeLarco, J.E. and Todoro, G.J. 1977. Nerve growth factor receptors on human melanoma cells in culture. Proc. Natl. Acad. Sci. USA 74, 565-569.

Ferlund, P. and Stenflo, J. 1982. Amino acid sequence of the light chain of bovine protein C. J. Biol. Chem. 257, 12170-12179.

Ferraiuolo, R., Stoppelli, M.P., Verde, P., Bullock, S., Lazzaro, P., Blasi, F and Pietropaolo, C. 1984. Synthesis and induction by tetradecanoyl phorbol acetate of urokinase in cultured human kidney careinoma cells. J. Cell. Physiol. 121, 368-374.

Fraki, J.E., Briggaman, R.A. and Lazarus, G.S. 1982. Uninvolved skin from psoriatic patients develops signs of involved psoriatic skin after being grafted onto nude mice. Science 215, 685-687.

Gilbert, W. 1978. Why genes in pieces? Nature 271, 501.

Gordon, S., Unkeless, J.C., Cohn, Z.A. 1974. Induction of macrophage plasminogen activator by endotoxin stimulation and phagocytosis. J. Exp. Med, 140, 995-1010.

Gross, J.L., Moscatelli, D and Rifkin, D.B. 1983. Increased capillary endothelial cell protease activity in response to angiogenic stimuli in vitro. Proc. Nat'l. Acad. Sci. USA 80, 2623-2627.

Guenzler, W.A., Steffens, G.J., Otting, F., Buse, G., Flohe, L. 1982a. Structural relationship between human high and low molecular mass urokinase. Hoppe Seyler's Z. Physiol. Chem. 363, 133-141.

Guenzler, W.A., Steffens, G.J., Otting, F., Kim, S.M., Frankus, E. and Flohe, L. 1982b. The primary structure of high molecular mass urokinase from human urine. The complete amino acid sequence of the A chain. Hoppe-Seyler's Z. Physiol. Chem., 363, 1155-1165.

Hamilton, J.A. 1983. Hypothesis: In vitro evidence for the invasive and tumor-like properties of the rheumatoid pannus. J. Rheumatol. 10, 845-851.

Hamilton, J.A. and Slywka, J. 1981. Stimulation of human synovial fibroblasts plasminogen activator production by mono nuclear cells supernatant. J. Immunol. 126, 851-855.

Hamilton, J.A., Zabriskie, J.B., Lachman, L.B. and Chen, Y. 1982. Streptococcal cell wall and synovial cell activation. Stimulation of synovial fibroblast plasminogen activator activity by monocytes treated with group A streptococcal cell wall sonicates and muramyl dipeptide. J. Exp. Med. 155, 1702-1718.

Hashimoto, K., Shafran, K.M., Webber, P.S., Lazarus, G.S. and Singer, K.H. 1983. Anti cell-surface pemphigus autoantibody stimulates plasminogen activator activity of human epidermal cells: a mechanism for the loss of epidermal cohesion and blister formation. J. Exper. Med. 157, 259-272.

Hecker, E. 1968. Carcinogenic principles from the seed oil of Croton tiglium and from other Euphorbiaceae. Cancer Res. 38, 2338-2349.

Heyneker, H.L., Holmes, W.E., and Vehar, G.A. 1983. Preparation of functional human urokinase proteins. Europ. Patent Applic. (no.-83103629.8) Publ. no. 0092182.

Hoylaerts, M., Rijken, D.G., Lijnen, N.R. and Collen, D. 1982. Kinetics of the activation of plasminogen by human tissue plasminogen activator. J. Biol. Chem. 257, 2912-2919.

Kelly, K., Cochran, B.H., Stiles, C.D. and Leder, P. 1983. Cell specific regulation of the c-myc gene by lymphocite mutagens and platelet derived growth factor. Cell 35, 603-610.

Kielberg, V., Andreasen, P.A., Grøndhal-Hansen, J., Nielsen, L.S., Skriver, L. and Danø, K. 1985. Proenzyme to urokinase-type plasminogen activator in the mouse in vivo. FEBS Letters. In press.

Kohno, T., Hopper, P., Lillquist, J.S., Suddith, R.L., Greenlee, R.E. and Moir, D.T. 1984. Kidney plasminogen activator. A precursor form of human urokinase with high fibrin affinity. Biotechnology, 2, 628-634, 1984.

Lee, L.S., Weinstein, I.B. 1978. Epidermal growth factor, like phorbol esters, induces plasminogen activator in Hela cells. Nature, 274, 696-697.

Lichti, U. Yuspa, S.M. and Hennings, H. 1978. Ornithine and
S-adenosylmethionine decarboxylases in mouse epidermal cell
cultures treated with tumor promoters. Carcinogenesis 2,
Mechanisms of tumor promotion and cocarcinogenesis, (Eds.
T.J.S. Slaga, A. Sivak and R.H. Boutwell). pp. 221-232.
Raven Press, New York.

Liotta, L.A., Goldfarb, R.H., Brundale, R., Siegal, G.P., Terranova,
V., Garbiza, S. 1981. Effect of plasminogen activator (urokinase),
plasmin and thrombin on glycoprotein and collagenous components of
basement membrane. Cancer Res. 41, 4629-4636.

Liotta, L.A., Kleinerman, J., Catanzaro, P. and Rynbrandt, D. 1977.
Degradation of basement membrane by murine tumor cells. J. Natl.
Canc. Inst. 58, 1427-1431.

Liotta, L.A., Tryggvason, K., Barbisa, S., Hart, I. and Foltz, C.M.
1980. Metastatic potential correlates with enzymatic degradation of
basement membrane collagen. Nature 284, 67-68.

Magnusson, S., Sottrup-Jensen, L., Petersen, T.E.,
Dudek-Wojciechowska, G. and Claeys, H. 1976. Homologous Kringle
structure common to plasminogen and prothrombin. Substrate
specificity of enzymes activating prothrombin and plasminogen.
Proteolysis and Physiological Regulation, (Eds. Ribbon, D.W. and
Brew), pp. 203-212 Academic Press, N.Y.

Mason, A.J., Evans, B.A., Cox, D.R., Shine, J. and Richards, R.I.
1983. Structure of the mouse kallikrein gene family suggests a role
in specific processing of biologically active peptides. Nature,
303, 300-307.

Matsuo, O., Rijken, D.C. and Collen, D. 1981. Thrombolysis by
human tissue plasminogen activator and urokinase in rabbits with
experimental pulmonary embolus. Nature 291, 590-591.

Michel, F. and Ko, C.S. 1974. Effect of pemphigus or bullous
pemphigoid sera and leukocytes on normal human skin in organ
culture: an in vitro model for the study of bullous diseases.
J. Invest. Dermatol. 62, 541-542.

Miskin, R. and Ben-Ishai, R. 1981. Induction of plasminogen
activators in normal and xeroderma pigmentosum fibroblasts. Proc.
Nat'l Acad. Sci. USA 78, 6236-6240.

Miskin, R. and Reich, E. 1980. Plasminogen Activator: Induction
of synthesis by DNA damage. Cell, 19, 217-224.

Miskin, R and Soreq, H. 1981. Microinjected Xenopus oocytes
synthesize active human plasminogen activator. Nucl. Acids Res. 9,
3355-3363.

Mullins, D.E. and Rohrlich, S.T.   1983.   The role of proteinase in cellular invasiveness.   Biochim. Biophys. Acta, 695, 177-214.

Nagamine, Y., Sudol, M. and Reich, E.   1983.   Hormonal regulation of plasminogen activator mRNA production in porcine kidney cells.   Cell 32, 1181-1190.

Nagamine, Y., Pearson, D., Altus, M.S. and Reich, E.   1984.   cDNA and gene nucleotide sequence of porcine plasminogen activator. Nucl. Acids Res. 12, 9525-9541.

Nielsen, L.S., Hansen, J.G., Skriver, L., Wilson, E.L., Kaltoft, K., Zeuthen, J. and Danø, K.   1982.   Purification of zymogens to plasminogen activator from human glioblastoma cells by affinity chromatography with monoclonal antibody.   Biochemistry (Wash.), 21, 6410-6415.

Ny, T., Elgh, F. and Lund, B.   1984.   The structure of the human tissue-type plasminogen activator gene.   Correlation of intron and exon structures to functional and structural domains.   Proc. Natl. Acad. Sci. USA., 81, 5355, 5359.

Opdenakker, G., Ashino-Fuse, H., Van Damie, J., Billiau, A. and DeSomer, P.   1983.   Effect of 12-0-tetradecanoylphorbol-13-acetate on the production of mRNAs for human tissue-type plasminogen activator.   Eur. J. Biochem. 131, 481-487.

Ossowski, L., Quigley, J.P., Kellerman, G.M. and Reich, E.   1973. Fibrinolysis associated with oncogenic transformation.   Requirement of plasminogen for correlated changes in cellular morphology, colony formation in agar and cell migration.   J. Exp. Med. 138, 1056-1064.

Ossowski, L. and Reich, E.   1983.   Antibodies to plasminogen activator inhibit human tumor metastasis.   Cell, 35, 611-619.

Ossowski, L., Unkeless, J.C., Tobia, A., Quigley, J.P., Rifkin, D.B., and Reich, E.   1973.   An enzymatic function associated with transformation of fibroblasts by oncogenic viruses.   J. Exp. Med. 137, 112-126.

Pennica, D., Holmes, W.E., Kohr, W.J., Harkins, R.N., Vehar, G.A., Ward, C.A., Bennett, W.F., Yelverton, E., Seeburg, P.H., Heyneker, H.L. Goeddel, D.V. and Collen, D.   1983.   Cloning and expression of human tissue-type plasminogen activator cDNA in E. coli.   Nature, 301, 214-221.

Piper, H.G., Hadlich, J. and Wurbach, G.   1967.   Fibrin-olytische Aktivitaet der Haut bei Psoriasis und einigen anderen Dermatosen.   Arch. Klin. Exp. Dermatol. 228, 249-258.

Quigley, J.P. 1976. Association of a protease (plasminogen activator) with a specific membrane fraction isolated from transformed cells. J. Cell. Biol. 71, 472-486.

Ralph, P., Williams, N., Moore, M.A.S. and Litcofsky, P.B. 1982. Induction of antibody-dependent and non-specific tumor Killing in human monocytic leukemia cells by nonlymphocyte factors and phorbol ester. Cell. Immunol. 71, 215-223.

Reich, E. 1978. Activation of plasminogen. A general mechanism for producing localized extracellular proteolysis. Molecular Basis of Biological Degradative Processes, Eds. Berlin R.D., Hermann, H., Lepow, I.H. and Tanzer, J.M.) pp. 155-169. Academic Press, New York.

Riccio, A., Grimaldi, G., Verde, P., Sebastio, P., Boast, S. and Blasi, F. 1985. The human urokinase-plasminogen activator gene and its promoter. Nucl. Acids Res. 13, 2759-2771.

RijKen, D.C. and Collen, D. 1981. Purification and characterization of the plasminogen activator secreted by human melanoma cells in culture. J. Biol. Chem. 256, 7035-7041.

Rovera, G., Santoli, D. and Damsky, C. 1979. Human promyelocytic leukemia cells in culture differentiate into macrophage-like cells when treated with a phorbol diester. Proc. Natl. Acad. Sci, USA. 76, 2779-2783.

Salerno, G., Verde, P., Nolli, M.L., Corti, A., Szots, H., Meo, T., Johnson, J., Bullock, S., Cassani, G. and Blasi, F. 1984. Monoclonal antibodies to human urokinase identify the single chain prourokinase precursor. Proc. Natl. Acad. Sci. USA, 81, 110-114.

Seifert, S.C. and Gelehrter, T.D. 1979. Mechanism of dexamethasone inhibition of plasminogen activator in rat hepatoma cells. Proc. Natl. Acad. Sci. USA, 75, 6130-6133.

Sheela, S., Barrett, J.C. 1982. In vitro degradation of radiolabelled, intact basement membrane mediated by cellular plasminogen activator. Carcinogenesis 3, 363-369.

Skriver, L., Nielsen, L.S., Stephens, R., and Danø. K. 1982. Plasminogen activator released as inactive proenzyme from murine cells transformed by sarcoma virus. Eur. J. Biochem. 124, 409-414.

Skriver, L., Larsson, L.I., Kielberg, V., Nielsen, L.S., Andreasen, P.B., Kristensen, P., and Danø, K. 1984. Immunocytochemical localization of urokinase-type plasminogen activator in Lewis lung carcinoma. J. Cell. Biol. 99, 752-757.

Sottrup-Jensen, L., Claeys, H. Zajdel, M., Petersen, T.E. and
Magnusson, S. 1978. The primary structure of human plasminogen:
Isolation of two lysine-binding fragments and one 'mini' plasminogen
(MW38,000) by elastase-catalyzed specific limited proteolysis.
Progress in Chemical Fibrinolysis and Thrombolysis (Eds. Davidson,
J.F., Rowan, R.M., Samama, M.M. and Desnoyers, P.C., eds.) vol. 3,
pp. 191-209, Raven Press, New York.

Sottrup-Jensen, L., Zajdel, M., Claeys, H., Petersen, T.E., and
Magnusson, S. 1975. Amino acid sequence of activation cleavage
site in plasminogen: homology with "pro" part of prothrombin.
Proc. Natl. Acad. Sci, USA, 72, 2577-2581, 1975.

Sporn, M.B. and Todaro, G.J. 1980. Autocrine Secretion and
malignant trasformation of cells. N. Engl. J. Med. 303,
878-880.

Steffens, G.J., Guenzler, W.A., Otting, F., Frankus, E., and Flohe,
L. 1982. The complete amino acid sequence of low molecular mass
urokinase from human urine. Hoppe-Seyler's Z. Physiol. Chem., 363,
1043-1058.

Stoppelli, M.P., Verde, P., Galeffi, P. Locatelli, E.K. and
Blasi, F. 1984. Regulation of the urokinase mRNA synthesis
by tumor promoters and epidermal growth factor. Peptide
Hormones, Biomembranes and Cell Growth (Eds. Bolis, G.C.,
Frati, L. and Verna, R.) pp. 245-252, Plenum Press.

Stoppelli, M.P., Corti, A., Soffientini, A., Cassani, G., Blasi, F.
and Assoian, R. 1985. Differentiation-enhanced binding of the
amino-terminal fragment of human urokinase plasminogen activator to
a specific receptor on U937 monocytes. Proc. Natl. Acad. Sci, USA,
in press.

Strassburger, W., Wollmer, A., Pitts, J.E., Glover, I.D.,
Tickle, I.J., Blundell, T.L., Steffens, G.J., Guenzler, W.A.,
Otting, F. and Flohe, L. 1983. Adaptation of plasminogen
activator sequences to known protease structures. FEBS
Letters. 157, 219-223.

Sumi, H. and Robbins, K.C. 1983. A functionally active heavy chain
derived from human high molecular weight urokinase. J. Biol. Chem.
258, 8014-8019.

Sutherland, E.W., Øye, I. and Butcher, R.W. 1965. The action of
epinephrine and the role of the adenylcyclase system in hormone
action. Rec. Progr. Horm. Res. 21, 623-642.

Swift, G.H., Craik, C.S., Stary, S.J., Quinto, C., Lahaie, R.G.,
Rutter, W.J. and McDonald, R.J. 1984. Structure of two related
elastase genes expressed in the rat pancreas. J. Biol. Chem. 259,
14271-14278.

Tripputi, P., Blasi, F., Verde., P., Cannizzaro, L.A., Emanuel, B.S. and Croce, C.M. 1985. Human urokinase gene is located on the long arm of chromosome 10. Proc. Natl. Acad. Sci. USA, 82, in press.

Unkeless, J.C., Gordon, S. and Reich, E. 1974. Secretion of plasminogen activator by stimulated macrophages. J. Exp. Med. 139, 834-850.

Unkeless, J.C., Tobia, A., Ossowski, L., Quigley, J.P., Rifkin, D.B. and Reich, E. 1973. An enzymatic function associated with transformation of fibroblasts by oncogenic viruses. J. Exp. Med. 137.

Valinsky, J.E., Reich, E. and LeDouarin, N.M. 1981. Plasminogen activator in the bursa of Fabricius. Correlation with morphogenetic remodeling and all migration. Cell. 25, 471-476.

Van de Werf, F., Ludbrook, P.A., Bergman, S.R., Tiefenbrunn, A.J., Fox K.A.A., DeGeest, H., Verstraete, M., Collen, D. and Sobel, B.E. 1984. Coronary thrombolysis with tissue-type plasminogen activator in patients with evolving myocardial infarction. N. Engl. J. Med. 310, 609-613.

Vassalli, J.D., Baccino, D., and Belin, D. 1985. A cellular binding site for the Mr55,000 form of the human plasminogen activator, urokinase. J. Cell. Biol. 100, 86-92.

Vassalli, J.D., Reich, E. 1977. Macrophage plasminogen activator: Induction by products of activated lymphoid cells. J. Exp. Med. 145, 429-437.

Verde, P., Stoppelli, M.P. Galeffi, P., DiNocera, P.P. and Blasi, F. 1984. Identification and primary sequence of an unspliced human urokinase poly-A+ RNA. Proc. Natl. Acad. Sci. USA, 81, 4727-4731.

Yang, J.K., Masters, J.N. and Attardi, G. 1984. Human dihydrofolate reductase gene organization. Extensive conservation of the G+C-rich 5' non coding sequences and strong intron size divergence from homologous mammalian genes. J. Mol. Biol. 176. 169-187.

Walker, G.C. 1984. Mutagenesis and inducible responses to deoxyribonucleic acid damage in Escherichia coli. Microbiology Rev. 48 60-93.

Walker, E.K. and Schleuning, W.D. 1985. Induction of fibrinolytic activity in HeLa cells by phorbol myristate acetate. Tissue-type plasminogen activator antigen and mRNA augmentation require intermediate protein biosynthesis. J. Biol. Chem. 260, 6354-6360.

Werb, Z., Mainardi, C.L., Vater, C.A. and Harris, E.D., Jr. 1977.
Endogenous activation of latent collagenase by rheumatoid synovial
cells. N. Engl. J. Med. 296, 1017-1023.

Wigler, M. and Weinstein, I.B. 1976. Tumor promoter induces
plasminogen activator. Nature, 259, 232-233.

Wiman, B. and Collen, D. 1978. Molecular mechanism of
physiological fibrinolysis. Nature, 272, 549-550.

Wun, T., Ossowski, L. and Reich, E. 1982. A proenzyme form of
human urokinase. J. Biol. Chem. 257, 7262-7268.

Human Genes and Diseases
Edited by F. Blasi
© 1986, John Wiley & Sons, Ltd.

IN VITRO TRANSFORMATION OF EPITHELIAL CELL BY ACUTE RETROVIRUSES

Giancarlo Vecchio, Pier Paolo Di Fiore, Alfredo Fusco, and
Giulia Colletta

Bernard E. Weissman and Stuart A. Aaronson

I. INTRODUCTION

In vitro cell transformation by oncogenic viruses, especially
by acute transforming retroviruses, has been extensively studied
by using fibroblastic cell systems. Although these studies have
greatly contributed to the understanding of problems related to
growth regulation in transformed cells, fibroblasts do not represent
the most suitable cell system to study questions related to the
relationships between cell transformation and differentiation. More
enlightening for this purpose have been the numerous studies performed with acute retroviruses and other cells of mesenchymal
origin, such as the cells of hematopoietic derivation (Graf and
Beug, 1978; Moscovici and Gazzolo, 1982; Beug et al., 1982). The
general picture which has emerged from the latter kind of studies is
that acute transforming retroviruses are capable of conferring the
in vitro malignant phenotype to cells of various hematopoietic
origins, while interfering with the differentiated pathway of the
particular cell type infected by blocking the expression of some
specific phenotypic markers (Graf and Beug, 1978; Beug et al.,
1982). The interpretation of such phenomena is that the cells
which have been the target of transformation induced by acute retro-

viruses have been hit at a particular stage of their differentiation
pathway and therefore express only the phenotypic markers which were
present prior to the infectious event and therefore prior to the
acquisition of the malignant properties.

Until recently, it was impossible to study the interrelationships
between malignant transformation and epithelial cell differentiation,
simply because of the lack of defined cell systems of epithelial
origin which maintain their specific differentiated parameters in
vitro. Such systems are indeed very valuable not only to study the
relationships between transformation and differentiation, but also
as general models for studying epithelial cell carcinogenesis, which
represents the most common type of carcinogenesis in humans.

Studies of epithelial cell transformation by acute retroviruses
have become even more interesting since the demonstration that
cellular homologs of acute transforming retrovirus oncogenes are
possibly involved in the neoplastic transformation of several human
epithelial cell histiotypes (for a review see Aaronson and Tronick,
1985). Also, in the light of these latter results and the lack
of a system of epithelial cells susceptible to transfection analysis
such as the NIH/3T3 cells, epithelial cell transformation by acute
retroviruses represents a good system for the evaluation of the
oncogenic potential, for epithelial cells, of the available oncogenes. Finally, since approximately 80 percent of human neoplasms
are carcinomas, i.e., malignant tumors originating from epithelial
cells, successful models to study in vitro the steps involved in
epithelial cell carcinogenesis are extremely useful.

The present review will briefly describe the results obtained by transformation of some epithelial cell systems by DNA tumor viruses, in consideration of the close similarities between DNA and RNA tumor viruses as transforming agents. In this context, the epithelial systems used will be also described and the main features of each system briefly analyzed. The review will then focus on studies of epithelial cell transformation obtained with RNA tumor viruses, emphasizing two systems in particular which have been studied recently in the Authors' laboratories and which may represent those best suited to study the two intimately related phenomena delineated above, i.e., the mechanisms of growth and differentiation in tissues that give rise to the majority of human malignant neoplasms.

II. TRANSFORMATION OF EPITHELIAL CELLS BY DNA TUMOR VIRUSES

Attempts to transform cells of epithelial derivation by DNA-containing tumor viruses have been tried numerous times using different epithelial cell systems as well as different DNA tumor viruses. Table I gives a summary of the systems used, as well as of the virus strains. Only some of the studies performed will be cited here in order to extrapolate general concepts from simple experimental approaches.

A. Studies with explant cultures. Most of the early studies performed utilized explant or primary cultures rather than continuous cell lines. All of those studies, therefore, suffered from the drawback of using mixtures of cell types rather than isolated epithelial cells. The possibility existed, therefore, that the

TABLE I.　TRANSFORMATION OF EPITHELIAL CELLS BY DNA TUMOR VIRUSES

Cell Type	Species	Virus Used	Differentiation Marker Examined	References
Thyroid gland	Hamster	SV40, Adeno-SV40 (LLE46)	None	Wells et al. (1966a)
Salivary gland	Hamster	Polyoma, SV40 Adenovirus, LLE46	Amylase	Wells et al. (1966b)
Pineal gland	Hamster	Polyoma, SV40 Adenovirus, LLE46	HIOMT (pineal enzyme)	Wells et al. (1966d) Orme et al. (1968)
Prostate gland	Hamster	SV40	Acid phosphatase	Paulson et al. (1968a) Paulson et al. (1968b)
Lens	Hamster	SV40	None	Albert et al. (1969)
Uvea, retina	Hamster	SV40, LLE46	None	Albert et al. (1968)
Lung	Hamster	SV40	None	Diamondopoulos and Enders (1965)
Liver	Hamster	SV40	None	Ibid
Kidney	Hamster	SV40	None	Rabson and Kirchstein (1962)
		Adeno-SV40 (LLE46)	None	Rabson et al. (1966)

Cell Type	Species	Virus Used	Differentiation Marker Examined	References
Salivary gland	Mouse	Polyoma	None	Dawe and Law (1959)
Mammary gland	Mouse	SV40	None	Anderson and Smith (1980) Butel et al. (1984)
Liver	Mouse	SV40	None	Anderson and Smith (1980)
Liver (Hepatocytes)	Rat	Adenovirus	None	Paraskeva and Gallimore (1980)
		SV40	α-fetoprotein albumin transferrin	Chou and Schlegl-Haueter (1981)
		SV40	Albumin α-fetoprotein tyrosine-amino-transferase	Isom et al. (1980); Isom et al. (1981)
Urothelium	Bovine	SV40	None	Elliott et al. (1974)
Lens epithelium	Bovine	SV40	α and γ crystallin	Weinstein et al. (1982)
Kidney	Monkey	Adenovirus 7 SV40	None	Wertz et. al. (1965)
		Epstein-Barr	None	Griffin and Karran (1984)
Thyroid gland	Human	SV40	Prostaglandins Calcitonin	Grimley et al. (1969) Rabson et al. (1962)

Cell Type	Species	Virus Used	Differentiation Marker Examined	References
Parathyroid gland	Human	SV40	Parathyroid Hormone	Deftos et al. (1968)
Keratinocytes	Human	SV40	Keratins Involucrin	Steinberg and Defendi (1979); Taylor-Papadimitriou et al. (1984) Banks-Schlegel and Howley (1983); Bernard et al. (1985)
Colon cells	Human	SV40	None	Moyer and Aust (1984)
Prostate gland	Human	SV40	None	Ohnuchi et al. (1982) Kaighn et al. (1980)
Amniotic fluid	Human	SV40	None	Walen (1982)
Breast	Human	SV40	Membrane Antigen	Chang et al. (1982)
Placenta	Human	SV40	Alkaline phosphatase, Human chorionic gonadotropin	Chou, (1978)
Retinal pigment epithelium	Human	SV40	None	Aronson, (1983)

cells which had been infected by the viruses used were not of
epithelial but of fibroblastic origin. The implication that some
of the cells originally transformed by DNA tumor viruses were indeed
of fibroblastic origin was obtained when the transformed cells in
some instances induced fibrosarcomas, not tumors of epithelial
origin, upon injection into syngeneic animals. This was true for
transformation of hamster pineal (Wells et al., 1966d; Orme et al.,
1968), as well as salivary glands (Wells et al., 1966b); for hamster
liver and lung cells (Diamandopoulos and Enders, 1965); for hamster
prostate gland cells (Paulson et al., 1968a; Paulson et al., 1968b)
and partially so for transformation of hamster thyroid tissue (Wells
et al., 1966a). In some instances the tumors induced have been
described as mixed tumors of epithelial and fibroblastic origin
(Wells et al., 1966a; Wells et al., 1966c). Only in a few cases
did the nature of the tumors arising in syngeneic animals indeed
appear to be epithelial (Rabson and Kirschstein, 1962; Rabson et
al., 1966). However, in many of the studies performed, the
epithelial nature of the transformed cells was not ascertained by
injecting the cells into syngeneic animals and looking at the
histological appearance of the tumors. For some of the reports,
therefore, the epithelial nature of the cells used for transformation
was based on different parameters. The best criterium for establishing the epithelial nature of a defined cell type is the availability
of specific markers for the particular epithelial cell studied. The
more specific the markers of epithelial differentiation, the more
reliable are the criteria used for identifying transformed cells
as truly epithelial. Specific markers of epithelial cells were

retained after transformation with SV40 of hamster pineal cells
(Wells et al., 1966d), hamster prostate tissue (Paulson et al.,
1968) and human parathyroid cells (Deftos et al., 1968).

The availability of specific epithelial differentiation markers
makes it also possible to investigate the interrelationships between
viral transformation and expression of the specific differentiated
phenotype of the particular epithelial cell studied. Such a possibility has been thoroughly exploited in the two types of epithelial
cells transformed by RNA tumor viruses on which the emphasis of this
review will be placed (see paragraph IV D - F). From early studies
performed with explant cultures, therefore, it appears that cells of
epithelial derivation can indeed be transformed by DNA tumor viruses
in vitro and, in the cases where biochemical parameters of specific
differentiation were available, transformation by viruses did not
seem to appreciably affect the expression of the differentiated
phenotype. In light of recent theories which distinguish between
"immortalization" of cells in culture and "full transformation", one
might wonder whether the maintenance of the differentiated phenotype
could be ascribed to simple immortalization (in the absence of true
transformation) of some of the cytotypes outlined above. This kind
of speculation seems particularly reasonable in those cases where
the transformed state was not tested by stringent criteria like
growth in soft agar or tumorigenicity in animals, as in the case of
human parathyroid cells infected with SV40 (Deftos et al. 1968). As
far as the viruses are concerned, another generalization which can
be drawn from these studies is that SV40 and the SV40-adenovirus
hybrid (LLE 46 strain) had the greatest oncogenic potential for

epithelial cell types, whereas polyoma and adenovirus had little
oncogenic potential for the same epithelial histiotypes.

B. Studies with pure epithelial cells. With the advent of
more sophisticated techniques for isolating purely epithelial cell
cultures, studies of viral transformation have acquired new interest
and have provided more accurate information. Among the best systems
studied are the ones utilizing human keratinocytes (Steinberg and
Defendi, 1979 and 1983; Defendi et al., 1982; Taylor-Papadimitriou
et al., 1982; Banks-Schlegel and Howley, 1983; Bernard et al.,
1985), human colon cells (Moyer and Aust, 1984), rat liver cells
(Paraskeva and Gallimore, 1980; Paraskeva et al., 1982; Isom et al.,
1980; Chou and Schlegel-Haueter, 1981), human placental cells (Chou,
1978), human breast cells (Chang et al., 1982), mouse mammary gland
cells (Butel et al., 1980), human prostate cells (Ohnuchi et al.,
1982; Kaighn et al., 1980), human amniotic cells (Walen, 1982) and
bovine lens epithelial cells (Weinstein et al., 1982). In particular, human keratinocytes, bovine lens epithelial cells and rat
hepatocytes are among the most suitable systems to study the interrelationships between transformation and expression of the differentiated phenotype. All three cell types, in fact, have several
well defined and biochemically characterized markers of differentiation, such as the family of the keratin proteins and involucrin
for keratinocytes, the lens proteins for lens cells and various
plasma proteins, such as α-fetoprotein, albumin and transferrin
for the liver cells.

The results obtained with these different cell histiotypes, all
transformed by SV40, are paradigmatically different. All three cell

types were phenotypically transformed by SV40 virus, as assayed by several criteria for the acquisition of the transformed phenotype. However, the transformed keratinocytes lost the ability to synthesize involucrin (Bernard et al., 1985) and two forms of keratins which are typical markers of epithelial stratification (the 56 and 58 KD keratins), whereas they retained the capability of synthesizing other forms of keratins (the ones migrating into the 48-52 KD range), which are probably associated with an earlier stage of differentiation (Okada et al., 1984). Furthermore, human keratinocytes transformed by DNA from a temperature sensitive mutant of SV40 expressed decreased amounts of keratins at the temperature permissive for transformation (33°) and the differentiated functions were restored at the non-permissive temperature (39°) (Banks-Schlegel and Howley, 1983). In contrast, transformed bovine lens epithelial cells retained the capacity, possessed by their normal counterparts, of synthesizing α and γ-crystallin (Weinstein et al., 1982). The results obtained by infecting rat liver cells with the wild type SV40 (Georgoff et al., 1984; Isom and Georgoff, 1984) showed that albumin synthesis was not affected by transformation; however, fetal rat liver cells infected by a temperature sensitive mutant (ts) of SV40 (Chou and Schlegel-Haueter, 1981) decreased their synthesis of albumin, transferrin and α-fetoprotein at the temperature permissive for transformation (33°). At the nonpermissive temperature (40°), the synthesis of the three proteins was reestablished while the cells lost the transformed phenotype (Chou and Schlegel-Haueter, 1981).

Most of the other studies performed with homogeneous isolated populations of epithelial cells did not place the emphasis on studying the relationship between transformation and differentiation, as much as in showing that SV40 exhibited oncogenic potential on cells of well-defined epithelial derivation.

III. PRESENCE OF CELLULAR HOMOLOGS OF RETROVIRAL ONCOGENES IN HUMAN AND EXPERIMENTAL TUMORS OF EPITHELIAL ORIGIN

Acute transforming retroviruses have arisen from recombination events occurred between the nontransforming retrovirus and the host cellular genome; they have, in fact, substituted viral genes necessary for replication with discrete segments of host genetic information. Upon transduction into the retroviral genome, such sequences derived from the normal cellular genes (proto-oncogenes) acquire the ability to induce neoplastic transformation (thus being defined as oncogenes). It has also been demonstrated that several virus isolates, from the same or even different species, have transduced the very same onc-sequence. This implies that the number of proto-oncogenes that can acquire transforming properties when transduced into the viral genome must be somehow limited.

Recently, a number of studies have indicated that certain human tumors (among which many of epithelial origin) contain dominant transforming genes capable of conferring the malignant phenotype to NIH/3T3 cells in vitro, in the DNA transfection assay (Cooper, 1982; Aaronson and Tronick, 1985). Most of the "human oncogenes" represent activated (in terms of malignant properties) alleles of the normal cellular counterparts (proto-oncogenes). So far, the majority of

tranforming genes detected in human neoplasia were found activated
by point mutations independently of any retroviral involvement and
related to a small group of retroviral oncogenes designated ras.

A general model can thus be envisioned in which the activation of normal cellular proto-oncogenes, either by transduction
into a retroviral host or by other mechanisms, such as point mutation, leads to the generation of transforming sequences (oncogenes)
capable of contributing to the malignant phenotype. Oncogenes have
been isolated so far from a variety of human malignancies, including
those of the hemopoietic lineage as well as sarcomas and carcinomas.
In particular, oncogenes have been isolated from many epithelial
malignancies including spontaneous carcinomas of ovary (Feig et al.,
1984), lung and pancreas (Pulciani et al., 1982a), colon carcinomas
(Pulciani et al., 1982a; McCoy et al., 1983), urinary tract (Fujita
et al., 1984) and from a variety of cell lines established from
carcinomas of the colon (McCoy et al., 1983), lung (Pulciani et al.,
1982b), urinary bladder (Pulciani et al., 1982b) and mammary gland
(Kraus et al., 1984). The frequency of detection of such oncogenes
from human tumors has ranged from 10-20% of the total number of
specimens tested. In a study conducted in one of our laboratories,
carcinomas were positive in 10-30% of the cases (see Table II).

It is critically important to establish the relevance of an
oncogene activated in a human tumor to the neoplastic process. At
the present this kind of evidence is still elusive. In the absence
of a direct approach linking oncogenes to the induction of epithelial
tumors, we think that infection of differentiated epithelial cells
with retroviruses bearing oncogenes belonging to different families

TABLE II. PRESENCE OF ACTIVATED ras ONCOGENES IN HUMAN TUMOURS

Type of Tumour	Percent positive	ras Oncogene Actived		
		H-ras	K-ras	N-ras
Carcinomas[a]	10-30	4/12	6/12	2/12
Sarcomas[b]	10	0/2	0/2	2/2
Haemotopoietic[c]	10-50	0/9	1/9	8/9

[a]Including lung, gastrointestinal and genitourinary carcinomas

[b]Including fibrosarcoma and rhabdomyosarcoma

[c]Including AML, CML, ALL, CLL.

is a unique tool to study both the interference of oncogenes in the differentiation program and the role that such genes play in epithelial cell transformation.

IV. TRANSFORMATION OF EPITHELIAL CELL SYSTEMS IN VITRO BY ACUTE RETROVIRUSES

A. Interaction of retroviruses with epithelial cytotypes in vitro. It is still widely accepted that fibroblasts and hematopoietic cells are the main targets for transformation by acute retroviruses. This conviction is based on the fact that in vivo retroviruses appear capable (with few exceptions) of inducing only malignancies of the mesenchymal cytotypes (for a complete review see Weiss et al., 1982). However, epithelial cells in vitro can indeed be target for acute transforming retroviruses. So far many different epithelial cytotypes have been successfully transformed in vitro by retroviruses. These include rat kidney epithelial cells (Ikawa et al., 1970), rat liver cells (Rhim et al., 1977; Ikawa,

1975; Altaner and Hlavayova, 1973), rat adrenocortical cells
(Auersperg, 1978; Auersperg and Calderwood, 1984, Auersperg et al.,
1977 & 1981), neuroepithelial cells from chick embryo lens (Jones et
al., 1981; Ephrussi and Temin, 1960) and rat lens (Trevithick et al.,
1979) rat thyroid cells (Fusco et al., 1981 & 1982; Colletta et al.,
1983; Ferrentino et al., 1984), mouse (Weissman and Aaronson, 1983;
Yuspa et al.,1983) or human (Rhim et al., 1985) keratinocytes, rat
ovarian surface epithelial cells (Adams and Auersperg, 1981), rat
ovarian granulosa cells (Harrison and Auersperg, 1981) and human
tracheobronchial cells (Yoakum et al., 1985) (see also Table III).

Some of these studies have suffered from the same criticism
of the earlier studies with DNA viruses in that markers of
epithelial cells were either not available or not utilized to
conclusively identify the transformed cells, nonetheless the
overwhelming weight of evidence (Table III) indicates that retroviruses can interact (at least in vitro) with epithelial cells,
causing alterations in their growth properties even including
malignant transformation. Possible explanations of the restricted
sarcomagenic/leukemogenic capability of these viruses in vivo may
be envisioned. One can speculate about the lower susceptibility
of the epithelial cells to the transforming event or, alternatively,
that the more rapidly growing sarcomas and/or leukemias can mask
in vivo initial epithelial malignant growths.

B. Relationships between transformed and differentiated
phenotype in transformed epithelial cells. Epithelial cells have
provided an excellent model system to study in vitro the relationships between the transformed and the differentiated phenotype,

since they display in vitro, in most of the cases, the differentiated
properties of the organs of origin in vivo. Moreover, it is known
that several epithelial neoplasias in vivo show alteration in the
maturation pathway and defects frequently occur in the differentiation of carcinoma cells grown in culture (Rheinwald and Beckett,
1980; Stanbridge et al., 1982). In addition, since many retroviral
oncogenes, analyzed in hematopoietic culture systems, altered differentiation, their use in epithelial cell types may lead to a
better understanding of the control of normal development and differentiation in epithelial cells.

The overall picture emerging from the available studies is that
transformation blocks the expression of the differentiated properties of epithelial cells. This conclusion is in close similarity
with that found in haematopoietic systems (Graf and Beug, 1978; Beug
et al., 1982). This is true for mouse keratinocytes (Weissman and
Aaronson, 1983), rat thyroid cells (Fusco et al., 1982; Colletta et
al., 1983), rat adrenocortical cells (Auersperg, 1978; Auersperg et
al., 1981), chicken epithelial cells (Jones et al., 1981) and human
tracheobronchial cells (Yoakum et al., 1985). We shall discuss
thoroughly the first two systems in a following section; here we
shall discuss briefly some of the properties of the three latter
systems with particular emphasis on the relationships between the
transformed and differentiated phenotypes.

Rat adrenocortical cell lines, in primary culture, grow in
vitro in two states of differentiation depending on culture conditions. When grown in 25% foetal bovine serum (FBS) they assume a
fibroblast-like morphology, although still retaining a limited

Table III: TRANSFORMATION OF EPITHELIAL CELLS BY RNA TUMOR VIRUSES

Cell Type	Type of Culture	Species	Virus Used	Differentiation Markers	Reference
Kidney	Continuous Cloned	Rat	Moloney MSV	None	Ikawa et al. (1970)
Liver	Primary	Rat	Kirsten MSV	None	Ikawa et al. (1975)
	Primary	Rat	B77ASV and Harvey MSV	None	Altaner and Hlavayova (1973)
	Primary	Rat	Kirsten MSV	None	Rhim et al. (1977)
Adrenocortex	Primary	Rat	Kirsten MSV	Enzymes of the steroid metabolic pathway, Production of tissue-specific hormones	Auersperg (1978) Auersperg et al. (1981)
Ovary (surface cells)	Primary	Rat	Kirsten MSV	17β-hydroxysteroid dehydrogenase	Adams and Auersperg (1981)
Ovary (granulosa cells)	Primary	Rat	Kirsten MSV	None	Harrison and Auersperg (1981)
Lens	Primary	Rat	Rous ASV	α,β,γ, crystallins	Trevithick et al. (1979)
	Primary	Chicken	Rous ASV	δ crystallin	Jones et al. (1981)

Cell Type	Type of Culture	Species	Virus Used	Differentiation Markers	Reference
Tracheo-bronchial cells	Primary	Human	v-Harvey oncogene[a]	Squamous differentiation upon TPA treatment	Yoakum et al. (1985)
Thyroid	Continuous cloned	Rat	Kirsten MSV Myeloproliferative SV	Thyroglobulin synthesis iodide uptake TSH dependence for growth	Fusco et al. (1982) Colletta et al. (1983)
Keratinocyte	Continuous cloned	Mouse	Kirsten MSV BALB MSV	Terminal differentiation upon Ca++ treatment	Weissman and Aaronson (1983)
	Primary	Mouse	Kirsten MSV	Terminal differentiation upon Ca++ treatment	Yuspa et al. (1983)
	Primary	Human	Ad-SV40 + Kirsten MSV	None	Rhim et al. (1985)

[a] v-Ha-ras was introduced into the cells by transfection.

expression of the adrenocortical differentiated functions; on the
other hand they display a typical epithelial morphology and retain
a more differentiated phenotype when cultured in 3-10% horse serum
(HS).

Adrenocortical cells in vitro retain many of the specific
markers of the glandular function in vivo, although with some
variations. Basically, either the fibroblast-like (FL) or the
epithelial-like cells (EL) are capable of accumulating lipids and
are positive for tissue-specific dehydrogenases like Δ5-3β hydroxy-
steroid dehydrogenase. Both of these activities are higher in the
morphologically more differentiated cytotype.

In addition these cells are capable of steroid metabolism, as
evidenced by the conversion of pregnenolone to 20α-hydroxypregnen-
olone. The most striking feature of adrenocortical differentiation
is, however, represented by the response to the adrenocorticotrophic
hormone (ACTH), the physiological stimulator of the gland in vivo.
Under ACTH stimulation, FL cells can metabolize pregnenolone to
11-deoxycorticosterone and 20α-dihydroprogesterone, in addition
to 20α-dihydroxypregnenolone. These two products are already
synthesised by EL cells in the unstimulated condition, but upon ACTH
stimulation EL cells can also produce corticosterone, which is a
tissue-specific hormone.

A continuous line of transformed rat adrenocortical cells
(TRA) has been successfully established following infection with the
Kirsten murine sarcoma virus (KiMSV). Interestingly, the transformation event eliminated the differential response of the cells to
the culture variables (HS/FBS) and resulted in similar growth rates

and high cellular density in all conditions. Transformation resulted also in a reduction of the differentiated phenotype: TRA, in
fact, did not produce deoxycorticosterone anymore nor responded
to culture conditions by qualitative changes in pregnenolone metabolism. Thus, it appears that transformation rendered these cells
unresponsive to the factors in the culture environment that cause
the modulation from the FL form to the more differentiated and
steroid secreting EL form.

Studies conducted on primary cultures of lens epithelial cells
infected with Rous sarcoma virus temperature-sensitive mutant for
transformation (ts RSV) seem to confirm the general pattern of in-
compatibility between the transformed state and the expression of
the differentiated phenotype (Jones et al., 1981). When uninfected
primary chicken lens cells are cultivated, they show a flattened
epithelial morphology and form "lentoid bodies" a characteristic
morphological feature of in vitro lens differentiation. Moreover,
they synthesize low amounts of δ crystallin, an absolutely specific
differentiation marker.

After infection with ts RSV, at the temperature permissive for
transformation, the cells start to grow faster. They become morphologically transformed, assuming a fibroblast-like morphology, and
"lentoid bodies" are no longer observable. In addition, the expression of the differentiated protein δ crystallin is lost. Since
also the specific mRNA for δ crystallin is no longer detectable, in
the transformed cells, it is likely that the block in the expression
of this differentiated function resides at the transcriptional
level. These results are similar to those reported for trans-

formed rat lens epithelial cells (Trevithick et al., 1979), where
α, β, and γ crystallin are no longer expressed after transformation with wild type Rous sarcoma virus.

Interestingly enough, in the case of transformed chicken lens
epithelial cells, the shift of the cells from the permissive to the
non-permissive temperature for tranformation results morphologically
in the reappearance of the "normal" epithelial phenotype. However,
the differentiation marker crystallin remains unexpressed, thus
suggesting an irreversible loss of at least one differentiated
function after retroviral transformation.

Recently, data on transformation and differentiation in epithelial cells have become available also in human systems in vitro,
particularly in the case of normal human bronchial epithelial cells
(NHBE) (Yoakum et al., 1985). When NHBE in primary culture are
cultivated in serum free medium (Lechner, 1982; Lechner, 1983) they
form an homogenous population which displays no sign of squamous
differentiation. When the same cells are cultivated in the presence
of small amounts (as low as 2%) of blood-derived serum or 12-0-
tetradecanoylphorbol-13acetate (TPA), they are capable of undergoing
terminal squamous cell differentiation.

Transfection of the v-H-ras oncogene in these cells, by means
of protoplast fusion, eventually results in the appearance of the
transformed phenotype as evidenced by the capability of growing as a
continuous cell line, growth in soft agar, and tumorigenicity upon
injection into athymic nude mice. Transformation is accompanied by
the loss of the capability to undergo terminal differentiation when
the cells are exposed to blood-derived serum or TPA (Yoakum et al.,

1985).

An exception to the general scheme outlined above (incompati-
bility between the transformed and the differentiated state) is,
however, represented by the transformation of rat ovarian surface
epithelial cells by the Kirsten MSV. In this case,the cells retained, following transformation, the expression of 17 β-hydroxysteroid
dehydrogenase activity, which the authors claimed to be a specific
differentiation marker for those cells (Adams and Auersperg, 1981).

C. Effects of oncogenes on homogeneous epithelial cell populations possessing differentiated functions and specific hormonal
and/or growth factor requirements. In recent years, it has become
apparent that a number of oncogenes encode proteins that interact
in the pathways by which growth factors stimulate normal cellular
proliferation (Waterfield et al., 1983; Doolittle et al., 1983;
Downward et al., 1984; Scherr et al., 1985). Optimal systems for
studying the effects of such genes on epithelial cells might involve
cell types which exhibit specific hormone or growth factor requirements as well as differentiated functions. In such systems, a given
oncogene might be analyzed for its ability both to alter differentiation and complement or abrogate growth factor requirements.

Our laboratories have, over the past several years, investigated
the interactions of oncogenes with two homogenous populations of
epithelial cells possessing specific markers of differentiation,
as well as, well-defined hormonal and growth factor requirements.
These systems, a rat thyroid and a mouse keratinocyte clonal-derived
cell lines, are representative of the two major types of epithelia,
the glandular and the lining epithelium, respectively. Besides pos-

sessing different functions, these epithelial cells differ substantially as far as growth potential and differentiation are concerned.
Lining epithelia show in vivo a typical kinetic behaviour consisting
of a continuous flux of cells from an undifferentiated highly replicative compartment (committed progenitors or stem cells) to other
compartments where the acquisition of a more differentiated phenotype
appears directly associated with a decreased proliferative capacity.
In contrast, glandular epithelia, which normally consist of highly
differentiated, quiescent cells, may, under appropriate stimuli,
undergo replication while retaining differentiated properties.
The fact that thyroid cells represent endodermal epithelium, while
keratinocytes, are of ectodermal origin, has further made it possible to study the interaction of retroviral transforming genes on a
broader spectrum of target cells than the previously most studied
targets of mesodermal derivations (Auersperg et al., 1977 and 1981;
Adams and Auersperg, 1981).

D. The rat thyroid epithelial cell system. Mammalian thyroid
glands consist of two main types of epithelial cells morphologically
and functionally different: follicular and parafollicular cells.
Follicular cells represent the vast majority of the thyroid tissue
and are under the control of the pituitary thyrotropic hormone
(TSH). They are deputed to the synthesis and secretion of the
thyroid hormones, thyroxine and triiodothyronine. Parafollicular
cells do not respond to the action of TSH and secrete a calcium
regulating hormone, calcitonin. Both types of cells are normally
nondividing (discontinuous replicators) even though they may in
pathological conditions, undergo replication. Efforts to grow

thyroid cells in culture led to the development of continuous cell lines of follicular origin (Ambesi-Impiombato and Coon, 1979). This continuous Fischer rat-derived cell line, FRT, was shown to be free of fibroblasts, possessed a typical epithelial morphology and grew relatively well without specific nutritional requirements. The only evidence that this cell line represented bona fide thyroid follicular epithelium was the specific binding by their plasma membranes of the thyrotropic hormone, the physiological thyroid stimulator (Yavin et al., 1981). However, this particular cell line did not exhibit either of the two typical markers of thyroid epithelium, i.e., synthesis of thyroglobulin, the glycoprotein within which the thyroid hormones thyroxine and thriiodothyronine are synthesized, as well as the capacity to concentrate iodide from the nutrient medium against a concentration gradient.

Since both of these properties are peculiar to thyroid cells, together with the presence of a membrane receptor for TSH, it seemed valuable to isolate cell variants which possessed each of these properties of thyroid epithelium.

In a further study, Ambesi-Impiombato et al. (1980) were able to isolate colonies and subsequently establish continuous cell lines from Fischer rat thyroid glands. These cells retained in vitro all three parameters of thyroid differentiated function (see Fig. 1). This cell line grew in a modified version of Ham's F-12 medium containing a low concentration of calf serum (0.5%) in the presence of six factors required for optimal growth: TSH, insulin, transferrin, somatostatin, hydrocortisone and the tripeptide glycyl-hystidyl-lysine (Ambesi-Impiombato et al, 1980). Those cells were

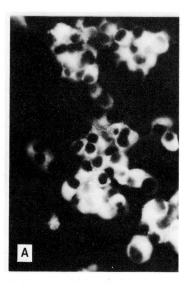 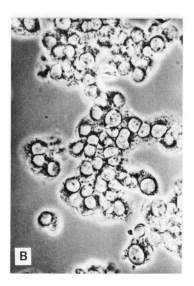

Figure 1:   Thyroglobulin Production by in vitro Cultured FRT-L
Cells as Detected by Indirect Immunofluorescence.

a) FRT-L cells were fixed and permeabilized in cold absolute methanol
and challenged with a monoclonal rabbit anti-rat-thyroglobulin IgG.
Second antibody was a rhodamine-labeled goat antirabbit IgG. (175X).
(courtesy of Dr. C. Tacchetti)

b) Phase contrast microphotograph of the same field shown in Fig.
1a.

designated FRT-L.

The original FRT-L clonal line has been maintained in culture
for several years and subclones with all the properties of the FRT-L
cell line have been derived as well. By adaptation of one of these
lines to growth in 5% serum, its doubling time was considerably
shortened to 48 hours as compared to 5-7 days for the parental FRT-L

line. The morphology of one such clonal line derived from the FRT-L
cell line is shown in Fig. 2a.

The availability of cloned populations of rat thyroid epithelial cells has made it possible to address the question of whether
this particular histiotype was indeed susceptible to transformation
by acute retroviruses. Early studies performed with the undifferentiated FRT cell line showed unequivocally that thyroid epithelial
cells could be malignantly transformed in vitro by KiMSV (Fusco et
al., 1981). Moreover, the fact that the transformed cells injected
into syngeneic animals induced tumors of epithelial morphology was
the definitive proof that the original FRT cell line was of epithelial derivation. More recent studies using rat thyroid cell lines
have confirmed the capability of at least three different strains of
the KiMSV to malignantly transform cells of thyroid epithelial
derivation (Fusco et al., 1982; Vecchio et al., 1982). In all cases
tested, the cells acquired typical morphological properties of
transformed cells, had a high colony-forming efficiency in agar and
grew as typical carcinomas in either newborn or adult syngeneic
Fischer rats (Vecchio et al., 1981). Fig. 2 b and c show the appearance of differentiated epithelial thyroid cells after transformation with two different strains of KiMSV.

Subsequent studies were aimed to ascertain whether the oncogenic potential towards thyroid epithelial cells was peculiar to the
KiMSV or was shared also by other acute transforming retroviruses.
As shown in Table IV the ability to transform the FRT-L cell clone
or other similar clones derived from the same cell line, was found
to be common to a variety of acute retroviruses, either containing

TABLE IV. TRANSFORMATION OF THE FRT-L CELL LINE BY DIFFERENT
RETROVIRAL ONCOGENES

Viral Strain	Oncogenes	Morphological Transformation	Growth in Agar
KiMSV (KiMuLV)	v-K-ras	+	+
KiMSV$_{wt}$ (Mol MuLV)	v-K-ras	+	+
KiMSV$_{ts}$ (Mol MuLV)	v-K-ras	+[a]	+[b]
MPSV$_{wt}$ (F MuLV)	v-mos	+	+
MPSV$_{ts}$ (F MuLV)	v-mos	+[a]	+[b]
HaMSV (AP 129)	v-H-ras	+	+
Rasheed MSV (RaLV)	v-H-ras	+	+
BALB/c MSV (AP-129)	v-H-ras	+	+
AF (F MuLV)	v-H-ras	+	+
MRSV (AP 170)	v-src	+	+

[a]Assayed at the temperature permissive for transformation, i.e. 33°.

[b]Assayed at 33°.

Figure 2: Morphological Appearance of FRT-L Cells Before and After
Transformation with Kirsten Murine Sarcoma Virus

a) FRT-L cells were grown in the presence of 5% calf serum and six
growth factors (see text) in Coon's modified Ham's F-12 medium. The
cells show a typical epithelial morphology growing in islets piling
up in the center of them. (450X).

b and c) Same cells after transformation with two strains of KiMSV.
Cells were capable of growing in the absence of the six growth
factors and lost their typical epithelial morphology. (450X).

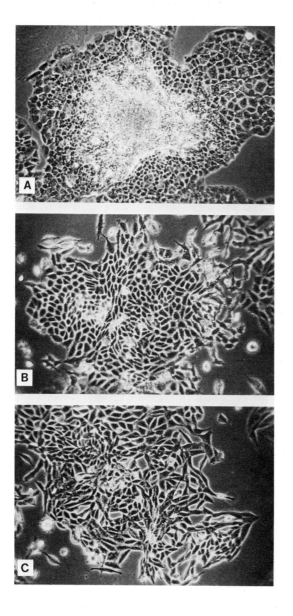

the ras-family of oncogenes, or containing other oncogenes, such as
mos and src. Transformation was assayed by morphological criteria
(see Fig. 2) as well as by the capacity of the virus-infected cells
to grow in semisolid medium. In several cases the transformed cells
were also able to induce tumors of epithelial morphology when injected into either newborn or adult syngeneic rats (see Fig. 3).

The availability of a cell system which expressed in vitro all
the properties of the differentiated thyroid phenotype has made it
possible to study the relationships between malignant cell trans-

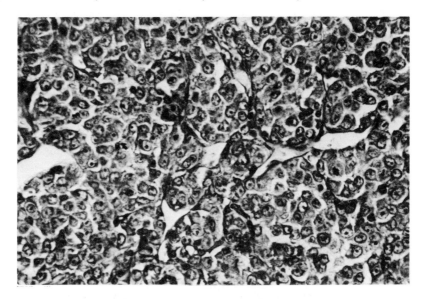

Figure 3:   Tumor Induced by Transformed FRT-L Cells Injected in
Syngeneic Adult Rat.
Transformed FRT-L cells (1 x 10^6) were injected subcutaneously
into newborn Fischer rats. Tumors were removed and fixed in cold
alcohol formalin and embedded in paraffin. Sections were stained
with hematoxylin and eosin. All of the tumors showed the histological appearance of adenocarcinomas. (X 270).

formation and the expression of thyroid differentiated functions. With all viruses tested an inverse correlation was found between the presence of the transformed and differentiated phenotypes. In fact, malignant transformation was always associated with the loss of dependence on TSH for the growth and also with the loss of the other two differentiated markers (thyroglobulin synthesis and secretion and iodide uptake). Among the three differentiated functions, synthesis of thyroglobulin was studied at a molecular level due to the availability of a c-DNA clone of rat thyroglobulin (Di Lauro et al., 1982). By the use of this probe, it was possible to demonstrate that the biosynthesis of thyroglobulin is blocked in the transformed cells at a transcriptional level, since no mRNA for thyroglobulin was found by hybridization to the cDNA clone either in the cytoplasm or nucleus of transformed cells. In contrast, large quantities of thyroglobulin mRNA were detected by this method in normal FRT-L cells (Colletta et al., 1983).

Studies performed with temperature sensitive mutants of two virus strains, i.e. the KiMSV (Colletta et al., 1983) and the myeloproliferative sarcoma virus (MPSV) (Fusco et al., 1985) have made it possible to elucidate in more detail the interrelationships between thyroid cell transformation and differentiation. After infection of the cells at the temperature permissive for transformation (33°C), one observes changes toward a transformed phenotype (see Table V). In addition, transformation is accompanied by a block in the expression of the differentiated phenotype. These results are analogous to those obtained by infecting the thyroid cells with the two wild type strains of KiMSV and MPSV.

TABLE V.   TRANSFORMATION AND DIFFERENTIATION MARKERS IN THRYOID
CELLS INFECTED WITH TEMPERATURE SENSITIVE MUTANTS OF KiMSV and MPSV.

	Transformation Markers		Differentiation Markers	
	Morphology	Growth in in Agar	Iodide Uptake	Tg synthesis
KiMSV$_{ts}$ 33°	+	+	-	-
KiMSV$_{ts}$ 39°	-	-	-	-
MPSV$_{ts}$ 33°	+	+	-	-
MPSV$_{ts}$ 39°	-	-	-	-

When the cells were shifted to the nonpermissive temperature
for transformation (39°C), a reversion of the transformed phenotype
toward the normal phenotype was observed. However, no reversion in
the expression of the thyroid specific differentiated functions was
detectable. Therefore, suppression of the differentiated phenotype
seems to be an irreversible phenomenon not continuously dependent on
the presence and activity of the viral oncogene product (Ferrentino
et al., 1984). The transforming proteins may trigger a pathway
which leads to the block of differentiation. However, this block
apparently becomes subsequently independent on the continuous
presence of the transforming proteins.

Temperature sensitive acute transforming retroviruses have
been reported to induce reversible blocks in differentiation in myoblasts (Fizman and Fuchs, 1975; Holtzer et al., 1975), retinal
melanoblasts (Boettiger et al., 1977) and chondroblasts (Pacifici
et al., 1977) of chicks. In other instances, results similar to
those obtained in the rat thyroid system (Gross and Rifkind, 1979;
Weber and Fris, 1979) have been observed. It is of interest that

in the only other study of the effects of ts viruses on epithelial
cells (Jones et al., 1981), also ts RSV was found to induce irreversible inhibition of the differentiated phenotype in chick lens
epithelial cells.

E. The mouse keratinocyte cell system. Cells in the epidermis
of mammalian species undergo a complicated yet well-characterized
program of terminal differentiation in vivo. The epidermis of the
skin can be divided into 4 layers; the basal cell layer, the stratum
spinosum, the stratum granulosum and the stratum corneum. The basal
cell layer is the only one which contains cells capable of dividing.
Once cells enter the stratum spinosum, they begin an irreversible
program of changes in their morphological, biochemical and immunological phenotype. These include the formation of desmosomes and
keratohyalin granules, an increase in the level of transglutaminase
and the crosslinking of keratin proteins. The cells eventually
enucleate and form cornified envelopes or squames which are shed
from the stratum corneum.

Initial attempts at establishing epidermal cells in tissue
culture were hindered by contamination with fibroblasts which
rapidly overgrew the slowly proliferating keratinocytes. However,
in 1975, Rheinwald and Green established primary cultures of human
foreskin keratinocytes by using irradiated mouse 3T3 cells as a
feeder layer. This feeder layer also provided a selection against
normal fibroblasts which could not attach to the dish under these
conditions. More recently, Peehl and Ham (1980) have developed a
growth medium which specifically supports the growth of human foreskin keratinocytes while inhibiting the growth of normal human

fibroblasts, thus eliminating the feeder layer requirement.

As early as 1974, investigators had tried to establish cultures
of mouse epidermal keratinocytes (Yuspa and Harris, 1974). Initial
results indicated that these primary cultures differentiate rather
than proliferate in culture. However, Hennings et al. (1980) have
since shown that the growth or differentiation of primary mouse
epidermal keratinocytes can be controlled by the level of extracellular calcium in the medium. When calcium levels are maintained
at 0.1 mM or less, the cells exhibited characteristics similar to
those of basal cells of the epidermis and could be maintained as
proliferating cultures. When the calcium level in the medium was
raised above this level, the cells ceased growing and began a
program of terminal differentiation similar to that observed in
normal epidermis.

The effects of infection of primary cultures of mouse epidermal
keratinocytes by KiMSV has recently been reported (Yuspa et al.,
1983). These studies indicated that the virus did not cause detectable alterations in the growth of the infected cells. However,
there was a dramatic effect on the ability of the cells to respond
to calcium induced terminal differentiation. While the uninfected
cells terminally differentiated within a period of 3-4 days, the
virus infected cells showed stratification but not terminal differentiation. Thus KiMSV appeared capable of blocking the terminal
differentiation program of primary mouse epidermal keratinocytes.

While primary cell cultures offer excellent model systems for
transformation studies, they also have several drawbacks. It is
often difficult to obtain large quantities of cells for study and

reproducibility can often be affected by the source of cells. In
addition, primary cultures may contain contaminating fibroblasts.
In order to overcome these difficulties, clonal cell lines, which
closely resemble primary keratinocytes in culture have been
established. These cell lines, designated BALB/MK, undergo a
program of terminal differentiation when exposed to extracellular
calcium levels greater than 1.0 mM. They are also dependent upon
the addition of epidermal growth factor (EGF) to serum-supplemented
medium for their continued proliferation. Although these cell lines
are aneuploid, they are non-tumorigenic when assayed in newborn
syngeneic mice. Therefore, BALB/MK cells provide the basis for a
cell transformation assay utilizing cells of epithelial keratinocyte
origin.

The dependency upon EGF for the continued growth of BALB/MK
cells appears to be a unique feature of these cells. As seen in
Figures 4 and 5, BALB/MK cells fail to grow in medium that is
supplemented with serum alone. However, addition of EGF to non-
proliferating BALB/MK cells leads to a burst of DNA synthesis 18
hours later (Figure 5). Thus the growth of the BALB/MK cells can
be controlled solely by the absence or presence of EGF in the
medium. In addition, BALB/MK cells are highly specific for EGF as
a mitogenic agent. A variety of other growth-promoting factors do
not support the proliferation of the BALB/MK cell line (Table VI).

Another feature of the BALB/MK cells is their ability to undergo a program of terminal differentiation similar to that observed in
vivo. The entry into this differentiation pathway is controlled by
the level of extracellular calcium in the medium in a manner similar

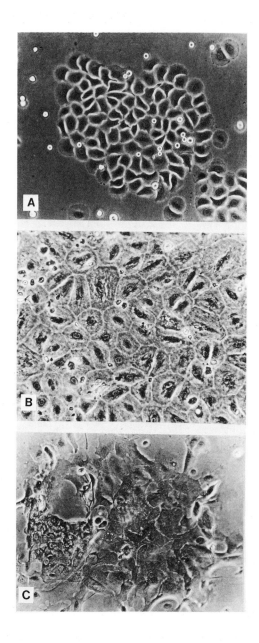

TABLE VI: GROWTH FACTORS TESTED ON BALB/MK CELLS

Factors which support growth	Factors which fail to support growth
Alpha transforming growth factor	Beta transforming growth factor
Endothelial-derived growth supplement	Dexamethasone
Epidermal growth factor	Fibroblast growth factor
	Insulin
	Interleukin 2
	Multiplication stimulating activity
	Platelet-derived growth factor

differentiation have been characterized and are easily assayed.

Therefore, the differentiation phenotype of BALB/MK cells can be

determined at any time in culture. Table VII summarizes the markers

to that observed in primary cultures. Several markers of epidermal

as well as the phenotype of BALB/MK cells.

As indicated in an earlier section, morphological changes have

not proven to be useful markers for transformation of epithelial

cells in culture. Indeed, infection of BALB/MK cells by BALB,

Harvey- or Kirsten-MSV lead to little or no change in morphology.

However, infection of these cells by such viruses invariably leads

Figure 4: Morphology of BALB/MK Cells Observed Under Different
Growth Conditions.

BALB/MK cells were seeded at a density of 1 x 10^5 cells/60 mm

petri dish in low calcium growth medium with 4 ng/ml EGF. (A) A

typical colony after two weeks in the presence of EGF. (B) A

typical colony two weeks after the removal of EGF. (C) A typical

colony after one week in the presence of EGF followed by a one week

exposure to high calcium.

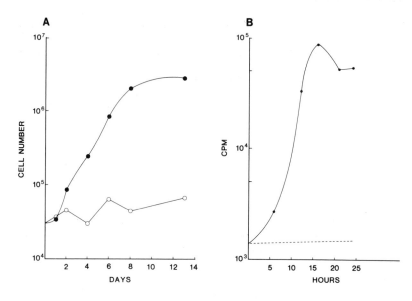

Figure 5: Growth and DNA Synthesis of BALB/MK cells in the Absence
or Presence of EGF.

A. BALB/MK cells were plated at a density of 1 x 10^5 cells/60 mm
petri dish in growth medium containing 0.09 mM calcium. Following
overnight incubation, one-half of the dishes received growth medium
supplemented with 4 ng/ml EGF. All cultures were fed with fresh
medium every three days. At various times, cells were removed and
counted in a hemacytometer. Each point represents the mean value of
triplicate dishes. Cell number is indicated in absence (O----O) or
presence (O----O) of EGF.

B. BALB/MK cells were plated at a density of 1 x 10^4 cells/35 mm
petri dish in growth medium alone. Following overnight incubation,
EGF was added to dishes at various times to a final concentration of
4 ng/ml while other dishes received growth medium alone. All dishes
were then treated with growth medium containing 10 μCi/ml of ^{3}H-
thymidine for a period of 2 hours. Dishes were then rinsed and

radioactivity incorporated radioactivity into DNA was precipitated with ice-cold 5% TCA and collected on filter paper. Filter papers were counted by standard liquid scintillation techniques. Each point is the average of triplicate dishes. CPM are indicated in the absence (----) or presence (0----0) of EGF.

TABLE VII: DIFFERENTIATION PHENOTYPE OF BALB/MK CELLS UNDER DIFFERENT EXTRACELLULAR CALCIUM CONCENTRATIONS

0.05 mM Calcium	1.5 mM Calcium
No cornified envelopes	Cornified envelopes produced
Desmosomes-few	Desmosomes-abundant
Filaggrin-absent	Filaggrin-present
Ornithine Decarboxylase-inducible	Ornithine Decarboxylase-not inducible
Pemphigoid antigen-present	Pemphigoid antigen-low or absent
Pemphigus antigen-absent or low	Pemphigus antigen-present
Transglutaminase-absent or low	Transglutaminase-high
Ulex Europaeus II lectin binding absent	Ulex Europaeus II lectin binding present

to an abrogation of EGF requirement for cell growth (Weissman and Aaronson, 1983). Thus the ability of the cells to grow in the absence of EGF provides a reproducible assay for infection of such cells by retroviruses. This altered phenotype can also be used as an assay for transformation by chemical carcinogens.

Another change which occurs following the infection of BALB/MK cells by acute transforming viruses is a block in the terminal differentiation pathway. When BALB/MK cells are exposed to high levels

of extracellular calcium, the vast majority undergo terminal differentiation within 4 to 7 days. However, either virally infected or
chemically altered BALB/MK cells are capable of growing equally well
at the same high level of extracellular calcium. This ability
appears to be due to the continued presence of a self-renewing basal
layer which is lost in BALB/MK cells exposed to high calcium conditions. It has previously been proposed that a block in the terminal
differentiation pathway of epidermal keratinocytes is an early
change in the process of malignant transformation (Yuspa and Morgan,
1981). Therefore, these results support the validity of the BALB/MK
system as a model for studing the effects of retroviruses on both
the growth control and differentiation of epithelial cells.

Since virus altered BALB/MK cells are blocked in their ability
to terminally differentiate in response to high calcium, it was of
interest to determine where in the differentiation pathway such
cells were blocked. The availability of differentiation markers of
the epidermis allowed us to determine the biochemical and immunological phenotype on these cells in both low and high calcium. A
summary of these results is shown in Table VIII. It appeared that
the KiMSV infected BALB/MK cells are blocked at a rather late
point in the differentiation pathway, since their differentiation
phenotype is similar to the BALB/MK cells in high calcium. This
would be consistent with the morphological stratification of these
cells observed in culture in the presence of high calcium.

When Kirsten-MSV infected BALB/MK cells were tested for tumorigenicity by growth in nu/nu (nude) mice, it was found that the

TABLE VIII: EXPRESSION OF EPIDERMAL DIFFERENTIATION MARKERS IN
UNINFECTED AND KIRSTEN-MSV INFECTED BALB/MK CELLS

Cell line and Calcium Conc.	Pemphigoid Ag	Pemphigus Ag	Ulex Europaeus II
BALB/MK-2 uninfected			
low Ca++	high	absent	absent
high Ca++	low	high	high
Kirsten-MSV infected			
low Ca++	high	low	absent
high Ca++	low	high	high

majority of cells are incapable of forming tumors. However, upon
injection of these cells, selection does occur in vivo which results
in the appearance of tumorigenic cell lines. This selection in vivo
is also accompanied by the acquisition of anchorage-independent
growth in vitro (Weissman et al., unpublished observations). These
results imply that there may be more than one step involved in the
malignant transformation of BALB/MK cells.

The effects of other retroviral onc genes have also been investigated in this system. Such studies have suggested that onc
genes from different functional families (i.e., ras family vs.
src family) have strikingly different effects on the growth and dif-

ferentiation of BALB/MK cells. While the block of differentiation induced by ras-oncogene containing retroviruses resides at a late stage of keratinocyte differentiation, the block associated with infection by retroviruses, containing tyrosine kinase related onco- genes, appears to reside at earlier stages (Weissman and Aaronson, in preparation). Therefore, the BALB/MK cell system should provide a means of contrasting the functions of different oncogenes in affecting the growth and differentiation of epidermal cells as well as studying the interaction of these genes in the process of neoplastic transformation.

F. Comparisons between the thyroid and the keratinocyte system. The results obtained by infection of clonal epithelial cells of either endodermic or ectodermic origin with retroviral oncogenes have established the ability of such genes to induce potent altera- tions in growth and differentiation in cells representing the vast majority of epithelial cells. Results previously obtained (Auersperg, 1978; Auersperg et al., 1981; Adams and Auersperg, 1981; Harrison and Auersperg, 1981) of transformation of epithelial cells of mesodermal derivation by acute retroviruses extend the general rule, that all existing epithelial histiotypes are susceptible to transformation by retroviral oncogenes.

One of the most striking features observed with either rat thyroid or mouse keratinocyte transformation is the complete loss of growth factor dependence upon the interaction with any of the retro- viral oncogenes utilized. Since this is a general phenomenon observed with all the tested oncogenes, it is tempting to speculate that viral oncogenes affect the control of cell proliferation either by interfering with a specific and possibly unique pathway which physiologically regulates cell growth, or by inducing the production of autocrine growth factors (such as TGF's) which substitute for the specific hormonal requirements. In the case of transformed keratinocytes it has not been possible to show the production of growth factors by the transformed cells. In fact, conditioned medium from virus-infected BALB/MK cells does not stimulate DNA synthesis of uninfected BALB/MK cells (Weissman, unpublished results). Thus, since the abolished hormonal requirement is that of a single growth factor, i.e., EGF, it is likely that the transforming genes interact at various steps in the same cellular growth regulatory pathway normally responsive to EGF. In the case of KiMSV transformed rat thyroid cells, preliminary results indicate that they produce a nondialyzable factor capable of stimulating the growth of normal FRT-L cells (Colletta et al., unpublished results). It seems, therefore, likely that oncogenes act, in this system, through the induction of growth factor molecules capable of substituting the physiological stimuli.

The interaction of most retroviral oncogenes with mouse keratinocytes or with rat thyroid epithelial cells, is associated with somewhat different transformed phenotypes. In the case of rat thyroid epithelial cells, complete neoplastic transformation is observed, as demonstrated by the acquired capability of the transformed cells to grow in agar and to form tumors in vivo. Such a pattern is not observed with mouse keratinocytes which, though altered in their basic growth properties, do not become capable of growth in soft agar and infrequently give rise to tumors in vivo.

A possible explanation for the differences in transformed phenotypes
is that the thyroid epithelial cell lines may already exhibit complementary alterations required for the full neoplastic phenotype.
In this regard, we have recently been able to obtain a new rat
thyroid epithelial cell line, designated PC-Cl 3, which behaves much
as mouse keratinocytes, in that it is not fully transformed by retroviral oncogenes. In fact, this cell line does not grow in agar
nor as a tumor in syngeneic animal or in nude mice after viral
infection, despite the abolishment of hormonal requirements and
effects on the cell morphology, (Fusco et al., manuscript in preparation). Thus, both the mouse keratinocyte and certain of the rat
thyroid cell lines may be useful as models to study multistep malignant transformation of epithelial cells.

Several investigators have reported fibroblast cell systems
(Ruley, 1983; Land et al., 1983) in which the cooperative action of
at least two oncogenes is required to achieve the expression of the
fully malignant phenotype. In an effort to establish such a model
for ascertaining the nature and minimum number of genes that could
induce malignant transformation for cells of epithelial origin,
human primary keratinocytes have been used in one of our laboratories (Rhim et al., 1985). Primary human keratinocytes could not
be propagated serially beyond two or three subcultures, nor did they
acquire indefinite life span when infected with the KiMSV. In
contrast, they became apparently "immortalized", but did not acquire
any phenotypic marker of malignant transformation when infected with
adenovirus 12-SV40 hybrid virus. Superinfection of the Ad12-SV40-
infected keratinocytes with KiMSV resulted in the acquisition of

the transformed properties as evidenced by morphological criteria, anchorage independent growth and ability to grow as carcinomas upon injection into nude mice. Human keratinocytes then appear to require a first "immortalization" step to become susceptible to further progression towards the malignant phenotype upon infection with KiMSV. Recent findings with human tracheal epithelium cells are also consistent with a multistep model for malignant conversion of epithelial cells in vitro (Yoakum et al., 1985).

By what mechanisms does the interaction of a viral oncogene product with epithelial cells bring about the shut off in expression of the differentiated functions? How can different metabolic pathways, such as those leading to production of widely different substances (such as hormones or structural proteins or enzymes) be negatively regulated by the viral oncogene products? There are three possible virus-cell interactions which can be hypothesized in this regard: 1) Oncogene products act on a "stem or progenitor cell." This cell has, by definition, both the capacity to divide and give rise to progeny cells capable of reaching an increasing degree of differentiation along a specific maturation pathway. Thus, the transformation event may either block the cell in the stem cell state or induce a later block in the maturation pathway. In the former case, the transformed population will consist of totally undifferentiated precursors. In the latter case, the population will consist of cells in which the block is unmasked as the cells reach a certain point in the differentiation pathway. 2) Oncogene products act on one of the several types of differentiated cells still capable of division along the maturation pathway. The degree

of differentiation reached by the transformed cells is therefore
"frozen" at the stage of differentiation which that cell had reached
before undergoing the transformation event. The prediction in this
case is that the differentiated phenotype will be that of the
specific precursor cell with which the viral oncogene interacted.
The cell targets of such interactions belong to the category of
cells which behave in vitro in a continuously replicating manner.
3) Oncogenes product act on a cell belonging to the in vivo category
of cells known as discontinuous replicators. Such cells are capable
of occasional division but have reached their highest degree of
differentiation. In this case the interaction might occur between
the oncogene product and a basic pathway controlling both growth and
differentiation, thus leading simultaneously to stimulation of
continuous and uncontrolled growth and to a block in the expression
of specific differentiation genes. Alternatively, the oncogene
product may interact separately with the pathways of growth and
differentiation, stimulating or deregulating the first and blocking
the second.

      The alterations of the differentiation pathway of BALB/MK seems
to be best explained by the first proposed mechanism. Since BALB/MK
cells consist of a homogeneous cloned population of committed progenitor cells (stem cells) they represent the only one available
target for the viral oncogene. Nevertheless, as predicted by the
hypothesis, differentiation of mouse keratinocytes can be blocked at
different stages, following the exposure of the cells to high Ca++
with the subsequent unmasking of blocks in the differentiation
pathway. It is still unknown what is the relationship between the

proliferative stimulus triggered by EGF and the differentiation state in mouse keratinocytes. If the two phenomena are linked, one would predict that the continuous proliferation induced by EGF might prevent the expression of the terminally differentiated phenotype and as a consequence the withdrawal of EGF should allow it. However, the fact that KiMSV transformed keratinocytes can undergo biochemical maturation, indicates that the two phenomena are not permanently coupled and stresses that differentiation is more complex than a mere cessation of cell growth.

Transformation of rat thyroid cells is well explained by the third proposed mechanism. These cells are, in fact, normally fully differentiated, although capable of replication. The complete block in expression of differentiated parameters in response to retroviral infection suggests that retroviral oncogenes interact to inhibit a fundamental pathway which controlls the expression of several different markers of differentiation. Indeed studies conducted with ts mutants for transformation suggest that viral oncogenes may interact separately, in this system, with the pathways of growth and differentiation. In fact, we have been able to demonstrate that the expression of the transformed phenotype can be modulated by the expression of the retroviral oncogene, whereas the block of the differentiation upon viral transformation is irreversible and cannot be modulated. We have therefore postulated (Colletta et al., 1983, Fusco et al., 1985) that the presence of the viral oncogene is necessary only to trigger the de-differentiation effect, whereas the persistence of the block with the time in culture is independent of the oncogene product itself.

In the thyroid system, moreover, the inhibition of differentiation, brought about by viral transformation, leads to a very complex
phenotype which is characterized by the loss of thyroglobulin biosynthesis and secretion, of the TSH responsiveness and of iodide
uptake (the latter one being by itself a complex phenomenon regulated by several enzymes). Therefore we tend to believe that
transformation affects an intracellular pathway which controls co-
ordinately the expression of all differentiated markers, rather
than interacting individually with each one of them.

A recent study (Setoyama et al., 1985) seems to support the
idea of the existence of such intracellular pathways. In their
model system utilized, mouse fibroblasts, transformation by MolMSV
is associated with suppression of the synthesis of type I collagen
and fibronectin, two mature products of normal fibroblasts. Chemical mutagenesis of the transformed cells can rescue (in some instances) the expression of the differentiation markers leaving the
expression and the function of the v-mos oncogene unchanged. Here,
in analogy with the thyroid cell system, one can postulate that the
mutagenic event has affected some pleiotropic or a series of trans-
acting factor(s) which can control the expression of type I collagen
and fibronectin. This result appears to be a mirror image of the
results obtained in one of our laboratories with the use of two ts
mutants for transformation on the thyroid rat system and also argues
in favour of the possibility of uncoupling differentiated and transformed phenotypes.

The regulation of cell growth and differentiation are two
phenomena extremely sensitive to the transformation event. Whether

they are permanently coupled or not is still a debatable question.
The in vivo findings that some tumors can be deregulated in growth
control and yet continue to retain a differentiated phenotype
favours the hypothesis that the two pathways can be uncoupled at
least in some situations. The results obtained in vitro with our
two epithelial systems confirm this concept. We believe that the
availability of in vitro models of hormone responsive, fully differentiated epithelial cells, will prove extremely valuable in
answering how oncogenes alter cell growth, cell differentiation, and
induce selective advantage of an undifferentiated cell in tumour
growth.

ACKNOWLEDGEMENTS

The authors are grateful to Drs. Matthias Kraus and C. Richter
King for helpful advices and discussions and to Frances Hyman and
Maude Atcheson for typing the manuscript. The work in the Laboratory of Naples was supported by the Progetto Finalizzato Oncologia
of the Consiglio Nazionale delle Ricerche.

REFERENCES

Aaronson, S.A., and Tronick, S.R. (1985). The role of oncogenes in
human neoplasia. "Important Advances in Oncology 1985" (Eds. V.T.
DeVita, S. Hellman, and S.A. Rosenberg), pp. 3-15, J.B. Lippincott
Co., Philadelphia, Pennsylvania.

Adams, A.T., and Auersperg, N. (1981). Transformation of cultured
rat ovarian surface epithelial cells by Kirsten Murine sarcoma
virus, Cancer Res., 41, 2063-2072.

Albert, D.M., Rabson, A.S., and Dalton, A.J. (1968). In vitro neoplastic transformation of uveal and retinal tissue by oncogenic DNA
viruses, Invest. Ophthalmonl., 7, 357-365.

Albert, D.M., Rabson, A.S., Grimes, P.A., and Von Sallmann, L. (1969). Neoplastic transformation in vitro of hamster lens epithelium by simian virus 40, Science, 164, 1077-1978.

Altaner, C., and Hlavayova, £. (1973). Transformation of rat liver cells with chicken sarcoma virus B77 and murine sarcoma virus, J. Virol., 11, 177-182.

Ambesi-Impiombato, F.S., and Coon, H.G. (1979). Thyroid cells in culture, Int. Rev. Cytol., Suppl., 163-172.

Ambesi-Impiombato, F.S., Parks, L.A., and Coon, H.G. (1980). Culture of hormone-dependent functional epithelial cells from rat thyroids, Proc. Natl. Acad. Sci. U.S.A., 77, 3455-3459.

Anderson, L.W., and Smith, H.S. (1980). Simian virus 40 and Moloney-murine sarcoma virus infection of bona fide mouse epithelium, J. Gen. Virol., 49, 443-446.

Aronson, J.F. (1983). Human retinal pigment cell culture, In Vitro, 19, 642-650.

Auersperg, N. (1978). Effects of culture conditions on the growth and differentiation of transformed rat adrenocortical cells, Cancer Res., 38, 1872-1844.

Auersperg, N., and Calderwood, G.A. (1984). Development of serum independence in Kirsten murine sarcoma virus-infected rat adrenal cells, Carcinogenesis, 5, 175-181.

Auersperg, N., Hudson, J.B., Goddard, E.G., and Klement, V. (1977). Transformation of cultured rat adrenocortical cells by Kirsten murine sarcoma virus (Ki-MSV), Int. J. Cancer, 19, 81-89.

Auersperg, N., Wan, M.W., Sanderson, R.A., Wong, K.S., and Mauldin, D. (1981). Morphological and functional differentiation of Kirsten murine sarcoma virus-transformed rat adrenocortical cell lines, Cancer Res., 41, 1763-1771

Banks-Schlegel, S., and Howley, P.M. (1983). Differentiation of human epidermal cells transformed by SV40, J. Cell Biol., 36, 330-337.

Bernard, B.A., Robinson, S.M., Semat, A., and Darmon, M. (1985). Reexpression of fetal characters in simian virus 40-transformed human keratinocytes, Cancer Res., 45, 1707-1716.

Beug, H., Hayman, M.J., and Graf, T. (1982). Leukemia as a disease of differentiation: retroviruses causing acute leukemias in chickens, Cancer Surveys, 1(2), 205-230.

Boettiger, D., Roby, K., Brumbaugh, J., Biehl, J., and Holtzer, H. (1977). Transformation of chicken embryo retinal melanoblasts by a temperature-sensitive mutant of Rous sarcoma virus, Cell, 11, 881-890.

Butel, J.S., Wong, C., and Medina, D. (1984). Transformation of mouse mammary epithelial cells by papovavirus SV40, Exp. Mol. Pathol., 40, 79-108.

Chang, S.E., Keen, J., Lane, E.B., and Taylor-Papadimitriou, J. (1982). Establishment and characterization of SV40-transformed human breast epithelial cell lines, Cancer Res., 42, 2040-2053.

Chou, J.Y. (1978). Human placental cells transformed by tsA mutants of simian virus 40: a model system for the study of placental functions, Proc. Natl. Acad. Sci. U.S.A., 75, 1409-1413.

Chou, J.Y., and Schlegel-Haueter, S.E. (1981). Study of liver differentiation in vitro, J. Cell Biol., 89, 216-222.

Colletta, G., Pinto, A., Di Fiore, P.P., Fusco, A., Ferrentino, M., Avvedimento, V.E., Tsuchida, N., and Vecchio, G. (1983). Dissociation between transformed and differentiated phenotype in rat thyroid epithelial cells after transformation with a temperature-sensitive mutant of the Kirsten murine sarcoma virus, Mol. Cell Biol., 3, 2099-2109.

Cooper, G.M. (1982). Cellular transforming genes, Science, 217, 801-806.

Dawe, C.J., and Law, L.W. (1959). Morphologic changes in salivary-gland tissue of the newborn mouse exposed to parotid-tumor agent in vitro, J. Natl. Cancer Inst., 23, 1157-1177.

Deftos, L.J., Rabson, A.S., Buckle, R.M., Aurbach, G.D., and Potts, J.T., Jr. (1968). Parathyroid hormone production in vitro by human parathyroid cells transformed by simiam virus 40, Science, 159, 435-436.

Defendi, V., Naimski, P., and Steinberg, M.L. (1982). Human cells transformed by SV40 revisited: the epithelial cells, J. Cell Physiol., 2, 131-140.

Diamandopoulos, G.T., and Enders, J.F. (1965). Studies on transformation of Syrian hamster cells by simian virus 40 (SV40): Acquisition of oncogenicity by virus-exposed cells apparently unassociated with the viral genome, Proc. Natl. Acad. Sci. U.S.A., 54, 1092-1099.

DiLauro, R., Obici, S., Acquaviva, A., and Alvino, C. (1982). Construction of recombinant plasmids containing rat thyroglobulin mRNA sequences, Gene, 19, 117-125.

Doolittle, R.F., Hunkapiller, M.W., Hood, L.E., Devare, S.G., Robbins, K.C., Aaronson, S.A., and Antoniades, H.N. (1983). Simian sarcoma virus onc gene, v-sis, is derived from the gene (or genes) encoding a platelet derived growth factor, Science, 221, 275-277.

Downward, J., Yarden, Y., Mayes, E., Scrace, G., Totty, N., Stockwell, P., Ullrich, A., Schlessinger, J., and Waterfield, M.D. (1984). Close similarity of epidermal growth factor receptor and v-erb-B oncogene protein seqences, Nature, 307, 521-527.

Elliott, A.Y., Stein, N., and Fraley, E.E. (1974). In vitro neoplastic transformatin of bovine embryonic urothelium by simian vacuolating virus 40 (SV40), Invest. Urol., 11, 411-413.

Ephrussi, B., and Temin, H.M. (1960). Infection of chick iris epithelium with the Rous sarcoma virus in vitro, Virology, 11, 547-552.

Feig, L.A., Bast, Jr., R.C., Knapp, R.C., and Cooper, G.M. (1984). Somatic activation of ras^K gene in a human ovarian carcinoma, Science, 223, 698-700.

Ferrentino, M., Di Fiore, P.P., Fusco, A., Colletta, G., Pinto, A., and Vecchio, G. (1984). Expression of the onc gene of the Kirsten murine sarcoma virus in differentiated rat thyroid epithelial cell lines, J. Gen. Virol., 65, 1955-1961.

Fizman, M.Y., and Fuchs, P. (1975). Temperature-sensitive expression of differentiation in transformed myoblasts, Nature, 254, 429-431.

Fujita, J., Yoshida, O., Yuasa, Y., Rhim, J.S., Hatanaka, M., and Aaronson, S.A. (1984). Ha-ras oncogenes are activated by somatic alterations in human urinary tract tumours, Nature 309, 464-466.

Fusco, A., Pinto, A., Ambesi-Impiombato, F.S., Vecchio, G, and Tsuchida, N. (1981). Transformation of rat thyroid epithelial cells by Kirsten murine sarcoma virus, Int. J. Cancer, 28, 655-662.

Fusco, A., Pinto, A., Tramontano, D., Tajana, G., Vecchio, G., and Tsuchida, N. (1982). Block in the expression of differentiation markers of rat thyroid epithelial cells by transformation with Kirsten murine sarcoma virus, Cancer Res., 42, 618-626.

Fusco, A., Portella, G., DiFiore, P.P., Berlingieri, M.T., DiLauro, R., Schneider, A.B., and Vecchio, G. (1985). A mos oncogene containing retrovirus (MPSV) transforms rat thyroid epithelial cells and blocks irreversibly their differentiation pattern, J. Virol., in press.

Georgoff, I., Secott, T., and Isom, H.C. (1984). Effect of simian virus 40 infection on albumin production by hepatocytes cultured in chemically defined medium and plated on collagen and non-collagen attachement surfaces, J. Biol. Chem., 259, 9595-9602.

Graf, T., and Beug, H. (1978). Avian leukemia viruses interaction with their target cells in vitro and in vivo, Biochim. Biophys. Acta, 516, 269-299.

Griffin, B.E., and Karran, L. (1984). Immortalization of monkey epithelial cells by specific fragments of Epstein-Barr virus DNA, Nature, 309, 78-82.

Grimley, P.M., Deftos, L.J., Weeks, J.R., and Rabson, A.S. (1969). Growth in vitro and ultrastructure of cells from a medullary carcinoma of the human thyroid gland: transformation by simian virus 40 and evidence of thyrocalcitonin and prostaglandins, J. Natl. Cancer Inst., 42, 663-680.

Gross, J.L., and Rifkind, D.B. (1979). The effect of avian retroviruses on limb bud chondrogenesis in vitro, Cell, 18, 707-718.

Harrison, J., and Auersperg, N. (1981). Epidermal growth factor enhances viral transformation of granulosa cells, Science, 213, 218-219.

Hennings, H., Michael, D., Cheng, C., Steinert, P., Holbrook, K., and Yuspa, S.H. (1980). Calcium regulation of growth and differentiation of mouse epidermal cells in culture, Cell, 19, 245-254.

Holtzer, H., Biehl, J., Yeoh, G., Meganathan, R., and Kaji, A. (1975). Effect of oncogenic virus on muscle differentiation, Proc. Natl. Acad. Sci. U.S.A., 72, 4051-4055.

Ikawa, Y. (1975). Transformation of cultured rat liver cells by murine sarcoma virus, Bibl. Haemat., 40, 165-177.

Ikawa, Y., Yoshijura, H., Sugano, M. (1970). Transformation of a clonal renal epithelial cell line by Moloney murine sarcoma virus (M-MSV), Bibl. Haematol., 36, 312-322.

Isom, H.C., and Georgoff, I. (1984). Quantitative assay for albumin-producing liver cells after simian virus 40 transformation of rat hepatocytes maintained in chemically defined medium. Proc. Natl. Acad. Sci. U.S.A., 81, 6378-6382.

Isom, H.C., Tevethia, M.J., and Kreider, J.W. (1981). Tumorigenicity of simian virus 40-transformed rat hepatocytes, Cancer Res. 41, 2126-2134.

Isom, H.C., Tevethia, M.J., and Taylor, J.M. (1980). Transformation of isolated rat hepatocytes with simian virus 40, J. Cell Biol., 85, 651-659.

Jones, R.E., DeFeo, D., and Piatigorsky, J. (1981). Initial studies on cultured embryonic chick lens epithelial cells infected with a temperature-sensitive Rous sarcoma virus, Vision Res., 21, 5-9.

Kaighn, M.E., Narayan, K.S., Ohnuchi, Y., Jones, L.W., and Lechner, J.F. (1980). Differential properties among clones of simian virus 40-transformed human epithelial cells, Carcinogenesis, 1, 635-645.

Kraus, M.H., Yuasa, Y., and Aaronson, S.A. (1984). A position 12-activated H-ras oncogene in all Hs578T mammary carcinosarcoma cells but not mammary cells of the same patient, Proc. Natl. Acad. Sci. U.S.A., 81, 5384-5388.

Land, H., Parada, L.F., and Weinberg, R. A. (1983). Tumorigenic conversion of primary embryo fibroblasts requires at least two cooperating oncogenes, Nature, 304, 592-602.

Lechner, J.F., Haugen, A., McClendon, I.A., and Pettis, E.W. (1982). Clonal growth of normal adult human bronchial epithelial cells in a serum-free medium, In Vitro, 18, 633-642.

Lechner, J.F., McClendon, I.A.,Laveck, M.A., Shemsuddin, A.M., and Harris, C.C. (1983). Differential control by platelet factors of squamous differentiation in normal and malignant human bronchial epithelial cells, Cancer Res., 43, 5913-5921.

McCoy, M.S., Toole, J.J., Cunningham, J.M., Chang, E.H., Lowy, D.R., and Weinberg, R.A. (1983). Characterization of a human/lung carcinoma oncogene, Nature, 302, 79-81.

Moscovici, C., and Gazzolo, L. (1982). Transformation of hemato-poietic cells with acute leukemia viruses, in Advances in Viral Oncology (Ed. G. Klein), pp. 83-106, New York, Raven Press.

Moyer, M.P., and Aust, J.B. (1984). Human colon cells: culture and in vitro transformation, Science 224, 1445-1447.

Ohnuchi, Y., Lechner, J.F., Bates, S.E., Jones, L.W., and Kaighn, M.C. (1982). Chromosomal instability of SV40 transformed human prostatic epithelial cell lines, Cytogen. Cell Genetics, 33, 170-178.

Okada, N., Steinberg, M.L., and Defendi, V. (1984). Re-expression of differentiated properties in SV40-infected human epidermal keratinocytes induced by 5-azacytidine. Exp. Cell Res., 153, 198-207.

Orme, S.K., Wells, S.A., Rabson, A.S., and Wurtman, R.J. (1968). In vitro neoplastic transformation of hamster pineal cells by three oncogenic DNA viruses. Cancer, 21, 477-482.

Pacifici, M., Boettiger, D., Roby, K., and Holtzer, H. (1977). Transformation of chondroblasts by Rous sarcoma virus and synthesis of the sulphated proteoglycan matrix, Cell, 11, 891-899.

Paraskeva, C., Brown, K.W., and Gallimore, P.H. (1982). Adenovirus cell interactions early after infection: in vitro characteristics and tumorigenicity of adenovirus type 2- transformed rat liver epithelial cells, J. Gen. Virol., 58, 73-81.

Paraskeva, C., and Gallimore, P.H. (1980). Tumorigenicity and in vitro characteristics of rat liver epithelial cells and their adenovirus-transformed derivatives, Int. J. Cancer, 25, 631-639.

Paulson, D.F., Fraley, E.E., Rabson, A.S., and Ketcham, A.S. (1968a). SV40-transformed hamster prostatic tissue: a model of human prostatic malignancy, Surgery, 64, 241-247.

Paulson, D.F., Rabson, A.S., and Fraley, E.E. (1968b). Viral neoplastic transformation of hamster prostate tissue in vitro, Science, 159, 200-201.

Peehl, D.M., and Ham, R.G. (1980). Growth and differentiation of human keratinocytes without a feeder layer or conditioned medium, In Vitro, 16, 516-525.

Pulciani, S., Santos, E., Lauver, A.V., Long, L.K., Aaronson, S.A., and Barbacid, M. (1982a). Oncogenes in solid human tumors, Nature, 300, 539-542.

Pulciani, S., Santos, E., Lauver, A.V., Long, L.K., Robbins, K.C., and Barbacid, M. (1982b). Oncogenes in human tumor cell lines, molecular cloning of a transforming gene from human bladder carcinoma cells, Proc. Natl. Acad. Sci. U.S.A., 79, 2845-2849.

Rabson, A.S., and Kirschestein, R.L. (1962). Induction of malignancy in vitro in newborn hamster kidney tissue infected with simian vacuolating virus (SV40), Proc. Soc. Exp. Biol. Med., 111, 323-328.

Rabson, A.S., Malmgren,R.A., and Kirschstein, R.L. (1966). Induction of neoplasia in vitro in hamster kidney tissue by adenovirus 7-SV40 hybrid strain (LLE46), Proc. Soc. Exp. Biol. Med., 121, 486-489.

Rabson, A.S., Malmgren, R.A., O'Conor, G.T., and Kirschstein, R.L. (1962). Simian vacuolating virus 40 (SV40) infection in cell cultures derived from adult human thyroid tissue, J. Natl. Cancer Inst., 29, 1123-1145.

Rheinwald, J.G.,and Beckett, M.A. (1980). Defective terminal differentiation in culture as a consistent and selectable character of malignant human keratinocytes, Cell, 22, 629-632.

Rheinwald, J.G., and Green, H. (1975). Serial cultivation of strains of human epidermal keratinocytes: the formation of keratinizing colonies from single cells, Cell, 6, 331-344.

Rhim, J.S., Jay, G., Arnstein, P., Price, F.M., Sanford, K.K., and
Aaronson, S.A. (1985). Neoplastic transformation of human epidermal
keratinocytes by AD12-SV40 and Kirsten sarcoma virus, Science, 227,
1250-1252.

Rhim, J.S., Kim, C.M., Okigaki, T., and Huebner, R.J. (1977).
Transformation of rat liver epithelial cells by Kirsten murine
sarcoma virus, J. Natl. Cancer Inst., 59, 1509-1516.

Ruley, H.E. (1983). Adenovirus early region 1A enables viral and
cellular transforming genes to transform primary cells in culture,
Nature, 304, 602-606.

Scherr, C.J., Rettenmier, C.W., Sacca, R., Roussel, M.F., Look,
A.T., Stanley, E.R. (1985). The c-fms proto-oncogene product is
related to the receptor for the mononuclear phagocyte growth factor,
CSF-1, Cell, 41, 665-676.

Setoyama, C, Liau, G., and DeCrombrugghe, B. (1985). Pleiotropic
mutants of NIH/3T3 cells with altered regulation in the expression
of both type I collagen and fibronectin, Cell, 41, 201-209.

Stanbridge, E.J., Der, C.J., Doersen, C.J., Nishimi, R.Y., Peehl,
D.M., Weissman, B.E., and Wilkinson, J.E. (1982). Human cell
hybrids analysis of transformation and tumorigenicity, Science, 215,
252-259.

Steinberg, M.L., and Defendi, V. (1979). Altered pattern of growth
and differentiation in human keratinocytes infected by simian virus
40, Proc. Natl. Acad. Sci. USA, 76, 801-805.

Steinberg, M.L., and Defendi, V. (1983). Transformation and
immortalization of human keratinocytes by SV40, J. Invest. Dermatol.
81, (1 suppl.), 131s-136s.

Taylor-Papadimitriou, J., Purkis, P., Lane, E.B., McKay, I.A., and
Chang, S.E. (1982). Effects of SV40 transformation on the cyto-
skeleton and behavioral properties of human keratinocytes, Cell
Differ., 11, 169-180.

Trevithick, J.R., Miller, G.G., Creighton, M.O., Hunter, E.H.,
Mousa, G.Y., and Blair, D.G. (1979). A rat epithelial cell line
RLE-R transformed with Rous sarcoma virus, Invest. Ophthalmol. Vis.
Sci., 18, 48.

Vecchio, G., Nitsch, L., Tajana, G., Pinto, A., and Fusco, A.
(1981). Experimental thyroid carcinogenesis by RNA tumor viruses,
in Advances in Thyroid Neoplasia 1981 (Eds. M. Andreoli, F. Monaco,
and J. Robbins), pp. 11-21, Field Educational Italia.

Vecchio, G., Pinto, A., Colletta, G., DiFiore, P.P., Fusco,A., Ferrentino, M., Grieco, M., and Tsuchida, N. (1982). Differentiation in retroviruses transformed cells: establishment of a new system of epithelial origin, in Expression of Differentiated Functions in Cancer Cells (Eds. R.F. Revoltella et al.), pp. 501-507, Raven Press, New York.

Walen, K.H. (1982). Anchorage independent growth of SV40 transformed human epithelial cells from amniotic fluids: differences within and among cell donors, In Vitro, 18, 203-212.

Waterfield, M.D., Scrace, G.T., Whittle, N., Stroobant, P., Johnsson, A., Wasteson, A., Westermark, B., Heldin, C.H., Huang, J.J., and Deuel, T.F. (1983). Platelet-derived growth factor is structurally related to the putative transforming protein, p28^{sis} of simian sarcoma virus, Nature, 304, 35-39.

Weber, M.J., and Friis R.R. (1979). Dissociation of transformation parameters using temperature-conditional mutants of Rous sarcoma virus, Cell, 16, 25-32.

Weinstein, B.I., Schwartz, J., Lonial, H., Dominguez, M.O., Gordon, G.G., Hochstadt, J., Southren, D.B., Dunn, M.W., and Southren, A.L. (1982). Normal and conditionally transformed bovine lens epithelial cell lines containing alpha- and gamma-crystallin, Exp. Eye Res., 34, 71-81.

Weiss, R., Teich, N., Varmus, H., and Coffin, J. (Eds.) (1982). RNA tumor viruses, molecular biology of tumour viruses, 2nd edition, pp. 1-1396, New York, Cold Spring Harbor Laboratory.

Weissman, B.E., and Aaronson, S.A. (1983). BALB and Kirsten murine sarcoma viruses alter growth and differentiation of EGF-dependent BALB/c mouse epidermal keratinocytes, Cell, 32, 599-606.

Wells, S.A., Jr., Orme, S.K., and Rabson, A.S. (1966a). Induction of neoplasia in vitro in hamster thyroid tissue by SV40 and adenovirus 7-SV40 hybrid (strain LLE46), Proc. Soc. Exp. Biol. Med., 123, 507-510.

Wells, S.A., Jr., Rabson, A.S., Malmgren, R.A., and Ketchmam, A.S. (1966b). In vitro neoplastic transformation of newborn hamster salivary-gland tissue by oncogenic DNA viruses, Cancer, 19, 1411-1415.

Wells, S.A., Rabson, A. S., and Malmgren, R.A. (1966c). Viral neoplastic transformation of endocrine tissues in the newborn hamster, Surg. Forum., 17, 105-106.

Wells, S.A., Jr., Wirtman, R.J., and Rabson, A.S. (1966d). Viral neoplastic transformation of hamster pineal cells in vitro: retention of enzymatic function, Science, 154, 278-279.

Wertz, R.K., O'Connor, C.C., Rabson, A.S., and O'Connor, G.T. (1965). Mixed infection of African green monkey kidney cells by adenovirus 7 and SV40, Nature, 208,1350.

Yuspa, S.H., and Harris, C.C. (1974). Altered differentiation of mouse epidermal cells treated with retinyl acetate in vitro, Exp. Cell Res., 86, 95-105.

Yuspa, S.H., and Morgan, D.L. (1981). Mouse skin cells resistant to terminal differentiation associated with initiation of carcino-genesis, Nature, 293, 72-74.

Yuspa, S.H., Vass, H., and Scolnick, E. (1983). Altered growth and differentiation of cultured mouse epidermal cells infected with oncogenic retroviruses: contrasting effects of viruses and chemicals, Cancer Res., 43, 6021-6030.

Yavin, E., Yavin, Z., Schneider, M.D., and Kohn, L.D. (1981). Mono-clonal antibodies to the thyrotropin receptor: implications for receptor structure and the action of antoantibodies in Graves disease, Proc. Natl. Acad. Sci. USA, 78, 3180-3184.

Yoakum, G.H., Lechner, J.F., Gabrielson, E.W., Korba, B.E., Malan-Shibley, L., Willey, J.C., Valerio, M.G., Shamsuddin, A.M., Trump, B.F., and Harris, C.C. (1985). Transformation of human bronchial epithelial cells transfected by Harvey ras oncogene, Science, 227, 1174-1179.

Human Genes and Diseases
Edited by F. Blasi
© 1986, John Wiley & Sons, Ltd.

MONOCLONAL ANTIBODIES TO THE INSULIN RECEPTOR AS PROBES

OF INSULIN RECEPTOR STRUCTURE AND FUNCTION

Ira D. Goldfine and Richard A. Roth

INTRODUCTION

In most tissues insulin is a major anabolic hormone. In general insulin promotes the metabolism of ions, sugars, lipids, and amino acids. In target cells insulin has three major sites of metabolic regulation (figure 1). At the plasma membrane insulin increases the transport of glucose and other substrates (1). Recent studies indicate that in adipocytes and other cells, insulin increases glucose transport by recruiting inert glucose trans-porters from the cell interior to the cell surface where they become active (2,3). In the cytoplasm, insulin activates a number of intracellular enzymes such as glycogen synthetase and pyruvate dehydrogenase by a dephosphorylation process (1). In the nucleus, insulin stimulates the synthesis of RNA and DNA (1).

In addition to these metabolic actions of insulin, insulin also activates a series of events that reduce the cellular response to the hormone (4,5). This phenomenon termed, "down regulation", is mediated in part by an insulin-induced loss of cell surface receptors (figure 2). Metabolic labeling studies of this pheno-menon indicate that insulin induces an acceleration of the degrada-tion of its own receptor (4,5).

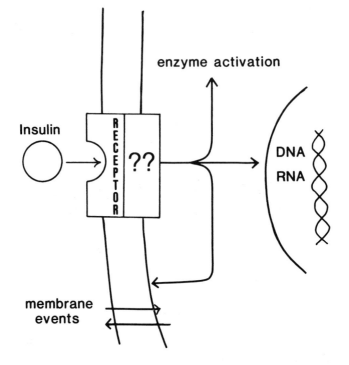

Fig. 1   Schematic diagram of the multiple metabolic actions of

           insulin on target cells.   Insulin first interacts with its

           receptor on the plasma membrane.   This interaction leads to

           regulation of three major types of metabolic actions; enhance-

           ment of membrane events such as glucose transport; enzyme

           activation in the cytoplasm and in cytoplasmic organelles;

           and DNA and RNA synthesis in nucleus.

How insulin carries out its metabolic and down regulatory

functions, however, is unknown.   Several mechanisms have been

proposed, including the generation of a unique second messenger at

the cell surface (6,7), initiation of a phosphorylation cascade

(8), and the direct interaction of insulin (or an insulin fragment)

with intracellular organelles (9,10). At present there is no

single comprehensive theory of insulin action that explains all of

the actions of insulin.

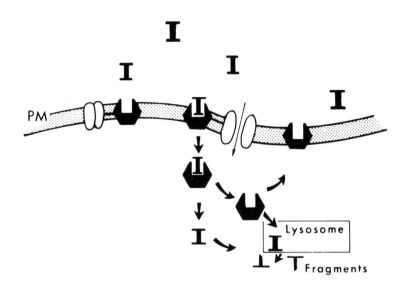

Fig. 2   Schematic diagram of the internalization of insulin and

its receptor. Insulin in the extracellular fluid binds to

its receptor and the insulin-receptor complex is then intern-

alized. Either insulin and its receptor can be recycled

back to the cell surface, or both of these structures can be

degraded in lysosomes and other cellular compartments.

INSULIN RECEPTOR STRUCTURE

The initial interaction of insulin with target cells is with

a receptor that is located on the plasma membrane. The insulin

receptor plays a critical role in both directing the hormone to a

specific target tissue and programming the response of the tissue

to the hormone (1). Thus, tissues with relatively high levels of

insulin receptor are more responsive to insulin than tissues with
low levels of receptor. In addition, as noted above, the mechanism
whereby insulin induces a specific response in a particular target
tissue is not known. It is likely, therefore, that a better
understanding of the molecular structure of the insulin receptor
will be critical in elucidating the biochemical mechanisms whereby
insulin controls cells.

Because of this critical role of the receptor, various
studies have focused on this molecule. The structure of the
insulin receptor has been elucidated in part by cross-linking
radioactive insulin to either whole cells or plasma membranes, and
then analyzing the resulting labeled proteins (11). Further
knowledge of the receptor has been obtained by purifying the
receptor and analyzing the structure of the purified protein (12).
Finally, the structure of the receptor has been studied by radioactively labeling cells and immunoprecipitating the receptor from
these cells via the use of specific antibodies to the receptor
(13). By all these techniques the insulin receptor has been shown
to be composed of two different subunits, an alpha subunit (Mr =
135,000) and a beta subunit (Mr = 95,000) (figure 3). Two of each
of these subunits have been hypothesized to be joined together via
disulfide bonds to give the final tetrameric structure of the
receptor. Both of these subunits are glycosylated (14) and exposed
to the extracellular environment. In contrast, only the beta
subunit appears to be exposed to the intracellular environment
(15). Recent work indicates that these two subunits are actually
synthesized from a single polypeptide precursor (16,17). After

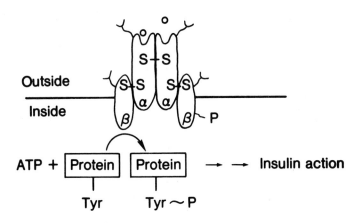

Fig. 3   Schematic diagram of the insulin receptor. In its basic

form, the insulin receptor consists of four subunits. There

are two alpha subunits and two beta subunits that are linked

by SS bonds. The two alpha subunits contain the hormone

binding sites and the two beta subunits contain an enzyme

activity. Both types of subunits are glycoproteins. When

insulin binds to the alpha subunit, the enzyme activity of

the beta subunit is stimulated. This enzyme activity induces

the phosphorylation both of itself and exogenous proteins.

This phosphorylation takes place at a tyrosine moiety and

is believed to be involved in many of the actions of insulin.

glycosylation, the precursor chain is cleaved to yield the two

chains.

In addition to the two structural chains, the insulin recep-

tor has two functional domains. These are the ability to bind

insulin, and the ability to phosphorylate other proteins. This

latter phosphorylating activity may regulate the subsequent res-

ponse of cells to insulin. Thus a model can be proposed where the binding of insulin to its receptor activates the intrinsic tyrosine specific kinase activity of the receptor. This activity could then phosphorylate either proteins or other molecules which mediate some or all of the actions of insulin.

The two functional domains of the receptor appear to reside in the aforementioned two structural domains. When radioactive insulin is crosslinked to its receptor, the alpha chain is primarily labeled (11). Also, when the beta chain of the receptor is proteolyzed, the insulin binding activity is not effected (18).

These results indicate that the insulin binding domain is localized to the alpha subunit. In contrast, the beta subunit appears to be involved in the kinase activity of the receptor. Covalent cross-linking of radioactive adenosine triphosphate to the receptor labels the beta subunit (19). In addition, proteolysis of this subunit destroys the kinase activity of the receptor (18). In addition to the kinase activity, the beta subunit also contains an autophosphorylation site (8). In both intact cells and purified receptors, predominately the beta subunit is phosphorylated (8,18). In purified receptors this phosphorylation occurs on tyrosine residues, but in the intact cell phosphorylations occur on other amino acids as well (20). Thus, the beta subunit may be a substrate for other kinases as well as the insulin receptor kinase (21). The significance of these phosphorylations are not clear, but most likely they are important in regulating the insulin responsiveness of cells (22).

MONOCLONAL ANTIBODIES

By virtue of their monospecificity, monoclonal antibodies
are different from other antibodies that are generated in intact
animals (polyclonal antibodies). The procedure described by
Kohler and Milstein allows one to grow lymphocytes in tissue
culture so that lymphocytes producing a single specific antibody
can be isolated (23). This procedure consists of immunizing mice,
fusing their splenic lymphocytes to cultured mouse myeloma cells,
and selecting a clone of cells producing only one type of antibody.
This approach has a number of important advantages over the classical approach of injecting whole animals with purified antigens.
First, even though a particular antigen cannot be purified, one
can immunize animals with the impure antigen and still obtain a
single antibody population that is directed against the particular
antigen one is interested in. This technique is particularly
useful for receptors that are hard to purify. Second, the antibody
that is produced is homogeneous; that is, the antibody is directed
only to a single antigenic site. In contrast, polyclonal antibodies contain mixtures of antibodies to many different regions of
the antigen. Finally, monoclonal antibodies can theoretically be
obtained in unlimited quantities. In contrast, polyclonal antibodies are often available in limited amounts, and will vary in
potency and specificity even with different bleedings of the same
animal. These advantages of the monoclonal antibodies have made
them extremely useful probes for many molecules, including receptors. A major probing technique for the insulin receptor has been
the use of antireceptor antibodies (24). The purpose of this

paper is to review our data concerning studies of the insulin

receptor using a monoclonal antibody.

PREPARATION OF A MONOCLONAL ANTIBODY TO THE INSULIN RECEPTOR

In order to study the interaction of antibodies with insulin

receptors in greater detail, we prepared a monoclonal antibody to

the binding site of the human insulin receptor (25). Prior studies

of the production of antireceptor antibodies in rabbits and mice

has only been partially successful (26,27). Although both poly-

clonal and monoclonal antibodies have been produced, these anti-

bodies are not able to inhibit insulin binding. In order to

produce an antibody that blocks insulin binding, we immunized mice

with IM-9 human lymphocytes (26,28). This human cell type has

numerous cell surface insulin receptors. When these cells were

injected into Balb C mice, greater than 80% of the animals produced

measurable titers of antibodies that inhibited insulin binding

(figure 4).

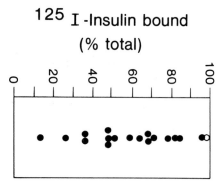

Fig. 4   Ability of sera from mice immunized with IM-9 cells to

    inhibit labeled insulin binding. The control with no serum

    is shown as an open circle and the sera from mice are shown

    with closed circles. Taken from reference 28.

The spleens of these mice were then removed and the cells fused to FO myeloma cells. After eight fusions, three monoclonal antibodies were identified. One monoclonal antibody was grown up and characterized. This antibody inhibited insulin binding to its receptor in IM-9 lymphocytes in a dose dependent manner (figure 5). The affinity of this antibody to its receptor in IM-9 cells

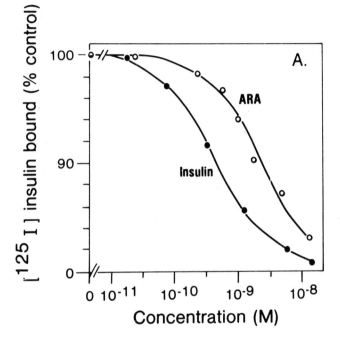

Fig. 5   The ability of insulin and antireceptor antibody (ARA) to inhibit insulin binding to IM-9 lymphocytes. Taken from reference 25.

was approximately 20% that of insulin (26). When cells were metabolically labeled with [S^{35}]methionine and solubilized with detergent, the antibody was able to immunoprecipitate the insulin receptor (figure 6). These data provided evidence, therefore,

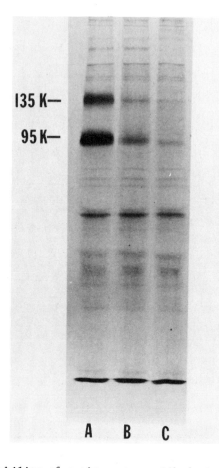

135 K—

95 K—

A   B   C

Fig. 6   The ability of antireceptor antibody to immuno-

precipitate insulin receptor from IM-9 lymphocytes.   IM-9

lymphocytes were incubated in methionine free media plus

[^{35}S]methionine for 16 hours.   The cells were then solubil-

ized with the detergent, Triton X-100.   Next, the receptor

was partially purified on a sepharose-wheat germ agglutinin

column and then immunoprecipitated with either antireceptor

antibody (A), antireceptor antibody plus 10^{-6}M insulin (B),

or normal antibody (C).   Polyacrylamide gels (7.5%) were run

and an autoradiograph is shown.   Adapted from reference 25.

that the antibody was interacting with the insulin receptor.

The ability of the antibody to compete for insulin binding with other human tissues were studied. In both placental membranes and human fat cells, the antibody progressively inhibited insulin binding, but in contrast with the IM-9 lymphocytes, in other cells the antibody was only approximately 1% as potent as insulin (figure 7) (26,29). These studies indicated, therefore, that the anti-receptor antibody had tissue specificity. When the antibody was tested for its ability to interact with non-human receptors, it had no activity on rat liver plasma membranes and rat adipocytes (25). These studies indicated, therefore, that the monoclonal antibody was species specific.

STUDIES WITH PROTEASE DIGESTION

As described above, the insulin receptor is composed of two subunits, α and β. It was of interest therefore to determine which subunit the antibody was binding to. From cross-linking studies, ^{125}I-insulin appears to be primarily linked to the α subunit (18,30). Therefore, one would predict that the antibody which inhibits insulin binding, would also bind to the α subunit. In contrast, the β subunit is primarily labeled by ^{32}P when the receptor is phosphorylated and when a photo-activatable radioactive ATP is crosslinked to the receptor (19). These results suggest that the β subunit is primarily involved in mediating the kinase activity of the receptor.

To test the roles of the two subunits of the insulin receptor in mediating these actions, one would have liked to separate the

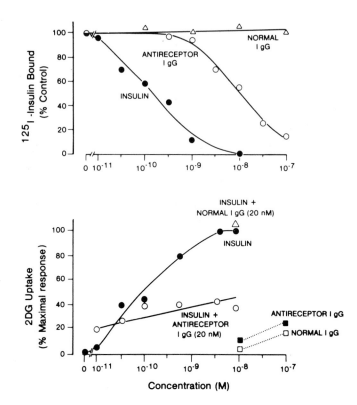

Fig. 7   The ability of antireceptor IgG to inhibit both insulin
binding (top panel) and insulin action (lower panel) in
human adipocytes.   In these cells either normal IgG or ARA
has a slight stimulatory effect on glucose transport.   In
these studies we used the model sugar, 2-deoxy-D-glucose.
This sugar is transported via the D glucose carrier, phosphorylated to 2-deoxy-D-glucose-6-P but then is not further
metabolized.   Normal IgG has no effect on either insulin
binding or insulin action.   Taken from reference 25.

two chains and test their activity alone. However, to separate

the two polypeptides requires the denaturation of the two chains.

Hence, another method was sought. Prior studies have indicated

that the β subunit is very susceptible to proteolysis (31).

Therefore, purified receptor preparations were treated with various

concentrations of different proteases. Crude collagenase prepara-

tions (the same mixtures that are used to prepare isolated rat

adipocytes) were found to degrade the β subunit into fragments of

less than 15,000 Mr without effecting the structure of the α

subunit (figure 8). Over the same concentration range, these

collagenase preparations were found to eliminate the kinase activ-

ity of the receptor preparations without affecting its insulin

binding activity (figure 9). Thus, these data further supported

the hypothesis that the α subunit of the insulin receptor is pre-

dominately involved in binding insulin whereas the β subunit is

predominantly responsible for the kinase activity. Furthermore,

this approach allowed a determination of which subunit the mono-

clonal antibody binds to. When the purified receptor preparations

were again digested with the collagenase preparations, the mono-

clonal antibody was found to precipitate the intact α subunit

(figure 10). These results therefore indicate that the monoclonal

antibody, like insulin, predominantly binds the α subunit.

BIOLOGICAL ACTIVITY OF THE MONOCLONAL ANTIBODY

The ability of the monoclonal antibody to mimic insulin

action was tested in several tissues. In adipocytes we studied

glucose transport with the model sugar, 2-deoxy-D-glucose. In

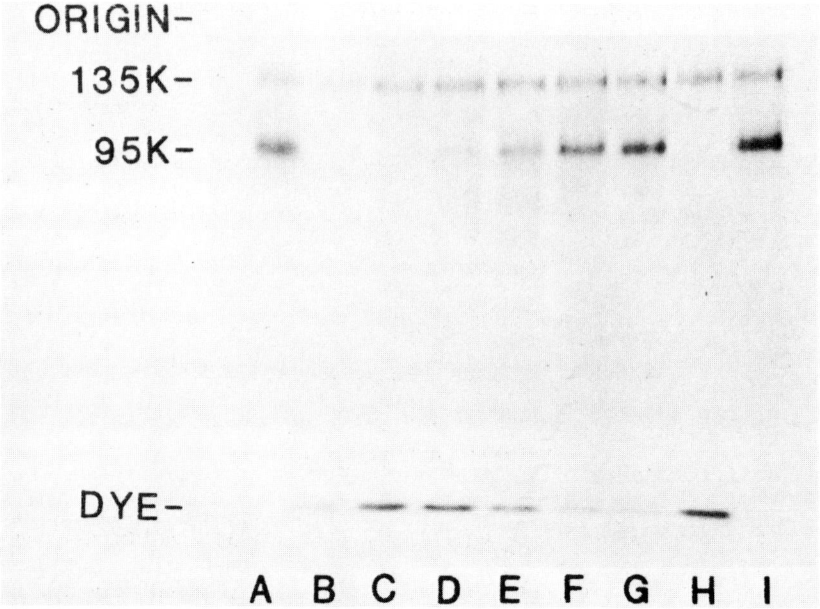

Fig. 8   Preferential degradation of the β subunit by collagenase
preparations. Metabolically ^{35}S-labeled, purified insulin
receptor was incubated with either 0 (A), 36 (B), 12 (C), 4
(D), 1 (E), 0.3 (F), or 0.1 (G) μg/ml of collagenase. The
samples were then reduced, denatured and analyzed on 7.5%
polyacrylamide gels. The autoradiograph of the dried gel is
shown. Taken from reference 18.

fibroblasts we studied amino acid transport with the model amino
acid, α-amino-isobutyrate (AIB). Both of these transport functions
are known to be regulated by insulin. The antibody had no activity
to stimulate glucose transport into adipocytes and amino acid
uptake into human fibroblasts (25). Since the antibody inhibited
the binding of insulin to its receptor, but did not mimic the

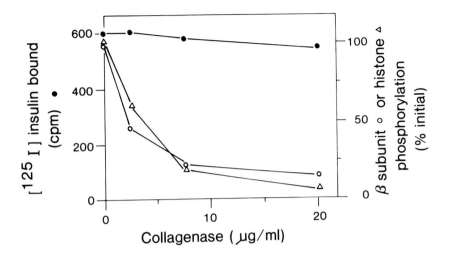

Fig. 9.   Effect of collagenase digestion on the insulin binding
and kinase activities of the receptor. Purified receptor
was incubated with the indicated concentrations of collagenase and then aliquots were tested for either insulin binding
or kinase activities. Taken from reference 18.

action of insulin in target cells, the possibility was considered
that the antibody could act as an antagonist of insulin action.
To study this possibility, cells are preincubated with monoclonal
antibody and insulin then added. In adipocytes (figure 7) and
fibroblasts (figure 11), the monoclonal antibody decreased the
action of insulin. The inhibitory activity of the antibody could
be overcome with higher concentrations of insulin indicating that
the antibody was a competitive antagonist (25).

THE EFFECT OF ANTIBODIES ON INSULIN RECEPTOR KINASE ACTIVITY

As mentioned previously, the alpha subunit of the insulin

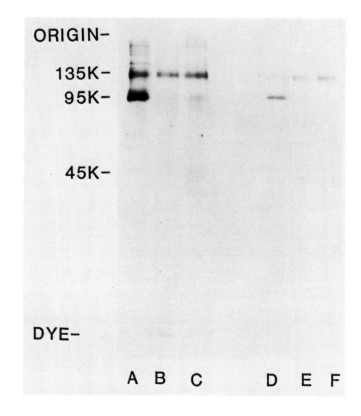

Fig. 10.   Effect of collagenase digestion on immunoprecipitation

of the receptor.   Purified receptor was incubated with

either buffer (A), 10 mg/ml crude collagenase (B), or 100

mg/ml purified collagenase (C).   Monoclonal was used to

precipitate either undigested receptor (D), receptor digested

with crude collagenase (E), or receptor digested with puri-

fied collagenase (F).   The precipitates were reduced, de-

natured and analyzed on gels.   The autoradiograph of the

dried gel is shown.   Taken from reference 18.

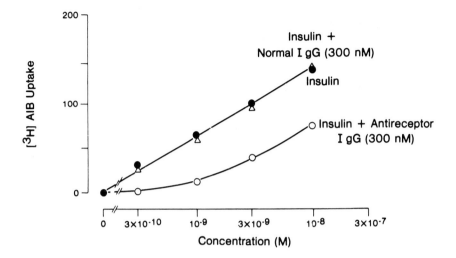

Fig. 11.   The ability of antireceptor IgG to inhibit α amino

isobutyric acid (AIB) uptake into human fibroblasts. AIB is

a model amino acid that is transported into cells via the

alanine transporter, but is not metabolized. From reference

25.

receptor binds the hormone, and the beta subunit has unique enzym-

atic activity. The beta subunit has the ability to phosphorylate

both itself and other proteins at a tyrosine moiety (figure 3).

Phosphorylation at tyrosines is relatively rare, and this property

has only been seen with the insulin, epidermal growth factor and

platelet derived growth factor receptors, and various oncogenes

(32-34). This phosphorylation is believed to mediate both normal

cell and tumor cell functions. When insulin is added to either

intact cells or purified receptors, it enhances phosphate incor-

poration into beta subunit tyrosine molecules (8,35) (figure 12).

We then tested the ability of the antireceptor antibody to either

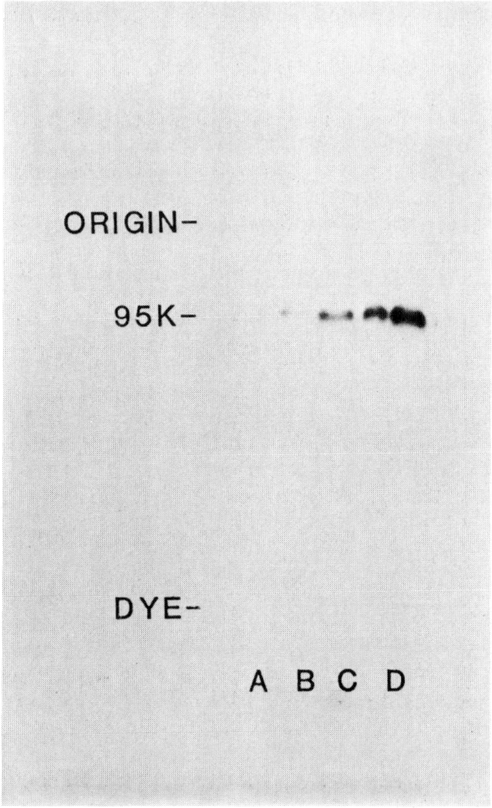

Fig. 12.   The effect of insulin on stimulation of tyrosine kinase
            activity of purified human insulin receptor. The effect of
            increasing insulin concentrations on stimulation of tyrosine
            kinase activity as measured by receptor beta subunit phosphor-
            ylation. An autoradiograph of SDS polyacrylamide gel (7.5%)
            is shown. A. control; B. 1.0 nM insulin; C. 10 nM insulin;
            D. 100 nM insulin. From reference 4.

mimic or inhibit this activity. The monoclonal antibody had no
effect on tyrosine phosphorylation either in whole cells or in
isolated, purified receptors. In contrast, the antibody progressively inhibited the effect of insulin in stimulating tyrosine
kinase activity (35) (figure 13). Therefore, the antibody, in
addition to blocking the metabolic actions of insulin, also blocked
insulin's ability to stimulate tyrosine kinase activity. These
data lend supportive evidence to the concept that certain metabolic
activities of insulin are mediated by the enzymatic activity of
the receptor.

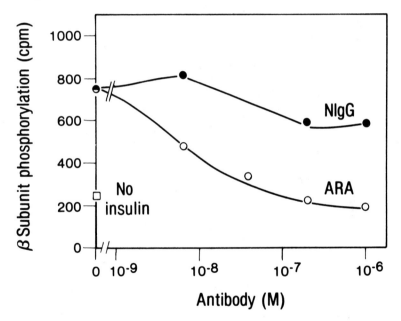

Fig. 13.   Effect of antireceptor antibody to inhibit insulin-
   stimulation of beta subunit phosphorylation of the human
   insulin receptor. Open square is the control with no insulin
   added and the open circle is phosphorylation in the presence
   of 100 nM insulin. ARA is antireceptor antibody, and NIgG
   is normal mouse IgG. From reference 35.

THE EFFECT OF MONOCLONAL ANTIBODY ON INSULIN RECEPTOR DOWN
REGULATION

It has been demonstrated in a number of cell types that
preincubation of cells with insulin leads to a progressive loss of
cellular insulin receptors (4,5). Careful examination of this
process reveals that insulin does not inhibit the synthesis of its
own receptor, but rather accelerates receptor degradation (5). We
then investigated whether the monoclonal antibody would mimic this
action of insulin (4). When the monoclonal antibody, at 10 nM,
was incubated with IM-9 lymphocytes, there was a time dependent
loss of insulin receptors (figure 14). Moreover, this loss was

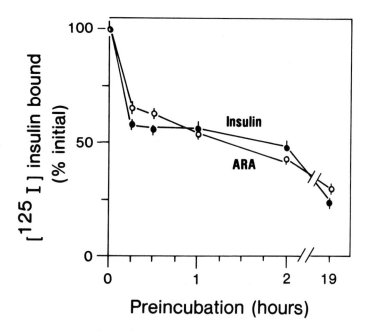

Fig. 14.   The effects of preincubation with 10 nM antireceptor
           antibody (ARA) and 1.0 μM insulin on subsequent ^{125}I insulin
           binding. From reference 4.

very similar to that seen with insulin itself. A rapid rate of
loss of binding was first seen after one hour, and a slower rate
of loss was then seen for up to 16 hours. The rapid loss of
receptors is believed to represent the internalization of the
hormone-receptor complex into the cell interior; the delayed loss
is believed to reflect the progressive degradation of the receptor
(4). When the dose dependence of the monoclonal antibody was
studied, it was seen that the monoclonal antibody caused from
regulation of the insulin receptors at lower molar concentrations
than insulin itself (figure 15). Metabolic labeling studies
indicated that the antibody, like insulin, induced an increase in
receptor degradation (4).

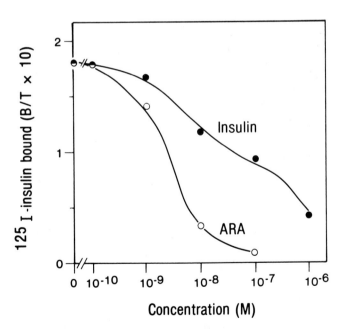

Fig. 15.   The dose response for insulin and antireceptor antibody
on cells preincubated 16 hours. From reference 4.

THE EFFECT OF MONOCLONAL ANTIBODY ON OTHER CELL SURFACE MOLECULES

Since the monoclonal antibody causes a rapid internalization of the cell surface insulin receptor and a subsequent acceleration of the degradation of the insulin receptor, it was possible that other cell surface molecules would also be effected. For example, an enzyme, called insulin degrading enzyme, is on the cell surface of rat hepatocytes (36), pancreatic acinar cells (37) and human IM-9 lymphocytes (38) and is hypothesized to be involved in an insulin degradation (38). It was therefore possible that this enzyme would be linked to the insulin receptor and also be decreased when the receptor was down regulated with the monoclonal antibody. However, when insulin receptors of IM-9 lymphocytes were decreased by 95% by down regulation with the monoclonal antibody, no change in surface levels of insulin degrading enzyme were detected (figure 16) (39). These data suggest that the majority of surface insulin degrading enzyme molecules in IM-9 cells are not linked to the insulin receptor.

THE EFFECT OF MONOCLONAL ANTIBODY ON THE INSULIN-LIKE GROWTH FACTOR RECEPTORS

Insulin is a member of a family of regulatory peptides which includes the closely related molecules, insulin-like growth factor I (IGF-I) and insulin-like growth factor II (IGF-II) (40). In contrast to insulin, which has a major role in regulating metabolic functions, the insulin-like growth factors (also called somatomedins) have a major role in regulating somatic growth. IGF-I is under the control of growth hormone; it is increased in acromegaly and gigantism, and decreased in hypopituitary dwarfism (40).

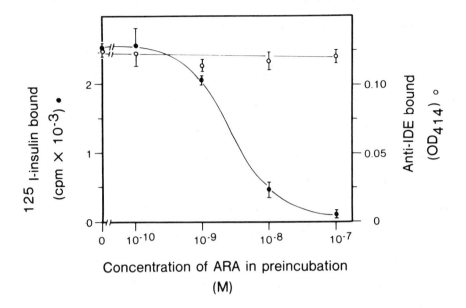

Fig. 16   Effect of down-regulation of the insulin receptor on

surface IDE.   IM-9 cells were down-regulated with the indi-

cated concentrations of monoclonal anti-receptor antibody

and then the surface insulin receptors and IDE were deter-

mined by ^{125}I-insulin binding and peroxidase-linked second

antibody, respectively.

There is a high degree of sequence homology between insulin and

these insulin-like growth factors (40).   In addition, the three

polypeptides exhibit several common biological activities, and at

high concentrations react to some extent with each other's receptor

(40).   As a consequence it has been difficult to clearly determine

which biological functions are associated with which hormone.

The monoclonal antibody was tested therefore for its ability to

interact with the receptors for the insulin-like growth factors.

The monoclonal antibody was found to only very weakly cross react
with the receptor for IGF-I and not to react with the receptor for
IGF-II (41) (figure 17).

Because of this specificity of the monoclonal antibody for
the insulin receptor, the monoclonal antibody could be used to
examine whether a particular effect of insulin is mediated via the
insulin receptor or via the insulin-like growth factor receptors.
Prior studies have indicated that cultured human hepatoma cells,
HEP-G2, respond to insulin by an increase in glycogen synthesis.

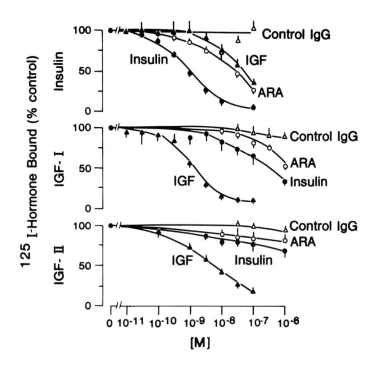

Fig. 17.   Effect of insulin, monoclonal antireceptor antibody
      (ARA).   IGF, and control IgG on ^{125}I-insulin (upper), [125]I-
      IGF-I (middle), and ^{125}I-IGF-II (lower) binding to human
      placenta membranes.   Taken from reference 41.

This increase reflects activation of the enzyme, glycogen syn-

thetase. In HEP-G2 cells we studied glycogen synthesis by mea-

suring labeled glucose incorporation into glycogen. The monoclonal

antibody was therefore used to determine whether the response of

these cells to insulin was mediated via the insulin receptor or

via the insulin-like growth factor receptor (42). The monoclonal

antibody was found to inhibit only approximately one third of the

response of these cells to insulin (figure 18). Moreover, the

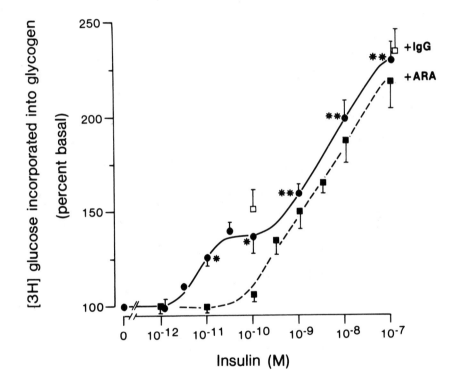

Fig. 18   Effect of increasing concentrations of insulin to

stimulate [^{3}H]glucose incorporation into glycogen of human

HEP-G2 hepatoma cells preincubated in the absence (O) and

presence ([]) of either 100 nM antireceptor antibody (ARA)

on normal IgG ([]). Taken from reference 42.

insulin-like growth factors were also capable of stimulating glycogen synthesis in these cells and this response was not inhibited at all with the monoclonal antibody (figure 19). These results indicated that the majority of the response to insulin in the HEP-G2 cells is mediated via the receptors for the insulin-like growth factors.

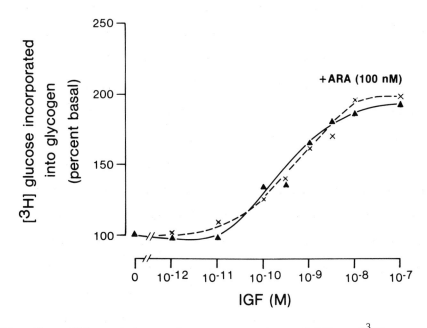

Fig. 19.   Effect of increasing concentrations of IGF on [3H] glucose incorporation into glycogen of human HEP G2 hepatoma cells preincubated in the absence (x) and presence (Δ) of antireceptor antibody (ARA). Taken from reference 42.

CONCLUSION

Studies with the monoclonal antibodies to the insulin receptor have revealed several important and interesting aspects of
receptor regulation. First, it is clear that interaction of an
antibody either at or very close to the insulin binding site is
not sufficient for generating the same transmembrane signals that
carry out the metabolic effects of insulin. Our monoclonal antibody antagonized the metabolic effects of insulin, and insulin
stimulation of receptor tyrosine kinase activity. These studies
lend support to the hypothesis, therefore, that tyrosine kinase
activity may mediate some of the metabolic effects of insulin. In
contrast, the antibody was an even more potent agonist than insulin
not sufficient for generating the same transmembrane signals that
carry out the metabolic effects of insulin. Our monoclonal antibody antagonized the metabolic effects of insulin, and insulin
stimulation of receptor tyrosine kinase activity. These studies
lend support to the hypothesis, therefore, that tyrosine kinase
activity may mediate some of the metabolic effects of insulin. In
contrast, the antibody was an even more potent agonist than insulin
itself in causing down regulation of the insulin receptor. These
data lead to two conclusions about the interaction of insulin with
its receptor. First, there appear to be two pathways of insulin
biological activity in target cells. One pathway is the metabolic
pathway, which is not mimicked by the monoclonal antibody. The
other pathway is the receptor down regulatory pathway, which is
mimicked by the antireceptor antibody. Moreover, since the
antibody does not stimulate tyrosine kinase activity, the data

suggest that tyrosine kinase activity is not necessary for this action of insulin. These studies indicate, therefore, that monoclonal antibodies are important probes of the action of insulin on target cells and suggest that additional studies with antibodies may further aid in elucidation of the mechanism of action of insulin.

REFERENCES

1.   Goldfine, I.D., Life Sci., 23, 2639 (1978).

2.   Cushman, S.W., and Wardzala, L.J., J. Biol. Chem., 255, 4758
      (1980).

3.   Ezaki, O., and Kono, T., J. Biol. Chem., 257, 14306 (1982).

4.   Roth, R.A., Maddux, B.A., Cassell, D.J., and Goldfine, I.D.,
      J. Biol. Chem., 258, 12094 (1983).

5.   Kasuga, M., Kahn, C.R., Hedo, J.A., Van Obberghen, E., and
      Yamada, K.M., Proc. Natl. Acad. Sci. USA, 78, 6917 (1981).

6.   Seals, J.R., and Jarett, L., Proc. Natl. Acad. Sci. USA, 77,
      77 (1980).

7.   Larner, J., Cheng, K., Schwartz, C., Kikuchi, K., Tamura,
      S., Creacy, S., Dubler, R., Galasko, G., Pullin, C., and
      Katz, M., Federation Proc., 41, 2724 (1982).

8.   Kasuga, M., Karlsson, F.A., and Kahn, C.R., Science, 215,
      185 (1982).

9.   Goldfine, I.D., Diabetes, 26, 148 (1977).

10.   Terris, S., and Steiner, D.F., J. Clin. Invest., 57, 885
      (1976).

11.   Pilch, P.F., and Czech, M.P., J. Biol. Chem., 254, 3375
      (1979).

12.   Jacobs, S., and Cuatrecasas, P., Endocrine Reviews, 2, 251
      (1981).

13.   Van Obberghen, E., Kasuga, M., Le Cam, A., Hedo, J.A., Itin,
      A., and Harrison, L.C., Proc. Natl. Acad. Sci. USA, 78, 1052
      (1981).

14.   Hedo, J.A., Kasuga, M., Van Obberghen, E., Roth, J., and

       Kahn, C.R., Proc. Natl. Acad. Sci. USA, 78, 4791 (1981).

15.   Hedo, J.A., and Simpson, I.A., J. Biol. Chem., 259, 11083

       (1984).

16.   Jacobs, S., Kull, F.C., and Cuatrecasas, P., Proc. Natl.

       Acad. Sci. USA, 80, 1228 (1983).

17.   Deutsch, P.J., Wan, C.F., Rosen, O.M., and Rubin, C.S.,

       Proc. Natl. Acad. Sci. USA, 80, 133 (1983).

18.   Roth, R.A., Mesirow, M.L., and Cassell, D.J., J. Biol. Chem.,

       258, 14456 (1983).

19.   Roth, R.A., and Cassell, D.J., Science, 219, 299 (1983).

20.   Kasuga, M., Zick, Y., Blithe, D.L., Crettaz, M., and Kahn,

       C.R., Nature, 298, 667 (1982).

21.   Jacobs, S., Sahyoun, N.E., Saltiel, A.R., and Cuatrecasas,

       P., Proc. Natl. Acad. Sci. USA, 80, 6211 (1983).

22.   Takayama, S., White, M.F., Lauris, V., and Kahn, C.R.,

       Proc. Natl. Acad. Sci. USA, 81, 7797 (1984).

23.   Kohler, G., and Milstein, C., Nature, 256, 495 (1975).

24.   Van Obberghen, E., and Kahn, C.R., Mol. Cell. Endocrinol.,

       22, 277 (1981).

25.   Roth, R.A., Cassell, D.J., Wong, K.Y., Maddux, B.A., and

       Goldfine, I.D., Proc. Natl. Acad. Sci. USA, 79, 7312 (1982).

26.   De Pirro, R., Lauro, R., Gelli, A.S., Bertoli, A., and

       Musiani, P., J. Clin. Lab. Immunol., 2, 27 (1979).

27.   Jacobs, S., Chang, K.-J., and Cuatrecasas, P., Science, 200,

       1283 (1978).

28.   Roth, R.A., Wong, K.Y., Maddux, B.A., and Goldfine, I.D.,
       Biochem. Biophys. Res. Commun., 101, 979 (1981).

29.   Roth, R.A., Maddux, B., Wong, K.Y., Styne, D.M., Van Vliet,
       G., Humbel, R.E., and Goldfine, I.D., Endocrinol., 112, 1865
       (1983).

30.   Massague, J., and Czech, M.P., J. Biol. Chem., 257, 6729
       (1982).

31.   Massague, J., Pilch, P.F., and Czech, M.P., J. Biol. Chem.,
       256, 3182 (1981).

32.   Waterfield, M.D., Scrace, G.T., Whittle, N., Stroobant, P.,
       Johnsson, A., Wasteson, A., Westermark, B., Heldin, C.-H.,
       Huang, J.S., and Deuel, T.F., Nature, 304, 35 (1983).

33.   Bishop, J.M., Cell, 32, 1018 (1983).

34.   Cooper, J.A., Reiss, N.A., Schwartz, R.J., and Hunter, T.,
       Nature, 302, 218 (1983).

35.   Roth, R.A., Cassell, D.J., Maddux, B.A., and Goldfine, I.D.,
       Biochem. Biophys. Res. Commun., 115, 245 (1983).

36.   Yokono, K., Roth, R.A., and Shigeaki, B., Endocrinol., 111,
       1102 (1982).

37.   Goldfine, I.D., Williams, J.A., Bailey, A.C., Wong, K.Y.,
       Iwamoto, Y., Yokono, K., Baba, S., and Roth, R.A., Diabetes,
       33, 64 (1984).

38.   Duckworth, W.C., and Kitabchi, A.E., Endocrine Reviews, 2,
       210 (1981).

39.   Roth, R.A., Mesirow, M.L., Cassell, D.J., Yokono, K., and
       Baba, S., Diabetes Research and Clinical Practice, in press.

40.   Zapf, J., Froesch, E.R., and Humbel, R.E., Curr. Top. Cell.
      Regul., 19, 257 (1981).

41.   Roth, R.A., Maddux, B., Wong, K.Y., Styne, D.M., Van Vliet,
      G., Humbel, R.E., and Goldfine, I.D., Endocrinol., 112, 1865
      (1983).

42.   Verspohl, E.J., Roth, R.A., Vigneri, R., and Goldfine, I.D.
      J. Clin. Invest., 74, 1436 (1984).

CELLULAR ONCOGENES AND THE PATHOGENESIS OF HUMAN CANCER

Riccardo Dalla-Favera and Ethel Cesarman

INTRODUCTION

Nearly a century of cancer research has provided unquestionable evidence that cancer derives from damage to the cell's genetic apparatus (Boveri, 1914). Since physical and chemical carcinogens as well as oncogenic viruses all have cell DNA as a common target, it had been repeatedly predicted that the pathogenesis of different tumors would involve the structural or functional alterations of one or more genes.

The long awaited identification of such altered genes has been obtained with the development of recombinant DNA technology, which has allowed the accumulation of considerable evidence for the involvement of a group of genes called oncogenes in the pathogenesis of cancer in different animal species including humans. The existence of cellular oncogenes was first suggested by studies on RNA tumor viruses or retroviruses. The tumorigenic properties of some of these viruses are due to the presence in their genome of viral oncogenes, which derive from normal cellular genes. More recently the existence of dominant-acting, cellular-transforming genes has been identified in the DNA of a variety of tumors. These genes are able to induce transformation when transferred into appropriate target cells in vitro. Both types of genes, those found in retroviruses and those detectable by DNA-mediated gene transfer, represent altered or "activated" versions of normal cellular genes that code for evolutionary conserved basic cellular functions. These genes are called cellular oncogenes (c-onc genes) or, less commonly yet more properly, proto-oncogenes.

The protein product of some viral (v-onc) or cellular oncogenes has been identified and from these studies evidence is emerging that these genes code for proteins that are involved in the control of cell proliferation and differentiation. It is therefore likely that the functional or structural alterations of these genetic elements may represent the molecular basis of the disruption of the cellular control mechanism that is typical of the transformed cell. Different alterations of these genes - mutations, amplifications, and chromosomal translocations - are, in fact, found with increasing frequency in different human tumors. A number of in vivo and in vitro observations suggest that each genetic alteration represents a step along the multistage process that leads to malignant transformation.

The objective of this chapter will be to review this rapidly evolving scenario. While detailed information on the more than twenty known oncogenes can be found in specialized reviews, we will focus on providing examples that are representative of general categories. Sections on the identification and normal function of cellular oncogenes will serve as a basis for a critical review of the different mechanisms of oncogene activation that are involved in human malignancies.

IDENTIFICATION OF CELLULAR ONCOGENES

a) Retroviral Oncogenes Identify Homologous Cellular Proto-Oncogenes.

The low complexity of the genetic information carried by tumorigenic retroviruses has made them an attractive and simple system to use in identifying the gene(s) responsible for malignant transformation (for reviews, see Bishop, 1983, 1985). Two main categories of retroviruses can be identified, based on the combined analysis of their biological properties and their genomic organization: (1) replication-competent, chronic leukemia viruses or helper viruses, and (2) replication-defective, acute leukemia or sarcoma viruses. Viruses of the first type contain all the viral genetic elements that are necessary for the virus replicative cycle in the cell (see Figure 1). While some of these viruses cause tumors in animals following a long latency period (chronic leukemia virus), their mechanism of action is still largely unknown and clearly does not directly involve any gene present in their genome. In general, they are not tumorigenic in vivo, do not transform cells in vitro, and do not carry any transforming sequence or viral oncogene. Viruses of the second type lack variable parts of the genetic information needed for their autonomous replication, and require the complementation of helper viruses of the first type. Structural genes are substituted by a transforming gene (see Figure 1), which is both necessary and sufficient for the in vivo and in vitro transforming capability of this type of virus. This gene is termed a viral oncogene (v-onc gene).

The elucidation of the origin of viral oncogenes (i.e., of the transition process between a replication-competent nontransforming retrovirus and a replication-defective transforming retrovirus) represented the starting point for our present knowledge of cellular oncogenes. It is now clear from DNA and protein homology data that all the known retroviral oncogenes are derived from homologous cellular oncogenes present in the genome of eukaryotic cells (see Figure 1). As a result of a poorly understood recombination event occurring at extreme low frequency, cellular oncogenes from different eukaryotic species have entered the genome of different retroviruses. During the process of viral

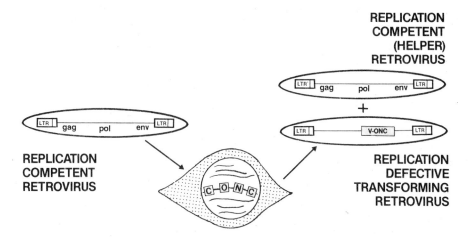

REPLICATION
COMPETENT
(HELPER)
RETROVIRUS

Fig. 1   Schematic representation of the origin of transforming
            retroviruses. GAG, POL and ENV indicates the three
            viral structural genes. LTR (long terminal repeats)
            represents the two noncoding regulatory sequences on
            the 5' and 3' ends of viral RNA. v-onc represents the
            viral oncogene. In the schematic representation of the
            cell (center), c-onc indicates a cellular oncogene.
            v-onc sequences are contained in one box and c-onc are
            split in several boxes to indicate that the former are
            arranged contiguously in the viral genome, while the
            latter have the classic exon-intron arrangement of most
            eukaryotic genes.

transduction, structural alterations have occurred in the cellular
oncogene that have turned it into a transforming viral oncogene.
        Cloned v-onc genes have been used to identify and clone
proto-oncogenes from different species. More than twenty v-onc
genes have been identified, all of which have been found to have
c-onc counterparts in cellular DNA. A list of known oncogenes,
their association with different types of tumors, and some known
biological and biochemical characteristics of their protein products
is presented in Table I.

        b) Oncogenes Identified by DNA-Mediated Gene
Transfer in NIH 3T3 Cells.

        The development of efficient DNA-mediated gene transfer
methods (Graham and Van Der Eb, 1973) has enabled researchers to
test for the presence of activated cellular oncogenes by introducing
total genomic DNA from tumor cells into appropriate recipient
cells. This approach was based on the assumption that activated

cellular oncogenes may act dominantly on the genetic background of
the recipient cell and induce cell transformation. The use of
NIH-3T3 mouse fibroblasts as highly sensitive recipient cells for
neoplastic transformation has led to the development of reliable and
consistent transfection assays in which transforming activity is
generally assayed by measuring the ability of tumor DNA to induce
foci of morphologically transformed cells on a background of
nontransformed NIH 3T3 cells. This system, which was initially used
to identify activated cellular oncogenes in chemically transformed
mouse cells (Shih et al., 1979; Cooper et al., 1980), has been
adapted for the identification and cloning of oncogenes from human
tumors following the strategy illustrated in Figure 2 (Shih et al.,
1981; Shih and Weinberg, 1982; Krontiris and Cooper, 1981; Murray et
al., 1981; Perucho et al., 1981). This strategy exploits
human-specific middle-repetitive Alu sequences that are contained
within or closely linked to many genes (Jelinek et al., 1980).
Because NIH 3T3 cells take up as much as $0.1^{\circ}/^{\circ}$ of a genome during
transfection, primary transfectants contain many DNA fragments with
human Alu sequences. However, serial passage of the transforming
gene by transformation of NIH 3T3 cells with DNA isolated from
transfectants generates secondary or tertiary transfectants
containing only the Alu sequences within or linked to the
transforming gene (Murray et al., 1981; Perucho et al., 1981). The
linkage of the Alu sequence to the transforming gene is then used to
clone the oncogene by screening libraries of transfectant DNA with
probes for Alu sequences (Pulciani et al., 1982b; Shih and Weinberg,
1982).

     Initial studies making use of either the sensitivity of
transforming activity to digestion with restriction endonucleases or
the pattern of middle-repetitive (Alu) sequences linked to or
contained within particular transforming genes suggested that
different tumor types might contain different transforming genes
(Lane et al., 1981, 1982b; Murray et al., 1981; Perucho et al.,
1981; Marshall et al., 1982; Pulciani et al., 1982a). Subsequent
studies employing clones of the transforming genes have demonstrated
that the transforming activity in many human tumors is due to genes
that are members of the ras family of oncogenes (i.e., the cellular
homologs of viral ras genes) (Shih and Weinberg, 1982; Der et al.,
1982, Der and Cooper, 1983). It is then intriguing to note that two
different approaches, one involving the study of viral oncogenes,
the second experimentally testing the transforming activity of tumor
DNA has lead to identification of the same genetic elements. As we
will discuss later in detail, a structural comparison of cloned
c-ras copies from tumor DNA and normal c-ras counterparts from
normal DNA shows that in tumor DNA one c-ras allele is activated by
specific point-mutations. Human transforming genes that are not
members of the ras gene family have been detected by transfection of
NIH 3T3 cells by Cooper and Lane and their coworkers (Lane et al.,
1981, 1982a,b) and more recently by several other groups that have

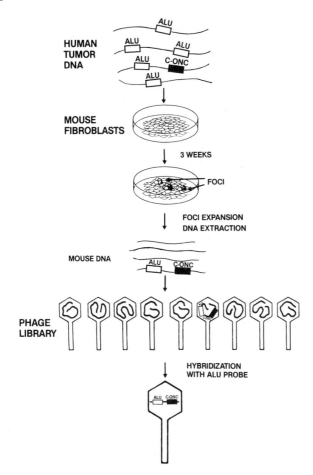

Fig. 2   Schematic representation of the assay used to isolate
            human transforming oncogenes by DNA-mediated gene
            transfer into NIH 3T3 mouse fibroblasts. See the text
            for a description of the procedure.

used a modification of the NIH 3T3 system assay involving the growth
of transfected cultures in semisolid medium (Cooper et al., 1980) or
in nude mice as a direct _in vivo_ selection (Blair et al., 1982;
Fasano et al., 1984a).

ROLE OF ONCOGENES IN NORMAL CELLS

With few exceptions, studies on the structure of viral or/and cellular oncogene sequences have preceded the ones dealing with the biochemical and functional characterization of their protein products. It is conceivable to assume that the high degree of conservation displayed by cellular oncogenes throughout evolution suggests that they must code for basic functions necessary for cell growth and/or development. Information regarding the nature of different onc-proteins has been derived from two main approaches. The first involves the study of v-onc proteins, which are more accessible to direct analysis than their cellular counterparts since they are generally produced in larger amounts and can be identified more easily. An alternative approach has been the use of molecular techniques to characterize the DNA sequence from which the protein sequence can be derived. While the function of most oncogenes is not known, for three oncogenes - sis, erb-B, and c-fms - the function of their corresponding protein product has already been substantially elucidated. These studies have allowed a preliminary classification of cellular oncogenes based on the distribution and/or biochemical function of their respective protein (see Table 1). Following is a review of the most important data on the different groups of oncogenes; we will describe the oncogene(s) that has been better characterized in each group as an example for a given class and as a basis for drawing conclusions on their general role in the control of cell proliferation (see Figure 3).

a) Growth Factors: PDGF and c-sis.

V-sis is the oncogene of the acutely transforming retrovirus simian sarcoma virus (SSV). Both v-sis and its human homolog, c-sis, have been cloned and sequenced (Dalla-Favera et al., 1981; Josephs et al., 1984). It is now known that the virus contains only a portion of the complete c-sis gene, and uses viral sequences to initiate transcription and translation of the onc-protein. Recently, two research teams found that when they sequenced some peptide residues derived from platelet derived growth factor (PDGF) and entered them into a computer database, a striking homology with the previously reported nucleotide sequence of the oncogene v-sis was found (Waterfield et al., 1983; Doolitle et al., 1983). In turn, this result suggested that c-sis is the gene encoding for PDGF or for a related molecule. Conclusive evidence supporting this hypothesis came from sequence analysis of the human c-sis gene (Chiu et al., 1984; Johnsson et al., 1984; Josephs et al., 1984). However, it is still not resolved whether this locus encodes both polypeptide chains PDGF-A and PDGF-B that constitute PDGF. PDGF is the major serum polypeptide mitogen for mesenchymal cells, and several observations concerning its mode of action are significant for the understanding of the molecular mechanism that

Table I Viral And Cellular Oncogenes Classified According to Their Known Properties

Group	Oncogene	V-ONC Species of Origin of Retrovirus	V-ONC Tumor Caused by Retrovirus	C-ONC Association With Human Tumors	PROTEIN (V-ONC OR C-ONC) Subcellular Location	PROTEIN (V-ONC OR C-ONC) Biochemical Function
I	sls (PDGF)	Wooley Monkey	Sarcoma	Glioblastoma(?)	Secreted	Growth Factor
II	erb-B	Chicken	Erythroleukemia Sarcoma	Carcinoma Glioblastoma	Plasma Membrane	Growth Factor Receptor (EGF) Tyrosine Specific Protein Kinase
	fms	Cat	Sarcoma	–	Plasma Membrane	Growth Factor Receptor (CSF-1) Tyrosine Specific Protein Kinase
	src	Chicken	Sarcoma	–	"	Tyrosine Specific Protein Kinase
	abl	Mouse	Lymphoma	Leukemia	"	"
	fes	Cat	Sarcoma	–	Cytoplasm Plasma Membrane(?)	"
	fps	Chicken				
	yes	Chicken	Sarcoma	–	Plasma Membrane(?)	
	fgr	Cat	Sarcoma	–	?	"
	ros	Chicken	Sarcoma	–	?	"
III	H-ras	Mouse	Sarcoma	Carcinoma	Inner Face of Plasma Membrane	GTP Binding GTPase
	K-ras	Mouse	Sarcoma	Carcinoma Leukemia Sarcoma	"(?)	"(?)
	N-ras	–	–	Neuroblastoma Leukemia Lymphoma Carcinoma	"(?)	"(?)
IV	myc	Chicken	Leukemia	Lymphoma Leukemia Carcinoma	Nucleus	DNA Binding
	myb	Chicken	Leukemia	Leukemia Carcinoma	"	"(?)
	fos	Mouse	Sarcoma	–	"	"(?)
	skl	Chicken	Sarcoma	Lymphoma(?)	Nucleus (?)	?
	N-myc	–	–	Neuroblastoma	?	?
V	mil	Chicken	Carcinoma	–	Cytoplasm	?
	raf	Mouse	Sarcoma	Carcinoma Glioblastoma	"	?
	mos	Mouse	Sarcoma	–	"	?
	rel	Turkey	Leukemia	–	?	?
	erb-A	Chicken	Erythroleukemia	–	?	?
	ets	Chicken	Leukemia Erythroblastosis	–	?	?

regulates cell growth (for reviews, see Stiles, 1983; Antoniades and Williams, 1983). PDGF stimulates a tyrosine phosphokinase activity (Nishimura et al., 1982) similar to the ones associated with several growth factor receptors. Furthermore, stimulation of quiescent 3T3 cells with PDGF leads to a rapid and transient increase in the expression of c-fos (Greenberg and Ziff, 1984) and to a temporary elevation of the levels of c-myc RNA (Kelly et al., 1983). This suggests that one role of PDGF in inducing proliferation of mesenchymal cells is to induce gene expression. Taken together these observations suggest that an oncogene, such as c-sis, can regulate the proliferation of specific responsive cells and that the ectopic expression of this gene in cells that do not make but are able to respond to PDGF could lead them to uncontrolled autonomous growth (Huang et al., 1984).

b)   Growth Factor Receptors: c-erbB and c-fms.

Growth factors act by binding to cell surface receptors, which represent the second step in a putative cascade of molecular events leading to the final signal for cell growth. It was a highly significant discovery that a receptor for a growth factor, namely the epidermal growth factor (EGF) receptor, showed significant homology with the sequence of v-erbB oncogene product (Downward et al., 1984). Analogous to the sis/PDGF system, this suggested that c-erbB proto-oncogene corresponded to the gene encoding the EGF receptor.

EGF is a potent mitogenic polypeptide that initiates a cellular response by binding specifically to its receptor, which is present on the surface of cells in several tissues. The EGF receptor is a membrane glycoprotein that is composed of an extracellular N-terminal region where the EGF binding site is localized, a transmembrane portion, and a C-terminal cytoplasmic domain that carries a tyrosine-specific protein kinase domain (for a review, see Hunter, 1984). Most notably, this internal domain is a common feature of a number of receptors and viral oncogene products (Yamamoto et al., 1983; see Table 1). Sequence comparison indicates that the v-erbB oncogene lacks the N-terminus of the EGF receptor (i.e., the EGF binding domain) yet it retains the cytoplasmic portion containing the common kinase domain (Downward et al., 1984). The possibility exists that the internal portion of the EGF receptor is shared by a family of receptors, and that c-erbB may code for a related molecule distinct in its extracellular domain. In turn, the lack of this domain in v-erb B suggests that the transforming potential of this onco-gene may be related to the capability of coding for a truncated receptor capable of transmitting a growth signal even in the absence of a ligand.

The discovery of v-erbB being related to a portion of the EGF receptor had implications for the function of several other oncogenes that are also associated with a tyrosine-specific protein

kinase activity (see Table 1). It is in fact possible that some of
these oncogenes code for other growth factor receptors. This widely
diffuse expectation found important confirmation in recent studies
indicating that the c-fms gene product is related to the receptor
for an hematopoietic growth factor, namely the macrophage growth
factor CSF-1. It has been shown that the c-fms gene product is
expressed at high levels in mature macrophages and that antibodies
against a recombinant v-fms coded polypeptide specifically react
with the murine CSF-1 receptor (Sherr et al., 1985). By analogy, it
is reasonable to expect that other oncogene proteins carrying
tyrosine kinase domains may be shown homologous to different growth
factor receptors.

c) Transducing Signals: The c-ras Family.

The oncogenes most frequently detected by DNA-mediated
transformation of NIH 3T3 cells belong to the family of c-ras genes
that code for a number of immunologically related proteins. Two
members of this family, H-ras and K-ras, are homologous to the
oncogenes of Harvey and Kirstein sarcoma viruses, respectively
(Ellis et al., 1981). The third known member, N-ras, has not been
found in any retrovirus so far and has been identified by
transfection of DNA from a neuroblastoma cell line into NIH 3T3
cells (Shimizu et al., 1983). These ras proteins are highly
homologous in their amino-terminal region, and are more divergent
toward·their carboxy-terminus (for a recent review, see Weinberg,
1984). Most of the characterization of these proteins involves
studies on the p21 H-ras protein, which has been shown to be
localized at the inner surface of the cell membrane (Willingham et
al., 1980; Furth et al., 1982). Its most notable biochemical
properties are its cability to bind GTP with high affinity and its
GTPase activity (Shih et al., 1980; McGrath et al., 1984; Sweet et
al., 1984). These properties, an identical subcellular localization
and a partial sequence homology, suggest that the H-ras protein may
belong to the family of nucleotide-binding G proteins, which are
involved in the transduction of signals from receptors on the
outside of the cell membrane to the adenylate cyclase system on the
inner side of the membrane (Gilman, 1984). G proteins can be
stimulatory (Gs) or inhibitory (Gi), each being formed by three
subunits (α, β, γ), and both containing a GTPase activity that
regulates their function. When a specific factor binds its
receptor, the Gs complex will bind GTP, leading to the activation of
adenylate cyclase and to further transmission of the stimulus
(Gilman, 1984). However, since the Gsα peptide contains an
intrinsic GTPase activity, it will hydrolyse GTP, resulting in the
Gsα protein ceasing its interaction with adenylate cyclase. These
observations suggest that the G proteins, and most likely the ras
proteins, are capable of both initiating and suppressing the
transduction of a signal originated in the membrane by the binding

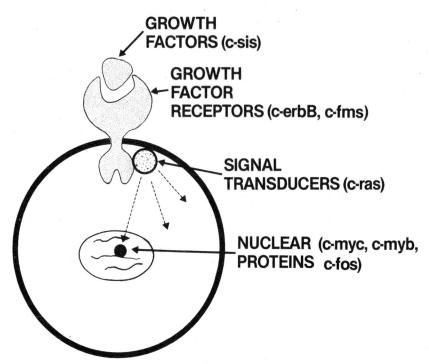

Fig. 3   Schematic representation of the main elements involved
              in the control of cell proliferation. In parenthesis
              are indicated some of the c-onc proteins belonging to
              the different functional classes.

of a ligand to its receptor. Consistent with these notions are the
observations that the addition of EGF to isolated cell membranes
results in enhanced GTP binding by the H-ras protein (Kamata and
Feramisco, 1984) and that microinjection of the c-ras protein into
quiescent cells induces DNA synthesis (Feramisco et al., 1984;
Stacey and Kung, 1984). Finally, activated c-ras genes from tumor
DNA code for proteins that have lost their regulatory GTPase
activity, which suggests that they may lead to constitutive rather
than regulated signal transduction (McGrath et al., 1984; Sweet et
al., 1984).

        d) Nuclear Proteins: c-myc.

        The c-myc proto-oncogene is the cellular homolog of the
oncogene, v-myc, of avian myelocytomatosis virus (Roussel et al.,
1979). The viral oncogene can cause tumors in a wide spectrum of
tissue in vivo, since it can cause sarcomas, carcinomas, and

hemotopoietic tumors in infected animals (Bishop, 1983). The analogous cellular myc gene does not have any tissue specificity in its pattern of expression since myc mRNA has been found in all tissue studied from different animal species (Sheiness and Bishop, 1979; Gonda et al., 1982). Rather than tissue specificity, c-myc gene expression displays an apparent cell-cycle specificity in several cell types tested. No myc mRNA is detectable in quiescent cells blocked in G_0 (Campisi et al., 1984), whereas induction of transcription can be obtained a few hours after mitogenic stimulus, in apparent correspondence with the G_0-G_1 transition of the cell cycle (Kelly et al., 1983; Greenberg and Ziff, 1984). Subsequently, myc mRNA and protein appear to be down-regulated to relatively lower levels that remain constant throughout the cell cycle (Thompson et al., 1985; Hann et al., 1985). Interestingly, together with a positive correlation with cell proliferation, c-myc expression displays a negative correlation with differentiation, at least as shown in a few experimental systems. In vitro induction of terminal differentiation in both the human promyelocytic leukemia (HL-60) (Westin et al., 1982) and in the Friend murine erythroleukemia cell lines (Lachman and Skoultchi, 1984) determines an early suppression of c-myc mRNA expression. Since in these cells terminal differentiation also involves an early and irreversible arrest of cell proliferation, it remains possible that this suppression may again reflect the proliferative status. Consistent with this idea, c-myc mRNA remains present after induction of differentiation in rat pheochromocytoma cells (PC12), which retain some proliferative capability in their partially differentiated state (Greenberg and Ziff, personal communication).

The induction of c-myc transcription appears to be instrumental for the transition through the G_0-G_1 phase of the cell cycle rather than simply related to it. This was demonstrated by showing that the expression of a transfected gene in mouse fibroblast can partially substitute for the mitogenic stimulus of PDGF (Armelin et al., 1984). In other words, c-myc expression appears to represent an additional step in the cascade of events that starts with growth factor/receptor interaction, extends through the intracellular signal transduction system, and leads to the activation that is thought to occur in the nucleus.

This hypothesis is confirmed by studies on the characterization of the c-myc protein and by preliminary data on its function. A 62-64Kd myc protein is recognized by different specific antisera in the nucleus of different cells (Ramsay et al., 1984; Donner et al., 1982), which has a consistent pattern of expression with the one observed for the c-myc gene (Persson and Leder, 1984). In vitro, this protein has been shown to bind to DNA. Furthermore, its subcellular location, as well as its biochemical properties and sequencing homology data, suggest that the c-myc protein shares some functional and structural similarities with proteins coded by the early region of genes of DNA tumor viruses (Ralston and Bishop,

1983). Most notably, the main functional characteristic of these nuclear DNA-binding proteins is their capability to regulate both positively and negatively the expression of other genes (Kingston et al., 1985). Such a trans-acting regulatory activity has been preliminarily reported for c-myc gene in experiments involving the cotransfection of c-myc gene clones and selected target genes (Kingston et al., 1984; our unpublished results). These observations point to the c-myc gene having a critical role in translating an extracellular mitogenic stimulus into a gene regulation process that is likely to be involved in the initiation of cell proliferation and possibly involved in the suppression of differentiation. While these hypotheses are now becoming amenable to direct experimental testing, it is important to note that other oncogenes, for example c-fos and c-myb, share some structural and/or functional characteristics with c-myc (Ralston and Bishop, 1983; Greenberg and Ziff, 1984), suggesting that they may belong to the same functional class.

MECHANISMS OF ONCOGENE ACTIVATION IN HUMAN MALIGNANCIES

Based on the general outline presented in the previous sections, it is theoretically possible that alterations involving the function of cellular oncogenes may be involved in the pathogenesis of cancer by disrupting the mechanisms regulating normal cell growth. Consistent with this theory, different alterations involving cellular oncogenes have been found in a considerable number of animal and human tumors. These alterations affect either the structure or the gene regulation pattern of a given oncogene and are commonly, yet perhaps improperly, referred to as "activation events." Different lines of experimentation have lead to the identification of three main mechanisms of activation in human tumors: (1) specific mutations have been identified by DNA-mediated gene transfer into NIH 3T3 cells, (2) tumor-specific chromosomal aberrations have been shown to be tied to the chromosomal position of various oncogenes, and (3) amplifications and rearrangements have been identified in various tumors by direct analysis of the genomic organization of different oncogenes.

a) Mutations.

We have previously described the transfection assay in NIH 3T3 cells that allows the detection of activated oncogenes from tumor DNA. In the initial studies, involving the analysis of DNA from the human bladder carcinoma line EJ, the human DNA sequences responsible for the transforming activity were shown to be homologous to the c-H-ras gene by hybridization and by heteroduplex analysis with a clone containing the human c-H-ras gene from normal DNA (Der et al., 1982; Parada et al., 1982; Santos et al., 1982).

Table II Examples of Mutations Found in Different
Transforming ras Oncogenes Compared to Normal Sequences
and Their Ability to Induce Transformation In Vitro.

	POSITION:	11	12	13		60	61	62	
		GCA	GGT	GGT		GGA	CAA	GAA	
Normal	K-ras	... ala	gly	gly	.....	gly	gln	glu	..
	v-K-ras	... ala	ser	gly	.....	gly	gln	glu	..
Transforming (Lung cancer)	K-ras	... ala	cys	gly	.....	gly	gln	glu	..
Normal	H-ras	... ala	gly	gly	.....	gly	gln	glu	..
	v-H-ras	... ala	arg	gly	.....	gly	gln	glu	..
Transforming (Bladder carcinoma)	H-ras	... ala	val	gly	.....	gly	gln	glu	..
Normal	N-ras	... ala	gly	gly	.....	gly	gln	glu	..
Transforming (Neuroblastoma)	N-ras	... ala	gly	gly	.....	gly	lys	glu	..
Transforming (Myeloid leukemia)	N-ras	...ala	gly	asp	.....	gly	gln	glu	..

Comparative sequence analysis between the EJ and the normal H-ras
clone demonstrated that the only difference was a single base change
of G to T in codon 12 of the first exon, resulting in the
replacement of a glycine residue by a valine in the corresponding
protein product (Tabin et al., 1982; Reddy et al., 1982; Taparowsky
et al., 1982). Therefore, the transforming activity of the EJ gene
was due to this single base pair mutation.

This early study served as the prototype for subsequent ones
that showed that approximately 20°/° of human tumors score
positive in the NIH 3T3 assay. In the great majority of these
positive cases a member of the ras family, - either H-ras, K-ras, or
N-ras - is involved, but there is no clear correlation between type
of tumors and a particular ras gene (Eva et al., 1983). Sequence
analysis of the biologically active clones shows that point
mutations are responsible for the activation (Reddy et al., 1982).
Additional studies, using either the NIH 3T3 assay or direct
restriction enzyme analysis of the involved DNA codons, have shown
that these mutations are somatic events detectable only in the
neoplastic cells of a given individual (Pulciani et al., 1982a;
Santos et al., 1984).

The sites in which the mutations occur are conserved among the three members of the ras family, involving either codon 12 or codon 61 (Taparowsky et al., 1983). Most notably, the same codons are found alternatively mutated in the v-H-ras (Dhar et al., 1982) and v-K-ras genes (Tsuchida et al., 1982) carried by their respective retroviruses (see Table II). In fact, in the case of codon 12, studies using c-H-ras mutants generated by in vitro mutagenesis, recombination, or spontaneous mutation have shown that the substitution of Gly 12 in the normal c-H-ras allele by any residue other than proline leads to transforming activity (Santos et al., 1983; Fasano et al., 1984b; Seeburg et al., 1984). Although the precise mechanism by which these minimal alterations so dramatically affect the biochemical function of these proteins is not known, several computer modeling studies have suggested that the loss of one glycine residue at position 12 could have profound effects on the three-dimensional conformation of the p21 proteins (Pincus et al., 1983; McCormick et al., 1985). Recently, it has been shown that the protein from the mutated gene has an impaired GTPase activity suggesting, as previously discussed, that the oncogenic protein would be incapable of down-regulating its own activity through the hydrolysis of GTP (McGrath et al., 1984; Sweet et al., 1984).

The repertoire of sites within c-ras genes that can be involved in point mutations of biological significance, as well as the repertoire of genes that can be activated by this mechanism, are likely to be larger than currently expected. Recently, using a modification of the NIH 3T3 assay that involves the direct in vivo selection of the transformed cells, transforming N-ras alleles carrying a mutation at codon 13 have been detected in a high percentage of the acute myelogenous leukemia (AML) DNA tested (Bos et al., 1985). These results are relevant since they indicate that: (1) several alternative mutation sites exist that are capable of conferring transforming activity to the N-ras gene, (2) different biologically active mutations may determine a different transformed phenotype that can be detected by different assays, e.g. by morphologic examination (codon 12 and 61) or by in vivo tumorigenicity (codon 13 in N-ras), and (3) some tumors may be associated with a constant activation event affecting the same gene (e.g., N-ras mutated at codon 13 in AML).

b) Chromosomal Aberrations.

Chromosomal aberrations of various types represent a distinctive feature of many human tumors. The consistent association between specific chromosomal markers and certain types of tumors has lead to the prediction that the chromosomal changes may be involved in carcinogenesis by altering the activity of genes important for the control of cell proliferation and differentiation (for reviews, see Klein, 1981; Rowley, 1982; Yunis, 1983). When cellular oncogenes were identified, they became immediate and obvious candidates for being the genetic elements involved in chromosomal aberrations. The strategy used for these studies involved the mapping of the normal chromosomal site for any of the given oncogenes. This was done by analysis of somatic cell hybrids constructed by fusing human cells with rodent cells. The hybrids retain all the rodent chromosomes but lose some of the human chromosomes. The sublocalization to specific chromosomal regions defined by banding has been obtained by in situ hybridization of a radioactive probe to denatured metaphase chromosome preparations.

Using these approaches, virtually all known human oncogenes have been mapped to specific chromosomal areas (see Table III and O'Brien, 1984). When these data are analyzed considering the chromosomal sites that are affected by tumor-specific chromosomal aberrations, a number of significant correlations emerge (Yunis, 1983). Since chromosomal regions recognized cytogenetically contain many thousands of kilobases of DNA, not all these correlations necessarily mean the involvement of the nearby oncogene. However, in the case of two aberrations the analysis of cloned chromosomal breakpoints has demonstrated the direct involvement of two oncogenes, namely the c-myc locus in the translocations specific of Burkitt lymphoma (BL) (Dalla-Favera et al., 1982b; Taub et al., 1982) and the c-abl locus in the translocation(s) typical of chronic myelogenous leukemia (CML) (de Klein et al., 1982). These two translocations have been analyzed in great detail at the molecular level and represent useful models for understanding the mechanisms of oncogene activation in other chromosomal aberrations.

Burkitt lymphomas are characterized by reciprocal translocations which, in $90°/°$ of the cases, involve chromosome 8 and chromosome 14 (Manolova et al., 1979) or, in $10°/°$ of cases, chromosome 8 and either chromosome 2 or 22 (Bernheim et al., 1981). The common feature of the three types of translocations is the recombination between the same region (band q24) of chromosome 8 and the chromosomal segments containing immunoglobulin (Ig) loci, namely the Ig heavy chain (IgH), light chain k (Igk), and light chain λ (Igλ) on chromosome 14, 2, and 22, respectively (Croce et al., 1979; Malcolm et al., 1982; Erikson et al., 1982). Two early findings together suggested that the c-myc oncogene could be involved in these translocations. First, the c-myc locus was located on chromosome 8 by analysis of somatic cell hybrids or by in situ hybridization (Dalla-Favera et al., 1982b; Taub et al., 1982; Neel et al., 1982). Second, rearrangements of this locus were detectable in a significant number of BL (Dalla-Favera et al., 1983; Taub et al., 1982). Direct evidence that the c-myc gene is translocated from chromosome 8 to chromosome 14 has been provided by analysis of somatic cell hybrid mouse cells and BL cells carrying the 8:14 translocation. Hybrids containing the aberrant chromsome $14q^+$ but not its reciprocal $8q^-$ contained the c-myc locus (Dalla-Favera et al., 1982b). Furthermore, the isolation from BL of clones

Table III   Chromosomal Locations of Human c-Oncogenes

Chromosome	c-onc	Band
1	N-ras Blym c-src c-ski	cen-p21 p32 p34-p36 q12-qter
2	N-myc	p23-pter
3	c-raf-1	p25
4	c-raf-2	-
5	c-fms	q34
6	c-K-ras-1 c-myb	p23-q12 q22-q24
7	c-erb-B met	pter-q22 p11.4-qter
8	c-mos c-myc	q22 q24
9	c-abl	q34
10	-	-
11	c-H-ras-1 c-ets	p15 q23-q24
12	c-K-ras-2	p12.05-pter
13	-	-
14	c-fos	q21-q31
15	c-fes	q25-q26
16	-	-
17	c-erb-A1 p53	p11-q21 q21-q22
18	-	-
19	-	-
20	c-src	q21-q13
21	-	-
22	c-sis	q12.3-q13.1
X	c-H-ras-2	

containing the c-myc gene linked to the IgH locus provided the
formal demonstration of the recombination between these two loci
(Hamlyn and Rabbitts, 1983; Gelmann et al., 1983; Battey et al.,
1983). The importance of these events for the pathogenesis of BL is
underscored by the finding that analogous chromosomal translocations
in mouse B-cell tumors are also characterized by recombinations
between c-myc and the various Ig loci.

After these initial observations, in a few selected cases
intensive investigation has been performed to examine the effects of
translocation on c-myc structure and function. It is now clear that
despite their apparent homogeneity at the cytogenetic levels, these
translocations display a considerable heterogeneity at the molecular
level since the different positions of the chromosomal breakpoints
in various BL cases lead to different structural alterations of both
the Ig and the c-myc loci. Within the IgH locus on chromosome 14,
the chromosomal breakpoints can be variably located within the
constant region genes Igμ (Dalla-Favera et al., 1982b; Taub et al.,
1982; Dalla-Favera et al., 1983; Erikson et al., 1983; Gelmann et
al., 1983; Adams et al., 1983), Igα (Showe et al., 1985), Igγ
(Hamlyn and Rabbitts, 1983), and also within the variable regions
(Erikson et al., 1982). In the variant (8:2) and (8:22)
translocations, the precise sites of breakpoints, within or in
proximity of the Igk (Taub et al., 1984b) and Igλ (Hollis et al.,
1984) loci respectively, have not been identified yet. Furthermore,
while in the t(8:14) at least part of the c-myc gene is translocated
from chromosome 8 to 14, it appears that in the variant
translocations c-myc remains on chromosome 8 and Ig genes move to
chromosome 8 (Magrath et al., 1983; Davis et al., 1984; Hollis et
al., 1984). Relatively to the c-myc locus, breakpoints have been
mapped at different sites, including 5' and 3' flanking cellular
sequences and within the 5' untranslated portion of the gene (Leder
et al., 1983). Furthermore, some studies have also indicated the
occurrence of small rearrangements, such as duplications,
insertions, deletions, or point mutations, both in noncoding and
coding portions of the translocated c-myc gene (Taub et al., 1984a;
Hayday et al., 1984; Rabbitts et al., 1983).

This heterogeneity of structural alterations finds a
counterpart in the pattern of myc in RNA expression in different BL
cases. While in some cases, relatively high levels of myc RNA can
be found, in some other cases these levels are lower than in normal
proliferating cells or non-tumorigenic B-lymphoblastoid lines which
are used as controls (Taub et al., 1984a; Lanfrancone et al.,
1984). These observations have lead to the presently widely
accepted notion that a deregulation rather than an absolute increase
of expression may be critical for c-myc activation in BL.

With respect to the precise mechanism by which this
deregulation occurs a clear understanding is still lacking, although
several models have been proposed depending upon the experimental
approach used. While we refer to another chapter of this book for a

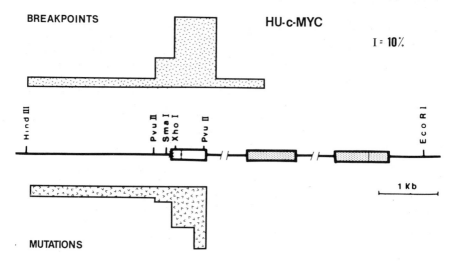

Fig. 4   Structural alterations of the c-myc locus in BL and
              their approximate location. Boxes within the c-myc
              partial restriction map represent the three c-myc exons
              (white box: noncoding exon; dotted boxes: coding
              exons). Boxed areas above and below the map indicate
              the relative frequency and approximate location of
              breakpoints and mutations within the c-myc locus.

detailed presentation of one of these lines of research, we
summarize here the different models as being based on two main
hypotheses. The first implies an active role for the Ig loci,
namely that transcriptional elements of Ig genes influence c-myc
expression after translocation. Abnormal transcriptional activation
of c-myc may be determined by the Ig enhancer elements within the Ig
locus (Hayday et al., 1984) or by putative "long distance" enhancer
elements that normally control Ig genes (Croce et al., 1984).
However, the known Ig enhancer element has been found in close
proximity of the c-myc gene in only a small minority of cases
(Hayday et al., 1984), and direct evidence for the existence of
other transcriptional regulatory elements is not yet available.
Alternatively, the finding of structural alterations in the c-myc
gene in the few BL analyzed, had pointed toward these events as the
primary cause of the deregulation of the translocated gene (Taub et
al., 1984a; Wiman et al., 1984).
        Along with this latter model we have recently attempted to
acquire an overall view of the type and frequency of structural
alterations affecting the c-myc locus by studying a relatively large

population of BL. The results of this study, schematically
illustrated in Figure 4, show that structural alterations of the myc
locus were detectable in all thirty BL tested which, however, can be
divided in two groups based on the type of alterations. The first
group is represented by cases in which the c-myc locus appears to be
truncated by the chromosomal breakpoint while in a second group, the
c-myc gene appears to be left intact by the translocations, yet a
significant number of mutations are detectable in the 5' portion of
the gene. Incidentally, these two groups appear to correlate with
the two epidemiological forms. of BL, i.e., all the sporadic
(American type) BL carry a truncated gene, while the majority of the
endemic (African type) BL carry an unrearranged mutated gene .
Taken together, these data suggest that structural alterations of
defined portions of the c-myc locus may represent a necessary, if
not sufficient, event in altering c-myc function following
translocation, while it remains possible that the Ig enhancer
elements may contribute to the novel regulation pattern of the
translocated gene in some cases.
         The clustering of either truncations or mutations shown in
Figure 4 identifies a region containing the entire first exon and a
few hundred base pairs of 5' flanking sequences which appear to be
constantly involved, suggesting that one or more regulatory
sequences may be present in this portion of the gene and that these
may be inactivated by any of the observed structural alterations.
The existence of a negative regulatory element in the c-myc locus
has been suggested by several groups, and a binding site for a
putative repressor protein has been mapped in a region approximately
1.5 kb from the 5' border of the first exon (Siebenlist et al.,
1984). More recently, in gene transfer experiments using 5' c-myc
constructs containing 5' or internal deletions in the 5' flanking
region, we have mapped a 200 bp region immediately adjacent to the I
exon, the removal of which results in an increase in the
transcription of the transfected myc clones. Furthermore, c-myc
expression may also be regulated at the post-transcriptional level,
as recent independent studies suggested that the removal of the
first non-coding exon from the c-myc gene may ffect myc expression
by increasing the stability of c-myc RNA (Piechaczyk et al., 1985)
or the efficiency of translation (Saito et al., 1983; Darveau et
al., 1985). It is then possible that different mechanisms may be
alternatively or concomitantly involved in different BL cases.
Additional ongoing studies comparing the levels of transcription,
mRNA stability and translation in different BL cases will elucidate
this complex mechanism of activation.
         In general, the case of c-myc translocation in BL stands as
an example of how chromosomal translocation can cause functional
abnormalities of an oncogene function without apparently altering
the protein product. Given the putative function of the c-myc
oncogene described in a previous section, it is obvious that the
constitutive rather than regulated expression of this gene may force

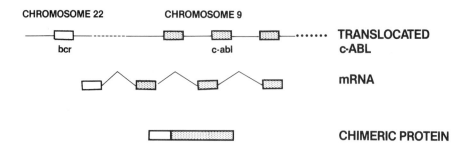

Fig. 5   Schematic representation of the recombination between
              the c-abl gene on chromosome 9 and the bcr locus on
              chromosome 22 in CML cells. See text for description
              and details.

the cell into a proliferative program. This is consistent with the
observed biological effect of transfected activated myc genes in
vitro as we will discuss in the following sections. Finally, the
example of the translocations involving the Ig and c-myc loci in BL
bear implications for analogous translocations involving Ig or
T-cell receptor gene loci in B- or T-cell lymphomas respectively.
The analysis of these translocations has already lead to the
preliminary identification of two new putative oncogenes (Tsujimoto
et al., 1984, 1985a, 1985b), as it is described in another chapter
of this book.
        In the second well-characterized chromosomal translocation
(i.e., the t[9:22] Philadelphia translocation typical of chronic
myelogenous leukemia), the involvement of the c-abl oncogene has
been conclusively demonstrated, yet the mechanism for its activation
appears to be different from the ones proposed for c-myc in BL.
Using a strategy similar to the one used for the mapping of c-myc in
BL, the c-abl locus has been mapped to the region of chromosome 9
which is translocated to chromosome 22 forming the Philadelphia
chromosome (de Klein et al., 1982; Heisterkamp et al., 1983; Bartram
et al., 1983). Using a "chromosome walking" procedure, clones
containing the breakpoints have been isolated from a few CML cases.
The breakpoint on chromosome 22 has been shown to cluster in a 5 kb
genomic area, termed "breakpoint clustering region" or bcr, in
different CML. This region recombines with variable segments
located immediately 5' to or within the c-abl locus on chromosome 9
(Heisterkamp et al., 1983). As illustrated in figure 5, the
recombination with the bcr locus appears to provide the c-abl gene
with new transcription initiation sites since an abnormal 8 kb mRNA
containing sequences from chromosome 22 fused to a c-abl transcript
is detectable in CML cells (Canaani et al., 1984; Collins et al.,

1984; Shtivelman et al., 1985). Correlating with this abnormal mRNA, an abnormally large c-abl protein is detectable in CML lines and pathologic samples (Konopka et al, 1984; Konopka et al., 1985). Significantly, this abnormal protein has protein kinase activity, like the transforming protein of the v-abl gene but unlike the normal c-abl protein. Additional studies will confirm whether the mechanism of c-abl activation in CML is mediated by these structural changes in its protein product.

c) Oncogene Amplification.

A number of examples of oncogene amplification in tumor cells have been reported. The first example was found in the human promyelocytic leukemia (APL) cell line HL-60, which was shown to contain 16-32 additional copies of c-myc (Collins and Groudine 1982; Dalla-Favera et al, 1982c) and relatively high levels of myc mRNA (Westin et al., 1982). c-myc amplification was also demonstrated in the leukemic cells from the same patient prior to their establishment as a cell line, indicating that the amplification had occurred in vivo, most likely as part of the pathogenesis of that particular leukemia case. The mechanism by which amplification may affect c-myc expression is not known. It is possible that an altered gene regulation pattern may result from the out-titration of the mechanism that normally regulates c-myc expression or that, as a consequence of the amplification, one or more myc copies may be structurally altered as it occurs in BL. Finally, since the chromosomal fragment appears to be mobile in the genome having been found both on chromosome 8 and in different marker chromosomes in different HL-60 clones (Nowell et al., 1983; Wollman et al., 1984; our unpublished results), the possible effects of these translocations on gene expression should also be considered.

Amplification does not appear to be a frequent event in AML since it has been found in only two of twenty-one cases tested. Interestingly, in one APL case the c-myb gene was found to be amplified (Pelicci et al., 1984) suggesting that alternative mechanisms, involving structurally and functionally similar genes, may be involved in the pathogenesis of phenotypically similar tumors.

Amplification of oncogenes other than c-myc and c-myb has been reported in several tumor types, (Table IV). Interestingly, in some tumors the amplification of a given oncogene appears to be a frequent event representing a molecular marker that correlates with some biological characteristics of the tumor. For instance, amplification of the c-myc gene is found with significant frequency in small cell lung cancer (SCLC) cell lines that display a more undifferentiated phenotype and a more aggressive growth pattern in vitro than SCLC lines lacking the amplification (Little et al., 1983). Analogously, an amplification of the N-myc gene (Schwab et al., 1983; Kohl et al., 1983), which is related to c-myc, is

Table 1V   Oncogene Amplifications in Human Tumors

Amplified Oncogene	Tumor Type	Comments	Reference
c-myc	Lung cancer	8/18 lung cancer cell lines 5/5 variant small cell lung carcinoma (SCLC-V)	Little et al., 1983
	Promyelocytic leukemia	HL60 cell line	Dalla-Favera et al., 1983 Collins and Groudine, 1982
	Colon carcinoma	COLO 320 cell line	Alitalo et al., 1983
	Breast carcinoma	1/5 cell lines	Kozbor and Croce, 1984
	Neuroblastoma	1/9 cell lines	Kohl et al., 1983
n-myc	Neuroblastoma	8/9 cell lines 1/1 tumor	Kohl et al., 1983 Schwab et al., 1983
c-myb	Acute myelogenous leukemia	ML cell line	Pelicci et al., 1984
	Colon carcinoma	COLO 201 and COLO 205 cell lines	Alitalo et al., 1984
c-erbB	Glioblastoma	4/10 tumors	Libermann et al., 1985
	Epidermoid Carcinoma	A431 cell line	Ullrich et al., 1984
c-abl	Chronic myelogenous leukemia	K562 cell line	Collins and Groudine, 1983

detectable in the more advanced metastatic stages of neuroblastoma (Brodeur et al., 1984; Schwab et al., 1984). These observations suggest that at least in some cases oncogene amplification may not be an early event during the pathogenesis of the tumor, but rather may represent a late step that confers an additional selective advantage to the cell.

In most cases, the presence of amplified oncogenes is accompanied by specific chromosomal alterations, namely homogeneous staining regions (HSR) and double minute (DM) chromosomes which in some instances have been shown to contain at least part of the amplified copies (Cowell 1982). Since both HSR and DM represent a frequent feature of many tumor types, it is possible that in many cases they contain amplified copies of presently unknown oncogenes, and that amplification may represent a mechanism frequently involved in the establishment or in the progression of the transformed phenotype.

NEOPLASTIC TRANSFORMATION AS A MULTISTEP PROCESS

The mechanisms of oncogene activation described in previous sections involve a single genetic event that may be necessary, yet not sufficient for determining a fully transformed phenotype. This observation becomes especially important when considering that much experimental and epidemiological evidence clearly demonstrates that neoplastic transformation is a multistage process (see Cairns, 1978). A critical question then arises about the individual role of different activated oncogenes in this process and the possible coexistence within a transformed cell of more than one activation event. Both in vivo and in vitro observations suggest that the genetic alterations that we have described may, in fact, represent defined, sometimes complementary, stages of tumor development.

Several studies have demonstrated that many tumors, both in humans and in animals, contain more than one activated oncogene. For instance, the APL cell line HL-60, which we have described as containing an amplified c-myc gene, also contains a mutated allele of the N-ras gene that is capable of transforming NIH 3T3 fibroblasts (Murray et al., 1983). Analogously, in the BL cell line, which carries an activated N-ras oncogene, the c-myc gene has undergone the chromosomal translocation typical of BL (Murray et al, 1983). These findings are organized in a model in Figure 6. While it is possible that these two events acting in concert are sufficient to produce the critical transformed phenotype, it is likely that a series of subsequent events contributes to the continuous selection of transformed cells with different biological properties. The observation that c-myc amplification may occur in cells that are already transformed suggests that some oncogenes may be involved in the establishment, maintenance, and progression of the transformed phenotype, while other genes may be involved in the development of additional biological properties specific for a given tumor type, such as, for instance, the capability to generate metastasis (Bernstein and Weinberg, 1985).

Further support for these theories and some functional information about the relationship between different oncogene-activation events derive from in vitro transfection experiments. These studies were prompted in part, by the observation, that while a single activated gene (e.g., c-ras) is sufficient to transform NIH 3T3 cells, the same gene is unable to transform primary cultures of normal mouse fibroblasts. This implies that NIH 3T3 cells had already undergone some changes on the malignancy pathway, and these changes are likely to be involved in the capability of NIH 3T3 cells to grow indefinitely in vitro as a permanent cell line. Based on the coexistence of activated myc and ras oncogene in vivo, experimental plans were devised involving the cotransfection of activated forms of these two genes into primary embryo fibroblasts (Land et al., 1983a, b). The success of these experiments in transforming normal cells provided direct evidence

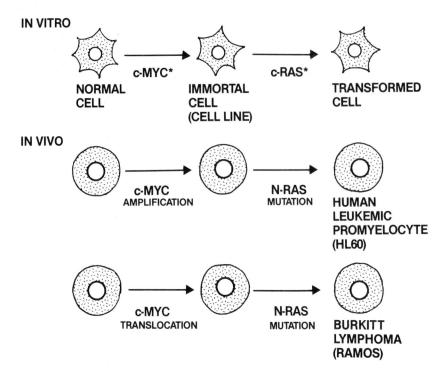

Fig. 6   Multistage transformation in vitro and in vivo. The in
                vitro section refers to transfection experiments
                involving activated(*) c-myc and c-ras genes as
                described in the text. The in vivo section represents
                a model of malignant transformation of promyelocytes or
                lymphocytes based on the findings in the two cell lines
                HL-60 and Ramos. This model does not necessarily imply
                a fixed temporal succession of the indicated
                alterations nor excludes the possibility that other
                genetic alterations may also be involved.

that at least two genetic changes are needed for tumorigenic
conversion in vitro (Figure 6). Furthermore, these experiments
suggested that different classes of oncogenes can be identified
based on their role in this assay.
        A first class is represented by genes that, when activated
are capable of making NIH 3T3 morphologically transformed and
tumorigenic in vivo, yet transfection of these genes into primary
cells does not lead to transformation. In addition to members of

the ras family of oncogenes, a number of cellular oncogenes (see Groups I and II in Table I) and oncogenes from the DNA tumor viruses (adeno, SV-40 and polyoma) share the same properties. Conversely, a second class of oncogenes, including c-myc are not able to transform NIH 3T3 cells, yet they can complement genes of the first class in the tumorigenic conversion of primary cells (Land et al., 1983a, b). By themselves these genes can lead to the outgrowth of cells that can overcome senescence and can, under appropriate conditions, be established as permanent nontumorigenic cell lines (Ruley et al., 1984). Most notably, the oncogenes that share these biological properties (namely, the large T-antigen genes of SV40 and polyoma, the EIA gene of adenovirus, and perhaps some cellular oncogenes of group IV in Table I) have other structural and functional similarities. They are all nuclear, DNA-binding proteins with putative trans-acting regulatory activity, as previously discussed.

Other studies suggest a more flexible model for the role of oncogenes belonging to different functional classes. For instance, using vectors in which the overexpression of a mutated c-ras gene is driven by a strong constitutive promoter, primary cells can be converted to immortalized tumorigenic cells without the complementation of a second oncogene (Spandidos and Wilkie, 1984). However, since in naturally occurring tumors mutations but not overexpression of the c-ras gene have been found, the significance of this finding may be strictly limited to its experimental context. Furthermore, the simultaneous alteration of both the structure and the expression may still be regarded as a dual genetic event.

Despite these unresolved issues, the studies described above must be regarded as providing both a technical and conceptual framework for experimental approaches aimed at dissecting the different genetic components involved in tumor development. The direct correlation between the observations made in vivo and experimental testing in vitro provides strong support for the validity of many of the theories discussed and points toward a number of potential developments along the same lines of research.

FUTURE PROSPECTS

The large body of experimental work briefly presented in this chapter strongly supports the general notion that cellular oncogenes are part of the genetic machinery controlling cell proliferation and differentiation and that various alterations affecting these genes are involved in tumor formation. We have attempted to stress the significant correlations between the data emerging from the genetic analysis and a number of well established observations on the epidemiology, biology, and cytogenetics of different tumors. It is important to mention that not even in a single type of tumor has the complete pathogenesis ever been elucidated and for the great

majority of tumors no cellular oncogene involvement has been
detected. It appears however that both the conceptual and technical
framework has been constructed and that a number of developments or
applications of these findings can be predicted in the near future.

The increasing number of correlations between the sites
chromosomal breakpoints and the location of oncogenes suggests that
the genetic architecture of several tumor-specific cytogenetic
aberrations will be elucidated. Moreover, it is important to note
that our present knowledge of chromosomal alterations and genomic
rearrangements is limited to microscopically detectable events which
are recognized by classic cytogenetic techniques. It is possible
that in tumor cells rearrangements or translocations may involve
sub-microscopic chromosomal fragments that would escape detection,
being too small to be detected by cytogenetic analysis and too large
to be detected by restriction enzyme Southern blot analysis. The
possibility of identifying such small rearrangements relies on the
improvemennt of chromosomal banding techniques as well as the
development of sensitive and reliable in situ hybridization methods
capable of detecting the eventual translocation of small genomic
fragments containing oncogenes.

The approach based on the identification of oncogenes by
DNA-mediated gene transfer also appears to be susceptible to future
technical and conceptual developments. The finding that only a
fraction of human tumors contain oncogenes which are capable of
transformation of NIH 3T3 cells, suggests that both the assay, based
on morphologic transformation, and the recipient cells, in terms of
tissue and perhaps species specificity, may be the limiting
factors. We have already mentioned that modifications of the assay
involving the direct selection for in vivo tumorigenicity have lead
to the identification of novel types of activation events and novel
oncogenes. Along the same lines, considering the role of oncogenes
in the control of cell proliferation, it is likely that different
assays based on specific growth conditions, for instance growth
factor independence or growth in appropriately defined media, will
lead to further developments in this area. Similarly, the definite
tissue specificity of expression displayed by certain oncogenes
suggests that some activation events may become phenotypically
detectable only in assays based on recipient cells of specific
lineages. While technical difficulties which are presently delaying
the practical utilization of these approaches do not appear to be
impossible to overcome, it is conceivable that these studies will
significantly enlarge the repertoire of known oncogenes and our
understanding of their relative mechanism of activation. The final
evidence for the causal, rather than circumstantial role (Duesberg,
1983; Rubin, 1984) of cellular oncogenes in human tumors will be
provided by showing that introducing the activated genes from a
given tumor in primary human cells of the same tissue results in
neoplastic transformation.

   While new experimental approaches will contribute to improve
our understanding of the pathogenesis of cancer, a number of general
observations of potential relevance for the diagnosis and
classification of different human tumors can be derived from our
present knowledge of oncogenes from several of the studies cited.
It is apparent that phenotypically similar tumors derived from the
same tissue do not always contain the same genetic alterations.
These differences may reflect the additive nature of the
transformation process, as in the case of the occurrence of oncogene
amplification in advanced stage tumors, or the alternative
involvement of functionally similar pathogenetic mechanisms, as it
is perhaps in the case of c-myc or c-myb amplification in AML.
Regardless of the precise mechanisms involved, this heterogeneity
suggests new criteria for the diagnosis and classification of
different tumor types which can be based on the type of pathogenetic
lesions. This type of classification, more than the ones relying on
classic, histologic, immunophenotypic or immunochemical criteria,
may serve as prognostic indicators of a number of biological and
clinical variables which often characterize phenotypically similar
tumors, such as invasiveness, hormone dependence and drug resistance.
   In a more distant future these same concepts may find
therapeutic applications. In this respect, the example of tumors
carrying altered growth factor receptors represents an example of
cases in which the identification of the genetic lesion suggest an
obvious and accessible target. In other cases the strategy may not
be as straightforward. It is however, encouraging to note that
since some genetic lesions are common to many different types of
tumors, very few, well-focused strategies may turn out to be useful
to attack many different enemies.

ACKNOWLEDGMENTS

   We are grateful to Pier Giuseppe Pelicci and Luisa
Lanfrancone for contributing some of the experimental data presented
in this chapter, to Francis Kern for a critical review and to Dianne
Nazario for expert editing. This work has been supported by Grant
No. CA 37165 from the National Cancer Institute and from the Cancer
Center Core Support Grant No. P30CA-16087. E.C. is supported by a
Fellowship from the Consejo Nacional de Ciencia y Tecnologia,
Mexico. R.D.F. is a Scholar of the Leukemia Society of America.

REFERENCES

Adams, J. M., Gerondakis, S., Webb, E., Corcoran, L. M. and Cory, S.
(1983). Cellular myc oncogene is altered by chromosome translocation
in an immunoglobulin locus in murine plasmacytomas and is rearranged
similarly in human Burkitt lymphomas. Proc. Nat. Acad. Sci. USA 80,
1982-1986.

Alitalo, K., Schwab, M., Lin, C. C., Varmus, H. E. and Bishop, J. M. (1983). Homogeneously staining chromosomal regions contain amplified copies of an abundantly expressed cellular oncogene (c-myc) in colon carcinoma. Proc. Nat. Acad. Sci. USA 80, 1707-1711.

Alitalo, K., Winqvist, R., Lin, C. C., de la Chapelle, A., Schwab, M. and Bishop, J. M. (1984). Aberrant expression of an amplified c-myb oncogene in two cell lines from a colon carcinoma. Proc. Nat. Acad. Sci. USA 81, 4534-4538.

Antoniades, H. N. and Williams, L. T. (1983). Human platelet-derived growth factor: Structure and function. Fed. Proc. 42, 2630-2634.

Armelin, H. A., Armelin, M. C. S., Kelly, K., Stewart, T., Leder, P., Cochran, B. H. and Stiles, C. D. (1984). Functional role for c-myc in mitogenic response to platelet-derived growth factor. Nature 310, 655-660.

Bartram, C. R., de Klein, A., Hagemeijer, A., van Agthoven, T., van Kessel, A. G., Bootsma, D., Grosveld, G., Ferguson-Smith, M. A., Davies, T., Stone, M, Heisterkamp, N., Stephenson, J. R. and Groffen, J. (1983). Translocation of c-abl oncogene correlates with the presence of a Philadelphia chromosome in chronic myelocytic leukaemia. Nature 306, 277-280.

Battey, J., Moulding, C., Taub, R., Murphy, W., Stewart, T., Potter, H., Lenoir, G. and Leder, P. (1983). The human c-myc oncogene: Structural consequences of translocation into the IgH locus in Burkitt Lymphoma. Cell 34, 779- 787.

Bernheim, A., Berger, R., and Lenoir, G. (1981). Cytogenetic studies on African Burkitt's lymphoma cell line t(8;14), t(2;8) and t(8;22) translocations. Cancer Genet. Cytogenet. 3, 307-315.

Bernstein, S. C. and Weinberg, R. A. (1985). Expression of the metastatic phenotype in cells transfected with human metastatic tumor DNA. Proc. Nat. Acad. Sci. USA 82, 1726-1730.

Bishop, J. M. (1983). Cellular oncogenes and retroviruses. Ann. Rev. Biochem. 52, 301-354.

Bishop, J. M. (1985). Viral oncogenes. Cell 42, 23-38.

Blair, D. G., Cooper, C. S., Oskarsson, M. K., Eader, L. A. and Vande Woude, G. F. (1982). New method for detecting cellular transforming genes. Science 218, 1122-1125.

Bos, J. L., Toksoz, D., Marshall, C. J., Verlaan-de Vries, M., Veeneman, G. H., van der Eb, A. J., van Boom, J. H., Janssen, J. W. G. and Steenvoorden, A. C. M. (1985). Amino-acid substitutions at codon 13 of the N-ras oncogene in human acute myeloid leukaemia. Nature 315, 726-730.

Boveri, T. (1914). Zur Frage der Erstehung Maligner Tumoren (Fischer, Jena).

Brodeur, G. G., Seeger, R. C., Schwab, M., Varmus, H. E. and Bishop, J. M. (1984). Amplification of N-myc in untreated human neuroblastomas correlates with advanced sisease stage. Science 224, 1121-1124.

Cairns, J. (1978). Cancer, science and society. W. H. Freeman, San Francisco.

Campisi, J., Gray, H. E., Pardee, A. B., Dean, M. and Sonenshein, G. E. (1984). Cell-cycle control of c-myc but not c-ras expression is lost following chemical transformation. Cell 36, 241-247.

Canaani, E., Steiner-Saltz, D., Aghai, E., Gale, R. P., Berrebia, A. and Januszewicz, E. (1984). Altered transcription of an oncogene in chronic myeloid leukaemia. Lancet I, 593-595.

Chiu, I. M., Reddy, E. P., Givol, D., Robbins, K. C., Tronick, S. R. and Aaronson, S. A. (1964). Nucleotide sequence analysis identifies the human c-sis proto-oncogene as a structural gene for platelet-derived growth factor. Cell 37, 123-129.

Collins, S. and Groudine, M. (1982). Amplification of endogenous myc-related DNA sequences in a human myeloid leukaemia cell line. Nature 298, 679-681.

Collins, S. and Groudine, M. (1983). Rearrangement and amplification of c-abl sequences in the human chronic myelogenous leukemia cell line K-562. Proc. Nat. Acad. Sci. USA 80, 4813-4817.

Collins, S. J., Kubonishi, I., Miyoshi, I., and Groudine, M. (1984). Altered transcription of the c-abl oncogene in K-562 and other chronic myelogenous leukemia cells. Science 225, 72-74.

Cooper, G. M., Okenquist, S., and Silverman, L. (1980). Transforming activity of DNA of chemically transformed and normal cells. Nature, 284, 418-421.

Cowell, J. K. (1982). Double minutes and homogeneously staining regions: Gene amplification in mammalian cells. Annu. Rev. Genet. 16, 21-59.

Croce, C. M., Shander, M., Martinis, J., Cicurel, L., D'Ancona, G., Dolby, T. W. and Koprowski, H. (1979). Chromosomal location of the genes for human immunoglobulin heavy chains. Proc. Nat. Acad. Sci. USA 76, 3416-3419.

Croce, C. M., Thierfelder, W., Erikson, J., Nishikura, K., Finan, J., Lenoir, G. M. and Nowell, P. C. (1983). Transcriptional activation of an unrearranged and untranslocated c-myc oncogene by translocation of a Cλ locus in Burkitt lymphoma cells. Proc. Nat. Acad. Sci. USA 80, 6922-6926.

Croce, C. M., Erikson, J., Ar-Rushdi, A., Aden, D., and Nishikura, K. (1984). Translocated c-myc oncogene of Burkitt lymphoma is transcribed in plama cells and repressed in lymphoblastoid cells. Proc. Nat. Acad. Sci. USA 81, 3170-3174.

Dalla-Favera, R., Gelmann, E. P., Gallo, R. C. and Wong-Staal, F. (1981). A human onc gene homologous to the transforming gene (v-sis) of simian sarcoma virus. Nature, 292, 31-35.

Dalla-Favera, R., Gelmann, E. P., Martinotti, S., Franchini, G., Papas, T. S., Gallo, R. C. and Wong-Staal, F. (1982a). Cloning and characterization of different human sequences related to the onc gene (v-myc) of avian myelocytomatosis virus (MC29). Proc. Nat. Acad. Sci. USA 79, 6497-6501.

Dalla-Favera, R., Bregni, M., Erikson, J., Patterson, D., Gallo, R. C. and Croce, C. M. (1982b). Human c-myc onc gene is located on the region of chromosome 8 that is translocated in Burkitt lymphoma cells. Proc. Nat. Acad. Sci. USA 79, 7824-7827.

Dalla-Favera, R., Wong-Staal, F., and Gallo, R. C. (1982c). Onc gene amplification in promyelocytic leukaemia cell kine HL-60 and primary leukaemic cells of the same patient. Nature 299, 61-63.

Dalla-Favera, R., Martinotti, S., Gallo, R. C., Erikson, J. and Croce, C. M. (1983). Translocation and rearrangements of the c-myc oncogene locus in human undifferentiated B-cell lymphomas. Science 219, 963-967.

Darveau, A., Pelletier, J., and Sonenberg, N. (1985). Differential efficiencies of in vitro translation of mouse c-myc transcripts differing in the 5' untranslated region. Proc. Nat. Acad. Sci. USA 82, 2315-2319.

Davis, M., Malcolm, S. and Rabbits, T. (1984). Chomosome translocation can occur on either side of the c-myc oncogene in Burkitt lymphoma cells. Nature 308, 286-288.

de Klein, A., Geurs van Kessel, A., Brosveld, G., Bartram, C. R., Hagemeijer, A., Bootsma, D., Spurr, N. K., Heisterkamp, N., Groffen, J. and Stephenson J. R. (1982). A cellular oncogene is translocated to the Philadelphia chromosome in chronic myelocytic leukaemia. Nature 300, 765-767.

Der, C. J., Krontiris, T. G. and Cooper, G. M. (1982). Transforming genes of human bladder and lung carcinoma cell lines are homologous to the ras genes of Harvey and Kirstein sarcoma viruses. Proc. Nat. Acad. Sci. USA 79, 3637-3640.

Der, C. J. and Cooper, G. M. (1983). Altered gene products are associated with activation of cellular ras^K genes in human lung and colon carcinomas. Cell 32, 201-208.

Dhar, R., Ellis, R. W., Shih, T. Y., Oroszlan, S., Shapiro, B., Maizel, J., Lowy, D., and Scolnick, E. M. (1982). Nucleotide sequence of the p21 transforming protein of Harvey murine sarcoma virus. Science 217, 934-937.

Donner, P., Greiser-Wilke, I., and Moelling, K. (1982). Nuclear localization and DNA binding of the transforming gene product of avian myelocytomatosis virus. Nature 296, 262-266.

Doolitle, R. F., Hunkapiller, M. W., Hood, L. E., Devare, S. G., Robbins, K. C., Aaronson, S. A., and Antoniades, H. N. (1983). Simian sarcoma virus onc gene, v-sis, is derived from the gene (or genes) encoding a platelet-derived growth factor. Science 221, 275-277.

Downward, J., Yarden, Y., Mayes, E., Scrace, G., Totty, N., Stockwell, P., Ullrich, A., Schlessinger, J. and Waterfield, M. D. (1984). Close similarity of epidermal growth factor receptor and v-erb-B oncogene protein sequences. Nature 307, 521-527.

Duesberg, P. H. (1983). Retroviral transforming genes in normal cellsμ Nature 304, 219-226.

Ellis, R., W. Defeo, D., Shih, T. Y., Gonda, M. A., Young, H. A., Tsuchida, N., Lowry, D. R. and Scolnick, E. M. (1981). The p21src genes of Harvey and Kirstein sarcoma viruses originate from divergent members of a family of normal vertebrate genes. Nature 292, 506-511.

Erikson, J. , Finan, J., Nowell, P. C. and Croce, C. M. (1982) Translocation of immunoglobulin V$_H$ genes in Burkitt lymphoma. Proc. Nat. Acad. Sci. USA 79, 5611-5615.

Erikson, J., ar-Rushdi, A., Drwinga, H. L., Nowell, P. C. and Croce, C. M. (1983). Transcriptional activation of the translocated c-myc oncogene in Burkitt lymphoma. Proc. Nat. Acad. Sci. USA 80, 820-824.

Eva, A., Tronick, S. R., Gol, R. A., Pierce, J. H. and Aaronson, S. A. (1983). Transforming genes of human hematopoietic tumors: Frequent detection of ras-related oncogenes whose activation appears to be independent of tumor phenotype. Proc. Nat. Acad. Sci. USA 80, 4926-4930.

Fasano, O., Birnbaum, D., Edlund, L., Fogh, J. and Wigler, M. (1984a). New human transforming genes detected by a tumorigenicity assay. Mol. Cell. Biol. 4, 1695-1705.

Fasano, O., Aldrich, T., Tamanoi, F., Taparowski, E., Furth, M. and Wigler, M. (1984b). Analysis of the transforming potential of human H-ras by random mutagenesis. Proc. Nat. Acad. Sci. 81, 4008-4012.

Feramisco, J. R., Gross, M., Kamata, T., Rosenberg, M. and Sweet, R. W. (1984). Microinjection of the oncogene form of the human H-ras (T-24) protein results in rapid proliferation of quiescent cells. Cell 38, 109-117.

Furth, M. E., Davis, L. J., Fleurdelys, B. and Scolnick, E. M. (1982). Monoclonal antibodies to the p21 products of the transforming gene of Harvey murine sarcoma virus and of the cellular ras gene family. J. Virol. 43, 294-304.

Gelmann, E. P., Psallidopoulos, M. C., Papas, T. S. and Dalla-Favera, R. (1983). Identification of reciprocal translocation sites within the c-myc oncogene and immunoglobulin μ locus in a Burkitt lymphoma. Nature 306, 799-803.

Gilman, A. G. (1984). G proteins and dual control of adenylate cyclase. Cell 36, 577-579.

Gonda, T. J., Sheiness, D. K. and Bishop, J. M. (1982). Transcripts from the cellular homologs of retroviral oncogenes: Distribution among chicken tissues. Mol. Cell. Biol. 2, 617-624.

Graham, F. L. and van der Eb, A. J. (1973). A new technique for the assay of infectivity of human adenovirus 5 DNA. Virology 52, 456-467.

Greenberg, M. E. and Ziff, E. (1984). Stimulation of 3T3 cells induces transcription of the c-fos proto-oncogene. Nature 311, 433-438.

Hamlyn, P. H. and Rabbits, T. H. (1983). Translocation joins the c-myc and the immunoglobulin γ1 genes in a Burkitt lymphoma revealing a third exon in the c-myc oncogene. Nature 304, 135-139.

Hann, S. R., Thompson, C. B., and Eisenman, R. N. (1985). c-myc oncogene protein synthesis is independent of the cell cycle in human and avian cells. Nature 314, 366-369.

Hayday, A. C., Gillies, S. D., Saito, H., Wood, C., Wiman, K., Hayward, W. S. and Tonegawa, S. (1984). Activation of a translocated human c-myc gene by an enhancer in the immunoglobulin heavy-chain locus. Nature 307, 334-340

Heisterkamp, N., Stephenson, J. R., Groffen, J., Hansen, P. F., De Klein, A., Bartram, C. R. and Grosveld, G. (1983). Localization of the c-abl oncogene adjacent to a translocation break point in chronic myelocytic leukaemia. Nature 306, 239-242.

Hollis, G. F., Mitchell, K. F., Battey, J., Potter, H., Taub, R., Lenoir, G. M. and Leder, P. (1984). A variant translocation places the λ immunoglobulin genes 3' to the c-myc oncogene in Burkitt's lymphoma. Nature 307, 752-755.

Huang, J. S., Huang, S. S. and Deuel, T. F. (1984).Transforming protein of simian sarcoma virus stimulates autocrine growth of SSV-transformed cells through PDGF cell-surface receptors. Cell 39, 79-87.

Hunter, T. (1984). The epidermal growth factor receptor gene and its product. Nature 311, 414-416.

Jelinek, W. R., Toomey, T. P., Leinwand, L., Duncan, C. H., Biro, P. A., Choudary, P. V., Weissman, S. M., Rubin, C. M., Houck, C. M., Deininger, P. L., and Schmid, C. W. (1980). Ubiquitous, interspersed repeated sequences in mammalian genomes. Proc. Nat. Acad. Sci. USA 77, 1398-1402.

Johnsson, A., Heldin, C. H., Wasteson, A., Westermark, B., Deuel, T. F., Huang, J. S., Seeburg, P. H., Gray, A., Ullrich, A., Scrace, G., Stroobant, P. and Waterfield, M. D. (1984). The c-sis gene encodes a precursos of the B chain of platelet-derived growth factor. EMBO J. 3, 921-928.

Josephs, S. F., Ratner, L., Clarke, M. F., Westin, E. H., Reits, M. S. and Wong-Staal, F. (1984). Transforming potential of human c-sis nucleotide sequences encoding platelet-derived growth factor. Science 225, 636-639.

Kamata, T. and Feramisco, J. R. (1984). Epidermal growth factor strimulates guanine nucleotide binding activity and phosphorylation of ras oncogene proteins. Nature 310, 147-150.

Kelly, K., Cochrn, B. H., Stiles, C. D. and Leder, P. (1983). Cell-specific regulation of the c-myc gene by lymphocyte mitogens and platelet-derived growth factor. Cell 35, 603-610.

Kingston, R. E., Baldwin Jr, A. S. and Sharp, P. A. (1984). Regulation of heat shock protein 70 expression by c-myc. Nature 312, 280-282.

Kingston, R. E., Baldwin, A. S. and Sharp, P. A. (1985). Transcription control by oncogenes. Cell 41, 3-5.

Klein, G. (1981). The role of gene dosage and genetic transpositions in carcinogenesis. Nature 294, 313-318.

Kohl, N. E., Kanda, N., Schreck, R. R., Bruns, G., Latt, S. A., Gilbert, F. and Alt, F. W. (1983). Transposition and amplification of oncogene-related sequences in human neuroblastomas. Cell 35, 359-367.

Konopka, J. B., Watanabe, S. M., and Witte, O. N. (1984). An alteration og the human c-abl protein in K562 leukemia cells unmasks associated tyrosine kinase activity. Cell 37, 1035-1042.

Konopka, J. B., Watanabe, S. M., Singer, J. W., Collons, S. J. and Witte, O. N. (1985). Cell lines and clinical isolates derived from Ph'-positive chronic myelogenous leukemia patients express c-abl proteins with a common structural alteration. Proc. Nat. Acad. Sci. USA 82, 1810-1814.

Kozbor, D. and Croce, C. M. (1984). Amplification of the c-myc oncogene in one of five human breast carcinoma cell lines. Cancer Res. 44, 438-441.

Krontiris, T. G. and Cooper, G. M. (1981). Transforming activity of human tumor DNAs. Proc. Nat. Acad. Sci. USA 78, 1181-1184.

Lachman, H. M. and Skoultchi, A. I. (1984). Expression of c-myc changes during differentiation of mouse erythroleukaemia cells. Nature 310, 592-594.

Land, H., Parada, L. F. and Weinberg, R. A. (1983a). Tumorigneic conversion of primary embryo fibroblasts requires at least two cooperating oncogenes. Nature 304, 596-602.

Land, H., Parada, L. F. and Weinberg, R. A. (1983b). Cellular oncogenes and multistep carcinogenesis. Science 222, 771-778.

Lane, M. A., Sainten, A., and Cooper, G. M. (1981). Activation of related transforming genes in mouse and human mammary carcinomas. Proc. Nat. Acad. Sci. USA 78, 5185-5189.

Lane, M. A., Neary, D. and Cooper, G. M. (1982a). Activation of a cellular transforming gene in tumours induced by Abelson murine leukaemia virus. Nature 300, 659-661.

Lane, M. A., Sainten, A., and Cooper, G. M. (1982b). Stage-specific transforming genes of human and mouse B- and T-lymphocyte neoplasms. Cell 28, 873-880.

Lanfrancone, L., Pelicci, P. G., Cesarman, E., Dalla-Favera, R. (1984). 'Mechanisms of oncogene activation in human hematopoietic tumors', in New Trands in Experimental Hematology, Oncogenes, Stem Cells, Bone Marrow Transplantation. (Eds. C. Peschle and C. Rizzoli), pp. 196-203. Aves Serono Symposia, Rome, Italy.

Le Beau, M. M., Westbrook, C. A., Diaz, M. O., Rowley, J. D. and Oren, M. (1985). Translocation of the p53 gene in t(15;17) in acute promyelocytic leukaemia. Nature 316, 826-828.

Leder, P., Battey, J., Lenoir, G., Moulding, C., Murphy, W., Potter, H., Stewart, T., and Taub, R. (1983). Translocations among antibody genes in human cancer. Science 222, 765-771.

Libermann, T. A., Nusbaum, H. R., Razon, N., Kris, R., Lax, I., Soreq, H., Whittle, N., Waterfield, M. D., Ullrich, A. and Schlessinger, J. (1985). Amplification, enhanced expression and possible rearrangement of EGF receptor in primary human brain tumours of glial origin. Nature 313, 144-147.

Little, C. D., Nau, M. M., Carney, D. N., Gazdar, A. F., and Minna, J. D. (1983). Amplification and expression of the c-myc oncogene in human lung cancer cell lines. Nature 306, 194-196.

Magrath, I., Erikson, J., Whang-Peng, J., Sieverts, H., Armstrong, G., Benjamin, D., Triche, T., Alabaster, O., and Croce, C. M. (1983). Synthesis of kappa light chains by cell lines containing an 8;22 chromosomal translocation derived from a male homosexual with Burkitt's lymphoma. Science 222, 1094-1098.

Malcolm, S., Barton, P., Murphy, C., Ferguson-Smith, M. A., Bentley, D. L. and Rabbitts, T. H. (1982). Localization of human immunoglobulin k light chain variable region genes to the short arm of chromosome 2 by in situ hybridization. Proc. Nat. Acad. Sci. USA 79, 4957-4961.

Manolova, Y., Manolov, G., Kieler, J., Levan, A. and Klein, G. (1979). Genesis of the 14q+ marker in Burkitt's lymphoma. Hereditas 90, 5-10.

Marshall, C. J., Hall, A., and Weiss, R. A. (1982). A transforming gene present in human sarcoma cell lines. Nature 299, 171-173.

Marshall, C. (1985) 'Human oncogenes', in Molecular Biology of Tumor Viruses, RNA tumor viruses, Second Edition, 2 / Supplements and Appendixes.(Eds. R. Weiss, N. Teich, H. Varmus and J. Coffin), pp 487-558, Cold Spring Harbor Laboratory, Cold Spring Harbour, New York.

McCormick, F., Clark, B. F. C., la Cour, T. F. M., Kjeldgaard, M., Norskov-Lauritsen, L. and Nyborg, J. (1985). A model for the tertiary structure of p21, the product of the ras oncogene. Science 230, 78-82.

McGrath, J. P., Capon, D. J., Goeddel, D. V. and Levinson, A. D. (1984). Comparative biochemical properties of normal and activated human ras p21 protein. Nature 310, 644-649.

Murray, M. J., Shilo, B. Z., Shih, C., Cowing, D., Hsu, H. W. and Weinberg, R. A. (1981). Three different human tumor cell lines contain different oncogenes. Cell 25, 355-361.

Murray, M. J., Cunningham, J. M., Parada, L. F., Dautry, F., Lebowitz, P. and Weinberg, R. A. (1983). The HL-60 transforming sequence: A ras oncogene coexisting with altered altered myc genes in hematopoietic tumors. Cell 33, 749-757.

Neel, B. G., Jhanwar, S. C., Chaganti, R. S. K., and Hayward, W. S. (1982). Two human c-onc genes are located on the long arm of chromosome 8. Proc. Nat. Acad. Sci. USA 79,7842-7846.

Nishimura, J., Huang, J. S., and Deuel, T. F. (1982). Platelet-derived growth factor stimulates tyrosine-specific protein kinase activity in Swiss mouse 3T3 cell membranes. Proc. Nat. Acad. Sci. 79, 4303-4307.

Nowell, P., Finan, J., Dalla-Favera, R., Gallo, R. C., ar-Rushdi, A., Romanczuk, H., Selden, J. R., Emanuel, B. S., Rovera, G. and Croce, C. M. (1983). Association of amplified concogene c-myc with an abnormally banded chromosome 8 in a human leukaemia cell line. Nature 306, 494-497.

O'Brien, S. J. (1984). Genetic Maps, pp.451, Cold Spring Harbour Laboratory, Cold Spring Harbour, New York.

Parada, L. F., Tabin, C. J., Shih, C., and Weinberg, R. A. (1982). Human EJ bladder carcinoma oncogene is homologue of Harvey sarcoma virus ras gene. Nature, 297, 474-478.

Pelicci, P. G., Lanfrancone, L., Brathwaite, M. D., Wolman, S. R. and Dalla-Favera, R. (1984). Amplification of the c-myb oncogene in a case of human acute myelogenous leukemia. Science 224, 1117-1121.

Persson, H. and Leder, P. (1984). Nuclear localization and DNA binding properties of a protein expressed by human c-myc oncogene. Science 225, 718-721.

Perucho, M., Goldfarb, M., Shimizu, K., Lama, C., Fogh, J. and Wigler, M. (1981). Biochemical transfer of single-copy eukaryotic genes using total cellular DNA as donor. Cell 14, 725-731.

Piechaczyk, M., Yang, J. Q., Blanchard, J. M., Jeanteur, P. and Marcu, K. B. (1985). Posttranscriptional mechanisms are responsible for accumulation of truncated c-myc RNAs in murine plasma cell tumors. Cell 42, 589-597.

Pincus, M. R., van Resenwoude, J., Harford, J. B., Chang, E. H., Carty, R. P. and Klausner, R. D. (1983). Prediction of the three-dimentional structure of the transforming region of the EJ/T24 human bladder oncogene product and its normal cellular homologue. Proc. Nat. Acad. Sci. USA 80, 5253-5257.

Pulciani, S., Santos, E., Lauver, A. V., Long, L. K., Aaronson, S. A. and Barbacid, M. (1982a). Oncogenes in solid human tumours. Nature 300, 539-542.

Pulciani, S., Santos, E., Lauver, A. V., Long, L. K., Robbins, K. C. and Barbacid, M. (1982b). Oncogenes in human tumor cell lines: Molecular cloming of a transforming gene from human bladder carcinoma cells. Proc. Nat. Acad. Sci. USA 79, 2845-2849.

Rabbitts, T. H., Hamlyn, P. H., and Baer, R. (1983). Altered nucleotide sequences of a translocated c-myc gene in Burkitt lymphoma. Nature, 306, 760-765.

Ralston, R. and Bishop, J. M. (1983). The protein products of the myc and myb oncogenes and adenovirus E1a are structurally related. Nature 306, 803-806.

Ramsay, G., Evan, G. I. and Bishop, J. M. (1984). The protein encoded by the human proto-oncogene c-myc. Proc. Nat. Acad. Sci. USA 81, 7742-7746.

Reddy, E. P., Reynolds, R. K., Santos, E. and Barbacid, M. (1982). A point mutation is responsible for the acquisition of transforming properties by the T24 human bladder carcinoma oncogene. Nature 300, 149-152.

Roussel, M., Saule, S., Lagrou, C., Rommens, C., Beug, H., Graf, T. and Stehelin, D. (1979). Three new types of viral oncogene of cellular origin specific for hematopoietic cell transformation. Nature 281, 452-455.

Rowley, J. D. (1982). Identification of the constant chromosome regions incolved in human hematologic malignant disease. Science 216, 749-751.

Rubin, H. (1984). Mutation and oncogenes-cause or effectμ Nature 309, 518.

Ruley, H. E., Moomaw, J. F., and Maruyama, K. (1984). Avian myelocytomatosis virus myc and adenovirus early region 1A promote the in vitro establishment of cultured primary cells. Cancer Cells 2, 481-486.

Saito, H., Hayday, A. C., Wiman, K., Hayward, W. S., and Tonegawa, S. (1983). Activation of c-myc gene by translocation: A model for translational control. Proc. Nat. Acad. Sci. USA 80, 7476-7480.

Santos, E., Reddy, E. P., Pulciani, S., and Barbacid, M. (1982). T24 human bladder carcinoma oncogene is an activated form of the normal human homoloque of BALB- and Harvey-MSV transforming genes. Nature 298, 343-347.

Santos, E., Reddy, E. P., Pulciani, S., Feldmann, R. J., and Barbacid, M. (1983). Spontaneous activation of a human proto-oncogene. Proc. Nat. Acad. Sci. USA. 80, 4679-4683.

Santos, E., Martin-Zanca, D., Reddy, Pierotti, M. A., Della Porta, G. and Barbacid, M. (1984). Malignant activation of a K-ras oncogene in lung carcinoma but not in normal tissue of the same patient. Science 223, 661-664.

Schwab, M., Alitalo, K., Klempnauer, K. H., Varmus, H. E., Bishop, J. M., Gilbert, F., Brodeur, G., Goldstein, M. and Trent, J. (1983). Amplified DNA with limited homology to myc cellular oncogene is shared by human neuroblastoma cell lines and a neuroblastoma tumour. Nature 305, 245-248.

Schwab, M., Ellison, J., Busch, M., Rosenau, W., Varmus, H. E. and Bishop, J. M. (1984). Enhanced expression of the human gene N-myc consequent to amplification of DNA may contribute to malignant progression of neuroblastoma. Proc. Nat. Acad. Sci. USA 81, 4940-4944.

Seeburg, P. H., Colby, W. W., Capon, D. J., Goeddel, D. V. and Levinson, A. D. (1984). Biological properties of human c-Ha-ras-1 genes mutated at codon 12. Nature 312, 71-75.

Sheiness, D., and Bishop, J. M. (1979). DNA and RNA from uninfected vertebrate cells contain sequences related to the putative transforming gene of avian myelocytomatosis virus. J. Virol. 31, 514-521.

Sherr, C. J., Rettenmier, C. W., Sacca, R., Roussel, M. F., Look, A. T. and Stanley, E. R. (1985). The c-fms proto-oncogene product is related to the receptor for the mononuclear phagocyte growth factor, CSF-1. Cell 41, 665-676.

Shih, C., Shilo, B. Z., Goldfarb, M. P., Dannenberg, A. and Weinberg, R. A. (1979). Passage of phenotypes of chemically transformed cells via transfection of DNA and chromatin. Proc. Nat. Acad. Sci. USA 76, 5714-5718.

Shih, T. Y., Papageorge, A. G., Stokes, P. E., Weeks, M. O. and Scolnick, E. M. (1980). Guanine nucleotide-binding and autophosphorylating activities associated with the p21src protein of Harvey murine sarcoma virus. Nature 287, 686-691.

Shih, C., Padhy, L. C., Murray, M. and Weinberg, R. A. (1981). Transforming genes of carcinomas and neuroblastomas introduced into mouse fibroblasts. Nature 290, 261-264.

Shih, C. and Weinberg, R. A. (1982). Isolation of a transforming sequence from a human bladder carcinoma cell line. Cell 29, 161-169.

Shimizu, K., Golfarb, M., Perucho, M and Wigler, M. (1983). Isolation and preliminary characterization of the transforming gene of a human neuroblastoma cell line. Proc. Nat. Acad. Sci. USA 80, 383-387.

Showe, L. C., Ballantine, M., Nishikura, K., Erikson, J., Kaji, H. and Croce, C. M. (1985). Cloning and sequencing of a c-myc oncogene in a Burkitt lymphoma cell line that is translocated to a germ line alpha switch region. Mol. Cell. Biol. 5, 501-509.

Shtivelman, E., Lifshitz, B., Gale, R. P. and Canaani, E. (1985). Fused transcript of abl and bcr genes in chronic myelogenous leukaemia. Nature, 315, 550-554.

Siebenlist, U., Hennighausen, L., Battey, J. and Leder, P. (1984). Chromatin structure and protein binding in the putative regulatory region of the c-myc gene in Burkitt lymphoma. Cell 37, 381-391.

Spandidos, D. A. and Wilkie, N. M. (1984). Malignant transformation of early passage rodent cells by a single mutated human oncogene. Nature 310, 469-475.

Stacey, D. W. and Kung, H. F. (1984). Transformation of NIH3T3 cells by microinjection of Ha-ras p21 protein. Nature 310, 508-511.

Stiles, C. D. (1983). The molecular biology of platelet-derived growth factor. Cell 33, 653-655.

Sukumar, S., Notario, V., Martin-Zanca, D. and Barbacid, M. (1983). Induction of mammary carcinomas in rats by nitroso-methylurea involves malignant activation of H-ras-1 locus by single point mutations. Nature 306, 658-661.

Sweet, R. W., Yokoyama, S., Kamata, T., Feramisco, J. R., Rosenberg, M. and Gross, M. (1984). The product of ras is a GTPase and the T24 oncogenic mutant is deficient in this activity. Nature 311, 273-275.

Tabin, C., Bradley, S., Bargmann, C., Weinberg, R., Papageorge, A., Scolnick, E., Dhar, R., Lowy, D., and Chang, E. (1982). Mechanism of activation of a human oncogene. Nature 300, 143-148.

Taparowsky, E., Suard, Y., Fasano, O., Shimizu, K., Goldfarb, M. and Wigler, M. (1982). Activation of the T24 bladder carcinoma transforming gene is linked to a single amino acid change. Nature 300, 149-152.

Taparowsky, E., Shimizu, K., Goldfarb, M. and Wigler, M. (1983). Structure and activation of the human N-ras gene. Cell 34, 581-586.

Taub, R., Kirsch, I., Morton, C., Lenoir, G., Swan, D., Tronick, S., Aaronson, S. and Leder, P. (1982). Translocation of the c-myc gene into the immunoglobulin heavy chain locus in human Burkitt lymphoma and murine plasmacytoma cells. Proc. Nat. Acad. Sci. USA 79, 7837-7841.

Taub, R., Moulding, C., Battey, J., Murphy, W., Vasicek, T., Lenoir, G. M. and Leder, P. (1984a). Activation and somatic mutation of the translocated c-myc gene in Burkitt lymphoma cells. Cell 36, 339-348.

Taub, R., Kelly, K., Battey, J., Latt, S., Lenoir, G. M., Tantravahi, U., Tu, Z. and Leder, P. (1984b). A novel alteration in the structure of an activated c-myc gene in a variant t(2;8) Burkitt lymphoma. Cell 37, 511-520.

Thompson, C. B., Challoner, P. B., Neiman, P. E. and Groudine, M. (1985). Levels of c-myc oncogene mRNA are invariant throughout the cell cycle. Nature 314, 363-369.

Tsuchida, N., Ryder, T., and Ohtsubo, E. (1982). Nucleotide sequence of the oncogene encoding the p21 transforming protein of of Kirstein murine sarcoma virus. Science 217, 937-939.

Tsujimoto, Y., Yunis, J., Onorato-Showe, L., Erikson, J., Nowell, P. C., and Croce, C. M. (1984). Molecular cloning of the chromosomal breakpoint of B-cell lymphomas and leukemias with t(11;14) chromosome translocation. Science 224, 1403-1406.

Tsujimoto, Y., Cossman, J., Jaffe, E., and Croce, C. M. (1985a). Involvement of the bcl-2 gene in human follicular lymphoma. Science 228, 1440-1446.

Tsujimoto, Y., Gorham, J., Cossman, J., Jaffe, E., and Croce, C. M. (1985b). The t(14;18) chromosome translocation involved in B-cell neoplasms result from mistakes in VDJ joining. Science 229, 1390-1393.

Ullrich, A., Coussens, L., Hayflick, J. S., Dull, T. J., Gray, A., Tam, A. W., Lee, J., Yarden, Y., Libermann, T. A., Schlessinger, J., Downward, J., Mayes, E. L. V., Whittle, N., Waterfield, M. D. and Seeburg, P. H. (1984). Human epidermal growth factor receptor cDNA sequence and aberrant expression of the amplified gene in A431 epidermoid carcinoma cells. Nature 309, 418-425.

Waterfield, M. D., Scrace, G. T., Whittle, N., Stroobant, P., Johnsson, A., Wateson, A., Westermark, B., Heldin, C. H., Huang, J. S. and Deuel, T. F. (1983). Platelet-derived growth factor is structurally related to the putative transforming protein p28sis of simian sarcoma virus. Nature 304, 35-39.

Weinberg, R. A. (1984). ras oncogenes and the molecular mechanisms of carcinogenesis. Blood 64, 1143-1145.

Westin, E. H., Wong-Staal, F., Gelmann, E. P., Dalla-Favera, R., Papas, T. S., Lautenberger, J. A., Eva, A., Reddy, E. P., Tronick, S. R., Aaronson, S. A. and Gallo, R. C. (1982). Expression of cellular homologues of retroviral onc genes in human hematopoietic cells. Proc. Nat. Acad. Sci. USA 79, 2490-2494.

Wigler, M., Pellicer, A., Silverstein, S. and Axel, R. (1978). Biochemical transfer of single-copy eukaryotic genes using total cellular DNA as donor. Cell 14, 725-731.

Willingham, M. C., Pastan, I., Shih, T. Y., and Scolnick, E. M. (1980). Localization of the src gene product of the Harvey strain of MSV to the plasma membrane of transformed cells by electron microscopic immunocytochemistry. Cell 19, 1005-1014.

Wiman, K. G., Clarkson, B., Hayday, A. C., Saito, H., Tonegawa, S. and Hayward, W. S. (1984). Activation of a translocated c-myc gene: Role of structural alterations in the upstream region. Proc. Nat. Acad. Sci. USA. 81, 6798-6802.

Wolman, S. R., Lanfrancone, L., Dalla-Favera, R., Ripley, S. and Henderson, A. S. (1985). Oncogene mobility in a huamn leukemia lina HL-60. Cancer Genet. and Cytogenet. 17, 133-141.

Yamamoto, T., Nishida, T., Miyajima, N., Kawai, S., Ooi, T. and Toyoshima, K. (1983). The erbB gene of avian erythroblastosis virus is a member of the src gene family. Cell 35, 71-78.

Yunis, J. J. (1983). The chromosomal basis of human neoplasia. Science 221, 227-236.

MOLECULAR GENETICS OF HUMAN B CELL NEOPLASIA

Carlo M. Croce

1. INTRODUCTION

Most human hematopoietic neoplasms carry non-random chromosome rearrangements, predominantly translocations and inversions (1-3). For example specific translocations between chromosomes 8 and 14, t(8;14)(q24;q32), between chromosomes 8 and 22, t(8;22)(q24;q11), and between chromosomes 2 and 8, t(2;8)(p11;q24) have been described in 75%, 16% and 9% of the cases of Burkitt lymphomas respectively (1-2,4-5). In more than 90% of cases of chronic myelogenous leukemias, a Philadelphia chromosome is present in the leukemic cells (3). This marker chromosome is the result of a reciprocal translocation between chromosomes 9 and 22, t(9;22)(q34;q11) (3). These observations suggested that chromosomal rearrangements may have a role in the pathogenesis of human hematopoietic tumors in particular and in human oncogenesis in general.

Recent findings provide strong support for the role of chromosomal translocations in the pathogenesis of human neoplasms. In Burkitt lymphoma it has been shown that the c-myc oncogene, that is the human homologue of a viral oncogene, v-myc, that can induce lymphomas in chickens is directly involved in the three different translocations observed in this disease (Fig. 1) (6-8). In Burkitt lymphomas with the t(8;14) chromosome translocation the c-myc oncogene translocates to the heavy chain locus on chromosome 14 (Figs. 1 and 2). While in the t(8;22) and t(2;8) variant chromosome translocations the c-myc oncogene remains

545

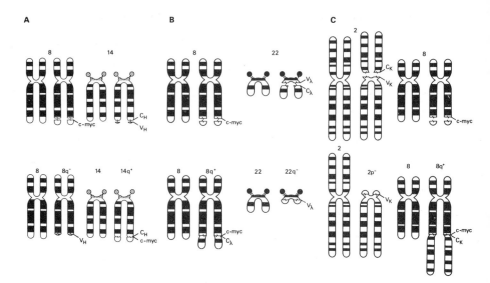

Fig. 1. In Burkitt's lymphomas with the t(8;14) translocation, the c-myc oncogene translocates to the heavy chain locus (A), and a portion of the immunoglobuin locus (V_H) is translocated to chromosome 8. In Burkitt's lymphomas with
the less frequent t(8;22) (B) and t(2;8) (C) translocations, the c-myc oncogene
remains on the involved chromosome 8, but the genes for the immunoglobulin
light chain constant regions (C_K and $C\lambda$) translocate to a region 3' (distal) to
the c-myc oncogene on the involved chromosome 8 (8q+). Again with these
translocations, the immunoglobulin loci are split so that sequences that encode
for the variable portion of the immunoglobulin molecule (V_K or $V\lambda$) remain on
chromosome 2 or 22, respectively.

on human chromosome 8, while the locus for either the lambda or kappa chains
translocates to a region that is distal to (3' to) the involved c-myc oncogene
(Figs. 1 and 2) (6-8). As shown in Fig. 2, there is considerable heterogeneity
of breakpoints in Burkitt lymphomas. In the cases with the t(8;14) chromosome

translocation the breakpoints are always 5' (proximal to) to the two coding

exons (exons 2 and 3) (9). The first exon of the c-myc oncogene is non-coding

(10-11). The breakpoints can occur 5' to th first exon of the c-myc oncogene or

can occur within the first intron (Fig. 2). In this case new cryptic promoters

are activated within the first intron, leading to novel transcripts that code,

however, for the same protein product (10-11) in the variant chromosome translo-

cations the breakpoints can occur at several sites 3' to the c-myc oncogene

(Fig. 2) (7-8). The biological consequences of these three different types of

translocations are the same: a transcriptional deregulation of the expression

of the c-myc oncogene that is transcribed constitutively at elevated levels

(12-13). The c-myc oncogene that can be regulated during normal B-cell dif-

ferentiation, fails to respond to normal transcriptional control and is

expressed constitutively at elevated levels as result of its proximity to gene-

tic elements within the three different immunoglobulin loci capable of acti-

vating gene transcription in cis (on the same chromosome). Thus, in Burkitt

lymphoma the translocations result in a quantitative difference in c-myc

expression in the involved B-cells (12-13).

In chronic myelogenous leukemia (CML) the human homologue, c-abl, of the

Abelson murine leukemia virus oncogene is translocated from its normal position

on chromosome 9 at band q34 to band q11 of chromosome 22. This translocation

results in the fusion of the c-abl oncogene with a gene on chromosome 22, named

bcr (14). Such fused gene is transcribed into an aberrant transcript 8.5 kb in

length and translated into an aberrant 210 kd c-abl protein that acquires

protein kinase activity (14). Thus the chromosomal alteration in CML results

in an aberrant gene product that expresses protein kinase activity similarly to

several retrovirus oncogene products such as v-src, c-erbB and c-abl.

Since specific chromosomal alterations involving the chromosomal regions

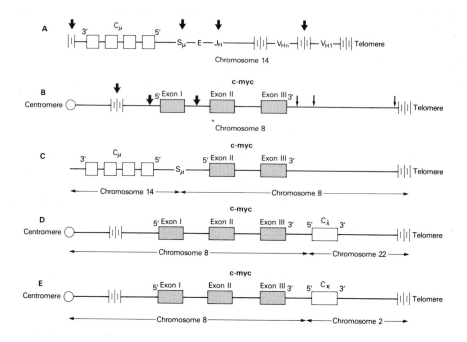

Fig. 2. (A) A general diagram of an immunoglobulin heavy chain gene. The
constant region (C) contains DNA segments that encode the common portion of
immunoglobulin molecules, while the variable region (V_H) contains DNA segments
that encode these portions of immunoglobulin molecules that differ from one
another. During B cell differentiation, immunoglobulin gene rearrangement
occurs so that one V_H gene segment comes to lie immediately adjacent to a D
segment (not shown) and the D segment lies adjacent to a short coding segment
(J_H). The element labeled E is a DNA sequence that enhances promoter function,
thereby increasing transcription of productively rearranged immunoglobuin gene.
The S region contains a DNA sequence that is involved in another type of DNA
rearrangement that is the basis for the class switch from secretion of one type
of heavy chain to another. The symbol (|ı||ı|) indicates that distances between

adjacent parts of the chromosome are not defined. In Burkitt's lymphoma with
the t(8;14) translocation the chromosome breakpoints within the heavy chain
locus may occur in the region carrying V_H genes, in the region between J_H and
V_H, in the heavy chain joining segment (J_H), in the switch region ($S\mu$), or may
involve a different constant region coding segment. The arrows indicate
possible sites for chromosomal breakpoints. (B) A general diagram of the c-myc
gene. In Burkitt's lymphomas with the t(8;14) translocation, the chromosomal
breakpoints on chromosome 8 are always 5' of the two coding exons (II and III)
of the c-myc oncogene (thick arrows). In some cases the c-myc oncogene is deca-
pitated by the chromosomal break, and the first exon of the gene remains on the
$8q^-$ chromosome, while the coding exons translocate to chromosome 14 (line C).
In Burkitt's lymphomas with the less frequent t(8;22) and t(2;8) translocations,
the breakpoints are distal to the c-myc oncogene (thin arrows). (C) Example of
a Burkitt's lymphoma with the t(8;14) translocation and a rearranged c-myc gene.
The Cu gene and the c-myc oncogene are inverted with respect to one another in
that the transcriptional orientation of the c-myc and of the Cu genes are in
opposite directions 5'->3'). (D) In Burkitt's lymphoma with the t(8;22)
translocation, the c-myc oncogene remains on chromosome 8, while the portion of
the lambda locus that encodes the constant region of a light chain translocates
to the chromosomal region 3' (distal) to the c-myc oncogene (line B). (E) In
Burkitt's lymphomas with the t(2;8) translocation, the c-myc oncogene also
remains on chromosome 8, while the constant region of the kappa locus translocates to a chromosomal region 3' (distal) to the c-myc oncogene (line B).

carrying the human immunoglobuin loci are frequently observed in human B cell malignancies we reasoned that it should be possible to isolate human genes involved in the pathogenesis of B cell neoplasms by taking advantage of their close proximity to the human heavy chain locus.

2. Cloning of chromosomal joinings between chromosomes 11 and 14 and between chromosomes 14 and 18 in human B cell neoplasms.

A reciprocal translocation between chromosomes 11 and 14, t(11;14)(q13;q32) has been observed in chronic lymphocytic leukemia (CLL) of the B cell type (15), in diffuse B cell lymphoma (16), and in multiple myeloma (17). By analyzing somatic cell hybrids between mouse myeloma cells and CLL cells carrying the t(11;14) (q13;q32) translocation, we found that the chromosome breakpoint on chromosome 14 in B cell malignancies with the t(11;14) chromosome translocation split the heavy chain locus (18). The constant region genes remain on chromosome 14, while the variable region genes translocate to the involved chromosome 14 (18). Then we used recombinant DNA technologies to clone the chromosomal joining between chromosomes 11 and 14 on the 14q+ chromosome of the CLL cells (CLL 271) we investigated (19). Human CLL DNA was partially digested and inserted into a phage vector (EMBL3A) (19). The recombinant clones were then screened by using a probe specific for the joining (J_H) segment of heavy chains and a probe for the u constant region (19). Two classes of recombinant clones were obtained; one class contained the productively rearranged u gene on normal chromosome 14, while the second class contained the unproductively rearranged u gene on the 14q+ chromosome (19). Analysis of the second class of recombinant clones indicated that they contained a DNA segment derived from chromosome 11, that rearranged with the J_H segment of the heavy chain locus (19). Unique DNA sequences derived from chromosome 11 flanking the breakpoint were then used to determine whether they could detect rearrangements of the homologues DNA sequences in other B cell noeplasms carrying the t(11;14) chromosome translocation. As shown in Fig. 3, a single DNA probe could detect rearrangment in a diffuse B

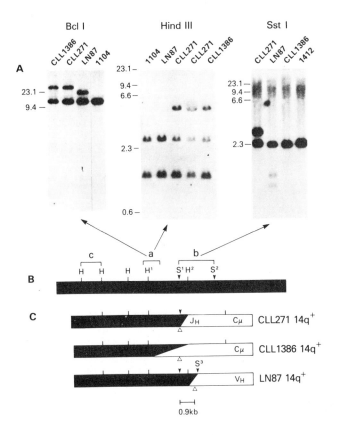

Fig. 3. Southern blot hybridization of CLL 271, CLL 1386 and LN87 DNA with

human chromosome 11-derived probes (A) and the maps of chromosome 14q+ in CLL

271, CLL 1386 and LN87 (C) and of normal chromosome 11 (B).

A, DNAs derived from neoplastic cells of CLL 271, CLL 1386 and LN87

were cut with the restriction enzymes shown, run on 0.7% agarose gels and trans-

ferred to nitrocellulose filters. The filters were hybridized with nic-

translated probe a (pRc8SmR) or b (shown in B). Hybridizations were carried out

(Fig. 3 caption cont.)

in 50% formamide/4 x SSC at 37°C and finally washed with 0.2 SSc at 65°C DNA 1104 and 1412 were obtained from T-cell lymphomas and used as germline control for chromosome 11. The size is shown in kb beside each panel. B, C, Structure of normal chromosome 11 in B is deduced by analysing the chromosome 11 sequence-containing recombinant clones. Filled box and open box show chromosomes 11 and 14, respectively. The cleavage sites by restriction enzyme SstI and HindIII are shown by S (▼) and H (|), respectively. All restriction sites are not shown. Open triangles represent the joining sites between chromosomes 11 and 14 on chromosome 14q+.

---

cell lymphoma (LN87) and in an additional CLL (CLL 1836) carrying a t(11;14) (q13;q32) chromosome translocation (19-20). Thus we were able to clone a specific segment of DNA that undergoes rearrangements in B cell tumors. We proposed to name this rearranging locus on chromosome 11 involved in B cell neoplasia, B cell lymphoma and leukemia 1 or bcl-1 (19-20).

We have also taken advantage of a cell line, 380, derived from a pre B cell leukemia carrying a t(8;14) and a t(14;18) chromosome translocation in the same cells (21) to attempt to clone the joining between chromosomes 14 and 18 (22). As previously mentioned, the t(14;18) translocation is observed in the majority of follicular lymphoma, one of the most common human B cell malignancies. Again, we partially cleaved the leukemic DNA, cloned it in a lambda phage vector and screened the recombinant phages for the presence of J_H and Cu sequences. Two classes of recombinant clones were observed. One class contained the joining between chromosomes 8 and 14 on one of the two 14q+ chromosomes, while the second class contained the joining between chromosomes 14 and 18 on the other 14q+ chromosome (22). Then the region of normal chromosome 14

that is involved in the rearrangements for which we proposed the name of bcl-2,
was entirely cloned and characterized (22-23). In addition we have used probes
from these regions to investigate the genomic rearrangements in a large collection of biopic samples from patients with follicular lymphoma (23). We
detected rearrangements of the bcl-2 locus in more than 60% of cases of follicular lymphomas and we found that such rearrangements occurred within two major
hot spots for rearrangements, where the hot spot defined by probe b is involved
much more frequently than the hot spot defined by probe c (Fig. 4)(23).
Therefore it is now possible to use two DNA probes to detect the occurrance of a
t(14;18) chromosome translocation in neoplastic B cells.

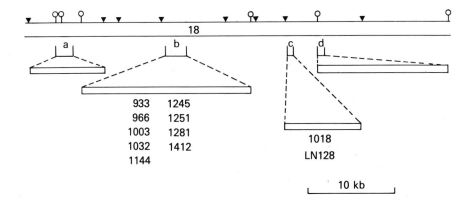

Fig. 4. Clustering of the breakpoints of t(14;18) on chromosome 18 in follicular lymphomas. The top bar represents normal chromosome 18. The open bar below
chromosome 18 indicates the germ line restriction fragment detected by each
probe. Sst I cleavage was used for probes a and c, and Bam HI digestion was
used for probes b and d. The numbers below open bars indicate follicular
lymphoma DNAs that showed rearrangement of the germ line restriction fragment.

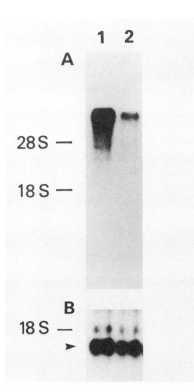

Fig. 5. RNA blot hybridization. Total cytoplasmic RNA was extracted from the cell lines shown, and poly(A)+ RNA was selected by oliogo(dT)-cellulose column chromatography. About 5 ug of poly(A)+ RNA from each cell ilne was glyoxalated, separated on a 1 percent agarose gel, and transferred to nitrocellulose filters. Each filter was hybridized with nick-translated probe in 50 percent formamide-4X SSC (standard saline citrate) at 37°C and washed finally in 0.5 x SSC at 55°C. (A) (Lane 1) line 380 RNA; (lane 2) line 697 RNA. The filter was hybridized with probe b. (B) the same filter as in (A) was rehybridized with a human phosphoglycerokinase complementary DNA probe, pHPGK-7e. The transcript is shown by the arrow.

We have also used probe b to detect bcl-2 transcripts (23). As shown in Fig. 5, this probe detected 6 kb transcripts. To determine whether the expression of the bcl-2 gene is enhanced by its proximity to the heavy chain enhancer localized between J_H and Su, we have compared the levels of bcl-2 transcripts in 380 pre B cells and in the 697 cells carrying a t(1;19) translocation, 697 cells also derive from a pre B cell leukemia (23). As shown in Fig. 5 the levels of bcl-2 transcripts in 380 was at least 10-fold higher than in 697 cells (ref). To confirm the validity of this finding we have washed the bcl-2 probe off the Northern blot, and have rehybridized it with a cDNA probe specific for an X-linked constitutive isozyme, phosphoglycerate kinase (PGK). As shown in Fig. 5 the levels of PGK transcripts in 380 and in 697 cells were approximately the same. Thus we have cloned a gene, bcl-2, that is involved in the majority of follicular lymphomas (23). In addition we have determined that the translocation results in an enhancement of bcl-2 expression, presumably because of its close proximity to an immunoglobulin heavy chain enhancer.

3. The t(11;14) and the t(14;18) chromosome translocations are the results of mistakes in V-D-J joining.

In order to understand the molecular mechanisms of chromosome translocations in B cell neoplasms, we have sequenced the joinings between chromosomes 11 and 14 on the 14q+ chromosomes of two independent cases of chronic lymphocytic leukemia (CLL) of the B cell type. As shown in Fig. 6 the two breakpoints on chromosome 11 are just 8 nucleotides apart and on chromosome 14 they involve the 5' region of the J_4 segment of the heavy chain locus (20). Interestingly we found stretches of extra nucleotides at the joinings between chromosomes 11 and 14 (Fig. 6), that did not derive from chromosomes 11 and 14 (20). Similarly extranucleotides at joining sites have been described in rearranged heavy chain genes and have been named N-region (20). These two observations, that the 5' end of one of the J_H segments is involved in the rearrangement and that N region are present at joining sites, suggested to us that the enzymatic system involved

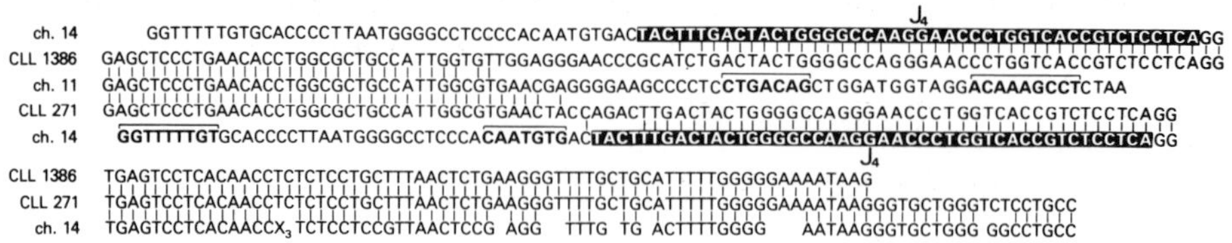

Fig. 6. DNA sequences of the joining sites between chromosomes 11 and 14 in CLL 271, CLL 1386 and of corresponding normal chromosome 11. Identical nucleotide sequences are shown by verticle lines. The boxed region indicates the J4 coding segment of the immunoglobulin heavy chain gene. The DNA sequences shown by brackets on chromosome 14 indicate the conserved sequence 7mer-9mer.

in V-D-J joining could be responsible for the chromosomal translocation (20). Since the V-D-J joining enzyme recognizes signal sequences (heptamer and nonamers) separated by a 12 and by a 23 nucleotide spacer respectively on the DNA segments to be joined, we have examined the region of normal chromosome 11 that is involved in the rearrangements for the presence of signal sequences for V-D-J joining (20). As shown in Fig. 6, we detected such sequences, heptamer and nonamer separated by a 12 nucleotide spacer, on chromosome 11, close to the breakpoints (Fig. 6). These results indicate that the t(11;14) chromosome translocation is the result of mistakes during V-D-J joining (20). The V-D-J joining enzyme instead of joining separated segments of DNA on the same chromosome joins separated segments of DNA on two different chromosomes (20).

As shown in Fig. 7, in follicular lymphomas, the rearrangements also occur in one of the J_H segments of the heavy chain locus. Sequence analysis of the breakpoints of 380 leukemic cells and of five independent cases of follicular lymphoma indicate that in every case the breakpoints on chromosome 14 were at the 5' end of one of the different J_H segments (Fig. 7) (24). As shown in Fig.

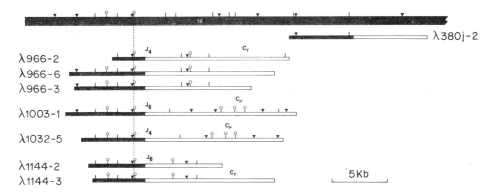

Fig. 7. Restriction maps surrounding the breakpoints of t(14;18) chromosome translocation in follicular lymphomas. Top bar shows chromosome 18. The horizontal lines under chromosome 18 represent insert DNAs of the recombinant clones containing the breakpoints of t(14;18) translocation from 4 follicular lymphoma DNAs. The filled and open bars represent chromosome 18 and 14 sequences respectively. The breakpoints were mapped by comparing the detailed restriction maps (data not shown) of breakpoint containing clones and those of corresponding normal chromosome 18 clones which were previously isolated. The restriction enzyme sites of HindIII, SstI and EcoRI are shown by |, ∇ and ⌽, respectively.

7 we detected the presence of N regions at joining sites in each of the five cases examined (24). Examination of the DNA sequence of normal chromosome 18 in the vicinity of the breakpoints indicates the presence of heptamers and nonamers with a 12 nucleotide spacer in close proximity to each breakpoint (Fig. 8). Thus we conclude that the t(14;18) chromosome translocation is also the result of a mistake in V-D-J joining during the pre B cell stage of differentiation.

It is of considerable interest that follicular lymphoma cells are at a more advanced stage of B cell differentiation than pre B cells, where the chro-

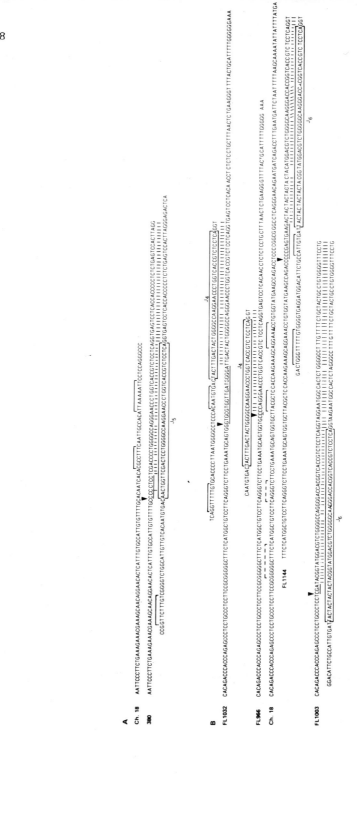

mosome translocation presumably occur. In addition, as shown in Fig. 7 the

bcl-2 gene can be associated to a γ chain gene. This indicates that at first

the bcl-2 gene translocates to a Cu region and then, during the process of

heavy chain switching, it translocates together with a J_H segment to a γ region

(24). These observations have important implications concerning the role of

proto-oncogene in human neoplasia. Since follicular lymphoma is a disease of

mature B cells and not a disease of pre B cells, it seems likely that the

expression of the malignant phenotype occurs only in more differentiated B

cells. The activation of the bcl-2 gene in pre B cells, however, might result

in a proliferative advantage of the involved pre B cells. Several possibilities

might explain these observations. One possibility is that pre B cells carrying

an activated bcl-2 gene can still respond to the microenvironment in which they

reside, while such ability might be lost by the more differentiated B cells.

Alternatively it is possible that the levels of enhancement of the bcl-2 gene

may differ according to the stage of B cell differentiation and that the immu-

noglobulin genes and the translocated bcl-2 gene are enhanced more in mature B

Fig. 8. Nucleotide sequences of the breakpoint of t(14;18) translocation and of

corresponding normal chromosome 18. DNA sequences were determined by chemical

degradation method as described. The breakpoint sequences are aligned so that

the chromosome 18-derived sequences match to the corresponding normal chromosome

18 sequences. The triangles represent the regions were DNA sequences of 380 and

4 FL's start to diverge from normal chromosome 18 sequences. The identical

nucleotides between the breakpoint sequences and J_H segment sequences (boxed)

are shown by vertical lines. The 7mer-9mer signal and signal-like sequences are

shown by the brackets. DNA sequences of chromosome 18 which are very similar to

the 7mer-9mer signal sequences 5' to D segment are shown by the broken brackets.

cells than in pre B cells. It is also possible that the bcl-2 product has to
interact with a cellular substrate that is expressed at later stages of B cell
differentiation in order to cause malignancy.

Finally it is possible that additional genetic alterations are required for
malignant transformation. This, however, does not seem very likely since pre B
cell leukemias with a t(14;18) chromosome translocation are not observed unless
they contain both a t(14;18) and a t(8;14) chromosome translocation. If additional genetic changes are required for malignant transformation, it is not
clear why these changes should occur in more mature B cells and not in pre B
cells.

Thus these findings open new avenues in our understanding of the relationships between cell differentiation and the expression of malignancy.

4. Molecular basis of T cell lymphoma and leukemias

Specific chromosome rearrangements predominantly translocations and inversions, are also observed in T cell malignancies (25-26). Interestingly most of
these rearrangements involve the region q11.2 of human chromosome 14 (25-26).
We thought that this chromosome region could have an important role in T cell
function and therefore attempted to determine whether one of the loci for the
human T cell recpetor resids on chromosome region 14q11.2. As shown in Table 1
a human cDNA clone for the alpha chain of the T cell receptor, detected the presence of the human gene for the alpha chain only when human chromosome 14 was
present in the hybrids (27). In situ hybridization to metaphase chromosomes
further localized the locus for the alpha-chain of the T cell receptor at chromosome band 14q11.2 (27). Thus it seems logical to speculate that the locus for
the alpha chain of the T cell receptor is involved in the chromosome rearrangements observed in T cell neoplasia and might have an important role in the
pathogenesis of this disease. Since in T cell CLL an inversion of chromosome 14

Table 1. Assignment of the gene for the α chain of the T cell receptor to human chromosome 14.

Hybrids	1	2	3	4	5	6	7	8	9	10	11	12	13	14	15	16	17	18	19	20	21	22	X	Tα Chain Gene
PXBIV-Cl 5	-	+	-	-	+	-	-	+	-	-	-	+	-	-	-	-	-	-	+	-	-	-	-	-
77 B10 Cl 28	+	-	+	-	+	-	-	-	-	-	-	-	-	+	-	-	-	+	-	-	-	-	+	+
DSK Cl 20-C	-	-	-	-	-	-	-	-	-	-	-	-	-	+	-	-	+	-	-	+	+	+	-	-
106 2B4 4C4	-	-	-	-	-	-	-	-	-	-	-	-	-	+	-	-	-	-	-	-	-	-	-	+
5263 Cl 7 S17*	-	-	-	-	-	-	-	-	-	-	-	-	-	+	-	-	-	-	-	-	-	-	-	+
D69 Cl 4S7	-	-	-	-	-	+	-	-	-	-	-	-	-	+	-	-	-	-	-	+	-	+	-	+
M44 Cl 2S5**	-	-	-	-	-	-	-	-	-	-	-	-	-	+	-	-	-	-	-	-	-	-	-	+
Nu9	-	-	-	-	-	+	+	-	-	-	-	-	-	-	-	-	-	-	-	-	-	-	-	-
D2 Cl 6SS	-	-	+	+	-	+	-	-	-	-	+	-	-	+	+	-	+	-	-	-	-	-	-	+
401 AD5 EF 3-1	-	-	-	-	-	-	-	+	-	-	-	-	-	-	-	-	-	-	+	-	+	+	+	-
77 B10 Cl 30	-	-	+	-	+	-	-	-	-	+	-	-	+	+	-	-	-	-	-	+	-	-	+	+
77 B10 Cl 31	+	-	+	-	+	-	-	+	+	+	-	-	+	+	-	-	-	+	-	+	-	-	+	+
53-87-3 Cl 10	-	-	-	-	-	-	+	-	-	-	-	-	-	-	-	-	-	-	-	-	-	-	-	-
GM54VA Cl 31	-	-	-	-	-	-	-	-	-	-	-	-	-	-	-	-	+	-	-	-	-	-	-	-
GM x LM Cl 5	-	-	-	+	-	-	+	+	-	-	-	-	-	-	+	-	-	-	-	+	-	+	-	
D2 Cl 6S3	-	-	+	+	-	+	-	-	-	-	+	-	-	-	+	+	+	-	-	-	-	+	-	-
77 B10 Cl 5	+	-	-	-	+	-	-	+	+	+	-	-	+	+	-	+	-	-	-	+	-	-	+	+
1P1	-	-	-	-	-	-	+	-	-	+	-	-	-	-	+	-	+	-	-	-	-	-	-	-
PT47 Cl 5	-	-	-	-	-	-	-	-	-	-	-	-	-	-	-	-	-	-	-	-	+	-		
5468 F1 Cl 1-11	-	-	-	-	-	-	-	-	-	+	+	+	-	-	-	-	-	+	-	-	+	-		
5468 F2 Cl 5	-	-	-	-	-	-	-	-	+	-	-	-	-	-	-	-	-	-	-	-	+	+		
DSK Cl 2	-	-	+	-	-	-	-	+	-	-	-	-	-	+	-	+	+	-	-	-	+	-		+
706B6-40 Cl 17	-	-	-	-	-	-	-	+	-	-	-	-	-	-	-	-	-	-	-	-	-	-		
640-63	-	-	-	-	-	-	-	-	+	-	-	-	-	+	-	-	-	+	-	-	+	-		
706-D1	-	+	-	-	-	+	-	-	-	-	-	+	+	-	-	+	+	+	-	-	-	+		

*Hybrid 52-63 Cl 7S17 carries the 14q+ chromosome of KOP-2 cells which have a t(14;X) chromosome translocation.

**Hybrid M44 Cl 2S5 contains only the 14q+ chromosome of P3HR-1 Burkitt lymphoma with the t(8;14) chromosome translocation.

with breakpoints at 14q11.2 and 14q32 is observed, we speculated that a
proto-oncogene may be located at band 14q32 and be activated by its close
proximity to the locus of the alpha chain of the T cell receptor (Fig. 9). To

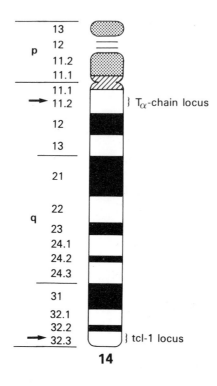

Fig. 9. Diagram of human chromosome 14. The arrows indicate the breakpoints
observed in CLL cells with an inversion of chromosome 14. The brackets indicate
the positions of the locus for the alpha-chain of the T-cell receptor (upper)
and of the putative tcl-1 oncogene (lower). The chromosomal inversion and the
t(14;14) (q11;q32) translocation should result in the juxtaosition of the T
alpha-chain locus and of the tcl-1 oncogene thereby activating tcl-1.

demonstrate conclusively that the locus for the alpha chain of the T cell recep-

tor is directly involved in the chromosome rearrangements observed in T cell

neoplasia, we have hybridized mouse leukemic T cells with two independent

leukemias carrying a t(11;14) (p13;q11.2) chromosome translocation and examined

the hybrid cells (Fig. 10) for the presence of the genes for the variable and

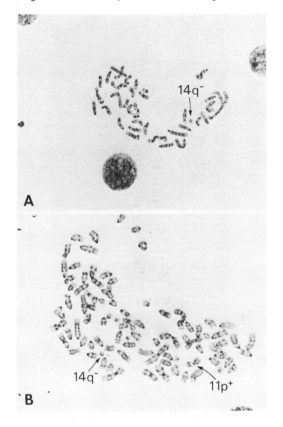

Fig. 10. (A) Partial trypsin/Giemsa banded metaphase from hybrud subclone 517
AA3-G7 containing the 14q⁻ chromosome (arrow), but not the normal 14, normal 11,
or the 11p+. (B) Partial trypsin/Giemsa banded metaphase from hybrid 517 B-D3
containing both the 11p+ and 14q⁻ translocation chromosomes (arrows), but not
the normal 11 or normal 14.

the constant regions of the alpha chain of the T cell receptor and of other
markers of chromosomes 11 and 14 (28). As shown in Table 2, the variable region
genes (Vα) remained on the involved chromosome 14 (14q⁻) while the gene for the
constant region translocated to the involved chromosome 11 (11p+) (Fig. 11).
Thus the chromosome breaks in these two leukemias split the locus for the alpha
chain of the T cell receptor between the genes for V α and C α (Fig. 9) (28). In

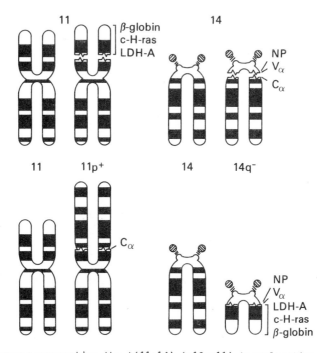

Fig. 11. Diagram representing the t(11;14) (p13;q11) translocation in acute
lymphocytic leukemia of the T cell type. The translocation breakpoint on chromosome 14 splits the locus for the alpha chain of the T cell receptor. The V
alpha genes remain on the 14q⁻ chromosome, while the C alpha gene translocates
to the involved chromosome 11 (11p+). The gene for human NP remains on the
involved chromosome 14 (14q⁻). The gene for LDH-A and the β-globin and c-H-ras
genes translocate to the involved chromosome 14 (14q⁻).

TABLE 2. PRESENCE OF THE Vα AND Cα GENES IN HYBRIDS BETWEEN MOUSE BW5147 CELLS AND HUMAN ALL CELLS

Hybrids	α-chain locus		Human Markers					Human Chromosomes+			
	Vα	Cα	NP*	LDH-A**	c-H-ras	β-globin	bcl-1	14	14q-	11	11p+
517 A-A3	+	+	+	+	+	+	-	++	+++	-	-
517 B-D3	+	+	+	+	+	+	+	-	+	-	+
517 B-B1	-	-	-	+	+	+	+	-	-	+	-
517 B-A3	+	-	+	+	+	+	-	-	+	-	-
517 B-D3-G8	+	-	+	+	+	+	-	-	++	-	-
517 B-D3-G9	+	+	+	-	+	+	+	-	++	-	+++
517 B-D3-D2A	-	+	-	-	-	-	+	-	-	-	++
517 A-A3-A10	+	+	+	+	+	+	-	+	++	-	-
517 A-A3-G7	+	-	+	+	+	+	-	-	+++	-	-
515 BD2-CF3	+	-	+	+	+	+	-	-	+	-	-
515 BD2-CF6	+	-	+	+	+	+	-	-	+++	-	-

*NP, nucleoside phosphorylase.

**LDH-A, lactic dehydrogenase A.

+Frequency of metaphases with relevant chromosomes: - = none; + = 10-30%; ++ = 30-50%; +++ = >50%. At least 25 metaphases were examined for each hybrid following trypsin/Giemsa staining. Selected metaphases were studied by the G11 technique to confirm the human origin of relevant chromosomes.

addition these results indicate that the Vα genes are proximal to the Cα gene at
band 14q11.2 (Fig. 11).

Thus, by taking advantage of leukemias and lymphomas carrying inversions and
translocations involving band 14q11.2 it should be possible to identify genes
that are involved in the pathogenesis of human T cell leukemias and lymphomas.

CONCLUSIONS

It is clear that specific chromosomal translocations are responsible for the
majority of human B cell neoplastic diseases. These translocations place proto-oncogenes in close proximity of genetic elements capable of activating gene
transcription in cis along considerable chromosomal distances. Thus the jux-
taposition of the protooncogenes and of the immunoglobulin loci result in the
deregulation of the involved proto-oncogene that is transcribed constitutively
at elevated levels leading to neoplasia (Fig. 12). Interestingly, the molecular
mechanisms involved in the t(11;14) and in the t(14;18) chromosome translocation
seems to involve the physiologic enzymatic system responsible for V-D-J joining.
Thus these translocations do not occur at random, but are the result of mistakes
in V-D-J joining (20,24), where the V-D-J joining enzyme joins separated
segments of DNA on two different chromosomes instead of joining two separated
segments of DNA on the same chromosome on the basis of signal sequences for
V-D-J joining (20,24). In view of these findings it seems logical to speculate
that the enzymes involved in immunoglobulin gene rearrangements may also be
involved in the t(8;14), t(2;8) and t(8;22) chromosome translocations observed
in Burkitt lymphomas. It would be of considerable interest to determine whether
other translocations might also be the result of enzymatic reactions.

Interestingly, a very similar situation seems to be occurring in T cell
neoplasms. The molecular analysis of T cell tumors indicates an important role

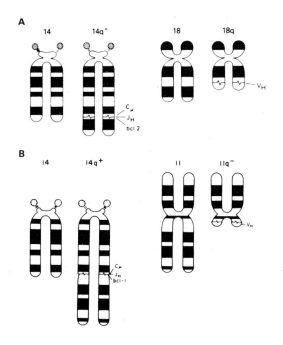

Fig. 12. Chromosome translocations in B cell lymphomas and leukemias of adults.
In follicular lymphomas with the t(14;18) translocation, the bcl-2 gene, normally located on band q21 of chromosome 18, translocates to the heavy chain
locus on chromosome 14 (A). In lymphomas and leukemias with the t(11;14)
translocation, the bcl-1 gene, normally located on band q13 of chromosome 11,
translocates to the heavy chain locus on chromosome 14 (B).

for the locus of the alpha chain of the T cell receptor in the pathogenesis of
these diseases. Since the locus for the alpha chain of the T cell receptor is
split by the chromosome break at band 14q11.2 and since the C locus translocates either to a different chromosome or to region 14q32, it is easy to predict
that by "chromosome walking" it will be possible to clone and identify the genes
that are involved in the pathogenesis of most T cell tumors.

Since most of human T cell malignancies in in areas endemic for the human T cell leukemia virus 1 (HTLV-1) also carry either translocations or inversions involving chromosome band 14q11.2, it seems likely that HTLV-1 _per se_ may not be a leukemogenic agent and might have an indirect role similar to that of EBV in the pathogenesis of African Burkitt lymphoma. Thus infection with HTLV-1 may increase the number of T cells susceptible to develop a chromosome translocation, possibly during T cell receptor gene rearrangements.

In conclusion the study of the molecular genetics of specific chromsoome translocations in human B and T tumors is providing a detailed explanation of the mechanisms and of the genes involved in these neoplasms.

REFERENCES

1. Manolov, G. & Manolova, Y. (1972) Nature (London) 237, 33-34.

2. Zech, L., Haglund, U., Nilsson, K. & Klein, G. (1976) Int. J. Cancer 17, 47-56.

3. Rowley, J.D. (1982) Science 216, 749-751.

4. Van denBerghe, H., Parloir, C., Gosseye, S., Englebienne, V., Cornu, G. & Sokal, G. (1979) Cancer Genet. Cytogenet. 1, 9-14.

5. Miyoshi, I., Hiraki, S., Kimura, I., Miyamato, K. & Sato, J. (1979) Experientia 35, 742-743.

6. Dalla Favera, R., Bregni, M., Erikson, J., Patterson, D., Gallo, R.C. & Croce, C.M. (1982) Proc. Natl. Acad. Sci. USA 79, 7824-7827.

7. Croce, C.M., Thierfelder, W., Erikson, J., Nishikura, K., Finan, J., Emanuel, B., Lenoir, G., Nowell, P.C. & Croce, C.M. (1983) Proc. Natl. Acad. Sci. USA 80, 6922-6926.

8. Erikson, J., ar-Rushdi, A., Drwinga, H.L., Nowell, P.C. & Croce, C.M. (1983) Proc. Natl. Acad. Sci. USA 80, 820-824.

9. Showe, L.C., Ballantine, M., Nishikura, K., Erikson, J., Kaji, H. & Croce, C.M. (1985) Molecular Cellular Biology 5, 501-509.

10. Watt, R., Nishikura, K., Sorrentino, J., ar-Rushdi, A., Croce, C.M. & Rovera, G. (1983) Proc. Natl. Acad. Sci. USA 80, 6307-6311.

11. Watt, R., Stanton, L.W., Marcu, K.B., Gallo, R.C., Croce, C.M. & Rovera, G. (1983) Nature 303, 725-728.

12. Nishikura, K., ar-Rushdi, A., Erikson, J., Watt, R., Rovera, G. & Croce, C.M. (1983) Proc. Natl. Acad. Sci. USA 80, 4822-4826.

13. ar-Rushdi, A., Nishikura, K., Erikson, J., Watt, R., Rovera, G. & Croce, C.M. (1983) Science 222, 390-393.

14. Shtivelman, E., Lifshitz, B., Gale, R.P. & Canaani, E. (1985) Nature (London) 315, 550-554.

15. Yunis, J.J., Oken, M.M., Kaplan, M.E., Ensrud, K.M., Howe, R.R. & Theologides, A. (1982) N. Engl. J. Med. 307, 1231-1236.

16. Nowell, P.C., Shankey, T.V., Finan, J., Guerry, D. & Besa, E. (1981) Blood 57, 444-451.

17. Van denBerghe, H., Parloir, C., David, G., Michaux, J.L. & Sokal, G. (1979) Cancer 44, 188.

18. Erikson, J., Finan, J., Tsujimoto, Y., Nowell, P.C. & Croce, C.M. (1984) Proc. Natl. Acad. Sci. USA 81, 4144-4148.

19. Tsujimoto, Y., Finger, L.S., Yunis, J., Nowell, P.C., & Croce, C.M. (1984) Science 226, 1097-1099.

20. Tsujimoto, Y., Jaffe, E., Cossman, J., Gorham, J., Nowell, P.C. & Croce, C.M. (1985) Nature (London) 315, 340-343.

21. Pegoraro, L., Palumbo, A., Erikson, J., Falda, M., Giovanozzo, B., Emanuel, B., Rovera, G., Nowell, P.C. & Croce, C.M. (1984). Proc. Natl. Acad. Sci. USA 81, 7166-7170.

22. Tsujimoto, Y., Yunis, J., Onorato-Showe, L., Nowell, P.C. & Croce, C.M. Science 224, 1403-1406.

23. Tsujimoto, Y., Cossman, J., Jaffe, E. & Croce, C.M. (1985) Science 228, 1440-1443.

24. Tsujimoto, Y., Gorham, J., Cossman, J., Jaffe, E. & Croce, C.M. (1985) Science, in press.

25. Zech, L., et al., (1981) Int. J. Cancer 32, 431.

26. Hecht, F., et al., (1984) Science 226, 1445.

27. Croce, C.M., Isobe, M., Palumbo, A., Puck, J., Ming, J., Tweardy, D., Erikson, J., Davis, M. & Rovera, G. (1985) Science 227, 1044-1047.

THE erbB-RELATED GROWTH FACTOR RECEPTORS

Tadashi Yamamoto, Kentaro Semba
and Kumao Toyoshima

INTRODUCTION

Acutely oncogenic retroviruses are capable of inducing leukemia and/or sarcomas in vivo after short latent
periods and of transforming hematopoietic cells and/or
fibroblasts in vitro. These viruses carry different
cell-derived sequences, known as oncogenes, responsible
for cell transformation. Cellular counterparts (or
proto-oncogenes) have been found in all metazoan species
examined(1) and one of the family of oncogenes, ras, is
also present in the genomes of the unicellular organisms
Saccharomyces cerevisiae(2,3) and Schizosaccharomyces
pombe(4). The remarkable conservation of proto-oncogenes
across vast distances of evolutionary time has led to the
speculation that proto-oncogenes play key roles in vital
functions in organisms. Genetic tools to study this
problem are now forthcoming from the discovery and
molecular cloning of cellular proto-oncogenes in
Drosophila(5,6,7,8,9,10) as well as yeast.

Molecular studies on retroviral oncogenes revealed
that they can be divided into at least 4 families: (i) The

src family-Their products have domains with significant homology to the sequence of pp60^{src}, an oncogene product of Rous sarcoma virus, responsible for protein-tyrosine kinase activity. (ii) The ras family-Their gene products have highly related sequences and show guanine nucleotide binding activity. In addition, the products of this family are suggested to function as "coupling factors" in a hormone receptor system that involves cyclic AMP-dependent protein kinase(11) and are suggested to be related with G-proteins. (iii) The myc family-The gene products have DNA binding activity and are expected to regulate gene expression. (iv) sis representing the growth factor related family. To date at least 19 genes have been identified as viral oncogenes and ten of them apparently belong to the src family. Most of them exhibit protein-tyrosine kinase activity, but some have protein-serine/threonine kinase activity(1,12,13).

Of particular interest is the fact that protein-tyrosine kinase activity is also associated with several receptors for polypeptide growth factors, such as epidermal growth factor (EGF)(14), platelet-derived growth factor (PDGF)(15), insulin(16) and insulin-like growth factor I(17). Upon binding of ligands to the respective receptors the tyrosine kinase activity of the receptors is believed to be important in transmission of mitogenic signals. This implies a possible link between the action of the growth factor-receptor complex and malignant cell growth induced by the oncogene product with protein-tyrosine kinase activity. In fact, recent analysis of the v-erbB gene, a member of the src family, and the EGF receptor gene indicated that the v-erbB gene is a part of the EGF receptor gene and codes for the carboxy-half of the receptor(18,19,20). Recently, the cellular homolog of the v-fms gene, an oncogene of the McDonough strain of feline sarcoma virus, was found to be a receptor for the mononuclear phagocyte growth factor, CSF-1(21). These findings, together with the extensive identity of the amino acid sequence of the v-sis gene product and PDGF(22,23), suggest that some v-onc gene products mimic the action of the polypeptide growth factor-receptor complex in activating a cellular pathway involved in cell proliferation. An attractive hypothesis is that deregulated expression of the growth factor receptor is important in the neoplastic process. In fact, recent studies showed amplification and over-expression of the EGF receptor gene in certain human tumors and tumor cell lines (see below).

THE erbB GENE IS AN ONCOGENE OF AVIAN ERYTHROBLASTOSIS VIRUS

Upon infection of birds with avian erythroblastosis virus (AEV), erythroleukemia is rapidly induced due to transformation of hematopoietic cells of erythroid lineage. The transformation of hematopoietic cells by AEV is suggested to begin with the infection of erythroid burst-forming unit (BFU-E) cells and to become evident only when the infected cells have matured to erythroid colony forming unit (CFU-E) cells(24).

Two independently isolated strains of AEV, the ES4(or R) strain and H strain, have been extensively characterized both virologically and molecularly. The genome of AEV-ES4 carries two cell-derived sequences, termed v-erbA and v-erbB, which are flanked by remnants of retroviral genes(25). In contrast, the genome of AEV-H consists of the v-erbB gene and the viral gag and pol genes, but not the erbA gene (26)(Fig. 1). The cellular homologs of the v-erbA gene and V-erbB gene are located on different human chromosomes and are not contiguous in the chicken genome. The generation of AEV-ES4 must, therefore, have required two independent recombination events.

Since the H strain and the ES4 strain of AEV are capable of inducing both erythroleukemia and fibrosarcomas in vivo, the v-erbB gene is evidently responsible for the induction of these tumors. The same conclusion was deduced by analyzing the transforming capacity of deletion mutants of the erbA gene and erbB gene in AEV-ES4. These deletion mutants, erbA$^-$B$^+$ and erbA$^+$B$^-$, were constructed by in vitro manipulation of the molecularly cloned AEV-ES4 DNA. The erbA$^-$B$^+$ mutants apparently transformed both erythroblasts and fibroblasts in tissue culture, whereas erbA$^+$B$^-$ mutants were unable to transform these cells. However, leukemic cells transformed by erbA$^-$B$^+$ mutants

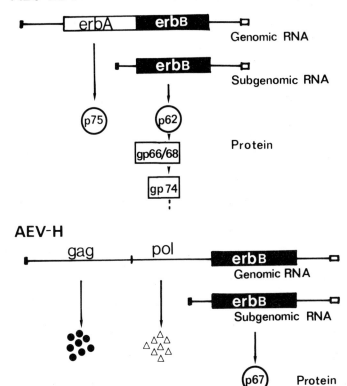

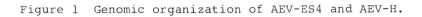

Figure 1 Genomic organization of AEV-ES4 and AEV-H.

frequently differentiated into mature cells, suggesting that the erbA gene may function in inhibiting differentiation of CFU-E transformed by erbB(27). Interestingly, some members of the src family, such as v-src and v-fps, are also reported to be capable of transforming erythroblasts in vitro, although less efficiently than v-erbB(28,29). Again the transformed cells could differentiate, probably due to lack of v-erbA function(29).

Another line of evidence for the transforming ability of the erbB gene emerged from studies on activation of the cellular oncogene (c-onc) by avian leukosis virus (ALV)(30,31). ALV is a replication-competent retrovirus that lacks an oncogene but induces a variety of tumors in susceptible chickens, including B-cell lymphomas, sarcomas and erythroblastosis after a long latent period. Analysis of genomic DNA prepared from neoplastic tissues of chickens in which erythroblastosis developed after infection with ALV revealed that ALV induces this neoplasm by activating the cellular proto-oncogene, c-erbB. This activation results from the insertion of ALV proviral DNA in the vicinity of the c-erbB gene. The strong promoter and enhancer functions of the LTR (long terminal repeat) sequence of ALV DNA are thought to induce enhanced transcription of the c-erbB gene.

THE erbB GENE PRODUCT: A MEMBER OF THE KINASE FAMILY

The v-erbB gene product of AEV-ES4 has been reported to be a 62 kd protein that becomes phosphorylated and glycosylated to yield a series of glycoproteins, gp66^{erbB}, gp68^{erbB}, gp74^{erbB} and gp82^{erbB} (32,33,34,35, see also Fig. 1). Both gp66^{erbB} and gp68^{erbB} are present on the rough endoplasmic reticulum, while highly glycosylated gp74^{erbB} is located in the plasmamembrane. Studies on temperature-sensitive mutants of AEV, which do not express gp74^{erbB} at the non-permissive temperature, suggested that cell surface expression of gp74^{erbB} is required for cell transformation (36).

Nucleotide sequence analysis showed that the v-erbB gene of AEV-H consists of 1812 nucleotides. The erbB protein, therefore, consists of 604 amino acid residues and has a calculated molecular weight of 67,638. Consistent with this, the primary product of the erbB gene in AEV-H infected chick embryo fibroblasts is a 67 kd protein (p67^{erbB}), which is then glycosylated to a glycoprotein of 72 kd (gp72^{erbB}). The erbB gene of AEV-H carries more c-erbB-derived sequence at its 3' end than AEV-ES4, which results in the production of an erbB protein of AEV-H with a larger molecular weight than that of AEV-ES4. There are apparently a series of glyco-proteins besides gp72^{erbB} in cells infected with AEV-H(37).

The predicted amino acid sequence of p67^{erbB} of AEV-H

consists of three functional domains, which we initially
called the G, S and E domains(18). Recent progress (see
below) led us to rename the G and S domains as the
extracellular domain and the kinase domain, respectively.
The extracellular domain and the kinase domain are
separated by a stretch of 23 hydrophobic amino acid
residues (see Fig. 2), which is believed to serve as a
membrane anchoring domain. The possible extracellular
domain is present on the amino-terminal side of the
membrane-spanning sequence and carries 2 possible sites of
N-linked glycosylation. A sequence of 285 amino acids in
the intracellular domain of $p67^{erbB}$, is significantly
homologous (38%) with the amino acid sequence of the
carboxyl-half of $pp60^{src}$, a transforming protein of Rous
sarcoma virus, in which protein-tyrosine kinase activity
is encoded. Moreover, the erbB protein has recently been
demonstrated to exhibit protein-tyrosine kinase activity
and its transforming ability is suggested to reside on the
kinase domain(38).

A mutant of AEV-H, td-130, has a deletion of 169
nucleotides and, therefore, produces a truncated erbB
protein of 42 kd which lacks the carboxyl one-third (E
domain) of the wild-type erbB protein. This mutant does
not induce erythroblastosis, but causes sarcomas in
chickens, suggesting that the E domain may be important in
transformation of erythroid cells but not of fibro-
blasts(26). This hypothesis is supported by the isolation

of a series of AEVs that carry deletions in the E domain and are unable to induce erythroblastosis (Robinson, H. personal communication).

Since the erbB protein is a membrane glycoprotein and is presumed to exert its protein-tyrosine kinase activity at the surface membrane, the erbB gene product has been thought to be related to a receptor that might control cell growth and/or cell differentiation.

THE erbB GENE PRODUCT IS A TRUNCATED VERSION OF THE EPIDERMAL GROWTH FACTOR RECEPTOR.

Epidermal growth factor (EGF) receptor, a 53-amino-acid-long mitogen, triggers a cascade of intracellular events by binding to specific receptors on the cell surface and eventually leads to initiation of the mitotic cycle(39). The mechanism by which the EGF-receptor complex exerts its effect on cell growth is currently being studied intensely. Ligand-receptor complexes have been shown to be concentrated in clathrin-coated pits, and to be internalized into receptosomes (or endosomes), transported to the golgi system and then degraded in lysozomes(40). The internalized receptor may activate intracellular mediators of the growth factor by some as yet unknown mechanism. Alternatively, internalization may merely lead to "down regulation", which possibly provides a termination signal for mitogenic activity. In addition, EGF binding induces a protein-tyrosine kinase activity

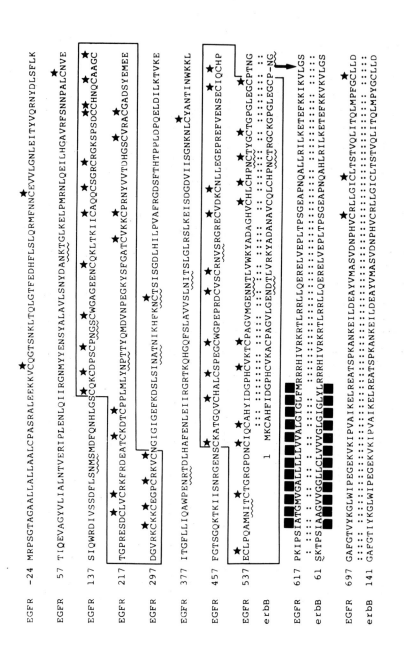

```
EGFR  777  YVRBHKDNIGSQYLLNWCVQIAKGMNYLEDRRLVHRDLAARNVLVKTPQHVKITDFGLAKLLGAEEKEYHAEGGKVPIKW
           : :::::::::::::::::::: :::::::::::: :::::::::::::::::::::::::: :::::::::::::::::::::
erbB  221  YIREHKDNIGSQYLLNWCVQIAKGMNYLEERRLVHRDLAARNVLVKTPQHVKITDFGLAKLLGADEKEYHAEGGKVPIKW

EGFR  857  MALESILHRIYTHQSDVWSYGVTVWELMTFGSKPYDGIPASEISSILEKGERLPQPPICTIDVYMIMVKCWMIDADSRPK
           ::::::::::::::::::::::::::::::::::::::::::::::::::::::::::::::::::::::::::::::::::::
erbB  301  MALESILHRIYTHQSDVWSYGVTVWELMTFGSKPYDGIPASEISSVLEKGERLPQPPICTIDVYMIMVKCWMIDADSRPK

EGFR  937  FRELIIEFSKMARDPQRYLVIQGDERMHLPSPTDSNFYRALMDEEDMDDVVDADEYLIPQQGFFSPSTSRTPLLSSLSA
           :::: :::::::::: :::::::::::::::::::: :::: : :: ::::::: :::::::::::::::::::::::::::
erbB  381  FRELIAEFSKMARDPPRYLVIQGDERMHLPSPTDSKFYRTLMEEEDMEDIVDADEYLVPHQGFFNSPSTSRTPLLSSLSA

EGFR  1017 TSNNSTVACIDRNGLQSCPIKEDSFLQRYSSDPTGALTEDSIDDTFLPVPEYINQSVPKRP-AGSVQNPVYHNQPLNPA-
           ::::: ::  : :::: : ::::::: ::::::: :: :::: ::::::::: :  :: ::  :::  :  :
erbB  461  TSNNSATNCIDRNG-QGHPVREDSFVQRYSSDPTGNFLEESIDDGFLPAPEYVNQLMPKKPSTAMVQNQIYNFISLTAIS

EGFR  1095 -PSRDPHYQDPHSTAVGNPEYLNTVQPTCVNSTFDSPAHWAQKGSHQISLDNPDYQQDFFPKEAKPNGIFKGSTAENAEY
           :  : :::::: :::::::::::: :::::::::: :  :::::::::: :: :  :::::::::: :
erbB  540  KLPMDSRYQNSHSTAVDNPEYLNTNQSPLAKTVFESSPYWIQSGNHQINLDNPDYQQDFLPTSCS  604

EGFR  1174 LRVAPQSSEFIGA  1186
```

Figure 2   Alignment of the amino acid sequences of the EGF receptor and erbB protein. Identities in sequences are marked by two dots between the two lines. The predicted transmembrane regions are shown by dotted bars, and the possible N-linked glycosylation sites by wavy lines. Stars show cysteine residues. Kinase domains are flanked by two horizontal arrows.

intrinsic to the EGF receptor(41,42,43). Tyrosine phosphorylation is thought to be an important initial event in the physiological proliferation cascade.

The EGF receptor is a 170 kd glycoprotein that spans the plasma membrane and its primary structure was recently deduced in studies on cDNA clones(20 and Fig. 2). The EGF receptor protein is made up of three distinct domains: an extracellular domain, which is present in the amino terminal moiety of the protein, a transmembrane domain, consisting of a short stretch of highly hydrophobic amino acids, and a cytoplasmic domain in the carboxy-half of the protein. The extracellular domain is highly glycosylated and is responsible for binding EGF. The cytoplasmic domain shows protein-tyrosine kinase activity and phosphorylates itself post-translationally at specific tyrosine residues.

The amino acid sequences of tryptic peptides generated from the EGF receptor have revealed that at least part of the receptor is highly homologous to the v-erbB protein(19). Direct comparison of the deduced amino acid sequence of the EGF receptor with that of the v-erbB protein indicates that the transforming protein, v-erbB, contains a short extracellular sequence, the transmembrane domain and the tyrosine kinase domain of the EGF receptor, but lacks most of the extracellular domain. The erbB protein also lacks 32 amino acid residues at the carboxy terminus of the receptor, which are replaced by 4 amino acids originating from the _env_ gene sequence of AEV-H. This replaced sequence of the receptor contains a tyrosine residue which is autophosphorylated(44). Thus, the _erbB_ protein represents the EGF receptor truncated at both the amino terminus and carboxy terminus represents the EGF receptor. The truncation at the carboxy terminus is not necessary for the transforming ability, since the ALV promotor-activated c-_erbB_ sequence, which is capable of transforming erythroid cells, codes for the v-_erbB_ protein-like EGF receptor, which lacks most of the extra-cellular domain but has no truncation at the carboxy terminus(45). These data suggest that transformation of cells by AEV may be the result of deregulated expression of a truncated EGF receptor molecule, which expresses protein-tyrosine kinase activity without EGF binding.

Another line of evidence that suggests a close relation between the EGF receptor and the _erbB_ protein is the assignment of the EGF receptor gene to human chromosome 7, since the human c-_erbB_ gene has also been mapped on the same chromosome(46,47,48,49).

STRUCTURAL FEATURES OF THE EGF RECEPTOR(20)

The primary structure of the EGF receptor is illustrated schematically in Figure 3 in comparison with _erbB_ protein and other receptor proteins; namely, the insulin receptor and the low density lipoprotein receptor, whose primary structures have been deduced from analysis

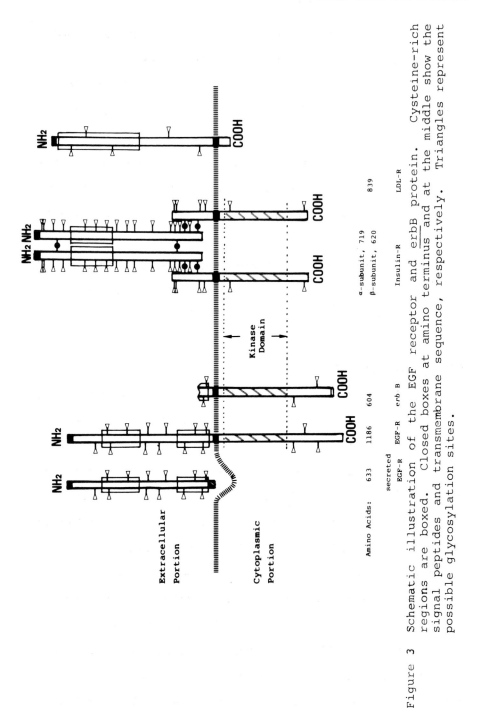

Figure 3   Schematic illustration of the EGF receptor and erbB protein. Cysteine-rich regions are boxed. Closed boxes at amino terminus and at the middle show the signal peptides and transmembrane sequence, respectively. Triangles represent possible glycosylation sites.

of their cDNA clones(50,51,52). The extreme amino terminus of the precursor of the EGF receptor consists of a hydrophobic sequence of 24 amino acids that is cleaved from the precursor and is not present in the mature receptor. This sequence is believed to function as a signal sequence to direct the nascent receptor to the endoplasmic reticulum. The mature EGF receptor consists of 1186 amino acids and has a predicted molecular weight of 131,360. This value is in good agreement with the molecular weight of the EGF receptor (138,000) synthesized in tunicamycin-treated cells, in which N-linked glycosylation is suppressed(53).

LIGAND BINDING DOMAIN

This domain, consisting of 621 amino-terminal amino acids, is located on the external surface of the plasma membrane and serves for EGF binding. An interesting feature of this domain is the abundance of cysteine residues (51 residues) most of which are concentrated within two regions of about 170 amino acids located at positions 134-313 and 446-612. A similar region is present in both the insulin receptor and the LDL receptor.

The cysteine-rich regions are rather hydrophylic and would facilitate the generation of a specific conformation for signal transmission through intramolecular or inter-molecular S-S bridges. These regions, however, may not be involved in the binding of ligands because of their highly hydrophilic character and because most ligand interactions with the receptor are thought to involve hydrophobic and electrostatic interactions.

The extra-cellular domain possesses 12 possible sites of N-linked glycosylation, and most of these are indeed glycosylated since 11 sites in the external domain have been shown to be glycosylated(53).

MEMBRANE-ANCHORING DOMAIN

A stretch of predominantly hydrophobic amino acids (residues 622-644), in an α-helical configuration, could span the lipid bilayer. The basic amino acid residues immediately carboxy terminal to the transmembrane sequence would interact with the phospholipid head groups and help in the correct allocation of the receptor at the cell surface.

KINASE DOMAIN

The cytoplasmic domain, consisting of 542 amino acids, undoubtedly contains a sequence (residues 690-940) that is shared by transforming proteins of the src family and exhibits protein-tyrosine kinase activity. Included in this kinase domain is the postulated ATP binding site(54) (Gly-X-Gly-X-X-Gly-X$_{14}$-Lys), which starts from Gly at position 695.

The carboxy-terminal sequence (residues 941 to 1186) corresponds to the E domain of the erbB protein, which is

thought to be important in interaction with the intracellular target protein. An interesting feature of this
domain is that it contains 3 major _in vitro_ phosphorylation sites (tyrosine residues at positions 1068, 1148 and
1173). The significance of phosphorylation of these
tyrosine residues is unclear. Initially, tyrosine
phosphorylation at position 1173 was thought to be
important for controll of cell growth, since the v-erbB
protein, which does not have this tyrosine residue causes
cell transformation. This possibility was, however,
disproved by the finding that the chicken c-erbB gene
activated by ALV promoter has transforming ability but has
no truncation of the 3' sequence.

It has been shown that tumor promoters such as the
phorbol diester 12-O-tetradecanoylphorbol-13-acetate
(TPA), stimulate protein kinase C-mediated phosphorylation
of the EGF receptor on threonine at position 654(55,56),
which in turn inhibits EGF-dependent induction of tyrosine
kinase activity of the receptor and decreases EGF binding
to the receptor(57,58,59). TPA also inhibits EGF-dependent proliferation of normal human embryo fibroblasts(60)
and growth of AEV transformed cells in soft agar(61).
Therefore, protein kinase C-mediated phosphorylation of
threonine 654 of the EGF receptor may provide a kind of
"feed back" regulation that decreases the effect of EGF
and the transforming ability of the erbB protein.

ANOTHER erbB-RELATED CELLULAR ONCOGENE DISTINCT FROM THE EGF RECEPTOR GENE.

The human epidermoid carcinoma cell line, A431, expresses elevated levels of EGF receptor as a consequence of amplification of the EGF receptor gene(see below), demonstrated by DNA blot hybridization using either v-erbB DNA or cloned EGF receptor cDNA as probe(20,62). The v-erbB DNA probe, but not the EGF receptor cDNA probe, detected, under non-strigent condition, an v-erbB-related gene that is not amplified in A431 cells and is named the c-erbB-2 gene. Part of the gene was cloned and the nucleotide sequence of the genomic clone revealed that the c-erbB-2 gene may encode protein-tyrosine kinase, whose amino acid sequence is deduced to be highly homologous (82 %) to the kinase domain of the EGF receptor. Moreover, the predicted amino acid sequence of the kinase domain of the c-erbB-2 product is highly homologous to v-erbB product and distantly related to the sequences of the products of other members of the src gene family(63).

A recently characterized neu gene, an oncogene detected in a series of rat neuro/glioblastomas by the gene-transfer technique, was also found to be related to the v-erbB gene and to encode a surface glycoprotein (gp185) with a molecular weight of 185,000(64,65). DNA blot hybridization analysis suggested that the rat c-erbB-2 gene is the neu gene. Furthermore, both the human counterpart of the neu gene and the human c-erbB-2 gene were independently mapped on human chromosome 17(66,67). From preliminary sequence data on the rat neu gene or the human c-erbB-2 gene, the c-erbB-2/neu gene is expected to encode a receptor protein that is very similar to the EGF receptor. Direct proof for this assumption awaits demonstration of the ligand binding site.

mRNA OF erbB-RELATED GENES

The c-erbB mRNA was detected initially in chicken embryo fibroblasts by RNA blot hybridization(68). Two species of transcripts, 9- and 12-kb mRNA, hybridized with the v-erbB DNA probe under conditions that did not allow cross-hybridization with related transcripts. The 12 kb mRNA is not the precursor form of the 9-kb mRNA, since both species are present in the cytoplasm as well as the nuclei. In addition, both species of RNA are synthesized throughout all developmental stages of chicken embryos. Two transcripts of 5.6- and 10-kb were also detected in human RNAs as c-erbB-related EGF receptor mRNAs(20,62 and Fig. 4). Structural analysis of human EGF receptor cDNA clones suggested that translation of the 5.6-kb mRNA results in production of the EGF receptor(20). The 5.6-kb mRNA consists of a short 5' non-coding sequence (about 200-300 nucleotides), 3.6 kb of coding sequence and a 3' non-coding sequence of about 1.7 kb. Although the 10-kb mRNA has not been analyzed in detail, it could result from transcriptions of another closely related gene, or,

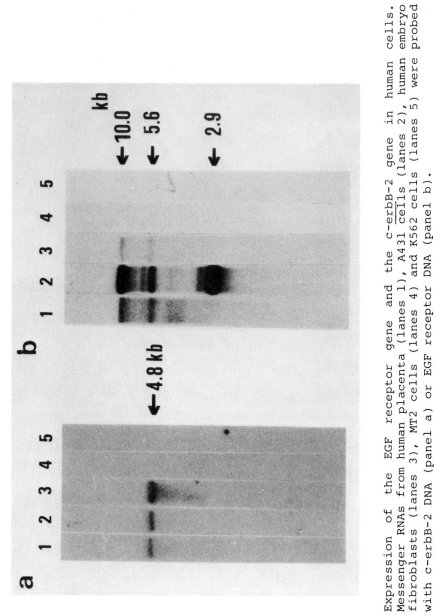

Figure 4   Expression of the EGF receptor gene and the c-erbB-2 gene in human cells. Messenger RNAs from human placenta (lanes 1), A431 cells (lanes 2), human embryo fibroblasts (lanes 3), MT2 cells (lanes 4) and K562 cells (lanes 5) were probed with c-erbB-2 DNA (panel a) or EGF receptor DNA (panel b).

alternatively, be generated from the same gene as that from which 5.8-kb mRNA is synthesized, by a different splicing mechanism or simply by carrying a longer non-coding sequence, analogous to other genes(69,70). Further studies are needed to elucidate the origin and function of the 10-kb mRNA. Minor mRNA species of 13-, 8.9-, 6.1- and 1.8 kb long were also reported to be detectable with the EGF receptor cDNA probe giving weak hybridization signals, but their nature are unknown. These RNAs might be synthesized from the yet unidentified erbB-related gene. However, extensive screening of the gene library indicated that there are not more than two v-erbB related genes in the human genome(49,63), suggesting that these minor species represent transcripts from a distantly related gene, such as a member of the src family or a gene encoding growth factor receptor, rather than the erbB-related gene.

The transcript of another v-erbB-related gene, c-erbB-2/neu, is a single species of 4.6-kb long in humans(Fig. 4). Analysis of c-erbB-2 cDNA clones showed that the 4.6-kb c-erbB-2 mRNA is similar to the 5.6-kb EGF receptor mRNA in that the coding sequence is flanked by short and long stretches of non-coding sequences at the 5' and 3' ends, respectively. The 3' non-coding sequence of the c-erbB-2 mRNA is rather shorter than that of the EGF receptor mRNA (Yamamoto, T. et al., in preparation).

The expressions of the two erbB-related genes are

rather cell-type specific. Neither the 10- and 5.6-kb EGF receptor mRNA nor the 4.6-kb c-erbB-2 mRNA is synthesized in leukemic cell lines K562 (chronic myelogenous leukemia cells) or MT2 (adult T-cell leukemia cells) or in a B-cell line, IM-9, established by EB virus infection. The two genes are transcribed at similar levels in keratinocytes, embryo fibroblasts, placenta and many cell lines established from carcinomas of epithelial cells, expect that epithelial cells often over-produce EGF receptor mRNA (see below).

Transformation of erythroid cells by AEV was suggested to be due to over-expression of part of the EGF receptor that includes the kinase domain. Quiescent expression of the EGF receptor gene in hematopoietic cells may be required for their normal growth, although we do not know the level of transcription of the EGF receptor gene in CFU-E or BFU-E cells. Detailed analysis of expression of the EGF receptor gene at each stage of differentiation of the hematopoietic cells of erythroid lineage may provide some clues on the mechanism of erythroblastosis.

AMPLIFICATION OF THE erbB-RELATED GENE IN HUMAN TUMORS.

There is accumulating evidence that over-expression of the cellular proto-oncogene can transform cells in vitro(1). For example, elevated expression of the proto-ras gene or proto-sis/PDGF-2 gene, which was placed under transcriptional control of the retroviral or SV40 promoter/enhancer, induces transformation of NIH3T3 cells(71,72,73). Gene amplification usually enhances their expression by increasing the number of DNA templates available for transcription. Therefore, analysis of the genes amplified in human tumors may provide clues to the neoplastic process.

Two karyotypic abnormalities, double minute chromosomes(DMs) and homogeneously staining regions of chromosomes (HSRs), signal the presence of amplified genes and are found in many tumor cells(74). DMs are small and paired chromatin bodies that lack centromeres and HSRs are regions of chromosomes where the unique binding pattern has been replaced by some chromatin materials. Is an amplified gene in DMs or HSRs related to the abnormal growth properties of tumor cells? Evidence for amplification of the cellular oncogenes c-myc and N-myc in HSRs found in a colon carcinoma cell line(COLO320)(75) and in a series of neuroblastoma cell lines(76), respectively, suggest that at least some genes amplified in DMs and HSRs may be causal in initial events or in biological progression of the malignant state. Generally, amplification in HSRs and DMs is of 100-fold magnitude, which implies that a gene amplified 5- to 10-fold with a small amplification unit would not be detected by current cytological techniques. In addition, possible involvement of a specific cellular oncogene in human tumors could be readily examined by the method of blot hybridization.

As mentioned above, A431 human carcinoma cells produce about 10-50 times more EGF receptors on their cell surface than most other cell types(77,78). This over-production was shown to due to about 30-fold amplification of the EGF receptor gene in A431 cells relative to that in placenta(20,62). Accordingly, A431 cells have about 50 times higher copy numbers of EGF receptor mRNAs(5.6-kb long and 10-kb long) than the placenta.

Not only A431 epidermoid carcinoma cells, but also 10 other cell lines established from human squamous cell carcinomas expressed high levels of EGF receptors(79). Other experimental data showed that 10 of 12 additional squamous carcinoma cell lines tested had an amplified EGF receptor gene(Fig. 5), but that the gene was amplified in only one of 6 primary tumors of squamous cells examined. The observed amplification of the receptor gene was in the range of 3- to 60-fold the that of keratinocytes and was associated with abundant expression, in proportion to the level of DNA amplification, of the RNAs and proteins. In contrast, no amplification of the EGF receptor gene was observed in 18 cell lines originating from leukemias and adenocarcinomas(80). Furthermore, 4 of 6 glioblastomas that express higher levels of the EGF receptor than normal were also found to have an amplified EGF receptor gene(81,82). These data suggest a high incidence of amplification of the EGF receptor gene in certain tumors.

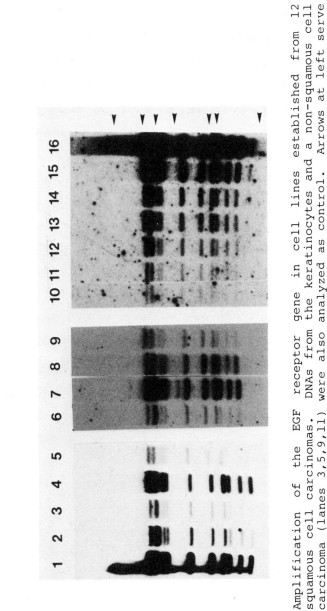

Figure 5   Amplification of the EGF receptor gene in cell lines established from 12 squamous cell carcinomas.   DNAs from the keratinocytes and a non-squamous cell carcinoma (lanes 3,5,9,11) were also analyzed as control.   Arrows at left serve as size marker (23,9.4,6.6,2.3,2.0 and 0.5kbp from top to bottom).

In addition, the data suggest that malignant squamous cells with an amplified EGF receptor gene may more readily be adapted to growth in tissue culture, than cells without an amplified EGF receptor gene, leading to their establishment as cell lines.

The neu gene is an activated oncogene detected by the gene-transfer technique in the genomes of several ethylnitrosourea-induced rat neuro/glioblastomas. As discussed before, the human counterpart of the neu gene is, probably, an erbB-related gene, which is distinct from the EGF receptor gene and has been termed the human c-erbB-2 gene, implying that disordered expression of the c-erbB-2 gene may be associated with human neoplasms. Indeed, amplification of the c-erbB-2 gene has been found in 3 adenocarcinomas; a salivary gland adenocarcinoma(63), a mammary carcinoma(83) and a cell line established from a well differentiated adenocarcinoma of the stomach(67). It is note worthy that amplification of the c-erbB-2 gene has been found only in adenocarcinomas, whereas amplification of the EGF receptor gene has frequently been found in squamous cell carcinomas. This suggests that the malignancies or tumor progressions of the two different types of epithelial cells are affected by unusual expressions of corresponding receptor proteins with similar, but distinct functions. DNAs from a number of tumors must be examined to clarify this point.

Provided that the c-erbB-2/neu gene encodes a

receptor protein similar to the EGF receptor, altered expression, either structural or quantitative, may be a frequent occurrence associated with tumorigenesis. Structural alteration of growth factor receptors would result in exertion of their intrinsic protein-tyrosine kinase activity continuously regardless of the presence or absence of the corresponding ligands. Enhanced expression of the receptor on the cell surface as a consequence of gene amplification would also result in constitutive expression of protein-tyrosine kinase activity. The mechanism of transmission by the receptor of extracellular mitogenic signals to the intracellular response is uncertain. Clarification of the molecular mechanism would provide an insight into the potential oncogenicity of the receptors.

By analogy to the activated ras oncogene in many human tumors(84), the activated neu gene probably produces a structurally altered protein product. Therefore, the growth properties of tumor cells could be modulated with antibodies that specifically recognize the altered protein conformation. To date, no comparison of the primary structure of the receptor protein produced from the amplified gene with that from the normal gene has been reported.

SYNTHESIS OF TRUNCATED EGF RECEPTOR

In addition to the two transcripts of normal size

(5.6- and 10-kb) that encode the complete EGF receptor, an unusual 2.9 kb mRNA was shown to be synthesized in A431 cells(20,62). This mRNA hybridizes with a DNA probe that carries the sequence coding for the extracellular domain of the receptor, but not with a probe containing the sequence for the cytoplasmic domain of the receptor. Nucleotide sequence analysis of cDNA clones derived from the 2.9 kb mRNA showed that this mRNA encodes a truncated EGF receptor protein of approximately 70 kd which, because it carries a signal sequence but not the membrane-anchoring sequence, should be secreted by A431 cells. In fact, monoclonal antibody, which recognizes the extra-cellular domain of the receptor, was found to immuno-precipitate a secreted 115 kd glycoprotein believed to be a glycosylated form of a 68 kd polypeptide(85).

In A431 cells, the level of expression of the 2.9-kb aberrant mRNA is about the same as that of the other two EGF receptor mRNAs (5.6- and 10-kb), which are over-expressed about 50-fold relative to their expressions in the placenta. The 2.9-kb mRNA is transcribed from a hybrid gene that was generated as a consequence of chromosomal translocation and consists of a coding sequence of unknown origin(86). Over-production of the 2.9-kb mRNA suggests that the hybrid gene is amplified to a similar extent to the gene for normal transcripts. To date, it has not been thought that the truncated receptor synthesized from the 2.9-kb mRNA would provide any clues to the neoplastic properties of A431 cells. However, we cannot exclude the possibility that extracellular accumulation of the truncated EGF receptor is associated with the appearance of the transformed phenotype. The truncated receptor may also provide a growth advantage to cells in culture, in which the EGF receptor is over-expressed.

CONCLUSIONS AND PROSPECTS

All metazoan organisms so far tested possess a family of proto-oncogenes, whose expressions are considered to be vitally important for the cells. It has been thought that studies on the altered activities of proto-oncogenes may eventually explain both the neoplastic origin and the properties of tumors. Transduction by retroviruses, induction of transcription by retroviral promoter/enhancer and DNA-mediated gene transfer provided evidence that proto-oncogenes are active in some tumors(84).

Although powerful selective pressure must have been at work to assure the conservation of proto-oncogenes during diversification of metazoan phyla, until recently we knew nothing about what kind of genes are proto-oncogenes. The c-sis gene, a cellular counterpart of the oncogene of simian sarcoma virus was first identified as the gene coding for a growth factor, PDGF. The second discovery was that the v-erbB gene is a truncated version of the EGF receptor gene. Subsequently, from studies in conjunction with those on the genetics of yeast, in which _ras_ homologs have been identified, the _ras_ proteins have been suggested to act as "coupling factors" that help to modulate the activity of adenyl cyclase. Analysis of the c-_erbB_-2/_neu_ gene provided further evidence that genes encoding growth factor receptor can be proto-oncogenes and that the receptor gene can be either activated to transform cells or amplified in certain tumors. In addition, the c-_fms_ gene was recently demonstrated, by biological analysis to encode the receptor for CSF-1. These findings suggest that altered expression of the protein components of an interdigitating network that controls cell growth during the course of differentiation is closely associated with neoplasms.

The precise role of altered proto-oncogenes in carcinogenesis remains largely uncertain. The trans-forming abilities of altered proto-oncogenes have been tested directly by gene-transfer. Among the proto-oncogenes demonstrated by transduction with retroviruses, c-_ras_ has been repeatedly identified as the active oncogene in human tumors. Initially it was thought that only this proto-oncogene of the _ras_ family is involved in tumorigenesis. However, extensive analysis of transform-ing genes from far more tumors suggested that other proto-oncogenes are also active oncogenes in tumor cells. The best example is the _neu_ gene, detected as an active oncogene in a series of rat neuro/glioblastomas, although we do not know whether the human version of the neu gene, c-erbB-2, is similarly activated in human tumors.

The possible involvement of proto-oncogenes of the src gene family was also demonstrated by the finding that human c-raf, the cellular homolog of the v-raf oncogene of 3611 murine sarcoma virus, is activated in an adeno-carcinoma of the stomach and in a glioblastoma(87,88). Although the neu gene product is believed to exhibit protein-tyrosine kinase activity, the v-raf gene and, therefore, possibly the c-raf gene, encodes protein kinase specific for threonine and serine residues, suggesting that altered expression of kinase activities, which must function in different pathways of normal cell growth, is involved in neoplasia.

It is believed that the neu gene is activated by damage to the coding domain, although no direct evidence for this is yet available. Does amplification of the proto-oncogene of neu, which is the rat c-erbB-2 gene, suffice to transform cells or maintain neoplastic properties? A point mutation, which results in a single amino acid change at residue 12 or 61 of the ras protein, has been shown to convert the proto-oncogene to the active oncogene with ability to transform NIH3T3 cells. Elevated expression of normal ras protein also transforms NIH3T3 cells. On the other hand, it has been suggested that enhanced expression of the c-src gene is not equivalent to that of the v-src gene, and that unique mutations on the

c-src gene are required for its acquisition of transforming ability(89,90).   Conceivably,   transformation   and
tumorigenesis can be induced by either increased expression or structural alteration of proto-oncogenes depending
on the proto-oncogene or tumor.   However, it should be
mentioned that qualitative change is often accompanied by
quantitative change of a single oncogene product or
several distinct oncogene products(84,91).   Therefore, it
is possible that at least in some cases, a combination of
over-expression and structural alteration of distinct sets
of   oncogenes   act   in   initiation   of   tumorigenesis   and
maintenance of neoplastic properties.   From this point of
view, it is worthwhile to examine whether the amplified
EGF receptor gene and c-erbB-2 gene carry mutations that
lead to structural changes of gene products.

Most gene products of the src gene family shows
protein-tyrosine kinase activity, which is expected to be
important in induction of neoplastic transformation.   As
discussed previously, this kinase activity is commonly
associated with receptor proteins of several polypeptide
growth factors and is believed to play a key role in the
control of cell growth.   A direct connection of the
oncogene product of the src family and the growth factor
receptor was shown by demonstration of a relation between
the EGF receptor and the erbB protein.   Other members of
the src gene family that encode the cytoplasmic kinase
domain as well as a possible transmembrane domain are the

fms and ros genes, the former, as mentioned previously, being linked with the CSF-1 receptor. The c-ros gene is closely related to the insulin receptor gene, although the two are not identical. Granted that cellular oncogenes of the src gene family whose products anchor into the cell-surface membrane encode receptors with protein-tyrosine kinase activity, is it unreasonable to speculate that the protein products of the src family that do not possess the membrane anchoring sequence function as subunits of a receptor complex, in concert with a protein that is devoid of the intracellular domain? Extensive efforts to analyze the primary structure and function of the proto-oncogenes of the src gene family are expected to clarify this problem.

ACKNOWLEDGEMENTS

We thank T. Akiyama for critical reading of the manuscript and S. Sasaki for help in preparing the manuscript.

REFERENCES

1. Bishop, J.M. and Varmus, H. "Functions and origins of retroviral transforming genes", in RNA Tumor Virus-2 (eds. Weiss, R., Teich, N., Varmus, H. and Coffin, J.) pp249-356, Cold Spring Harbor Laboratory, New York (1985)

2. Defeo-Jones, D., Scolnick, E.M., Koller, R. and Dhar, R. Nature **306**, 707-709 (1983)

3. Powers, S., Kataoka, T., Fasano, O., Goldfarb, M., Stratherm, J., Broach, J. and Wigler, M. Cell **36**, 607-612 (1984)

4. Fukui, Y. and Kaziro, Y.,   EMBO J. **4**, 687-691 (1985)

5. Shilo, B.-Z. and Weinberg, R.A.   Proc. Natl. Acad. Sci. U.S.A. **78**, 6789-6792 (1981)

6. Neuman-Silberberg, F.S., Schejter, E., Hoffman, F.M. and Shilo, B.-Z. Cell **37**, 1027-1033 (1984)

7. Simon, M.A. Kornberg, T.B. and Bishop, J.M.   Nature **302**, 837-839 (1983)

8. Hoffman, F.M., Fresco, L.D., Hoffman-Falk, H. and Shilo, B.-Z. Cell **35**, 393-401 (1983)

9. Livneh, E., Glazer, L., Segal, D., Schlessinger, J. and Shilo, B.-Z. Cell **40**, 599-607 (1985)

10. Katzen, A.L., Kornberg, T.B. and Bishop, J.M.   Cell **41**, 449-456 (1985)

11. Toda, T., Uno, I., Ishikawa, T., Powers, S., Kataoka, T., Broek, D., Cameron, S., Broach, J., Matsumoto, K. and Wigler, M. Cell **40**, 27-36 (1985)

12. Mark, G.E. and Rapp, U.R., Science **224**, 285-289 (1984)

13. Sutrave, P., Bonner, T.I., Rapp, U.R., Jansen, H.W., Patschinsky, T. and Bister, K.   Nature **309**, 85-88 (1984)

14. Cohen, S., Carpenter, G. and King, L.   J. Biol. Chem. **255**, 4834-4842 (1980)

15. Nishimura, J., Hung, H.S. and Deuel, T.F.   Proc. Natl. Acad. Sci. U.S.A. **79**, 4303-4307 (1982)

16. Kasuga, M., Zick, Y., Blithe, D.L., Crettaz, M. and Kahn, C.R. Nature **298**, 667-669 (1982)

17. Rubin, J.B., Shia, M.A. and Pilch, P.F.   Nature **305**, 438-440 (1983)

18. Yamamoto, T., Nishida, T., Miyajima, N., Kawai, S., Ooi, T. and Toyoshima, K. Cell **35**, 71-78 (1983)

19. Downward, J., Yarden, Y., Mayes, E., Scrace, G., Toffy, N., Stockwell, P., Ullrich, A., Schlessinger, J. and Waterfield, M.D. Nature **307**, 521-527 (1984)

20. Ullrich, A., Coussens, L., Hayblick, J.S., Dull, T.J., Gray, A. Tam, A.W., Lee, J., Yarden, Y., Libermann, T.A., Schlessinger, J., Downward, J., Mayes, E.L.V., Whittle, N., Waterfield, M.D. and Seeburg, P.H.

Nature **309**, 418-425 (1984)

21. Sherr, C.J., Rettenmier, C.W., Sacca, R., Roussel, M.F., Look, A.T. and Stanley, E.R. Cell **41**, 665-676 (1985)

22. Waterfield, M.D., Scrace, G.T., Whittle, N., Stroobant, P., Johnsson, A., Wasteson, A., Westermark, B., Heldin, C.-H., Huang, J.S. and Deuel, T.F. Nature **304**, 35-39 (1983)

23. Doolittle, R.F., Hunkapiller, M.W., Hood, L.E., Devare, S.G., Robbins, K.C., Aaronson, S.A. and Antoniades, H.N. Science **221**, 275-277 (1983)

24. Samarut. J. and Gazzolo, L. Cell **28**, 921-929 (1982)

25. Graf, T. and Stehelin, D. Biochim Biophys Acta **651**, 245-271 (1982)

26. Yamamoto, T., Hihara, H., Nishida, T., Kawai, S. and Toyoshima, K. Cell **34**, 225-232 (1983)

27. Frykberg, L., Palmieri, S., Beug, H., Graf, T., Hayman, M.J. and Vennstrom, B. Cell **32**, 227-238 (1983)

28. Pierce, J.H., Aaronson, S.A. and Anderson, S. Proc. Natl. Acad. Sci. U.S.A. **81**, 2374-2378 (1984)

29. Kahn, P., Adkins, B., Beug, H. and Graf, T. Proc. Natl. Acad. Sci. U.S.A. **81**, 7122-7126 (1984)

30. Fung, Y.K., Lewis, W.G., Crittenden, L.B. and Kung, H.J. Cell **33**, 357-368 (1983)

31. Raines, M.A., Lewis, W.G., Crittenden, L.B. and Kung, H.J. Proc. Natl. Acad. Sci. U.S.A. **82**, 2287-2291 (1985)

32. Privalsky, M., Sealy, L., Bishop, J.M., McGrath, J. and Levinson, A. Cell **32**, 1257-1267 (1983)

33. Hayman, M.J., Ramsay, G.M., Savin, K., Kitchener, G., Graf, T. and Beug, H. Cell **32**, 579-588 (1983)

34. Hayman, M.J. and Beug, H. Nature **309**, 460-462 (1984)

35. Decker, S.J. J. Biol. Chem. **260**, 2003-2006 (1985)

36. Beug, H. and Hayman, M.J. Cell **36**, 963-972 (1984)

37. Nishida, T., Sakamoto, S., Yamamoto, T., Hayman, M., kawai, S. and Toyoshima, K. Gann **75**, 325-333 (1984)

38. Kris, R.M., Lax, I., Gullick, W., Waterfield, M.D., Ullrich, A., Fridkin, M. and Schlessinger, J. Cell **40**, 619-625 (1985)

39. James, R. and Bradshaw, R.A. "Polypeptide growth-factors" in Ann. Rev. Biochem. (eds. Snell, E.E., Boyer, P.D., Meister, A. and Richardson, C.C.) Vol 53 pp259-292 Annual Reviews Inc. California (1984)

40. Pastan, I. and Willingham, M.C. Trends in Biochem. Sci. **8**, 250-254 (1983)

41. Carpenter, G., King, L. and Cohen, S. Nature **276**, 409-410 (1979)

42. Hunter, T. and Cooper, J.A. Cell **24**, 741-752 (1981)

43. Basu, M., Biswas, R. and Das, M. Nature **311**, 477-480 (1984)

44. Downward, J., Parker, P. and Waterfield, M.D. Nature **311**, 483-485 (1984)

45. Nilsen, T.W., Maroney, P.A., Goodwin, R.G., Rottman, F.M., Crittender, L.B., Raines, M.A. and Kung, H.-J. Cell **41**, 719-726 (1985)

46. Jansson, M., Philipson, L. and Vennstrom, B. EMBO J. **2**, 561-565 (1983)

47. Shimizu, N., Behzadian, M.A. and Shimizu, Y. Proc. Natl. Acad. Sci. U.S.A. **77**, 3600-3604 (1980)

48. Kondo, I. and Shimizu, N. Cytogenet. Cell. Genet. **35**, 9-14 (1983)

49. Spurr, N.K., Solomon, E., Jansson, M., Sheer, D., Goodfellow, P.N., Bodmer, W.F. and Venstrom, B. EMBO J. **3**, 159-163 (1984)

50. Ullrich, A., Bell, J.R., Chen, E.Y., Herrera, R., Petruzzelli, L.M., Dull, T.J., Gray, A., Coussens, L., Liao, Y.-C., Tsubokawa, M., Mason, A., Seeburg, P.H., Grunfeld, C., Rosen, O.M. and Ramachandran, J. Nature **313**, 756-761 (1985)

51. Ebina, Y., Ellis, L., Jarnagin, K., Edery, M., Graf, L., Clauser, E., Ou, J.-h., Masiarz, F., Kan, Y.W., Goldfine, I.D., Roth, R.A. and Rutter, W.J. Cell **40**, 747-758 (1985)

52. Yamamoto, T., Davis, C.G., Brown, M.S., Schneider, W.J., Casey, M.L., Goldstein, J.L. and Russel, D.W. Cell **39**, 27-38 (1984)

53. Mayes, E.L.V. and Waterfield, M.D. EMBO J. **3**, 531-537 (1984)

54. Kamps, M.P., Taylor, S.S. and Sefton, B.M. Nature **310**, 589-592 (1984)

55. Nishizuka, Y. Nature **308**, 693-698 (1984)

56. Hunter, T., Ling, N. and Cooper, J.A. Nature **311**, 480-483 (1984)

57. Cochet, C., Gill, G.N., Meisenhelder, J., Cooper, J.A. and Hunter, T. J. Biol. Chem. **259**, 2553-2558 (1984)

58. Iwashita, S. and Fox, C.F. J. Biol. Chem. **259**, 2559-2567 (1984)

59. Friedman, B., Frackelton, A.R., Ross, A.H., Connors, J.M., Fujiki, H., Sugimura, T. and Rosner, M.R. Proc. Natl. Acad. Sci. U.S.A. **81**, 3034-3038 (1984)

60. Decker, S. Mol. Cell. Biol. **4**, 1718-1723 (1984)

61. Decker, S.J. J. Biol. Chem. **260**, 2003-2006 (1985)

62. Merlino, G.T., Xu, Y.-H., Ishii, S., Clark, A.J.L., Semba, K., Toyoshima, K., Yamamoto, T. and Pastan, I. Science **224**, 417-419 (1984)

63. Semba, K., Kamata, N., Toyoshima, K. and Yamamoto, T. Proc. Natl. Acad. Sci. U.S.A. in press (1985)

64. Schechter, A.L., Stern, D.F., Vaidyanathan, L., Decker, S.J., Drebin, J.A., Green, M.I. and Weinberg, R.A. Nature **312**, 513-516 (1984)

65. Drebin, J.A., Stern, D.F., Link, V.C., Weinberg, R.A. and Green, M.I. Nature **312**, 545-548 (1984)

66. Weinberg, R.A. personal communication

67. Fukushige, S., Yoshida, M.C., Suzuki, T., Semba, K. and Yamamoto, T. to be published

68. Venstron, B. and Bishop, J.M. Cell **28**, 135-143 (1982)

69. Setzer, D.R., McGrogan, M., Nunberg, J.H. and Schimke, R.T. Cell **22**, 316-370 (1980)

70. Tosi, M., Young, R.A., Hagenbuchle, O. and Schibler, U. Nucleic Acid Res. **9**, 2313-2323 (1981)

71. Chang, E.H., Furth, M.E., Scolnick, E.M. and Lowy, D.R. Nature **297**, 479-483 (1982)

72. Gazit, A., Igarashi, H., Chiu, I.-M., Srinivasan, A., Yaniv, A., Tronick, S.R., Robbins, K.C. and Aaronson, S.A. Cell **39**, 89-97 (1984)

73. Clarke, M.F., Westin, E., Schmidt, D., Josephs, S.F., Ratner, L., Wong-Steal, F., Gallo, R.C. and Reitz Jr., J.S. Nature **308**, 464-467 (1984)

74. Barker, P.E. Cancer Genet. Cytogenet. **5**, 81-94 (1982)

75. Alitaro, K., Winqvist, R., Lin, C.C., de la Chapelle, A., Schwab, M. and Bishop, J.M. Proc. Natl. Acad. Sci. U.S.A. 4534-4538 (1984)

76. Schwab, M., Alitaro, K., Klempnauer, K.-H., Varmus, H.E. Bishop, J.M., Gilbert, F., Brodeur, G., Goldstein, M. and Trent, J. Nature **305**, 245-248 (1983)

77. Fabricant, R.N., DeLarco, J.E. and Todaro, G.J. Proc. Natl. Acad. Sci. U.S.A. **74**, 565-569 (1977)

78. Haigler, H., Ash, J.F., Singer, S.J. and Cohen, S. Proc. Natl. Acad. Sci. U.S.A. **75**, 3317-3321 (1978)

79. Cowley, G., Smith, J.A., Gusterson, B., Hendler, F. and Ozanne, B. "The amount of EGF receptor is elevated on squamous cell carcinoma" in Cancer Cells (eds. Levine, A.J., Vande Woude, G.F., Topp, W.C. and Watson, J.D.) Vol 1 pp5-10, Cold Spring Harbor Lab. New York (1984)

80. Yamamoto, T., Kamata, N., Kawano, H., Shimizu, S., Kuroki, T., Toyoshima, K., Rikimaru, K., Nomura, N., Ishizaki, R., Pastan, I., Gamou, S. and Shimizu, N. Cancer Res. in press (1985)

81. Libermann, T.A., Razon, N., Bartal, A.D., Yarden, Y., Schlessinger, J. and Sorew, H. Cancer Res. **44**, 753-760 (1984)

82. Libermann, T.A., Nusbaum, H.R., Razon, N., Kris, R., Lax, I., Soreq. H., Whittle, N., Waterfield, M.D., Ullrich, A. and Schlessinger, J. Nature **313**, 144-147 (1984)

83. King, C.R., Kraus, M.H. and Aaronson, S.A. Science in press (1985)

84. Land, H., Parada, L.F. and Weinberg, R.A.   Science
    222, 771-778 (1983)

85. Weber, W., Gill, G.N. and Spiess, J.   Science **224**,
    294-297 (1984)

86. Merlino, G.T., Ishii, S., Whang-Peng, J., Knutsen, T.,
    Xu, Y.-H., Clark, A.J.L., Stratton, R.H., Wilson,
    R.K., Ma, D.P., Roe, B.A., Hunts, J.H., Shimizu, N.
    and Pastan, I. Mol. Cell. Biol. **5**, 1722-1734 (1985)

87. Fukui, M., Yamamoto, T., Maruo, K., Kawai, S. and
    Toyoshima, K. Proc. Natl. Acad. Sci. U.S.A. in press
    (1985)

88. Shimizu, K., Nakatsu, Y., Sekiguchi, M., Hokamura, K.,
    Tanaka, K., Terada, M. and Sugimura, T. Proc. Natl.
    Acad. Sci. U.S.A. in press (1985)

89. Parker, R.C., Varmus, H.E. and Bishop, J.M. Cell **37**,
    131-139 (1984)

90. Iba, H., Takeya, T., Cross, F.R., Hanafusa, T. and
    Hanafusa, H.   Proc. Natl. Acad. Sci. U.S.A. **81**,
    4424-4428 (1984)

91. Murray, M.J., Cunningham, J.M., Parada, L.F., Dautry,
    F., Lebowitz, P. and Weinberg, R.A. Cell **33**, 749-757
    (1983)

Human Genes and Diseases
Edited by F. Blasi
© 1986, John Wiley & Sons, Ltd.

ALDOLASE GENE AND PROTEIN FAMILIES: STRUCTURE, EXPRESSION AND

PATHOPHYSIOLOGY

Francesco Salvatore, Paola Izzo and Giovanni Paolella

1. INTRODUCTION

The aldolase family of enzymes is essential for the degradation of glucose-6-phosphate. This crucial role has prompted numerous investigations aimed at defining the structure and the catalytic aspects of these enzymes. More recently, the structure and expression of the aldolase genes have also been the object of intensive study. The clinical implications of this family of enzymes became apparent when Hers and Joassin (1961) associated a decrease in the catalytic activity of one isoenzyme of the aldolase family, liver aldolase B, to the presence of an autosomal recessive human disease, hereditary fructose intolerance (HFI). Thanks to the availability of gene probes for aldolase B (Besmond et al., 1983; Paolella et al., 1984; Rottmann et al., 1984), it is now possible to apply DNA recombinant techniques to the study of the disease at the gene level.

Following the lines of the comprehensive review by Horecker et al. (1972), the present work is an attempt to provide a general overview of the topics concerning the aldolase protein family, with particular reference to more recent developments in studies on their molecular and functional properties. In addition,

611

we shall give an account of recent investigations, up to 1984, on aldolase gene structure and expression in both physiological and abnormal conditions.

2. THE ALDOLASE ISOENZYME PROTEIN FAMILY

(a) Brief overview

The aldolase family consists of three isoenzymes: muscle fructose 1,6-bisphosphate aldolase (Meyerhof et al. 1936), hepatic fructose-1-phosphate aldolase (Leuthardt et al. 1952; Hers and Kusaka, 1953), and brain aldolase (Penhoet' et al., 1966). The three aldolase isoenzymes, now called aldolase A, B and C, respectively, occur in all vertebrates (Lebherz and Rutter, 1969). A fourth aldolase variant, aldolase D, has been identified; however, as yet it has only been found in a salmonoid fish (Lebherz and Rutter, 1969).

The main properties of the three aldolases can be summarised as follows:

- The native enzymes exist as a tetrameric combination of monomers of approximately the same size ($M_r \cong 40,000$) (Kawahara and Tanford, 1966; Masters and Winzor, 1971; Penhoet et al., 1967; Penhoet et al., 1969b).

- The constituent monomers are homologous in isoenzymes A and B, (four monomers A in aldolase A, and four monomers B in aldolase B), and essentially heterologous in the C type (three hybrid members of the AC set, A_3C, A_2C_2, AC_3, in addition to the homologous A_4 and C_4 species); another AB hybrid set (A_4, A_3B, A_2B_2, AB_3, B_4) has been found in kidney (Penhoet et al., 1966;

Penhoet et al., 1967; Lebherz and Rutter, 1969).

- Only three structural genes code for the three types of protein monomers described so far (Rutter et al., 1968).

- The three monomers have independent catalytic activity within the tetramer, and the functional role of the tetrameric nature of the aldolases has yet to be completely clarified (Meighen and Schachman, 1970; Penhoet and Rutter, 1971).

- The general similarities and some differences of the three monomer variants may be summarised as follows: a) similar, although not identical, conformation at the active site, since the three variants catalyze the cleavage or the formation of phosphorylated fructose via a Schiff-base that is formed between the substrate and the NH_2 group of a lysyl residue of the enzyme (Horecker et al., 1963; Penhoet et al., 1969b; Morse and Horecker, 1968); b) similar subunit association sites as revealed by in vitro and in vivo random tetramer production (Penhoet et al., 1966; Penhoet et al., 1967; Lebherz and Rutter, 1969); c) distinct epitopes for the three subunits (no antiserum against one cross-reacts with the other two) (Penhoet et al., 1969a; Penhoet and Rutter, 1971); d) different primary structures, as demostrated by amino acid composition and fingerprint pattern after trypsin digestion: this is consistent with the differences in several catalytic properties (Penhoet et al., 1969a).

The overall reaction catalyzed by the aldolases (E.C.4.1.2.13) through a lyase-type mechanism is illustrated below.

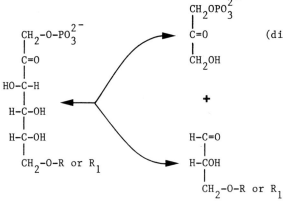

R = -H (fructose-1-phosphate and glyceraldehyde)

R_1 = $-PO_3^{2-}$ (fructose-1,6-bisphosphate and glyceraldehyde-3-phosphate)

        The metabolic pathways, along which the aldolases operate, are shown in Fig. 1.

        Aldolase A degrades fructose-1,6-bisphosphate, and thus is involved principally in the main catabolic pathway for glucose-6-phosphate. Aldolase B degrades liver fructose-1-phosphate, the main source of which is dietary fructose. When aldolase B activity is drastically diminished, as in hereditary fructose intolerance (HFI), dietary fructose accumulates, and provokes liver damage which can lead to cirrhosis and to malfunctioning of the gastrointestinal tract (Gitzelmann et al., 1983). That the primary defect in HFI is a decrease or a complete lack of liver fructose-1-phosphate aldolase, or aldolase B, has been demonstrated by direct enzyme assay on liver biopsy from HFI patients. Since antibodies against the enzyme show a 30% reaction with a liver protein from HFI patients, it has been suggested that a mutation of the structural gene is responsible for the defect of

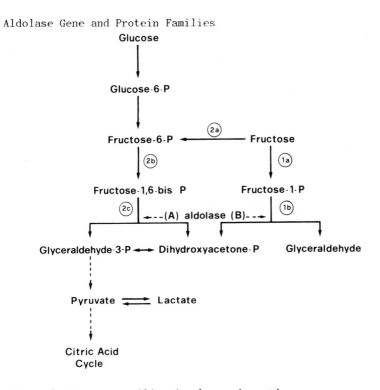

Figure 1. <u>Fructose utilization by various tissues.</u>
1a and 1b   = via F-1-P (intestine, liver, kidney)
2a, 2b and 2c = via F-1,6-bisP (muscle, adipose tissue)
Aldolase B deficiency (reaction 1b) in man causes hereditary
fructose intolerance. Fructokinase deficiency (reaction 1a) in man
causes essential fructosuria.

aldolase B (Nordmann et al., 1968). However, there is evidence of a

genetic inhomogeneity at the level of the hepatic aldolase, i.e.,

there may be more than one type of genetic defect at the level of

the enzyme (Koster et al., 1975; Gitzelmann et al., 1974).

Therefore, this is evidence that the family of aldolase proteins

can be exploited for the study of the molecular basis of a disease.

Moreover, also aldolase A deficiency in red cells has been

associated to hemolytic anemia, mental retardation, and glycogen

accumulation in the liver (Beutler et al., 1973; Lowry and Hanson,

1977).

The isoenzyme protein family of human aldolases is also important in the study of tissue- and organ-specific expression, as the electrophoretic patterns derived from the various tissues have shown. Moreover, the isoenzyme pattern of aldolase appears to change during development from the foetal stage to full-growth, and the foetal-type pattern has been observed to reappear in various tumor liver cells and in the course of liver chemical carcinogenesis. All these findings, together with results of RNA studies, will be discussed later in this Chapter.

Lastly, the question as to the kind of regulatory processes responsible for the selective expression of each enzyme subunit may be approached by the study at the molecular level of this gene family.

In the following sections we shall summarize the main findings that have been obtained at the protein and gene level. We shall also describe the implications that can be envisaged in the field of molecular pathology, i.e., the diagnosis of molecular diseases at DNA level, and the impact that the study of this family could have on the more general topics of cell differentiation and de-differentiation.

(b) Isolation and purification

The name aldolase was given to the enzyme when Meyerhof et al. (1936) found that it catalyzed the reversible condensation of dihydroxyacetone-P (DHAP) with several aldehydes. Shortly after, Meyerhof showed that aldolase was able to cleave fructose 1,6-bisphosphate (then known as fructose-1,6-diphosphate) in one mole each of DHAP and glyceraldehyde-3-phosphate (G3P) (Meyerhof,

1938a; 1938b). The enzyme was easily purified and crystallized from mammalian muscle (Warburg and Christian, 1943). The rather simple isolation procedures yielded large quantities of the enzyme that could be used for mechanicistic and molecular investigations (Warburg and Christian, 1943; Taylor et al., 1948; Lai, 1968). Also liver aldolase has been purified to homogeneity and isolated in crystalline form (Penhoet et al., 1969a; Göeschke and Leuthardt, 1963; Morse and Horecker, 1968; Gracy et al., 1969; Gürtler and Leuthardt, 1970; Ikehara et al., 1970). Lastly, aldolase C has been isolated in pure form from chicken brain (Marquardt, 1970) and beef brain (Penhoet et al., 1969b). In humans, aldolase A has been purified from skeletal muscle (Freemont et al., 1984), aldolase B from liver (Gürtler and Leuthardt, 1970), and aldolase C from brain (Willson and Thompson, 1980).

(c) Kinetics and molecular properties

Aldolases have been divided in two classes according to their mechanism of action: class I consists of Schiff-base-forming aldolases, which are widely distributed in animals and higher plants; class II consists of metallo-aldolases, which are practically limited to bacteria and moulds, and which implicate the requirement for a metal ion cofactor (Horecker et al., 1972).

Class I aldolases implicate the formation of a carbanion intermediate (enamine) after the Schiff-base formation, which allows the condensation of the two triosephosphates. The Schiff-base is formed between the lysyl amino group (aa 227) of the active site and the face si of the C-2 group of fructose bisphosphate (Trombetta et al., 1977). There is another intermediate of the reaction, the pyruvaldehyde-aldolase-orthophosphate complex (Grazi and Trombetta, 1978) that is in equilibrium with the aldolase-DHAP complex.

The specificity of the three aldolase forms towards the substrates, as demonstrated by apparent K_M, is very different. Aldolases A and C show a greater specificity than aldolase B for fructose-bisphosphate, whereas aldolase B has a greatèr specificity than aldolases A and C for fructose-1-phosphate. The activity ratio towards the two substrates (Fru-P_2/Fru-1-P) also varies greatly among the three forms: for aldolase A it is 50; for aldolase C, 25; and for aldolase B, 1 (Penhoet et al., 1966; 1969a).

The variation in activity among the various aldolases is of great physiological significance, in that it is consistent with the essential glycolytic nature of aldolase A as compared to aldolase B. The latter is devoted mainly to the utilization of exogenous fructose after its phosphorylation to fructose-1-phosphate (see Fig. 1). Furthermore, the K_M for G3PD is lower for aldolase B than it is for aldolase A, which is consistent with the fact that aldolase B is involved essentially in neoglucogenesis (Rutter et al., 1963).

Each of the four subunits of the enzyme has a binding site for the substrate, and each is catalytically active (Meighen and Schachman, 1970; Penhoet and Rutter, 1971).

Aldolase A, B and C active tetramers have a similar molecular weight. It ranges from 160,000 to 148,000 (Kawahara and Tanford, 1966; Masters and Winzor, 1971; Penhoet et al., 1967), and the C type is relatively smaller than the other two (about 30 residues shorter, for each polypeptide chain) (Lee and Horecker, 1974).

The three isoenzymes have very different amino acid compositions. The main differences are in aspartic acid, alanine and methionine between A and B; and in lysine, cysteine, glutamic acid, methionine and leucine between C and the other two forms (Penhoet et al., 1969b). The immunological reactivity of the three subunits is different, as demonstrated by the absence of cross-reactivity with specific antibodies against each of the other subunits (Penhoet et al., 1969a; Penhoet and Rutter, 1971).

The electrophoretic mobility is noticeably different among the three isoenzymes; the hybrid species, made up of mixed populations of A and B, or of A and C, having intermediate mobility. The C form migrates towards the anode, while A and B migrate towards the cathode (Penhoet et al., 1966; Lebherz and Rutter, 1969).

Rabbit liver aldolase and rabbit liver fructose-1,6-bisphosphatase form a one-to-one molecule complex. The complex formation is very specific, since muscle aldolase cannot replace liver aldolase. These findings support the possibility that the complex can play a role in gluconeogenesis (MacGregor et al., 1980).

(d) Primary structure

In this section we describe findings on the primary protein structure of the three aldolase forms. In a later section we shall report on studies of gene structure, and shall compare the sequences obtained with those obtained by direct protein analysis.

The homologies at the protein and gene level will also be discussed later in this paper.

The structure of aldolase A has been completely sequenced by Lai (1975) from rabbit muscle. Independently, Sajgò and Hajòs (1974) also determined the complete amino acid sequence of the protein from rabbit tissue. However, there were many discrepancies between the two sequences. Work from a third laboratory (Benfield et al., 1979) solved several of these differences, but at the same time it showed other discrepancies. These differences were to be solved only much later with the advent of mRNA studies (see below). Most of differences found were at the level of the Glu-Gln and Asp-Asn couples, where the analytic procedures at the protein level can lead to ambiguities.

In addition, evidence for a microheterogeneity of aldolase A monomers emerged from studies by Kochman et al. (1968), Susor et al. (1969), and Anderson et al. (1969). They found two forms that differ only in one residue of aspargine (contained in the α-form) which is deamidated to aspartic acid in the β-form. The difference has been shown to be age-dependent and presumably due to post-translational modification of residue 358 (now 360) of the sequence, as determined by Lai (1975).

The β-subunit starts to appear when the animal (rabbit) is three-five months old and it reaches a steady-state level in the adult animal, where 50% of subunits are in the deamidated form (Koida et al., 1969). A second type of microheterogeneity was suggested from the presence of organic phosphorus within the molecule (Kobashi et al. 1966).

Recently, a human skeletal muscle aldolase has been sequenced for 2/3 of its length (Freemont et al., 1984). With this sequencing it became possible to: i) resolve many of the discrepancies existing for rabbit muscle skeletal aldolase; ii) establish that fructose-1,6-bisphosphate aldolase is highly conserved with only 4 differences per 100 residues between human and rabbit sequences, and 20 differences per 100 residues between human and Drosophila aldolases; iii) interpret the three-dimensional structure of this human enzyme, which is amenable to X-ray studies.

Before the advent of gene structure studies, the only sequences available for aldolase B, were limited sections of ox liver aldolase from residues 1-39 and 173-248 (Benfield et al., 1979), and sequences near the active site lysine (aa.215-241) (Morse and Horecker, 1968; Ting et al., 1971). The only available aldolase C sequence from protein structure studies is reported in Horecker et al. (1972), as unpublished data from Felicioli and Horecker.

(e) Active site of the enzyme: structure and function

Class I aldolases, which form Schiff-base intermediates in their reaction mechanism, contain a specific lysine residue at the active site that reacts with the carbonyl group of the substrate.

Besides the lysine residue at position 227, other lysine, cysteine, histidine and tyrosine residues appear to be involved in the active site of the fructose-1,6-bisphosphate aldolase. These residues, though far apart along the polypeptide chain, lie on the surface of the molecule and they must converge in sufficient proximity to form the active site of the enzyme. Both sequence studies and the chemical reactivity of the functional group in aldolase, strongly suggest that the middle part of the sequence is probably required for the formation of a folding domain, which is responsible for the preservation of the three-dimensional structure (Stellwagen, 1976; Lambert et al., 1977).

The mechanism of action of the enzyme has been the subject of a series of elegant investigations, which have led to the elucidation of several aspects of enzyme-substrate interactions and to the understanding of the sequential binding of substrates (Grazi and Trombetta, 1980; Lai et al., 1974; Grazi et al., 1983; Grazi and Trombetta, 1984). A detailed discussion of these studies is beyond the scope of this review. However, a study performed by Grazi and Trombetta (1979) is particularly pertinent to the physiological role of the aldolase isoenzymes. The mechanism of action of the liver enzyme described by them accounts for its prevalent function in gluconeogenesis as opposed to glycolysis.

(f) Effects of proteases

Fructose-1,6-bisphosphate aldolase A and aldolase B appear to be, in vivo, a likely substrate for a limited proteolysis by lysosomal cathepsins, particularly cathepsins B, M and D. The action of these carboxy-terminal exopeptidases may produce: (i) a change in the enzyme specificity: a large decrease of activity toward fructose-1,6-bisphosphate and little or no decrease toward fructose-1-phosphate (Bond and Barret, 1980); (ii) inactivation of the enzyme as the first step of its catabolic fate (Petell and Lebherz, 1979); (iii) resistance to total proteolysis with maintenance of immunological cross-reactive material (Offermann et al., 1983); (iv) loss of complex formation between fructose-1,6-bisphosphate aldolase and fructose-1,6-bisphosphatase, when attacked by cathepsin M (Pontremoli et al., 1982a).

3. THE ALDOLASE ISOENZYME GENE FAMILY

The cloning and sequencing of DNA coding for aldolase isoenzymes, both cDNAs and structural genes, promises to yield a wealth of information on the aldolase system. Four groups have independently cloned and sequenced aldolase isoenzyme cDNAs in mammals (Costanzo et al., 1983; Besmond et al., 1983; Tsutsumi et al., 1983; Putney et al., 1983). Basically, two approaches have been used to obtain DNA sequences coding for aldolase isoenzymes. The more traditional method is based on in vitro translation of mRNA prepared from aldolase-producing cells followed by specific aldolase mRNA isolation and cDNA synthesis (Besmond et al., 1983; Tsutsumi et al., 1983). The second method allows the identification of aldolase cDNAs by means of directly sequencing M13 clones from a cDNA library and comparing the sequences obtained to the known protein sequences (Costanzo et al., 1983; Putney et al., 1983). The small probes so obtained are used to identify full-length cDNAs and genomic clones.

(a) mRNA isolation and cDNA fragment synthesis

mRNA from muscle tissue (Lebherz and Doyle, 1976) and liver tissue (Oda and Omura, 1980; Tsutsumi and Ishikawa, 1981; Grégori et al., 1982) has been used to obtain in vitro synthesis of aldolase isoenzymes. Using mRNA from chicken muscle, Lebherz and

Doyle (1976) obtained the synthesis of aldolase tetramers in a
wheat germ extract. The newly synthesized aldolase polypeptide
immunoprecipitated with anti-aldolase antibodies and was very
similar in size to native aldolase subunits. Moreover, the subunits
formed tetramers, which indicated that the tetrameric process is
formed without an intermediate processing step. Oda and Omura
(1980) obtained consistent results for aldolase B synthesized from
rat liver RNA. They also showed that aldolase B mRNA is mainly
synthesized on free ribosomes. Tsutsumi and Ishikawa (1981)
purified aldolase B mRNA by immunoabsorption of the polysomes to
anti-aldolase B antibodies and subsequent sucrose gradient,
obtaining a 127-fold enrichment. Aldolase B has also been
synthesized from human liver RNA. Grégori et al. (1984)
demonstrated that translatable aldolase B mRNA is 1.3% of the total
mRNA in adult liver, but it is 0.1% in a six months old fetus, and
0.01% in the liver of a 4.5 months old fetus.

Aldolase isoenzyme cDNA fragments have been isolated by
cloning the cDNA obtained from purified specific mRNA (Tsutsumi et
al., 1983). Simon et al. (1983) used a similar approach, but they
took advantage of the differential expression of aldolase B mRNA in
fasted rats and in rats fed a carbohydrate-rich diet. cDNAs
prepared from the two RNAs were used as positive and negative
probes, respectively, to screen an aldolase B-enriched cDNA
library.

(b) Cloning of cDNA fragments by shot-gun sequencing

We screened a cDNA library (Costanzo et al., 1983) by
sequencing M13 recombinant clones and identifying them by

comparison with the known protein sequences. By this method we identified several cDNAs coding for human proteins encoded by liver mRNAs. Most of them were cDNAs for secretory proteins, but some, such as ferritin and aldolase B, coded for intracellular proteins. Glyceraldehyde-3-phosphate dehydrogenase has been identified in the same way (Arcari et al., 1984), and the same approach was used by Putney et al. (1983) to screen rabbit muscle cDNAs. Aldolase A cDNA has been identified in this way, as have many actins and other muscle-specific mRNAs.

(c) Identification of full-length cDNAs

The identification and sequencing of full-length cDNAs has led to the elucidation of both the structure of aldolase mRNA and the complete sequencing of several aldolase proteins. Moreover, with full-length cDNAs it became possible to identify the correspondent mRNAs and genes, and therefore to start to study their expression in vivo.

Aldolase A cDNA was first described by Tolan et al. (1984), who identified in a rabbit muscle library, a set of overlapping clones that almost completely covered the mRNA. The remaining portion was determined by primer extension experiments. The total length of the mRNA seems to be 1375 bp plus the poly-A tail. The 5'-untranslated region is 62 nucleotides long. The 3'-non-coding region is 218 nucleotides and shows the canonical poly-A addition signal. The coding region codes for a single polypeptide chain of 364 amino acids, one more than the previously known protein sequence. The sequence described by Tolan et al. (see also Fig. 2) contributed to solving the ambiguities between the two

PAT ALDOLASE B cDNA
HUMAN ALDOLASE B cDNA
RABBIT ALDOLASE A cDNA
MOUSE ALDOLASE A cDNA
MOUSE ALDOLASE C cDNA

different protein sequences reported in literature (Lai et al., 1974; Sajgò and Haigòs, 1974; see also Section 2(d) above). Small sequence variations were found between the different clones, thus suggesting either allelic variations or, possibly, more active genes per cell.

The human aldolase B cDNA sequence was independently determined by two groups (Rottmann et al., 1984; Paolella et al., 1984). The structure of rat aldolase B cDNA was also elucidated (Tsutsumi et al., 1984a). Both the mRNAs code for a 364 amino acid protein that shows strong similarities with aldolase A (see the following Section). The mRNA is larger than aldolase A mRNA. The largest clone reported is 1600 nucleotides long (Paolella et al., 1984), but it may still lack some nucleotides at the 5'-end. The 3'-non coding region is larger than in aldolase A mRNA, being 387 bases in rat and 434 bases in man. Both mRNAs have a canonical poly-A addition signal, but the human mRNA has a second addition signal in the middle of the 3'-untranslated sequence. However, it is not known where this site is used, if at all. cDNA sequencing also provided the first complete sequence for aldolase B (see Fig.

Figure 2. Comparison of the known cDNA sequences coding for aldolase isoenzymes. Human aldolase B is taken as a reference. All the other sequences have been aligned with it in order to maximize homology. Equal nucleotides are indicated by a hyphen; different nucleotides are shown. A full stop indicates an unknown nucleotide. To maximize homology in the untranslated regions of rat aldolase B cDNA, some nucleotides have been removed and are indicated above the line, and gaps, represented by spaces, have been introduced. Initiation and termination triplets are boxed.

2 for a comparison among the various cDNA aldolase sequences).

Recently, in our laboratory, a partial aldolase C cDNA was isolated from a mouse library, and this allowed us to assign the sequence of the first 227 amino acids of aldolase C (Paolella et al., 1985a). Also aldolase C is highly homologous to aldolases A and B (Fig. 2) and is colinear with them, at least in the sequenced region. The amino terminal end is also coincident. As the aldolase C subunit can substitute aldolase A subunits in forming active tetramers, it is possible that the three isoenzymes have the same length. However, Lee and Horecker (1974) found that aldolase C isolated from rabbit brain was somewhat smaller (about 30 amino acids) than the other two enzymes.

(d) Comparison of mRNA sequences

A comparison of aldolase mRNA sequences is shown in Fig.2. Taking human aldolase B as a reference, all the known cDNA sequences have been aligned to obtain the maximum homology. Three cDNA sequences are full-length or almost full-length. Rat aldolase B was sequenced by Tsutsumi et al. (1984); human aldolase B by Paolella et al. (1984); and rabbit aldolase A by Tolan et al. (1984). These contain a complete coding sequence with two extensions, in the 5'- and 3'-untranslated regions. All of them reach the poly-A addition site. Partial sequences from mouse aldolase A and C cDNAs are available (Paolella et al., 1985a), and these have also been aligned with the corresponding regions of aldolase B cDNA.

The coding region is very similar in all the sequences reported. All of them may be aligned without gaps or insertions, starting from a single methionine codon (positions 67-69) and ending with a termination codon in position 1159-61. The stop codon is TAG in both aldolase B cDNAs, and TAA in aldolase A cDNA. Of course, the two sequences for aldolase B in man and rat show the highest homology, 88%, but aldolase A cDNA is also very similar, showing a 69 and 70% homology, respectively, for rat and human aldolase B cDNAs. Most of the differences consist of changes in the third position of the codons and often result in no amino acid change. Mouse aldolase A cDNA is also very similar to rabbit aldolase A cDNA.

Both the 5'- and 3'-untranslated regions show a lower degree of homology. The 3'-untranslated regions are highly variable in length, ranging from 218 to 414 bases. Rat aldolase B cDNA sequence is similar to human aldolase B cDNA: if a few gaps and insertions are allowed, the overall homology is 55%. Aldolase A cDNA differs from both, and no significant homology could be found. Similarly, in the 5'-non coding region the two aldolase B cDNAs are 50% homologous, whereas aldolase A is different. In this region, also the mouse aldolase C sequence is known, but again no extensive homology is evident. However, all the sequences show small conserved patterns of a few nucleotides between positions 55 and 66, which is just upstream from the first ATG codon.

The codon usage in the three complete sequences is reported in Fig. 3. The most striking feature in aldolase A codon usage is the prefential use of codons ending in C or G (Tolan et al., 1984). This trend is also present in mouse aldolase A and, to a lesser extent, in mouse aldolase C (not shown). On the other hand, rat aldolase B does not show any such preference (Tsutsumi et al., 1984a), and does not use the CG doublet. Also in human aldolase B cDNA (Paolella et al., 1984) CG-ending codons are very rarely used: TCG, ACG are used once each, and CCG is never used.

HUMAN ALDOLASE B

```
TTT Phe   4   1 %    TCT Ser   6   1 %    TAT Tyr   5   1 %    TGT Cys   3   0 %
TTC Phe   6   1 %    TCC Ser   3   0 %    TAC Tyr   6   1 %    TGC Cys   6   1 %
TTA Leu   1   0 %    TCA Ser   2   0 %    TAA  *    0   0 %    TGA  *    0   0 %
TTG Leu   1   0 %    TCG Ser   1   0 %    TAG  *    1   0 %    TGG Trp   3   0 %

CTT Leu   4   1 %    CCT Pro   6   1 %    CAT His   4   1 %    CGT Arg   3   0 %
CTC Leu  11   3 %    CCC Pro   2   0 %    CAC His   5   1 %    CGC Arg   3   0 %
CTA Leu   4   1 %    CCA Pro   5   1 %    CAA Gln   3   0 %    CGA Arg   2   0 %
CTG Leu  12   3 %    CCG Pro   0   0 %    CAG Gln  19   5 %    CGG Arg   3   0 %

ATT Ile   6   1 %    ACT Thr   6   1 %    AAT Asn   4   1 %    AGT Ser   4   1 %
ATC Ile  13   3 %    ACC Thr  11   3 %    AAC Asn  13   3 %    AGC Ser   5   1 %
ATA Ile   0   0 %    ACA Thr   3   0 %    AAA Lys   4   1 %    AGA Arg   1   0 %
ATG Met   7   1 %    ACG Thr   1   0 %    AAG Lys  20   5 %    AGG Arg   2   0 %

GTT Val   8   2 %    GCT Ala  19   5 %    GAT Asp   5   1 %    GGT Gly   7   1 %
GTC Val   2   0 %    GCC Ala  18   4 %    GAC Asp   8   2 %    GGC Gly   5   1 %
GTA Val   5   1 %    GCA Ala   6   1 %    GAA Glu  12   3 %    GGA Gly  10   2 %
GTG Val   7   1 %    GCG Ala   1   0 %    GAG Glu  11   3 %    GGG Gly   7   1 %
```

RAT ALDOLASE B

```
TTT Phe   5   1 %    TCT Ser   4   1 %    TAT Tyr   2   0 %    TGT Cys   2   0 %
TTC Phe   5   1 %    TCC Ser   5   1 %    TAC Tyr  10   2 %    TGC Cys   6   1 %
TTA Leu   0   0 %    TCA Ser   4   1 %    TAA  *    0   0 %    TGA  *    0   0 %
TTG Leu   4   1 %    TCG Ser   2   0 %    TAG  *    1   0 %    TGG Trp   3   0 %

CTT Leu   7   1 %    CCT Pro   8   2 %    CAT His   5   1 %    CGT Arg   2   0 %
CTC Leu  12   3 %    CCC Pro   1   0 %    CAC His   4   1 %    CGC Arg   4   1 %
CTA Leu   5   1 %    CCA Pro   3   0 %    CAA Gln   5   1 %    CGA Arg   4   1 %
CTG Leu   7   1 %    CCG Pro   1   0 %    CAG Gln  17   4 %    CGG Arg   1   0 %

ATT Ile   6   1 %    ACT Thr   3   0 %    AAT Asn   5   1 %    AGT Ser   4   1 %
ATC Ile  10   2 %    ACC Thr   9   2 %    AAC Asn  10   2 %    AGC Ser   6   1 %
ATA Ile   1   0 %    ACA Thr   4   1 %    AAA Lys   3   0 %    AGA Arg   3   0 %
ATG Met   6   1 %    ACG Thr   4   1 %    AAG Lys  21   5 %    AGG Arg   4   1 %

GTT Val   7   1 %    GCT Ala  23   6 %    GAT Asp   5   1 %    GGT Gly   2   0 %
GTC Val   4   1 %    GCC Ala  12   3 %    GAC Asp   8   2 %    GGC Gly  11   3 %
GTA Val   1   0 %    GCA Ala   7   1 %    GAA Glu   6   1 %    GGA Gly  10   2 %
GTG Val  10   2 %    GCG Ala   1   0 %    GAG Glu  16   4 %    GGG Gly   4   1 %
```

RABBIT ALDOLASE A

```
TTT Phe   0   0 %    TCT Ser   3   0 %    TAT Tyr   2   0 %    TGT Cys   1   0 %
TTC Phe   7   1 %    TCC Ser   9   2 %    TAC Tyr  10   2 %    TGC Cys   7   1 %
TTA Leu   0   0 %    TCA Ser   2   0 %    TAA  *    1   0 %    TGA  *    0   0 %
TTG Leu   2   0 %    TCG Ser   0   0 %    TAG  *    0   0 %    TGG Trp   3   0 %

CTT Leu   0   0 %    CCT Pro   4   1 %    CAT His   2   0 %    CGT Arg   5   1 %
CTC Leu   7   1 %    CCC Pro   9   2 %    CAC His   9   2 %    CGC Arg   6   1 %
CTA Leu   0   0 %    CCA Pro   1   0 %    CAA Gln   3   0 %    CGA Arg   0   0 %
CTG Leu  25   6 %    CCG Pro   5   1 %    CAG Gln  13   3 %    CGG Arg   3   0 %

ATT Ile   6   1 %    ACT Thr   3   0 %    AAT Asn   3   0 %    AGT Ser   3   0 %
ATC Ile  15   4 %    ACC Thr  16   4 %    AAC Asn  11   3 %    AGC Ser   4   1 %
ATA Ile   0   0 %    ACA Thr   2   0 %    AAA Lys   1   0 %    AGA Arg   0   0 %
ATG Met   3   0 %    ACG Thr   1   0 %    AAG Lys  25   6 %    AGG Arg   1   0 %

GTT Val   3   0 %    GCT Ala   6   1 %    GAT Asp   7   1 %    GGT Gly   4   1 %
GTC Val   6   1 %    GCC Ala  27   7 %    GAC Asp   7   1 %    GGC Gly  12   3 %
GTA Val   3   0 %    GCA Ala   4   1 %    GAA Glu   4   1 %    GGA Gly   4   1 %
GTG Val  10   2 %    GCG Ala   5   1 %    GAG Glu  20   5 %    GGG Gly  10   2 %
```

Figure 3. Codon usage in the three known cDNA sequences coding for aldolase isoenzymes.

(e) Comparison of the known protein sequences

Table 1 shows the homology percentages obtained by comparing all the known aldolase sequences, whether obtained by direct protein sequencing or by translating the cDNA sequence. Rabbit aldolase A has been sequenced with both procedures, using the cDNA-derived sequence, which seems the most reliable. It is clear that the sequences are all related, showing a minimal homology of 69%. The homology is clearly higher when type A aldolases or type B aldolases are compared. It is between 97–98% for aldolase A, and 95% for aldolase B. An exception is aldolase B from ox, which shows values around 82–83% when compared to the other type B aldolases. However, the sequence for ox aldolase B is much smaller (one-third of the protein) and was obtained by protein sequencing only. On the other hand, when type A aldolases are compared to type B aldolases, the homology falls to 69–76%.

	Ox (B)	Rat (B)	Human (B)	Mouse (C)	Mouse (A)	Rabbit (A)	Human (A)
Ox (B)	100	82 (117)	83 (117)	71 (93)	76 (78)	72 (117)	69 (117)
Rat (B)	82 (117)	100	95 (364)	73 (226)	73 (257)	70 (364)	71 (273)
Human (B)	83 (117)	95 (364)	100	71 (226)	74 (257)	69 (364)	71 (272)
Mouse (C)	71 (93)	73 (226)	71 (226)	100	93 (128)	84 (226)	78 (61)
Mouse (A)	76 (78)	73 (257)	74 (257)	93 (128)	100	97 (257)	97 (208)
Rabbit (A)	72 (117)	70 (364)	69 (364)	84 (226)	97 (257)	100	98 (272)
Human (A)	69 (117)	71 (273)	71 (272)	78 (61)	97 (208)	98 (272)	100

Table 1. Percentage of homology after comparison between all the known aldolase sequences. The numbers in parentheses indicate the overlapping sequences (as number of amino acids). A, B and C indicate aldolase A, B and C, respectively.

Aldolase C shows similar values to aldolase B, between 71 and 73%, but it is significantly more similar to type A aldolase, showing a homology ranging between 78 and 93%.

The aminoacid sequences of all the known aldolases are compared in Fig. 4. Taking rabbit aldolase A as a reference, all the known aldolase sequences are aligned accordingly. It is immediately evident that the homology is spread along the whole protein molecule, rather than being clustered in a few regions. The largest completely identical region is 61 amino acids and lies between position 166 and 226 of mouse aldolases A and C. The major differences between aldolases A and B may be observed in the carboxyterminal end, where 25 amino acids out of 47 are different. Around the active site, aa. 215-241, there is a general conservation of the sequence. The more striking differences appear in positions 218 and 236, where asparagine and alanine residues are conserved in aldolases B, and are substituted by serine and proline, respectively in aldolases A. Similar, but perhaps less significant, is a valine-isoleucine substitution in position 222. The sequence at the active site of mouse aldolase C is only partially known, and is so far identical to aldolase A. The rabbit aldolase C sequence, originally determined around the active site, has many ambiguities, but it also seems to be similar to aldolase A. Some amino acids are characteristic of aldolase C. Examples are Leu-132, Leu-104 and Phe-79, which substitute for Ser, Val and Leu, respectively, that are conserved in both A- and B-type aldolases.

```
                    10        20        30        40        50        60        70        80        90       100
-YQY-
MPHSHPALTPEQKKELSDIAHRIVAPGKGILADESTGSIAKRLQSIGTENTEENRRFYRQLLTADDRVNPCIGGVILFHETLYQKADDGRPFPQVIKS
--Y--SA----------L--T--------------M--SQ--V------------------------F----------D-N-V--VRT-QD
--A-RF--Q--------E--QS--N-----------V-TMGN--R-KV---------------QF-EI-FSV-SSI-QS------DSQ-KL-RNIL-E
--A-RF--S--E--Q--N------------------V-TMGN--R-KV---------------QF-E--FSV-NSISQS------DSQ-KL-RNIL-E
--EAAF--S-E--A--ET-R--D-------------V-TM.

                   110       120       130       140       150       160       170       180       190       200
KGGVVGIKVDKGVVPLAGTNGETTTQGLDGLSERCAQYKKDGADFAKWRCVLKIGEHTPSALAIIENANVLARYASICQQNGIVPIVEPEILPDGDHDLK
--IL---L---D----------L-----------------------------------------DR---L-----------M------------VI-------E
--I---L-Q-GA----K--I-------------------------V-G---A-R-ADQC--S--Q----A-------------M---L-------V------E
--I---L-Q-GA----K--I-------------------------V-G---A-R-SDQC--S---Q----A------------------L------VI--S--ME
................................................................EED-L----------------------VI--S--ME

                   210       220       230       240       250       260       270       280       290       300
RCQYVTEKVLARVYKALSDHHIYLEGTLLKPNMVTPGHACTQKYSHEEIAMATVTALRRTVPPAVTGVTFLSGGQSEEERSINLNAINKCPLLKPWALTF
--H-----------V-------N---V----------------A----K--TP-QV----------------H---A-P-IC------M--D-TL-----L--P--K-S
--H---S------------------N---V---------L--A----K--TP-QV----------------H---A-PSIC------M--D-TL-----YR--PR-K-S
--H-E--------------------N---V------------A----K--TPQ-V

                   310       320       330       340       350       360       370       380       390       400
SYGRALQASALKAWGGKKENLKAAQEEYVKRALANSLACQGKYTPSGQAGAASESLFISNHAY          HUMAN ALDOLASE A
                                                -I----------S           MOUSE ALDOLASE A
--A----AA-KE-T--AFM---M--CQ-AK-Q-VHT-SS---STQ---TACYT-                   RABBIT ALDOLASE A
--A----AA-K--T--AFM---V--CQ-A--Q-VHT-SS---STQ---TASYT-                   MOUSE ALDOLASE C
                                                                          HUMAN ALDOLASE B
                                                                          RAT ALDOLASE B
                                                                          OX ALDOLASE B
```

Figure 4. Comparison of the known aldolase protein sequences. Rabbit aldolase A, sequenced both at the protein and the cDNA level, is taken as a reference: the cDNA-derived sequence (Tolan et al., 1984) has been used. All the other sequences have been aligned with the rabbit aldolase A: identical amino acids are indicated by a hyphen; a full stop indicates an unknown amino acid. Amino acids are indicated by the one-letter code.

A typical region that differentiates aldolase isoenzymes is from residue 88 to 100. In this region 10 residues out of 13 are different between human aldolase B and rabbit aldolase A. Mouse aldolase C shows a third completely different sequence, which has 8 substitutions with respect to aldolase A, and 10 with respect to aldolase B. On the contrary, this region is highly conserved among type B aldolases, as shown by the 100% identity between rat and human proteins. The aldolase A sequence is also conserved: mouse aldolase A is identical, where the sequence is known, to the rabbit enzyme. Human aldolase A, whose gene was recently sequenced in our laboratory (Izzo et al., 1985), also shows the same sequence.

Another region of great variability is between positions 4-5 at the amino terminal region. Glu, Arg, Ser or Gln may be found in position 4 and similarly Ala, Phe, Tyr or Hys may be found in position 5. Despite this variability, the polypeptide chain length is highly conserved. All the cDNA-derived sequences start with a methionine in position 1, which is absent from the sequences determined by direct protein sequencing. This methionine may be lost during the maturation process, and so be absent from the mature protein. Alernatively, the amino terminal might be lost during the purification procedure or, being modified it could interfere with the first steps of the sequence determination. This possibility is supported by a report of a single derivatized methionine in aldolase proteins from some tissues (Lebherz et al., 1984).

(f) Evolutionary derivation of the aldolase isoenzyme genes

From the data discussed so far it is evident that at least

three different genes must exist to generate the different mRNAs. Further confirmation of this comes from Southern blot experiments that show a complex pattern with bands of different intensities (Paolella et al., 1985b). However, the genes are clearly related and probably derive from a single ancestral gene. It appears that the isoenzymes must have diverged well before the various mammalian species separated. This is supported by the presence of the three isoenzymes in all the mammalian species tested and by the much stronger homology between isoenzymes of the same type in different species. Moreover, the aldolase B gene, which codes for a more specialized protein, and which shows a lower degree of homology for the other isoenzymes, probably diverged from the common branch at an earlier stage. Aldolase A and aldolase C genes, which show a greater similarity, probably separated more recently. This interpretation is also in agreement with the catalytic properties of the isoenzymes. In fact, aldolase A and aldolase C are both active on fructose-1,6-bisphosphate, but have a very low affinity for fructose-1-phosphate. The two substrates are instead well recognised by aldolase B, which is mainly involved in fructose metabolism in liver.

(g) Cloning of the structural genes

Two aldolase genes have been recently isolated. The aldolase B gene was isolated from rat (Tsutsumi et al., 1985), whereas the aldolase A gene was isolated from man (Izzo et al., 1985). The two genes have a similar intron-exon organization, although the introns are much shorter in the aldolase A gene than in the aldolase B gene.

The aldolase B gene in rat is quite large, spanning 14 kb from the "cap-site" to the poly-A site. The aldolase A gene, on the other hand, is smaller. Both genes are split in 9 exons by 8 introns, ranging in size between 0.4 and 4.7 kb in the rat gene, with an average size around 1 kb; the human gene introns are somewhat smaller ranging between 0.1 and 1.0 kb. Seven introns interrupt the coding region and are located exactly in the same position on both genes. Six exons (III-VIII) of 212, 55, 161, 84, 175, 200 bp are completely contained in the coding region. The remaining nucleotides derive from exons IX and II. Exon IX contains 93 bases of the coding sequence and the entire 3'-untranslated region in both genes. In the aldolase B gene, it is 387 or 388 nucleotides, in aldolase A it is 200 nucleotides. As expected from the cDNA data, the 3'-non coding regions do not show any significant homology in the two genes. Exon II contains the first 112 bp of the coding region, and is only slightly larger than 112 bp in both genes: it is 122 bp in the aldolase B gene and 132 bp in the aldolase A gene. In both genes the 5'-untranslated region is split by an intron located just a few bases upstream from the starting ATG codon. Exon I is located rather far away (4.6 kb) in the rat aldolase B gene and it is 71 bp long. This exon is not yet known for aldolase A gene.

The transcription initiation site has been determined in the aldolase B gene by S1 mapping (Tsutsumi et al., 1985). It corresponds to a four-nucleotide sequence, AGAT, which generates multiple RNAs, starting at each of the four bases. The most frequent initiation corresponds to the T nucleotide, which has been taken as the starting point of exon I. Upstream from this point, 870 nucleotides have been sequenced in the 5'-flanking region (Tsutsumi et al., 1985). A typical TATA box (Goldberg, 1979) is located 25 base-pairs before the initiation site, and a CCAAT sequence (Efstratiadis et al., 1980) is also present at position -126. Further upstream, three A-rich sequences are present. Two of them are Alu-family repeats. The third sequence reported by Tsutsumi et al.(1985) shows some homology with the identifier (ID) sequences (Sutcliffe et al., 1982) described in brain specific genes, which seem to be involved in the regulation of their expression.

4. OCCURRENCE, EXPRESSION AND REGULATION OF ALDOLASE ISOENZYMES

(a) Tissue specificity and developmental expression

All cells appear to contain aldolase in their cytosol; however, various tissues show a preponderance of one particular isoenzyme over the others (Penhoet et al., 1966; Lebherz and Rutter, 1969). Aldolase A is considered to be the muscle specific form; it is the first of the three forms to be produced in the developing embryo; it persists as the sole form in most adult cell types, including muscle cells, blood cells and fibroblasts (Lebherz and Rutter, 1969). However, aldolase A is also expressed with other isoenzymes in kidney, liver and brain (Masters, 1968; Schapira et al., 1975). Aldolase B is expressed in the liver and small intestine. Aldolase C, the brain specific form, is much less known. In mammals it is selectively expressed in the brain where it is located in the astrocytes of the cerebral cortex (Thompson et al.,

1982). Low levels of immunoreactive aldolase C have been demonstrated in erythrocytes, liver and adrenal gland (Willson and Thompson, 1980).

The tissue expression of the three isoenzymes varies according to the stage of development (Rutter et al., 1963). All three isoenzymes are found in foetal liver, however the level of aldolase A is much higher than that of the other two aldolase forms. As the foetus matures, the amount of A progressively diminishes and the amount of B increases (Hommes and Draisma, 1970). At birth, aldolase A production is switched off, and aldolase B becomes the predominant isoenzyme. In adult rat liver aldolase B constitutes approximately 0.5% of the total liver protein. Aldolase C also progressively diminishes as the foetus matures, and at birth its amount in the liver is negligible. Such ontogenic changes have not been found in rat brain. Both isoenzyme A and isoenzyme C are present in foetal rat brain, and their relative expression is not modified in the adult tissues (Masters, 1968). There are no extensive studies of ontogenic changes in aldolase expression in muscle. In chicken, the A and C isoenzymes are contained in foetal muscle. The C form gradually diminishes during development and is not detected in the adult. The amount of A increases so that it becomes the sole isoenzyme in the adult muscle, where it constitutes up to 5% of the total muscle protein (Lebherz and Rutter, 1969). The tissue specific distribution of the three isoenzymes is very similar in almost all examined vertebrate species.

The possible subunit structures of mammalian aldolases are

as follows (Masters, 1981):

$$A_4 \quad A_3B \quad A_2B_2 \quad AB_3 \quad B_4$$
$$A_3C \qquad\qquad CB_3$$
$$A_2C_2 \qquad C_2B_2$$
$$AC_3 \quad C_3B$$
$$C_4$$

The tissue distribution of various aldolases has been examined in man, monkey, bovine, swine, rat, rabbit, chicken, turtle, frog, perch and shark (Lebherz and Rutter, 1969).

The structure of the tetramers in these species may be recapitulated as follows:

- muscle: A_4 in all species;

- liver : A_4 and B_4 in all species;

- brain : AC hybrids in all species;

- heart : AC hybrids in all species, but A_4 in bovine, rat and rabbit;

- spleen: A_4 in all species, but AC hybrids in bovine, swine, chicken and perch;

- kidney: AB hybrids in all species, but A_4 and B_4 in frog and shark; also AC hybrids in chicken, turtle and perch.

Trout and salmon have a quite different pattern, showing a new D subunit in a CD hybrid combination in brain, and C_4 tetramers in spleen and kidney, while liver, heart and muscle show the same aldolase structure as the other species.

The distribution of aldolase A and aldolase B in vertebrate tissues is consistent with the specialized physiological functions exerted by the two proteins. In fact, the catalytic properties of aldolase A_4 indicate a glycolytic capacity, while the catalytic properties of aldolase B_4 suggest that this particular enzyme is mainly active in gluconeogenesis and in the metabolism of fructose through the formation of fructose-1-phosphate (Rutter et al., 1963). It is well known that skeletal muscle is rich in glycolytic activity, whereas liver and kidney cortex are the site of gluconeogenesis activity, and liver is the primary site of fructose utilization.

(b) Intracellular distribution and function

Glycolytic enzymes are conventionally considered soluble cell components. However, several reports indicate that some enzymes, in some tissues, are bound to particulate cell fractions.

In skeletal muscle, up to 40% of aldolase activity is not solubilized under conditions that usually solubilize proteins (Arnold and Pette, 1968). The bound enzymes appear to be associated to the contractile muscle protein, actin, and can be solubilized only under particular experimental conditions (Arnold and Pette, 1968; 1970). It would appear that the 'free' and the 'bound' forms found within the same cell type are distinct enzymes, at least as far as localization is concerned. However, evidence suggests that there is no difference in functional or in molecular properties between the two enzymes (Penhoet et al., 1969). Therefore, the change in localization may not be a reflection of a difference in their physiological functions, other than what would be required for storage and for the availability of enzyme molecules for cell metabolism.

However, aldolase isoenzymes differ in their tendency to

bind to muscle particulate fractions. As an example, the transition
of C to A during skeletal muscle development results in an
increased number of A subunits bound to particulate fractions (C_4
aldolase, was not detected in the bound fraction) (see Lebherz,
1975a). This differential behavior could be because an increasing
number of negative charges in the C subunits determines
electrostatic repulsion between C subunits and actin, whereas the
positively charged aldolase A subunits are attracted to actin
(Arnold and Pette, 1968).

There is no known catalytic basis upon which to suggest
different specialized functions for aldolase A and aldolase C.
However, they do have opposite net electrostatic charges at
physiological pH, which is consistent with evidence that indicates
their association with different cellular structures ('free' and
'bound' forms). Consequently, the two enzymes may be distinguished
functionally, not on the basis of their molecular function, but by
their activity in different regions of the cell (Clarke and
Masters, 1973).

(c) Subunit assembly and turnover

Assembly and turnover of aldolases has been studied
in vivo by subunit exchange (Lebherz 1975b). In vitro studies had
suggested that the quaternary structures of aldolases were so
stable that subunit exchange between aldolase tetramers did not
occur (Lebherz, 1972). Further investigations indicated that all
aldolase tetramers at the time of initial assembly are constructed
of newly synthetized subunits, which is additional evidence that
subunits exchange does not occur (Lebherz, 1975b). Moreover, post-

translational processing of aldolase polypeptide chains did not appear to be involved in the formation of functional tetramers (Lebherz and Doyle, 1976). In addition, the degradation of all subunit forms of an aldolase tetramer appear to be coupled because they are not reincorporated into other tetramers. Thus, there is a coordinate turnover of aldolase tetramer subunits (Lebherz, 1975b).

(d) Variation of aldolase levels under physiological conditions

Aldolase B and its mRNA levels vary in the liver according to dietary status. In fasted rabbit, the levels of both decrease to one-half the values found in livers of non fasted animals. However, the concentration of aldolase proteins, measured with a specific antibody, remains unchanged. This implies that fasting induces a modification of liver aldolase, which results in the accumulation of an inactive, immunologically cross-reactive form (Pontremoli et al., 1979; Pontremoli et al., 1982b). The modification has been related to limited proteolysis at the -COOH terminus (Pontremoli et al., 1982a). This modified form is produced in vivo by catepsin M, as reported in Section 2(f).

Inactive enzyme molecules have been described in aged animals for several proteins including aldolase B and aldolase A. Petell and Lebherz (1979) suggested that the defective aldolase molecules, arising from a splitting of a -COOH terminal tyrosine residue, are due to an artefact during the protein isolation procedures. However, Reznick et al. (1981), on the basis of a reduced ratio of catalytic activity versus units of enzyme antigen, suggested that the reduced catalytic activity of aldolase molecules in aged animals was due to a decrease in the rate of protein available for disposal by the protein degradation system.

The role of hormones in aldolase B gene expression has been studied to elucidate the mechanisms that regulate the induction of mRNA to dietary stimuli. Liver, kidney and small intestine of normal animals, of adrenal- and thyroid-deprived animals, and of glucagon- and cAMP-treated animals were examined for their RNAs levels (Munnich et al., 1985). It was found that in vivo hormonal control in liver differs significantly from that occurring in kidney and in the small intestine. In the liver, mRNA synthesis requires dietary carbohydrates, no glucagon release, and the presence of permissive hormones, such as insulin for glucose- and maltose-fed rats. In the small intestine, the presence of dietary carbohydrates and insulin induces mRNA synthesis, whereas glucagon and cAMP exert no effect. In the kidney, the synthesis of mRNA is practically unaffected by diet and hormones.

(e) Tissue and developmental changes evaluated by mRNA analyses

A recent series of studies that measure the concentrations of mRNAs coding for both aldolase A and aldolase B have begun to provide information on the tissue-specific expression of aldolase isoenzymes, and the regulation of their expression during development (Grégori et al., 1982; Numazaki et al., 1984; Mukai et al., 1984).

To determine the tissue specific concentrations of aldolase isoenzymes, mRNA levels have been analyzed by Northern blot hybridization (Tsutsumi et al., 1984b; Numazaki et al., 1984). Aldolase A was detected in brain, muscle and hepatoma cells, but not in liver cells, whereas aldolase B was detected in liver cells, but not in the other tissues examined. The correspondance of these results with those obtained at the protein level gives an indication that the control of protein concentration is related mainly to the mRNA content.

Another interesting observation is the presence of two different mRNA species of aldolase A in different rat tissues; one in muscle (1,650 bp) and another in brain and hepatoma cells (1,550bp) (Tsutsumi et al., 1984b). The difference should be due to the untranslatable 5'- and/or 3'-regions since the coded proteins are identical, although microheterogeneity in the coding regions cannot be excluded until the complete sequences of these mRNAs are known. The mechanism for the selection of the size of the expressed mRNA has not been defined in this case. Therefore, usage of alternative promoters or different mRNA processing (splicing) could be important in the regulation of mRNA expression.

mRNA analyses performed in foetal and adult rat liver have yielded information on mRNA regulation during development. It was confirmed that aldolase B increased during the foetal period, and it was twelve times greater at birth than in the 14-day old foetus (Numazaki et al., 1984). In contrast, aldolase A content decreased during the foetal period, and at birth it was one-eighth of that found in 14-day foetal liver. Also in this case the level of isoenzyme expression is controlled by the level of mRNA.

Again, as far as the mechanisms that control the steady-state of mRNA are concerned, the evidence is such that it is still impossible to discriminate between transcription, processing of precursor mRNAs, and degradation of mature and precursor mRNAs.

Another important problem is presented by the type of liver cell that produces the various aldolase mRNAs. It seems that the same cells produce aldolase B and the foetal type of aldolases A and C, although intervention by hematopoietic cells cannot be ruled out (Hatzfeld et al., 1977).

In conclusion, all available data support the hypothesis that changes in the levels of aldolase A and B mRNA lie behind the modification of the isoenzyme pattern during development.

(f) Variations in tumors

Studies already reviewed by Horecker et al. (1972) indicated that in hepatomas aldolase B is replaced by aldolase A and that the extent of the transition is related to the growth of the tumor. In summary, the analyses of aldolase isoenzymes, and kinetic data obtained by measuring the FDP/F1P activity ratio supported the resurgence of a foetal-type aldolase pattern in a liver that is in transition from a normal to a cancerous state. There was also evidence that the appearance of aldolase A preceded the appearance of tumors.

Several investigators have confirmed the resurgence of the foetal-type pattern of aldolases in cancerous liver cells (Schapira et al., 1963; Hatzfeld et al., 1977; Matsushima et al., 1968; Gracy et al., 1970; Ikehore et al., 1970; Weber et al., 1980).

More recently, specific DNA probes have been used to estimate the concentrations of mRNA in various tissues. The levels of aldolase A and B mRNAs in the liver have been studied (Daimon et al., 1984) during hepatocarcinogenesis induced by 3'-methyl-4-dimethyl-aminobenzene (3'-Me-DAB). The result showed an increased expression of aldolase A mRNA and a decreased expression of aldolase B, after 4 week's treatment with the azo-dye as compared to normal liver. Moreover, in this experiment the level of aldolase A mRNA in the livers of azo-dye-fed rats was at least two times higher than that expected from the kinetic data of the enzyme activity. This observation may also indicate the presence of some regulatory process at the translational or post-translational steps during the synthesis of functional aldolase (Daimon et al., 1984).

Asaka et al. (1983a; 1983b) developed a radioimmunoassay method for the quantitative determination of aldolase A in human serum and tissue. They showed that serum aldolase A levels increased in liver cell carcinoma tissue, while in almost all of the nonmalignant liver diseases studied the aldolase A levels remained at normal values.

To determine whether foetal or unusual forms of the isoenzyme were present in cells from preneoplastic livers of rats fed a diet lacking choline and containing 0.1% DL-ethionine, Hayner et al. (1984) and Yaswen et al. (1984) used a centrifugal elutriation to isolate oval, parenchymal and biliary cells. They showed that oval cells, but not hepatocytes, express foetal forms of aldolase. The results indicate that in animals receiving the carcinogenic diet, isoenzyme alterations associated with neoplasia result from the proliferation of a new cell population that contains these enzymes, and not from de-differentiation of mature hepatocytes.

5. MOLECULAR PATHOLOGY OF ALDOLASES

A deficiency in any one of the three aldolase isoenzymes could, in theory, have clinical implications. A total deficit of aldolase A is probably lethal, but a reduced activity has been described in some conditions (Beutler et al., 1973). A deficiency of the B form causes hereditary fructose intolerance (Gitzelmann et al., 1983). No deficit has yet been described for aldolase C.

(a) Aldolase B deficiency (hereditary fructose intolerance)

(i) Symptoms and clinical course

Hereditary fructose intolerance (HFI), first described three decades ago, is a recessive autosomic disease (Chambers and Pratt, 1956). The disease usually starts as the child is weaned from the mother's milk, and fructose or sucrose is introduced into the diet. Soon after fructose ingestion, diarrhea and vomiting appear. Continued fructose administration results in severe liver damage, and the patient develops hyperbilirubinemia, hypoalbuminemia and a marked elevation of liver enzymes in the plasma. A typical symptom is hypoglycemia. The disease is controlled by a fructose-free diet; uncontrolled it progresses to biliary cirrhosis.

In older children and adults, the syndrome is less marked. Abdominal pain and diarrhea follow fructose ingestion, but often the patient tends to avoid fructose-containing food, and the disease may go undiagnosed. Of course, the risk remains that an intake of fructose could induce a severe attack, which may result in death.

(ii) Pathophysiology and enzyme defect

Hers and Joassin (1961) first described a deficit in aldolase B activity in the liver. Since aldolase B uses fructose-1-phosphate as a substrate, it is the enzyme responsible for the catabolism of exogenous dietary fructose. In fact, the absence of this activity makes it impossible for the affected liver to metabolize exogenous fructose, and causes an increase in fructose-1-phosphate concentration. Fructose-1-phosphate is directly responsible for the liver damage. Moreover, it inhibits a number of liver enzymes which, in turn, could cause the other symptoms of the disease. Inhibition of fructokinase causes the increased fructose levels both in plasma and in urine. Inhibition of glycogen degradation may cause the hypoglycemia. Aldolase B itself is an important enzyme in gluconeogenesis (Grazi and Trombetta, 1979) and its deficit may lead to a reduced glucose synthesis.

Aldolase B activity in the liver of patients with HFI was studied in an attempt to understand the molecular basis of the enzyme deficiency. The activity towards fructose-1-phosphate is generally reduced (Schapira et al., 1974) or absent in HFI patients (Steinmann and Gitzelmann, 1981). The activity towards fructose-1,6-bisphosphate, on the other hand, is usually conserved. Probably the best marker for diagnosis is the ratio between the two activities, FDP/F1P, which is generally about 1 in the liver of normal individuals. This value is increased to 2-3, and sometimes more in individuals affected by HFI (Schapira et al., 1962). The origin of the activity towards fructose-1,6-bisphosphate is not clear. It has been suggested that it is due to an increased

production of aldolases A and C in the liver (Schapira, 1975).
However, aldolase B itself is usually immunologically present in
the liver, and shows some residual activity towards fructose-1,6-
bisphosphate.

Aldolase B has been purified from liver tissue of HFI
patients and well characterized (Cox et al., 1983). Kinetically,
its K_M towards fructose-1-phosphate is invariably increased, which
indicates a drastically reduced affinity. However, the underlying
molecular defect can differ from patient to patient. In some cases,
the enzyme is abnormally unstable when heated. In one family, a
slightly larger protein, with higher molecular weight, has been
demonstrated (Cox et al., 1983). Reactivation of the enzyme has
also been described after treatment with anti-aldolase B antibodies
(Gitzelmann et al., 1974) or reducing agents (Schapira et al.,
1974). It is noteworthy that treatment with reducing agents does
not modify the activity of the normal enzyme (Lemonnier et al.,
1974). In cases in which the protein is present at normal levels,
the mutation is probably a variation in the aminoacid sequence,
which results in an abnormal instability of its tertiary structure.
During treatment with reducing agents, the protein may assume its
normal structure and the activity may be recovered. However, in all
these cases, information on the primary structure of the mutant
protein will help to clarify the molecular basis of the defect.

(iii) Clinical and laboratory diagnosis

The approach to the diagnosis of HFI is conditioned by the
age of the patient. Since the symptoms of HFI are not specific,
diagnosis in children is not an easy task. The history of the

patient should be investigated for recurrent diarrhea after the ingestion of sugar or fruits. Laboratory data are again not specific for the disease, but hypoglycemia and fructosuria are suggestive of HFI. A definitive diagnosis can only be obtained through a fructose load test or by directly measuring the enzyme activity in liver specimens.

The fructose tolerance test (Steinmann and Gitzelmann, 1981) is performed by administering 200 mg/Kg of fructose and measuring several parameters (blood glucose, phosphorus, urate, magnesium and fructose) at standard intervals. In the affected child, glucose concentration is characteristically lowered, plasma phosphorous is also decreased, whereas magnesium and urates are increased. In adult patients phosphoremia is also decreased, but the other parameters remain within normal limits. A typical phenomenon is the increase of plasma potassium.

To measure the activity of the enzyme in the liver, a liver specimen should be collected by agobiopsy or by laparoscopy. Aldolase B activity towards fructose-1-phosphate is reduced or absent, and the ratio between the two activities is generally 2-3. The test is the best currently available, but it is still indecisive: some patients with severe liver disorders may present values similar to fructose intolerant patients (Steinmann and Gitzelmann, 1981).

The prenatal diagnosis of HFI in at-risk families and the screening of heterozygotic carriers deserve a special mention. It is not yet possible to identify the heterozygotic state. The fructose challenge is unhelpful, and of course invasive techniques, such as liver biopsy, cannot be used in healthy individuals. Prenatal diagnosis is equally impossible because it can be done only on liver fragments, which, at this stage of development, do not produce aldolase B activity in detectable amounts, even in healthy individuals, since the predominant species are aldolase A and/or C (Hommes and Draisma, 1970).

(iv) Perspectives for diagnosis by gene probes

Now that cDNA and gene probes are available for human aldolase B, the techniques of genetic analysis may be applied to the study of fructose intolerant individuals. In our laboratory we used aldolase B cDNA as a probe to study the aldolase B gene in healthy individuals (Paolella et al., 1985b). The hybridization patterns are reported in Fig. 5, where a Southern transfer of human DNA, digested with several common restriction enzymes, and hybridized to aldolase B cDNA is reported. The patterns are quite conserved in healthy individuals, except for an extra Pvu II band which, in our laboratory, was observed in 3 unrelated subjects out of 10 tested. This band represents a possible polymorphism and is shown in Fig. 6.

A report of a difference in the hybridization pattern in patients suffering from HFI comes from Grégori et al. (1984), who demonstrated a variation in a Bam HI band in one patient out of 11 tested. However, he was a compound heterozygote, since the extra band was present only in the allele derived from the father. The other allele, although defective, had a normal hybridization pattern.

Although from the information obtained so far by using

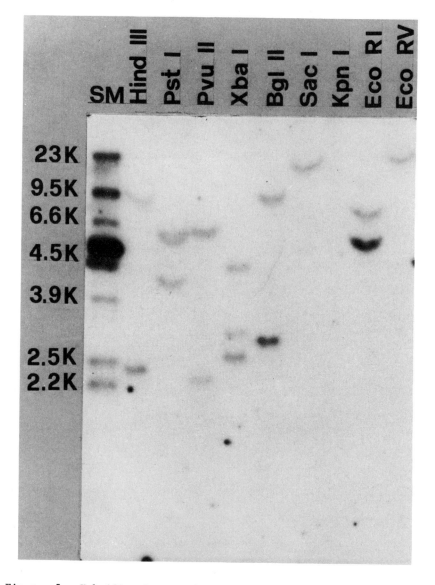

Figure 5. Hybridization patterns to human genomic DNA from peripheral blood of a healthy individual. Aldolase B cDNA, was used as a probe (Paolella et al., 1984). Hybridization conditions were stringent (0.1 x SSC, 65°C). The first slot contains the size markers from 2.2 to 23 kb, the other slots contain the restriction fragments after digestion with the enzymes indicated.

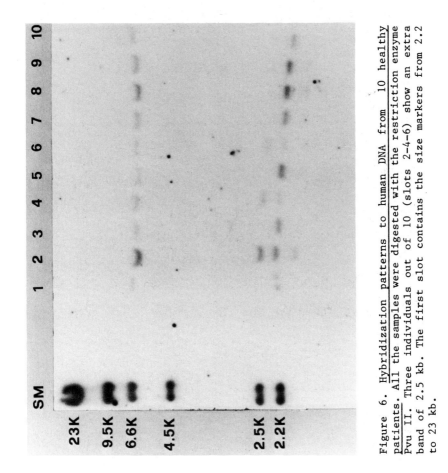

Figure 6. Hybridization patterns to human DNA from 10 healthy patients. All the samples were digested with the restriction enzyme Pvu II. Three individuals out of 10 (slots 2-4-6) show an extra band of 2.5 kb. The first slot contains the size markers from 2.2 to 23 kb.

gene probes it is still not possible to diagnose the disease, this
will certainly be possible as soon as more data on normal and ill
subjects become available. In fact, the availability of the genes
coding for aldolase B may soon reveal the mutant genes in patients
without the need for invasive techniques. This will not only allow
diagnosis, but will hopefully lead to prenatal diagnosis and
heterozygous identification.

(b) Aldolase A deficiency

A nonspherocytic hemolytic anemia associated to mental
retardation and increased hepatic glycogen, apparently due to a
deficiency of red cell aldolase, type A, has been described in a
patient by Beutler et al. (1973). In the same patient, besides the
previous symptoms, Lowry and Hanson (1977) found a growth
retardation, midfacial hypoplasia and hepatomegaly. As far as we
are aware there are no recent investigations on aldolase A
deficiency. However, also in this condition, alterations of gene
structure and expression are now amenable to a more direct
approach.

6. CONCLUDING REMARKS

The aldolase isoenzyme family has been studied at protein
level and, more recently, at DNA level. While the protein approach
has revealed the main features of the molecular and catalytic
properties of the enzyme, the molecular genetic studies have begun
to provide insights into the molecular structure of the gene and,
in addition, to give some indication as to the nature of its defect
in the human disease, fructose hereditary intolerance. Cloning and

characterization of human genes will also be instrumental in the understanding of the molecular mechanism of gene expression and regulation in both normal and pathological situations.

Several groups of investigators, including ours, are actively working along these lines, and it is not inconceivable that the aldolase isoenzyme family will become a useful model with which to study the relationships between gene expression and modulation during development and tissue differentiation.

ACKNOWLEDGEMENTS

        The work performed in the authors' laboratory was supported by grants from the Ministero della Pubblica Istruzione (Rome, Italy), and from the Consiglio Nazionale delle Ricerche, with special reference to the "Progetto finalizzato Ingegneria Genetica" (Rome, Italy). We are indebted to Jean Gilder, Rosaria Calabria and Ferdinando Dello Stritto for their help in preparing the manuscript.

REFERENCES

Anderson, P.J., Gibbons, I., and Perham, R.N. (1969).'A comparative study of the structure of muscle fructose 1,6-diphosphate aldolases', Eur.J.Biochem., 11, 503-509.

Arcari, P. Martinelli, R., and Salvatore, F. (1984). 'The complete sequence of a full-length cDNA for human liver glyceraldehyde-3-phosphate dehydrogenase: Evidence for multiple mRNA species', Nucl.Acids Res., 12, 9179-9189.

Arnold, H., and Pette, D. (1968). 'Binding of glycolytic enzymes to structure proteins of the muscle', Eur.J.Biochem., 6, 163-171.

Arnold, H., and Pette, D. (1970). 'Binding of aldolase and triosephosphate dehydrogenase to F-actin and modification of catalytic properties of aldolase', Eur.J.Biochem., 15, 360-366.

Asaka, M., Nagase, K., Miyazaki, T., and Alpert, E. (1983a). 'Radioimmunoassay of aldolase A. Determination of normal serum levels and increased serum concentration in cancer patients', Cancer, 51, 1873-1878.

Asaka, M., Nagase, K., Miyazaki, T., and Alpert, E. (1983b). 'Aldolase A isoenzyme levels in serum and tissues of patients with liver diseases', Gastroenterology, 84, 155-160.

Benfield, P.A., Forcina, B.G., Gibbons, I., and Perham, R.N. (1979). 'Extended amino acid sequences around the active-site lysine residue of class-I fructose 1,6-bisphosphate aldolases from rabbit muscle, sturgeon muscle, trout muscle and ox liver', Biochem.J., 183, 429-444.

Besmond, C., Dreyfus, J.C., Grégori, C., Frain, M., Zakin, M.M., Sala Trepat, J., and Kahn, A., (1983). 'Nucleotide sequence of a cDNA clone for human aldolase B', Biochem.Biophys.Res. Commun., 117, 601-609.

Beutler, E., Scott, S., Bishop, A., Margolis, N., Matsumoto, F., and Kuhl, W. (1973). 'Red cell aldolase deficiency and hemolytic anemia: a new syndrome', Trans.Assoc.Am.Phys., 86, 154-166.

Bond, J.S., and Barrett, A.J. (1980). 'Degradation of fructose-1,6-bisphosphate aldolase by cathepsin B. A further example of peptidyldipeptidase activity of this proteinase', Biochem.J., 189, 17-25.

Chambers, R.A., and Pratt, R.T.C. (1956). 'Idiosyncrasy to fructose', Lancet, 11, 340.

Clarke, F.M., and Masters, C.J. (1973). 'On the distribution of aldolase isoenzymes in subcellular fractions from rat brain', Arch.Biochem.Biophys., 156, 673-683.

Costanzo, F., Castagnoli, L., Dente, L., Arcari, P., Smith, M., Costanzo, P., Raugei, G., Izzo, P., Pietropaolo, T.C., Bougueleret, L., Cimino, F., Salvatore, F., and Cortese, R. (1983). 'Cloning of several cDNA segments coding for human liver proteins', EMBO J., 2, 57-61.

Cox, T.M., O'Donnell, M.W., Camilleri, M., and Burghes, A.H. (1983). 'Isolation and characterization of a mutant liver aldolase in adult hereditary fructose intolerance', J.Clin. Invest., 72, 201-213.

Daimon, M., Tsutsumi, K., Sato, J, Tsutsumi, R., and Ishikawa, K. (1984). 'Changes of aldolase A and B messenger RNA levels in rat liver during azo-dye-induced hepatocarcinogenesis', Biochem.Biophys.Res.Commun., 124, 337-343.

Efstratiadis, A., Posakony, J.W., Maniatis, T., Lawn, R.M., O' Connel, C., Spritz, R.A., De Riel, J.K., Forget, B.G., Weissmann, S.M., Slightom, J.L., Blechl, A.E., Smithies, O., Baralle, F.E., Shoulders, C.C., and Proudfoot, N.J. (1980). 'The structure and evolution of the human β-globin gene family', Cell, 21, 653-668.

Freemont, P.S., Dunbar, B., and Fothergill, L.A. (1984). 'Human skeletal-muscle aldolase: N-terminal sequence analysis of CNBr- and o-iodosobenzoic acid-cleavage fragments", Arch. Biochem.Biophys., 228, 342-352.

Gitzelmann, R., Steinmann, B., Bally, C., and Lebherz, H.G. (1974). 'Antibody activation of mutant human fructosediphosphate aldolase B in liver extracts of patients with hereditary fructose intolerance', Biochem.Biophys.Res.Commun., 59, 1270-1277.

Gitzelmann, R., Steinmann, B., and Van Den Berghe, G. (1983).
'Essential fructosuria, hereditary fructose intolerance, and
fructose-1,6-diphosphatase deficiency' in The Metabolic Basis
of Inherited Disease (Eds. Stanbury, J.B., Wyngaarden, J.B.,
Fredrickson, D.S., Goldstein, J.L., and Brown, M.S.), 5th Ed.,
McGraw Hill, New York, pp.118-140.

Göeschke, H., and Leuthardt, F. (1963). 'Kristallisation einer
Aldolase aus Kaninchenleber', Helv.Chim.Acta, 46, 1791-1792.

Goldberg, M. (1979). 'Sequence analysis of Drosophila histone
genes', Ph.D. thesis, Stanford University, Palo Alto, Calif.

Gracy, R.W., Lacko, A.G., and Horecker, B.L. (1969). 'Subunit
structure and chemical properties of rabbit liver aldolase'
J.Biol.Chem., 244, 3913-3919.

Gracy, R.W., Lacko, A.G., Brox, L.W., Adelman, R.C., and Horecker,
B.L. (1970). 'Structural relations in aldolases purified from
rat liver and muscle and Novikoff hepatoma', Arch.Biochem.
Biophys., 136, 480-490.

Grazi, E., and Trombetta, G. (1978). 'A new intermediate of the
aldolase reaction, the pyruvaldehyde-aldolase-orthophosphate
complex', Biochem.J, 175, 361-365.

Grazi, E., and Trombetta, G. (1979). 'Fructose-1,6- bisphosphate
aldolase from rabbit liver. Reaction mechanism and
physiological function', Eur.J.Biochem., 100, 197-202.

Grazi, E., and Trombetta, G. (1980). 'The aldolase-substrate
intermediates and their interaction with glyceraldehyde-3-
phosphate dehydrogenase in a reconstructed glycolytic system',
Eur.J.Biochem., 107, 369-373.

Grazi, E., Trombetta, G., and Lanzara, V. (1983). 'Fructose-
1,6-bisphosphate aldolase from rabbit muscle. Kinetic
resolution of the enamine phosphate from the enamine-aldehyde
intermediate at low temperature', Biochemistry, 22, 4434-4437.

Grazi, E., and Trombetta, G. (1984). 'Fructose-1,6-bisphosphate
aldolase from rabbit muscle: Different catalytic behavior of
the dihydroxyacetone phosphate binding sites at low
temperature', Arch.Biochem.Biophys., 233, 595-602.

Grégori, C., Besmond, C., Kahn, A., and Dreyfus, J.C. (1982).
'Characterization of messenger RNA for aldolase B in adult and
fetal human liver', Biochem.Biophys.Res.Commun., 104, 369-375.

Grégori, C., Besmond, C., Odievre, M., Kahn, A., and Dreyfus, J.C.
(1984). 'DNA analysis in patients with hereditary fructose
intolerance', Ann.Hum.Genet., 48, 291-296.

Gürtler, B., and Leuthardt, F. (1970). 'Veber die Heterogeneität
der Aldolasen', Helv.Chim.Acta, 53, 654-658.

Hayner, N.T., Braun, L., Yaswen, P., Brooks, M., and Fausto, N. (1984). 'Isoenzyme profiles of oval cells, parenchymal cells, and biliary cells isolated by centrifugal elutriation from normal and preneoplastic livers', Cancer Res., 44, 332-338.

Hatzfeld, A., Elion, J., Mennecier, F., and Schapira, F. (1977). 'Purification of aldolase C from rat brain and hepatoma', Eur.J.Biochem., 77, 37-43.

Hers, H.G., and Joassin, G. (1961). 'Anomalie de l'aldolase hépatique dans l'intolérance au fructose', Enzymol.Biol.Clin., 1, 4-14.

Hers, H.G., and Kusaka, T. (1953). 'Le metabolisme du fructose-1-phosphate dans le foie', Biochim.Biophys.Acta, 11, 427-437.

Hommes, F.A., and Draisma, M.I. (1970). 'The development of L- and M-type aldolases in rat liver' Biochim.Biophys.Acta, 222, 251-252.

Horecker, B.L., Rowley, P.T., Grazi, E., Cheng, T., and Tchola, O. (1963). 'The mechanism of action of aldolases. IV. Lysine as the substrate-binding site', Biochem.Z., 338, 36-51.

Horecker, B.L., Tsolas, O., and Lai, C.Y. (1972). 'Aldolases' in The Enzymes (Ed. Boyer, P.D.) pp. 213-258. Academic Press, New York.

Ikehara, Y., Endo, H., and Okada, Y., (1970). 'The identity of the aldolases isolated from rat muscle and primary hepatoma'. Arch.Biochem.Biophys., 136, 491-497.

Izzo, P., Costanzo, P., Lupo, A., Paolella, G., Cimino, F., and Salvatore, F. (1985). 'The nucleotide sequence of human aldolase A gene', submitted for publication.

Kawahara, K., and Tanford, C. (1966). 'The number of polypeptide chains in rabbit muscle aldolase', Biochemistry, 5, 1578-1584.

Kobashi, K., Lai, C.Y., and Horecker, B.L. (1966). 'Organic phosphate groups in native and borohydride-reduced aldolase', Arch.Biochem.Biophys., 117, 437-444.

Kochman, M., Penhoet, E., and Rutter, W.J. (1968). 'Characterization of the subunits of aldolases A and C'. Federation Proc., 27, 590, abstr. 2101.

Koida, M., Lai, C.Y., and Horecker, B.L. (1969). 'Subunit structure of rabbit muscle aldolase: extent of homology of the α and β subunits and age-dependent changes in their ratio', Arch. Biochem.Biophys., 134, 623-631.

Koster, J.F., Slee, R.G., and Fernandez, J. (1975). 'On the biochemical basis of hereditary fructose intolerance'. Biochem.Biophys.Res.Commun., 64, 289-294.

Lai, C.Y. (1968). 'Studies on the structure of rabbit muscle aldolase. I. Cleavage with cyanogen bromide: An approach to the determination of the total primary structure', Arch. Biochem.Biophys., 128, 202-211.

Lai, C.Y. (1975). 'Studies on the structure of rabbit muscle aldolase. Determination of the primary structure of the COOH-terminal BrCN peptide: The complete sequence of the subunit polypeptide chain', Arch.Biochem.Biophys., 166, 358-368.

Lai, C.Y., Nakai, N., and Chang, D. (1974). 'Amino acid sequence of rabbit muscle aldolase and the structure of the active center', Science, 183, 1204-1206.

Lambert, J.M., Perham, R.N., and Coggins, J.R. (1977). 'Intramolecular ionic interactions of lysine residues and a possible folding domain in fructose diphosphate aldolase', Biochem.J., 161, 63-71.

Lebherz, H.G. (1972). 'Stability of quaternary structure of mammalian and avian fructose diphosphate aldolases', Biochemistry, 11, 2243-2250.

Lebherz, H.G. (1975a). 'On the regulation of fructose diphosphate aldolase isozyme concentrations in animals cells', in Isoenzymes (Ed. Markert, C.L.), Academic Press, New York, vol.3, pp. 253-279.

Lebherz, H.G. (1975b). 'Evidence for the lack of subunit exchange between aldolase tetramers in vivo', J.Biol.Chem., 250, 7388-7391.

Lebherz, H.G., Bates, O.J., and Bradshaw, R.A. (1984). 'Cellular fructose-P_2 aldolase has a derivatized (blocked) NH_2 terminus', J.Biol.Chem., 259, 1132-1135.

Lebherz, H.G., and Doyle, D. (1976). 'Synthesis of functional aldolase tetramers in a heterologous cell-free system', J.Biol.Chem., 251, 4355-4358.

Lebherz, H.G., and Rutter, W.J. (1969). 'Distribution of fructose diphosphate aldolase variants in biological systems', Biochemistry, 8, 109-121.

Lee, Y., and Horecker, B.L. (1974). 'Subunit structure of rabbit brain aldolase', Arch.Biochem.Biophys., 162, 401-411.

Lemmonier, F., Grégori, C., Schapira, F., and Szajnet, M.F. (1974). 'Possible role of thiol groups in the abnormal kinetics in hereditary fructose intolerance', Biochem.Biophys.Res.Commun., 61, 306-312.

Leuthardt, F., Testa, E., and Wolf, H.P. (1952). 'Über den stoff-wechsel des fructose-1-phosphäts in der leber', Helv. Physiol.Pharmacol.Acta, 10, c57-c59.

Lowry, R.B., and Hanson, J.W. (1977). 'Aldolase A deficiency with syndrome of growth and developmental retardation, midfacial hypoplasia, hepatomegaly and consanguineous parents', Birth Defects Orig.Art.Ser.XIII (3B), 223-228.

MacGregor, J.S., Singh, V.N., Davoust, S., Melloni, E., Pontremoli, S., and Horecker, B.L., (1980). 'Evidence for formation of a rabbit liver aldolase-rabbit liver fructose-1,6-bisphosphatase complex', Proc.Natl.Acad.Sci.USA, 77, 3889-3892.

Marquardt, R.R. (1970). 'Multiple molecular forms of avian aldolases. IV. Purification and properties of chicken (Gallus domesticus) brain aldolase', Can.J.Biochem., 48, 322-333.

Masters, C.J. (1968). 'The ontogeny of mammalian fructose-1,6-diphosphate aldolase', Biochim.Biophys.Acta, 167, 161-171.

Masters, C.J. (1981). 'Interactions between soluble enzymes and subcellular structure', CRC Critical Reviews in Biochemistry, 105-143

Masters, C.J., and Winzor, D.J. (1971). 'The molecular size of enzymically active aldolase A', Biochem.J., 121, 735-736.

Matsushima, T., Kawabe, S., Shibuya, M., and Sugimura, T. (1968). 'Aldolase isozymes in rat tumor cells', Biochem.Biophys.Res. Commun., 30, 565-570.

Meighen, E.A., and Schachman, H.K. (1970). 'Hybridization of native and chemically modified enzymes. I. Development of a general method and its application to the study of the subunit structure of aldolase', Biochemistry, 9, 1163-1176.

Meyerhof, O. (1938a). 'Sur l'isolement de l'acide 3-glycéroaldéhyde phosphorique biologique au cours de la dégradation enzymatique de l'acide hexose diphosphorique', Bull.Soc.Chim.Biol., 20, 1033-1042.

Meyerhof, O. (1938b). 'Sur l'isolement de l'acide 3-glycéroaldéhyde phosphorique biologique au cours de la degradation enzymatique de l'acide hexose diphosphorique' Bull.Soc.Chim.Biol., 20, 1345-1358.

Meyerhof, O., Lohmann, K., and Schuster, Ph. (1936). 'Über die Aldolase, ein Kohlenstoff-verknupfendes Ferment. I.Mitteilung: Aldolkondensation von Dioxyacetonphosphorsäure mit Acetaldehyd', Biochem.Z., 286, 301-343.

Morse, D.E., and Horecker, B.L. (1968). 'Rabbit liver aldolase: Determination of the primary structure at the active site', Arch.Biochem.Biophys., 125, 942-963.

Mukai, T., Joh, K., Miyahara, H., Sakakibara, M., Arai, Y., and Hori, K. (1984). 'Different expression of rat aldolase A mRNA in the skeletal muscle and ascites hepatoma cells', Biochem. Biophys.Res.Commun., 119, 575-581.

Munnich, A., Besmond, C., Darquy, S., Reach, G., Vaulont, S., Dreyfus, J.C., and Kahn, A. (1985). 'Dietary and hormonal regulation of aldolase B gene expression', J.Clin.Invest., 75, 1045-1052.

Nordmann, Y., Schapira, F., and Dreyfus, J.C. (1968). 'A structurally modified liver aldolase in fructose intolerance: immunological and kinetic evidence', Biochem.Biophys.Res. Commun., 31, 884-889.

Numazaki, M., Tsutsumi, K., Tsutsumi, R., and Ishikawa, K. (1984), 'Expression of aldolase isozyme mRNAs in fetal rat liver', Eur.J.Biochem., 142, 165-170.

Oda, T., and Omura, T. (1980). 'Biosynthesis of aldolase B by free ribosomes in rat liver', J.Biochem., 88, 437-442.

Offermann, M.K., Chlebowski, J.F., and Bond, J.S. (1983). 'Action of cathepsin D on fructose-1,6-bisphosphate aldolase', Biochem.J., 211, 529-534.

Paolella, G., Santamaria, R., Izzo, P., Costanzo, P., and Salvatore, F. (1984). 'Isolation and nucleotide sequence of a full-length cDNA coding for aldolase B from human liver', Nucl.Acids Res., 12, 7401-7410.

Paolella, G., Buono, P., Mancini, F.P., Izzo, P., and Salvatore, F. (1985a). 'Structure and expression of mouse aldolase genes: Brain specific aldolase C amino acid sequence is closely related to aldolase A', submitted for publication.

Paolella, G., Buono, P., Santamaria, R., and Salvatore, F. (1985b). 'Aldolase B gene: Study of hybridization patterns in healthy individuals', It.J.Biochem., in press.

Penhoet, E.E., Kochman, M., and Rutter, W.J. (1969a). 'Isolation of fructose diphosphate aldolases A, B and C', Biochemistry, 8, 4391-4395.

Penhoet, E.E., Kochman, M. and Rutter, W.J. (1969b). 'Molecular and catalytic properties of aldolase C', Biochemistry, 8, 4396-4402.

Penhoet, E.E., Kochman, M., Valentine, R., and Rutter, W.J. (1967). 'The subunit structure of mammalian fructose diphosphate aldolase', Biochemistry, 6, 2940-2949.

Penhoet, E.E., Rajkumar, T., and Rutter, W.J. (1966). 'Multiple forms of fructose diphosphate aldolase in mammalian tissues', Proc.Natl.Acad.Sci.USA, 56, 1275-1282.

Penhoet, E.E., and Rutter, W.J. (1971). 'Catalytic and immunochemi-
    cal properties of homomeric and heteromeric combinations of
    aldolase subunits', J.Biol.Chem., 246, 318-323.

Petell, J.K., and Lebherz, H.G., (1979). 'Properties and metabolism
    of fructose diphosphate aldolase in livers of old and young
    mice', J.Biol.Chem., 254, 8179-8184.

Pontremoli, S., Melloni, E., Michetti, M., Salamino, F., Sparatore,
    B., and Horecker, B.L. (1982a). 'Limited proteolysis of liver
    aldolase   and   fructose-1,6-bisphosphatase   by   lysosomal
    proteinases: Effect on complex formation', Proc.Natl.Acad.Sci.
    USA, 79, 2451-2454.

Pontremoli, S, Melloni, E., Michetti, M., Salamino, F., Sparatore,
    B., and Horecker, B.L. (1982b). 'Characterization of the
    inactive form of fructose-1,6-bisphosphate aldolase isolated
    from livers of fasted rabbits', Proc.Natl.Acad.Sci. USA, 79,
    5194-5196.

Pontremoli, S., Melloni, E., Salamino, F., Sparatore, B., Michetti,
    M., and Horecker, B.L. (1979). 'Changes in activity of
    fructose-1,6-bisphosphate aldolase in livers of fasted rabbits
    and   accumulation   of   crossreacting   immune   material',
    Proc.Natl.Acad.Sci.USA, 76, 6323-6325.

Putney, S.D., Herlihy, W.C., and Schimmel, P. (1983). 'A new
    troponin T and cDNA clones for 13 different muscle proteins,
    found by shotgun sequencing', Nature, 302, 718-721.

Reznick, A.Z., Lavie, L., Gershon, H.E., and Gershon, D. (1981).
    'Age-associated accumulation of altered FDP aldolase B in
    mice', FEBS Letters, 128, 221-224.

Rottmann, W.R., Tolan, D.R., and Penhoet, E.E. (1984). 'Complete
    amino acid sequence for human aldolase B derived from cDNA and
    genomic clones', Proc.Natl.Acad.Sci.USA, 81, 2738-2742.

Rutter, W.J., Blostein, R.E., Woodfin, B.M., and Weber, C.S.
    (1963). 'Enzyme variants and metabolic diversification', Adv.
    Enz.Regul., (Ed. Weber, G.), vol. 1, pp. 39-56.

Rutter, W.J., Rajkumar, T., Penhoet, E., Kochman, M., and
    Valentine, R. (1968). 'Aldolase variants: Structure and
    physiological significance', Ann.N.Y.Acad.Sci., 151, 102-117.

Sajgò, M., and Hajòs, G. (1974). 'The amino acid sequence of rabbit
    muscle   aldolase',   Acta Biochim.Biophys.Acad.Sci.Hung.,   9,
    239-241.

Schapira, F. (1975). 'Kinetic and immunological abnormalities of
    aldolase B in hereditary fructose intolerance', Biochem.
    Soc.Trans., 3, 232-234.

Schapira, F., Dreyfus, J.C., and Schapira, G. (1963). 'Anomaly of aldolase in primary liver cancer', Nature, 200, 995-997.

Schapira, F., Hatzfeld, A., and Grégori, C. (1974). 'Studies on liver aldolases in hereditary fructose intolerance', Enzyme, 18, 73-83.

Schapira, F., Hatzfeld, A., and Weber, A. (1975). 'Resurgence of some fetal isozymes in hepatoma',in Isoezymes (Ed.Markert, C.L.), vol. III, pp. 987-1003, Academic Press, New York.

Schapira, F., Schapira, G., Dreyfus, J.C. (1962). 'La lésion enzymatique de la fructosurie bénigne', Enzymol.Biol.Clin., 1, 170-175.

Simon, M.P., Besmond, C., Cottreau, D., Weber, A., Chaumet-Riffaud, P., Dreyfus, J.C., Sala Trepat, J., Marie, J., and Kahn, A. (1983). 'Molecular cloning of cDNA for rat L-type pyruvate kinase and aldolase B', J.Biol.Chem., 258, 14576-14584.

Steinmann, B., and Gitzelmann, R. (1981). 'The diagnosis of hereditary fructose intolerance', Helv.Paediat.Acta, 36, 297-316.

Stellwagen, E. (1976). 'Predicted structure for aldolase', J.Mol. Biol., 106, 903-911.

Susor, W.A., Kochman, M., and Rutter, W.J. (1969). 'Heterogeneity of presumably homogeneous protein preparations', Science, 165, 1260-1262.

Sutcliffe, J.G., Milner, R.J., Bloom, F.E., and Lerner, R.A. (1982). 'Common 82-nucleotide sequence unique to brain RNA', Proc.Natl.Acad.Sci.USA, 79, 4942-4946.

Taylor, J.F., Green, A.A., and Cori, G.T. (1948). 'Crystalline aldolase', J.Biol.Chem., 173, 591-604.

Thompson, R.J., Kynoch, P.A.M., and Willson, V.J.C. (1982). 'Cellular localization of aldolase C subunits in human brain', Brain Res., 232, 489-493.

Ting, S.M., Lai, C.Y., and Horecker, B.L. (1971). 'Primary structure at the active sites of beef and rabbit liver aldolases', Arch.Biochem.Biophys., 144, 476-484.

Tolan, D.R., Amsden, A.B., Putney, S.D., Urdea, M.S., and Penhoet, E.E. (1984). 'The complete nucleotide sequence for rabbit muscle aldolase A messenger RNA', J.Biol.Chem., 259, 1127-1131.

Trombetta, G., Balboni, G., di Iasio, A., and Grazi, E. (1977). 'On the stereospecific reduction of the aldolase-fructose 1,6 bisphosphate complex by $NaBH_4$', Biochem.Biophys.Res.Commun., 74, 1297-1301.

Tsutsumi, K., and Ishikawa, K. (1981). 'Purification of messenger RNA coding for rat liver aldolase B subunit', Biochem.Biophys.Res.Comm., 100, 407-412.

Tsutsumi, K., Mukai, T., Hidaka, S., Miyahara, H., Tsutsumi, R., Tanaka, T., Hori, K., and Ishikawa, K. (1983). 'Rat aldolase isoenzyme gene. Cloning and characterization of cDNA for aldolase B messenger RNA', J.Biol.Chem., 258, 6537-6542.

Tsutsumi, K., Mukai, T., Tsutsumi, R., Hidaka, S., Arai, Y., Hori, K., and Ishikawa, K. (1985) 'Structure and genomic organization of the rat aldolase B gene', J.Mol.Biol., 181, 153-160.

Tsutsumi, K., Mukai, T., Tsutsumi, R., Mori, M., Daimon, M., Tanaka, T., Yatsuki, H., Hori, K., and Ischikawa, K. (1984a). 'Nucleotide sequence of rat liver aldolase B messenger RNA', J.Biol.Chem., 259, 14572-14575.

Tsutsumi, R., Tsutsumi, K., Numazaki, M., and Ishikawa, K. (1984b). 'Two different aldolase A mRNA species in rat tissues', Eur.J.Biochem., 142, 161-164.

Warburg, O., and Christian, W. (1943). 'Isolierung und Kristallisation des Gärungsferments Zymohexase" Biochem.Z., 314, 149-176.

Weber, A., Le Provost, E., Boisnard-Rissel, M., Berges, J., Schapira, F., and Guillouzo, A. (1980). 'Localization of fetal aldolases during early stages of azo-dye hepatocarcinogenesis in rat', Biochem.Biophys.Res.Commun., 92, 591-597.

Willson, V.J.C., and Thompson, R.J. (1980). 'Human brain aldolase C$_4$ isoenzyme: purification, radioimmunoassay, and distribution in human tissues', Ann.Clin.Biochem., 17, 114-121.

Yaswen, P., Hayner, N.T., and Fausto, N. (1984) 'Isolation of oval cells by centrifugal elutriation and comparison with other cell types purified from normal and prenoeplastic livers', Cancer Res., 44, 324-331.

Index

Acquired immunodeficiency
  syndrome 54
Acute transforming
  retroviruses 415
Acyl-CoA-cholesterol transferase
  (ACAT) 305, 306
Adenocarcinomas 594
Adenosine deaminase (ADA)
  deficiency 159
Adenosine receptors 131
Adolase A deficiency 654
Adrenocortical cells 429, 432
Adrenocorticotrophic hormone
  (ACTH) 432
Adrenoleukodystrophy 3, 51, 52,
  65
Albumin 231
Aldolase
    class I 617, 621
    class II 617
Aldolase A 612, 614, 615, 618,
  620, 622, 625, 628–32, 634–39,
  641–46
Aldolase B 611, 612, 614, 618,
  621, 622, 624, 627–32, 634–39,
  642, 643, 645–54
Aldolase C 612, 618, 628, 631,
  635, 637, 638, 641, 645
Aldolase D 612
Aldolase gene and protein
  families 611–65
Aldolase isoenzyme gene family
  623–37
Aldolase isoenzyme genes,
  evolutionary derivation of
  634–35
Aldolase isoenzyme protein family
  612–23

Aldolase isoenzymes
    intracellular distribution
      and function 640–41
    molecular pathology 647–54
    occurrence, expression and
      regulation of 637–46
    subunit assembly and
      turnover 641–42
    tissue and developmental
      changes 643–45
    tissue specificity and
      development expression
      637–40
    variations in tumors 645–46
Aldolase levels under physio-
  logical conditions 642–43
Aldolase protein sequences 632
Aldolase sequences 631
Allopurinol 128, 130
α-amino-isobutyrate (AIB) 484,
  487
α-Fetoprotein 231
Aγ -globin gene 280
Amino acids 56, 57, 135, 136,
  140, 342, 344, 345, 363, 389,
  390, 392, 476, 484, 577, 581,
  582, 585, 586, 619, 625, 632
Amino oxidases 228, 229
Aminopterin 133
Amniocentesis 67
AMP 127
Anemia 263
Anti-aldolase antibodies 624
Anti-receptor antibodies 477,
  479, 480, 487, 489–91
apoAI/CIII gene complex 328
Apolipoprotein AI and CIII
  genes 326–28

Apolipoproteins 300
    functions of 302
    importance of 303
Arginosuccinate synthetase 21
Avian erythroblastosis virus
    (AEV) 573-76
Avian leukosis virus (ALV) 576
8-Azaguanine 134

B cell leukemias 567
B cell lymphoma 550, 567
B cell neoplasia 545-70
B cell neoplasms
    chromosomal joinings in
        550-55
    chromosome translocations
        in 555
BALB/MK cells 447, 449-55, 458
Becker muscular dystrophy 4, 18,
    101-2
β-globin gene 280, 282
    mutations unlinked to 267
    organization and structure
        of 258-59
β-globin gene disorders, 257-98
    diagnosis and prevention of
        284-87
    inherited 260-61
    natural history of 257
    scientific history of 257
β-globin synthesis 275, 276-77,
    280
β-glucuronidase 100
β-thalassaemia 257, 261-75
    analysis of 284
    antenatal diagnosis of 275
    characterization of mutations
        and haplotypes 271-75
    diagnosis of 284-87
    hematological and biochemical
        phenotypes caused by 261
    molecular mechanisms of
        264-67
    mutation linkage to haplo-
        types 268-71
    mutations affecting mRNA
        translation 267
    mutations affecting RNA-
        processing 264
    mutations affecting splice

    sites 264
    mutations affecting trans-
        cription 264
    mutations creating alter-
        native splice sites 265-66
    mutations in Mediterranean
        area 271-75
    mutations unlinked to β-globin
        gene 267
    prenatal diagnosis 287
Biotin analogues 15
BMD 106
Bridging markers 17, 20
Burkitt lymphoma 517, 545, 546,
    548, 568

C-propeptide domain 350-51
C-terminal telopeptide domain
    350-51
Cadmium in human tissues 207
Calcium 451, 452
Cancer pathogenesis 503-44
Carcinoma cells 399, 400
Carrier detection 67, 71-73
Carrier status diagnosis 64
Cartilage 346
Cathepsins B, M and D 622
cDNA, full-length 625
cDNA fragment synthesis 623
cDNA libraries 319, 321, 624
cDNA sequences 627, 630
Cell migration 378, 379, 381, 386
Cellular homologs of retroviral
    oncogenes 425-27
Cellular oncogenes 503-44
    amplification in tumor cells
        523-24
    identification of 504-7
    mechanisms of activation in
        human malignancies 514-24
    role in normal cells 508-14
Centromeric sequence 22
Ceruloplasmin 56, 228
Cholesteryl esters 303
Chondrodystrophies 367
Choreoathetosis 149
Chorionic villi biopsy 67, 71
Chromatin 133
Chromatin conformation 186-90
Chromosomal aberrations 516-23

Chromosomal balance 169
Chromosomal joinings in B cell
    neoplasms 550-55
Chromosome breakage 53
Chromosome sorting 10-11
Chromosome translocations 545,
    553, 555, 566, 567
Chromosomes, carrying -
    thalassaemia mutations 272
Chronic lymphocytic leukemia
    (CLL) 550, 555, 562
Chronic myelogenous leukemia
    (CML) 545, 549
Chylomicrons (CM) 300
CIS
    non-deletional forms with
    β-globin synthesis in
        280-81
    non-deletional forms without
    β-globin synthesis in 281
Cloning techniques 77
Clotting factors 4
c-myc proto-oncogene 512-14
Coagulation factor genes 52
Coagulation factors 55
Collagen 341-75
    fibrillar 344-46
    genetic disorders of 356
    interstitial 344
    nonfibrillar 344-46
    subunits of 346
    Type I 342, 345, 364
    Type II 345,367
    Type III 345
    Type IV 345
    Type V 345, 346
    types of 342, 344
Collagen defects 342
Collagen disease 341
Collagen exons 351
Collagen fibers 347
Collagen genes
    analysis of 352
    chromosomal location 354-56
    RFLPs associated with 367-70
    structure of 349-71
Collagen metabolism 341, 346-47
Collagen molecule 342-351
Collagen structure 342-46
Collagenase digestion 485, 486
Collagenase preparations 483

Colour blindness 3, 4, 51-53
Concurrent A-thalassaemia 277-78
Connective tissue 341, 342, 344
Cooley's anemia 257
Copper distribution in Menkes'
    disease 232-33
Copper metabolism in Menkes'
    disease 227-28, 233-41
Copper proteins in Menkes'
    disease 228-32
Copper toxity 239
c-ras family 511-12
Creatine-phosphokinase (CPK) 102
Crossover points 91-121
Cytochrome c oxidase 228, 229
Cytogenetic tests 71

$s\beta$ - thalassaemia 278-83
s-globin gene 280
Dermatological diseases 403
Diaminooxidase 230
Diazepam 131
Dihydroxyacetone-P (DHAP) 616
DNA analysis 284, 285
DNA damage 404
DNA hypomethylation 191
DNA markers 6
DNA mediated gene 599
DNA mediated gene transfer
    505-7
DNA methylation 191
DNA microcloning 80
DNA modification studies 191-94
DNA mutations 146
DNA polymorphic sequences 324
DNA polymorphisms 91-121
DNA precursors 76
DNA recombinant technology 4
DNA restriction techniques 286
DNA segments 77
DNA sequence polymorphism 197-97
DNA sequences 7, 12, 24, 80,
    93-95, 547, 559, 623
DNA synthesis 447, 450, 472
DNA tumor viruses 417-25
DNase I levels 189-90
DNase I-sensitive site 190-91
Dopamine 131
Dopamine β-hydroxylases 230

Dopamine metabolism 132
Dosage hybridisation 12
Down regulation 471, 490-91, 579
Drosophilia 27, 571, 621
Duchenne muscular dystrophy 2, 4,
    12, 14, 16, 101-2, 112
DX13 probe 60-62, 77

Ehlers-Danlos Syndromes 365-67
    Type IV 365
    Type VII 366, 367
    Type VIIa 366
    Type IX 226
End-organ unresponsiveness 4
Enzyme deficiency 127
Epidermal growth factor (EGF)
    447, 449, 451, 459, 510, 572
Epidermal growth factor (EGF)
    receptor 579-85
    synthesis of truncated 597-99
Epidermoid carcinoma cells 594
Epithelial cells
    comparisons between thyroid
        and keratinocyte systems
        454-61
    effects of oncogenes on
        homogeneous 435-36
    in vitro transformation by
        acute retroviruses 415-70
    rat thyroid system 436-45
    relationships between trans-
        formed and differentiated
        phenotype 428-35
    studies with pure 423-25
Epithelial cytotypes, interaction
    of retroviruses in vitro
    427-28
erbB gene 573-76
erbB gene product 577-83
erbB-related cellular oncogene
    588-89
erbB-related genes
    amplification in human
        tumors 592-97
    mRNA of 589-97
Erythroleukemia 573
Escherichia coli 404
Extracellular proteolysis 377-80

Fabry disease 98
Factor V 56
Factor VIII 20, 23, 53-59, 61,
    77, 100, 180
Factor IX 23, 52-55, 57-59, 62,
    67, 73, 100, 180
Factor X 57
Familial combined hyperlipidemia
    317-18
Familial hypercholesterolemia 310
Familial hypertriglyceridemia
    313-15
Fatty acid levels 65
Fibrillogenesis 342
Fibrin 377
Fibrinoid degeneration 341
Fibroblastic cell systems 415
Flow cytometry 8
Flow karyotypes 9
Flow sorting 8
Foetal blood sampling 67
Follicular lymphoma 552, 556,
    557, 567
Fragile X mental retardation
    syndrome 51, 66-67, 100, 112
    analysis with DNA markers
        67-74
    future prospects 76-77
    mutation effect on recombin-
        ation 74
    unanswered questions 76-77
FRT-L cells 438-40, 442
Fructose 1, 6-bisphosphate
    616, 622
Fructose-1-phosphate 648

GABA (γ-amino butyric acid) 132
γ-globin synthesis, increased
    level of 278
Gene analysis 2
Gene cloning 2
Gene deletions 58
Gene libraries 2
Gene mapping 91-93
Gene sequences 2
Gene specific probes 20
Gene therapy 124, 151, 159
Genetic analysis 71, 92

Genetic counseling 13, 67
Genetic disease 124, 125, 132
Genetic disorders 20
Genetic distances 92
Genetic map of q27-q28 region 77-80
Genetic mapping 25, 112
    single-copy sequences for 9-10
Genetics 2
Genome mapping 6-7
Genomic cloning 179-82, 190-91
Genomic digest patterns 175
Genomic DNA 111, 322, 323
Genomic libraries 80, 319
Globin gene cluster, non-alpha 268
Globin gene expression 259-60
Globin gene scheme 270
Glucose-6-phosphate 611
Glucose-6-phosphate dehydrogenase 51, 192
Glutamine phosphoribosyl amido transferase 127
Glyceraldehyde-3-phosphate (G3P) 616, 621
Glycerol 3'-phosphate dehydrogenase 21
Glycerol 6-phosphate dehydrogenase (G6PD) 3, 52, 75, 77, 92, 98, 100
Glycine 344
GMP 127
Gout 123-26, 128
Gouty arthritis 127, 128, 130, 144
Gouty bone disease 123
Grandfather's rule 92
Growth factor domain 384, 392
Growth factor receptors 510-11, 571-609
Growth factors 508-9
Growth hormone 492

Haplotype identification 111, 113
Haplotype reconstruction 93
    through cell hybrids 109-11
Haplotypes
    in chromosomes carrying

β-thalassaemia mutations 272
of β-thalassaemia mutations 268-71
    in Mediterranean area 271-75
HAT (Hypoxanthine, Aminopterin, Thymidine) 133
Hb Lepore 282
Heavy metals 207
Hematopoietic cells 573, 592, 645
Hematopoietic neoplasms 545
Hemophilia 2
Hemophilia A 3, 20, 51-90, 112
Hemophilia B 3, 51-90, 112
Hemophilia mutations 58-64
    direct detection 58-59
    indirect detection 59-62
Hep-3 cells 379, 380
Hepatitis 54
Hepatocarcinogenesis 645
Hereditary fructose intolerance (HFI) 611, 614, 647-54
Hereditary Persistence of Fetal Hemoglobin (HPFH) 278-83
    deletional forms 282-83
    heterocellular 279-80
    non-deletional forms 280-81
Herpes virus timidine kinase 217
Heterogeneity 69, 71
Heterozygosity 71
High density lipoprotein-2 (HDL2) 300
High density lipoprotein-3 (HDL3) 300
HPRT 3, 19, 53, 80, 106, 124
HPRT deficiency 125-27, 132, 145-47, 149, 160
HPRT gene 123-68, 198
    polymorphic variants in 142
HPRT gene expression 152, 157
HPRT gene fragments 141
HPRT gene point mutations 160
HPRT infected cells 155, 156
HPRT locus 143
HPRT mutations 144, 147, 150, 151
HPRT particles 156
HPRT protein 135, 145-47, 154
HPRT retroviral vector 154
HPRT sequence 136
HPRT transcription unit 150
HPRT virus 157,158

Human T cell leukemia virus 1
  (HTLV-1) 568
Hybridization patterns 651
3-Hydroxy-3 methylglutaryl CoA
  reductase synthesis 306
Hydroxylysine 346
Hydroxyproline 344
4-Hydroxyproline 346
Hyperbilirubinaemia 647
Hypercholesterolemia, familial
  310
Hyperlipidemia 299-340
  classification of 306
  familial 306
  familial combined 317-18
  familial type 1 307-8
  familial type 3 312-13
  genetic nature of 306
  isolation of genes 319
  pedigree analysis 318
  prevalent forms of 318-33
Hyperlipoproteinemia, type 5
  315-17
Hypertriglyceridemia 328-33
  apoAl/3.2 allele association
    with 328
  familial 313-15
  type IV/V 328
Hyperuricemia 128
Hypoalbuminemia 647
Hypoglycemia 647
Hypopituitary dwarfism 492
Hypoxanthine guanine phosphoribosyl transferase.
  See HPRT

Immunoglobulin gene rearrangement 547
Immunoglobulin heavy chain
  gene 547
IMP 127, 134
In situ hydridization 14-16, 133
In vitro cell transformation by
  oncogenic viruses 415
Insulin 471-502
  internalization of 473
  metabolic actions of 471
Insulin degrading enzyme 492
Insulin-like growth factors
  492-96
Insulin receptor
  down regulation 471
  preparation of monoclonal
    antibody to 478-81
  structure and role 473-76
Insulin receptor down regulation
  490-91
Insulin receptor kinase 476,
  485-89
Intermediate density lipoproteins
  (IDL) 300, 305
Intracellular organelles 473
Intragenic markers 59,60
Intravascular fibrinolysis 377

Jumping cloning vectors 77

Keratinocyte cell system in
  mouse 445-54
Kidney stones 130
Kinase domain 586-87
"Kinky hair" disease 223
Kirsten murine sarcoma virus
  (KiMSV) 432, 435, 439, 440,
  443, 444, 446, 449, 452-53,
  455-57, 459
Klinefelter's syndrome 21

λ EMBL-3PGK-M 181
λ PGK-G1 181
Lecithin cholesterol acyl
  transferase (LCAT) 305
Lentoid bodies 433
Lepore globin chains 282
Lesch-Nyhan disease 125, 126,
  128, 130-32, 144-50, 155, 159
Lesch-Nyhan syndrome 98, 112, 124
Leukemias 594
  molecular basis of 560-66
Leukemic cells 545
Ligand binding domain 585-86
Linear sequences 92
Linkage analysis 9, 60, 65, 69,
  79, 92, 93

Linkage computer program 67
Linkage mapping 1-50
Linkage markers 3-4
Linkage relationships 92
Linked markers 60-63
Lipid analysis 315
Lipids transport in man 303-6
    endogenous 305-6
    exogenous 303-5
Lipoprotein analysis 315
Lipoprotein lipose (LPL) 307
Lipoprotein transport cycle 304
Lipoproteins
    components of 301
    physical properties of 299
    principal function of 299
    proportion of protein in 300
    structure of 299-303
Liver cells 54
Low density lipoprotein (LDL)
    57, 300, 305, 306, 310
Lysine residues 347
Lysyl oxidase 230

Macroorchidism 75
Mammalian cytogenetics 24
Marfan syndrome 363-65
Meiotic recombination 74
Membrane-anchoring domain 586
Menkes' cells 235, 237-39, 242,
    244
Menkes' disease 20, 222-43
    animal model 225-26
    clinical picture 223-24
    copper distribution in 232-33
    copper metabolism in 227-28
    copper proteins in 228-32
    genetics 224-25
Menkes' syndrome 101
Metallothionein
    chromosomal location 216
    expression in cultured
        cells 233-41
    gene expression 217-22
    gene organization 213-16
    gene regulation 207-56
    history of discovery of 207
    inducibility of 211-13
    protein chemistry 208-22

    pseudogenes 215
Metaphase chromosomes 8
Microcytic anemia 257
Microdissection of mitotic
    chromosomes 77
Mild β-thalassaemia gene 277
Molecular biology 112
Molecular genetics 27
Monoclonal antibodies 477-81
    biological activity 483-85
    effect on insulin receptor
        down regulation 490-91
    effect on insulin-like growth
        factor receptors 492-96
    effect on insulin receptor
        kinase activity 485-89
    effect on other cell surface
        molecules 492
    preparation to insulin
        receptor 478-81

Mouse bone marrow infection 158
mRNA 321, 589-98, 623
mRNA analyses 644
mRNA levels 643
mRNA sequences 628-30
mRNA synthesis 400
mRNA translation 267
Multipoint linkage 76, 93
Multipoint linkage analysis 114
Murine leukemia virus (MLV) 152
Mutant loci, localisation of
    16-18
Mutation detection 62-64, 73
Mutations 514-16
Myeloproliferative sarcoma
    virus (MPSV) 443, 444

N-propeptide domain 353-54
N-terminal telopeptide domain
    353-54
Neoplastic transformation as
    multistep process 525-27
Neurospora crassa 209
Nick translation studies 188-89
NIH3T3 cells 416, 425, 505-7,
    511, 515-16, 601
Non-alpha globin gene cluster
    258

Non-disjunction families 21
Nonreutilizable oxypurine
  hypoxanthine 127
Norepinephrine 132
Nuclear proteins 512-14
Nuclease sensitivity 186,
  188, 194
Nucleic acid sequence 138
Nucleic acid studies 147
Nucleotide sequences 136, 559

Ocular albinism 4
Oligonucleotides 274, 285
Ornithine transcarbamylase
  (OTC) 100
Osteogenesis imperfecta 356-63
  analysis of atypical variant
    363
  classification of syndromes
    357
  mild dominant forms of 370
  Type I 358
  Type II 358
  Type III 361, 368
  Type IV 362

Pemphigus 401-2
Pgk cDNA probes 174-78, 181,
  183
Pgk deficiency 98
Pgk gene 169-205
  cloning 173-74
  dispersion of 183-85
  molecular mapping 174-78
  sensitivity to restriction
    endonucleases 194-96
  structure of 178-79
Pgk gene expression 186
Pgk locus 196-98
Pgk molecular genetics 185-86
Pgk pseudogenes 184
Phase relationships 92
Phenotypes 2
Phenylalanine hydroxylase 5
Philadelphia chromosome 545
Phorbol esters 398
Phosphoglycerate kinase.

See Pgk
Physical mapping 10-16, 25
Plasma lipoproteins.   See
  Lipoproteins
Plasmin 377
Plasmin inhibitors 378, 401
Plasminogen 377, 388
Plasminogen activator 377-414
  and disease 400-5
  genes 394-98
  genetic regulation 398-400
  level of activity 379
  molecular types 382
Platelet-derived growth factor
  (PDGF) 508, 572, 599
Point mutations 58
Polymorphic alleles, identifying
  322-26
Polymorphic DNA markers 59-62
Polymorphic sequences 324
Polymorphisms 93
Polypeptide growth factors 572
Premutation hypothesis 77
Prenatal diagnosis 18, 65, 67,
  71
Procollagens 344
Progenitor cell 457
Proline 344
Promotor mutations 264
Protease digestion 481-83
Protein C 57
Protein sequences 631-34
Proteins 382
Proto-oncogenes 504-5, 571,
  599-601
PRPP 127
Pseudogenes 20
Psoriasis vulgaris 403
Purine biosynthesis 127, 129
Purine metabolism 124, 125, 130
Purine nucleoside phosphorylase
  (PNP) deficiency 159

q27-q28 region 52-53
  genetic map of 77-80
Rare disorders 93
Recombinant DNA 54, 94-98, 179
Recombinant DNA molecule 153
Recombinant DNA technology 319,

322–26
Recombination fraction 69–71, 74
Recombination frequency 71
Recombination hot-spots 93
Registration peptides 347, 350
Relative probability of linkage
  (RPL) 70
Renal disease 128
Restriction endonucleases
  treatments 194–96
Restriction fragment length
  polymorphism (RFLP) 4–6, 17,
  24, 51, 63, 80, 93, 95, 322
  associated with collagen
    genes 367–70
  linked to apo AI/CIII
    complex 328–33
  markers 61–62
Retinitis pigmentosa 20
Retinoschisis 20, 101
Retroviral oncogenes
  cellular homologs of
    425–27
  identify homologous cellular
    proto-oncogenes 504–5
Rheumatoid arthritis 403, 405
RNA blot hybridization 554
RNA processing 264
RNA synthesis 472
RNA tumor viruses 422, 430
RNA viruses 153

Saccharomyces cerevisiae 571
Salvage pathways 129
Schizosaccharomyces pombe 571
Segregation analysis 64–66,
  71, 73, 75
Sequence analysis 179–182
Severe combined immunodeficiency
  4
Sex chromosomes 27
Sex-linked disease 2
Sex-linked inheritance 2
Shot-gun sequencing 624–25
Sicilian sβ-thalassaemia 282
Signal peptide 384
Signal propeptide domain 353–54
Silent β-chain variant 276
Silent β-thalassaemia gene 276

Silent β-thalassaemia mutations
  262
Silent heterocellular HPFH
  gene 280
Simian sarcoma virus (SSV) 508
Single-copy DNA sequences 22
Single-copy sequences for
  genetic mapping 9–10
Sintenic maps 92
Skin cancers 402
Small cell lung cancer (SCLC)
  523
Somatic cell genetics 91
Somatic cell hybrids 8,
  12–14, 91, 92, 105, 111
Spanish sβ-thalassaemia 283
Spasticity 149
Splice sites
  mutations affecting 264
  mutations creating alter-
    native 265–266
St14 probe 7, 60–62, 64, 65, 67
Stem cell 457
Superoxide dismutase 231
SV40 virus 422–24

T cell malignancies, molecular
  basis of 560–66
T cell receptor 560–68
T cell tumors 566
Thalassaemia 149, 261
Thalassaemia intermedia 263,
  275–78
  mild defects in β-globin
    chain synthesis 276–77
6-Thioguanine 134
Thymidine 134
Thymidylate 76
Thyroglobulin production 438
Thyroid epithelial cell system
  in rat 436–45
Thyrotrophic hormone (TSH) 436,
  437, 443, 460
Tissue plasminogen activator 57
Tissue-type PA (tPA) 386–88
  β-chains of 393
  chromosomal location of
    genes 398
  genes 394–98

physiological functions
  383-94
secondary structure of 387
structure function relation-
  ship 389-94
TMP (thymidine monophosphate)
  134
Transcriptional activity 191
Transducing signals 511-12
Triglycerides 303, 314
Triple-helical domain 351-53
Triple-helix formation 347
Trophoblast cells 67
Tumor cells 379
Tumor metastasis 380
Tumor promotion 398
Tyrosinase 230
Tyrosine kinase activity 488,
  489, 572, 579
Tyrosine phosphorylation 489

Uric acid 128
Uric acid nephrolithiasis 149
Urinary PA (uPA) 382-86
  β-chains of 393
  biosynthesis 382-83
  chromosomal location of
    genes 398
  genes 394-98
  nucleotide structure of
    porcine and human 397
  physiological functions
    393-94
  structure and properties 383
  structure function relation-
    ship 389-94
  subcellular localization
    385-86
  synthesis 398
Urokinase 382
Urokinase PA (uPA) 379, 381

Variant hemoglobins 260
V-D-J joining 555-60, 566
Very low density lipoproteins
  (VLDL) 300, 305, 314
Viral oncogenes 503, 504
von Willebrand factor 54

Wilson's disease 243

Xanthine 128
Xanthine oxidase 128
X-autosomal translocations
  18, 98, 109
X-chromosome
  cloned DNA probes 26
  fragmentation of 13
  rate of assignment of new
    genes 92
  repetitive sequences on
    21-22
  total linkage map 23-25
X-chromosome 1-50, 132-33,
  143, 144, 173, 174, 193
X-chromosome centromere 22
X-chromosome DNA probes 5
X-chromosome haplotypes 107
  mapping 101-9
X-chromosome inactivation
  169-205
  developmental biology of 170
  molecular biology of 171-73
X-chromosome long arm 51-90
X-chromosome sequences,
  isolation of 7-9
Xeroderma pigmentosum (XP)
  402-4
Xg cluster 3, 92
X-linked disorders 20-21, 27, 54,
  98, 99, 100, 111-13
X-linked genes 91-121
X-linked ichthyosis 98, 101
X-linked markers 21, 95
X-linked mental retardation
  non-specific 75
  without fragile site 75
X-linked muscular dystrophies
  16-18
X-linked muscular retardation
  19-20
X-linked phenotype 24
X-linked polymorphisms 111, 114
X-linked recombinant clones 96
X-linked retinitis pigmentosa
  (RP) 101
X-syntenic groups 25
X/Y homologous sequences 22-23